STUDY GUIDE AND SELECTED SOLUTIONS MANUAL

Karen C. Timberlake
Los Angeles Valley College

BASIC CHEMISTRY

Timberlake & Timberlake

THIRD EDITION

Prentice Hall
Boston Columbus Indianapolis New York San Francisco Upper Saddle River
Amsterdam Cape Town Dubai London Madrid Milan Munich Paris Montréal Toronto
Delhi Mexico City São Paulo Sydney Hong Kong Seoul Singapore Taipei Tokyo

Acquisitions Editor: Terry Haugen
Marketing Manager: Erin Gardner
Assistant Editor: Jessica Neumann
Managing Editor, Chemistry and Geosciences: Gina M. Cheselka
Project Manager: Ed Thomas
Operations Specialist: Maura Zaldivar
Supplement Cover Designer: Paul Gourhan
Cover Illustrator: Blakely Kim
Cover Photograph: magnetcreative/iStockphoto.com

Pearson Prentice Hall
Upper Saddle River, NJ 07458

Printed in the United States of America

10 9 8 7 6 5 4 3

ISBN-13: 978-0-321-67626-9
ISBN-10: 0-321-67626-2

Prentice Hall
is an imprint of

PEARSON

www.pearsonhighered.com

Preface

This Study Guide is intended to accompany *Basic Chemistry, Third Edition.* The purpose of this Study Guide is to provide students with additional learning resources to increase their understanding of the key concepts in the text. Each chapter in the Study Guide correlates with a chapter in the text. Within each chapter, there are Learning Exercises that focus on problem solving, which in turn promote an understanding of the chemical principles of that Learning Goal. Following the Learning Exercises, a Check List of Learning Goals and a multiple-choice Practice Test provide a review of the chapter content. Finally, the Answers and Solutions to Selected Text Problems give solutions for all the odd-numbered problems in each chapter.

I hope that this Study Guide will help in the learning of chemistry. If you wish to make comments or corrections, or ask questions, you can send me an e-mail message at *khemist@aol.com*.

Karen Timberlake
Los Angeles Valley College

To the Student

One must learn by doing the thing;
though you think you know it, you
have no certainty until you try.
—*Sophocles*

Here you are in a chemistry class with your textbook in front of you. Perhaps you have already been assigned some reading or some problems to do in the book. Looking through the chapter, you may see words, terms, and pictures that are new to you. This may very well be your first experience with a science class like chemistry. At this point you may have some questions about what you can do to learn chemistry. This *Study Guide* is written with those considerations in mind.

Learning chemistry is similar to learning something new such as tennis or skiing or driving a car. If I asked you how you learn to play tennis or ski or drive a car, you would probably tell me that you would need to practice often. It is the same with learning chemistry. Learning the chemical concepts and learning to work the problems depends on the time and effort you invest in it. If you practice every day, you will find that learning chemistry is an exciting experience and a way to understand the current issues of the environment, health, and medicine.

Manage Your Study Time

I often recommend a study system to students in which you read one section of the text and immediately practice the questions and problems that go with it. In this way, you concentrate on a small amount of information and actively use what you learned to answer questions. This helps you to organize and review the information without being overwhelmed by the entire chapter. It is important to understand each section, because they build like steps. Information presented in each chapter proceeds from the basic to the more complex. Perhaps you can only study three or four sections of the chapter. As long as you also practice doing some problems at the same time, the information will stay with you.

Form a Study Group

I highly recommend that you form a study group in the first week of your chemistry class. Working with your peers will help you use the language of chemistry. Scheduling a time to meet each week helps you study and prepare to discuss problems. You will be able to teach some things to other students in the group, and sometimes they will help you understand a topic that puzzles you. You won't always understand a concept right away. Your group will help you see your way through it. Most of all, a study group creates a strong support system whereby students like you get together to help each other complete the class successfully.

Go to Office Hours

Try to go to your tutor's and/or professor's office hours. Your professor wants you to understand and enjoy learning this material. Often a tutor is assigned to a class or there are tutors available at your college. Don't be intimidated. Going to see a tutor or your professor is one of the best ways to clarify what you need to learn in chemistry.

I hope that this *Study Guide* will help in the learning of chemistry. If you wish to make comments or corrections, or ask questions, you can send me an e-mail message at khemist@aol.com.

Using This Study Guide

Now you are ready to sit down and study chemistry. Let's go over some methods that can help you learn chemistry. This *Study Guide* is written specifically to help you understand and practice the chemical concepts that are presented in your class and in your text. Some of the exercises teach basic skills; others encourage you to extend your scientific curiosity. The following features are part of this *Study Guide:*

1. Study Goals
The Study Goals give you an overview of what the chapter is about and what you can expect to accomplish when you complete your study and learning of a chapter.

2. Think About It
Each chapter in the *Study Guide* has a group of questions that encourage you to think about some of the ideas and practical applications of the chemical concepts you are going to study. You may find that you already have knowledge of chemistry in some of the areas. That will be helpful to you. Other questions give you an overview of the chemistry ideas you will be learning.

3. Key Terms
Each chapter in the *Study Guide* introduces Key Terms. As you complete the description of the Key Terms, you will have an overview of the topics you will be studying in that chapter. Because many of the Key Terms may be new to you, this is an opportunity to review their meaning.

4. Chapter Sections
Each section of the chapter begins with the Key Concepts to illustrate the important ideas in that section. The summary of concepts is written to guide you through each of the learning activities. When you are ready to begin your study, read the matching section in the textbook and review the sample exercises in the text. Included in each chapter are the Guides to Problem Solving with step-by-step directions for working out a problem as well as the icons for Tutorials or Self Study Activities that are found in *MasteringChemistry*.

5. Learning Exercises
The Learning Exercises give you an opportunity to practice problem solving related to the chemical principles in the chapter. Each set of exercises reviews one chemical principle. There is room for you to answer the questions or complete the exercise in this *Study Guide*. The answers are found immediately following each exercise. (Sometimes they will be located at the top of the next page.) Check your answers right away. If they don't match the answer in the study guide, go back to the textbook and review the material again. It is important to make corrections before you go on. In learning tennis, you hit the ball a lot from the base line before you learn to volley or serve. Chemistry, too, involves a layering of skills such that each one must be understood before the next one can be learned.

At various times, you will notice some essay questions that illustrate one of the concepts. I believe that writing out your ideas is a very important way of learning content. If you can put your problem-solving techniques into words, then you understand the patterns of your thinking and you will find that you have to memorize less.

6. Checklist
Use the Checklist to check your understanding of the Study Goals. This gives you an overview of the major topics in the section. If something does not sound familiar, go back and review. One aspect of being a strong problem-solver is the ability to check your knowledge and understanding as you go along.

7. Practice Test

A Practice Test is found at the end of each chapter. When you have learned the material in a chapter, you can apply your understanding to the Practice Test. If the results of this test indicate that you know the material, you are ready to proceed to the next chapter. If the results indicate more study is needed, you can repeat the Learning Exercises you need to work on. Answers are found at the end of the Practice Test.

8. Answers and Solutions to Selected Problems in the Text

The complete solutions to all of the odd-numbered problems for each chapter can be found at the end of this Study Guide. As you work the odd-numbered questions and problems in the text, you can check your answers and see the solutions.

Contents

Answers to Selected Problems

1

Chemistry in Our Lives

Study Goals

- Define the term chemistry and identify substances as chemicals.
- Describe the activities that are part of the scientific method.
- Develop a study plan for learning chemistry.

Think About It

1. Why can we say that the salt and sugar we use on food are chemicals?

2. How can the scientific method help us make decisions?

3. What are some things you can do to help you study and learn chemistry successfully?

Key Terms

Match each of the following key terms with the correct definition:

1. scientific method **2.** experiment **3.** hypothesis
4. theory **5.** chemistry

a. _____ an explanation of nature validated by many experiments

b. _____ the study of substances and how they interact

c. _____ a possible explanation of a natural phenomenon

d. _____ the process of making observations, writing a hypothesis, and testing with experiments

e. _____ a procedure used to test a hypothesis

Answers **a.** 4 **b.** 5 **c.** 3 **d.** 1 **e.** 2

1.1 Chemistry and Chemicals

- A chemical is any material used in or produced by a chemical process.
- A chemical is a substance containing one type of material that has the same composition and properties.

◆ Learning Exercise 1.1

Indicate if each of the following is a chemical or not:

a. _____ aluminum **b.** _____ heat

c. _____ sodium fluoride in toothpaste **d.** _____ ammonium nitrate in fertilizer

e. _____ time

Answers **a.** yes **b.** no **c.** yes **d.** yes **e.** no

1.2 Scientific Method: Thinking Like a Scientist

- The scientific method is a process of making observations, writing a hypothesis, and testing the hypothesis with experiments.
- A theory develops when experiments that validate a hypothesis are repeated by many scientists with consistent results.

MasteringChemistry
Tutorial: The Scientific Method

◆ Learning Exercise 1.2

Identify each of the following as an observation (O), a hypothesis (H), or an experiment (E).

a. _____ Sunlight is necessary for the growth of plants.

b. _____ Plants in the shade were shorter than plants in the sun.

c. _____ Plant leaves are covered with aluminum foil and their growth measured.

d. _____ Fertilizer is added to plants.

e. _____ Ozone slows plant growth by interfering with photosynthesis.

f. _____ Ozone causes brown spots on plant leaves.

Answers **a.** H **b.** O **c.** E **d.** E **e.** H **f.** O

1.3 A Study Plan for Learning Chemistry

- Components of the text that promote learning include: *Looking Ahead, Learning Goals, Concept Checks, Sample Problems, Study Checks, Guides to Problem Solving (GPS), Chemistry and Health, Chemistry and the Environment, Chemistry and History, and Chemistry and Industry, Questions and Problems, Concept Maps, Chapter Reviews, Key Terms, Understanding the Concepts, Additional Questions and Problems, Challenge Problems, Glossary/Index, Answers to Selected Questions and Problems, and Combining Ideas.*

- An active learner continually interacts with chemical concepts while reading the text and attending lecture.
- Working with a study group clarifies ideas and illustrates problem solving.

◆ Learning Exercise 1.3

Which of the following activities would be included in a successful study plan for learning chemistry?

a. _____ attending lecture once in a while

b. _____ working problems with friends from class

c. _____ attending review sessions

d. _____ planning a regular study time

e. _____ not doing the assigned problems

f. _____ going to the instructor's office hours

Answers **a.** no **b.** yes **c.** yes **d.** yes **e.** no **f.** yes

Checklist for Chapter 1

You are ready to take the Practice Test for Chapter 1. Be sure you have accomplished the following learning goals for this chapter. If you are not sure, review the section listed at the end of the goal. Then, apply your new skills and understanding to the Practice Test.

After studying Chapter 1, I can successfully:

_______ Describe a substance as a chemical (1.1).

_______ Identify the components of the scientific method (1.2).

_______ Design a study plan for successfully learning chemistry (1.3).

Practice Test for Chapter 1

1. Which of the following would be described as a chemical?
   A. sleeping **B.** salt **C.** singing
   D. listening to a concert **E.** energy
2. Which of the following is not a chemical?
   A. wool **B.** sugar **C.** feeling cold
   D. salt **E.** vanilla

For questions 3 through 7, identify each statement as an observation (O), a hypothesis (H), or an experiment (E):

3. _______ More sugar dissolves in 50 mL of hot water than in 50 mL of cold water.

4. _______ Samples containing 20 g of sugar each are placed separately in a glass of cold water and a glass of hot water.

5. _______ Sugar consists of white crystals.

6. _______ Water flows downhill.

7. _______ Drinking ten glasses of water a day will help me lose weight.

For questions 8 through 12, answer yes or no.

To learn chemistry, I will:

8. ________ Work the problems in the chapter and check answers.

9. ________ Attend some lectures, but not all.

10. ________ Form a study group.

11. ________ Set up a regular study time.

12. ________ Wait until the night before the exam to start studying.

Answers to the Practice Test

1. B	**2.** C	**3.** O	**4.** E	**5.** O
6. O	**7.** H	**8.** yes	**9.** no	**10.** yes
11. yes	**12.** no			

2
Measurements

Study Goals

- Write the names and abbreviations for the metric or SI units used in measurements of length, volume, mass, temperature, and time.
- Distinguish between measured numbers and exact numbers.
- Convert a standard number to scientific notation.
- Determine the number of significant figures in a measurement.
- Round off a calculator answer to the correct number of significant figures.
- Write conversion factors from quantities and units in an equality.
- Use prefixes to change a unit to a larger or smaller unit.
- Use metric and SI units, U.S. units, a percentage, ppm or ppb, and density as conversion factors.
- In problem solving, convert the initial unit of a measurement to another unit.
- Calculate the density of a substance; use density to determine the mass or volume of a substance.

Think About It

1. What kind of device would you use to measure each of the following: your height, your weight, and the quantity of water to make soup?

2. When you do a measurement, why should you write down a number and a unit?

3. Why does oil float on water?

Key Terms

Match each of the following key terms with the correct definition:

a. metric (SI) system **b.** exact number **c.** significant figures
d. conversion factor **e.** density **f.** scientific notation

1. _____ all the numbers recorded in a measurement including the estimated digit

2. _____ a fraction that gives the quantities of an equality in the numerator and denominator

3. _____ the relationship of the mass of an object to its volume, usually expressed in g/mL

4. _____ a number obtained by counting items or from a definition

5. _____ a decimal system of measurement used throughout the world

6. _____ a form of writing a number using a coefficient and a power of ten

Answers **1.** c **2.** d **3.** e **4.** b **5.** a **6.** f

2.1 Units of Measurement

- In the sciences, physical quantities are described in units of the metric or International System (SI).
- In the metric system, length or distance is measured in meters (m), volume in liters (L), mass in grams (g), time in seconds (s), and temperature in degrees Celsius (°C) or kelvins (K).

◆ Learning Exercise 2.1

Indicate the type of measurement in each of the following measurements:

1. length **2.** mass **3.** volume **4.** temperature **5.** time

a. _____ 45 g	**b.** _____ 8.2 m	**c.** _____ 215 °C	**d.** _____ 45 L
e. _____ 825 K	**f.** _____ 8.8 s	**g.** _____ 140 g	**h.** _____ 50 s

Answers **a.** 2 **b.** 1 **c.** 4 **d.** 3 **e.** 4 **f.** 5 **g.** 2 **h.** 5

2.2 Scientific Notation

- A numerical value written in scientific notation has two parts: a number from 1 to 9 called a *coefficient,* followed by a power of ten.
- For numbers greater than ten, the decimal point is moved to the left to give a positive power of ten.
- For numbers less than 1, the decimal point is moved to the right to give a negative power of ten.

MasteringChemistry

Tutorial: Scientific Notation
Tutorial: Using Scientific Notation

Study Note

1. For a number greater than 10, the decimal point is moved to the left to give a coefficient 1 to 9 and a positive power of ten. For a number less than 1, the decimal point is moved to the right to give a coefficient 1 to 9 and a negative power of ten.

2. The number 2.5×10^3 means that 2.5 is multiplied by 10^3 (1000).

$$2.5 \times 1000 = 2500$$

3. The number 8.2×10^{-2} means that 8.2 is multiplied by 10^{-2} (0.01).

$$8.2 \times 0.01 = 0.082$$

◆ Learning Exercise 2.2A

Write each of the following measurements in scientific notation:

a. 240 000 cm ______________ **b.** 825 m ______________

c. 230 000 kg ______________ **d.** 53 000 y ______________

e. 0.002 m ______________ **f.** 0.000 001 5 g ______________

g. 0.08 kg ______________ **h.** 0.000 24 s ______________

Answers **a.** 2.4×10^5 cm **b.** 8.25×10^2 m **c.** 2.3×10^5 kg **d.** 5.3×10^4 y
e. 2×10^{-3} m **f.** 1.5×10^{-6} g **g.** 8×10^{-2} kg **h.** 2.4×10^{-4} s

◆ Learning Exercise 2.2B

Indicate the larger number in each pair.

a. 2500 or 2.5×10^2 ______________ **b.** 0.04 or 4×10^{-3} ______________

c. 65 000 or 6.5×10^5 ______________ **d.** 0.000 35 or 3.5×10^{-3} ______________

e. 300 000 or 3×10^6 ______________ **f.** 0.002 or 2×10^{-4} ______________

Answers **a.** 2500 **b.** 0.04 **c.** 6.5×10^5
d. 3.5×10^{-3} **e.** 3×10^6 **f.** 0.002

◆ Learning Exercise 2.2C

Write each of the following in standard form:

Examples: 2×10^2 m = 200 m and 3.2×10^{-4} g = 0.000 32 g

a. 4×10^3 m ______________ **b.** 5.2×10^4 g ______________

c. 1.8×10^5 g ______________ **d.** 8×10^{-3} L ______________

e. 6×10^{-2} kg ______________ **f.** 3.1×10^{-5} g ______________

Answers **a.** 4000 m **b.** 52 000 g **c.** 180 000 g
d. 0.008 L **e.** 0.06 kg **f.** 0.000 031 g

2.3 Measured Numbers and Significant Figures

- A measured number is obtained when you use a measuring device to determine a quantity.
- An exact number is obtained by counting items or from a definition that relates units in the same measuring system.
- There is uncertainty in every measured number, but not in exact numbers.
- Significant figures in a measured number are all the reported figures including the estimated digit.
- A number is a significant figure if it is

  a. not a zero

  b. a zero between nonzero digits

  c. a zero at the end of a decimal number
- Zeros written at the beginning of a number with a decimal point or zeros in a large number without a decimal point are not significant figures.

◆ Learning Exercise 2.3A

Are the numbers in each of the following statements measured (M) or exact (E)?

a. ____ There are 7 days in 1 week.
b. ____ A concert lasts for 73 min.
c. ____ There are 1000 g in 1 kg.
d. ____ The potatoes have a mass of 2.5 kg.
e. ____ A student has 26 DVDs.
f. ____ A snake is 1.2 m long.

Answers **a.** E (counted) **b.** M (use a watch) **c.** E (metric definition) **d.** M (use a balance) **e.** E (counted) **f.** M (use a metric ruler)

> *MasteringChemistry*
> **Self Study Activity: Significant Figures**
> **Tutorial: Counting Significant Figures**

> **Study Note**
>
> Significant figures (abbreviated SFs) are all the numbers reported in a measurement including the estimated digit. Zeros are significant unless they are placeholders appearing at the beginning of a decimal number or in a large number without a decimal point.
>
> 4.255 g (4 SFs) 0.0042 m (2 SFs) 46 500 L (3 SFs)

◆ Learning Exercise 2.3B

State the number of significant figures in each of the following measured numbers:

a. 35.24 g ____
b. 0.000 080 m ____
c. 55 000 m ____
d. 805 mL ____
e. 5.025 L ____
f. 0.006 kg ____
g. 268 200 mm ____
h. 25.0 °C ____

Answers **a.** 4 **b.** 2 **c.** 2 **d.** 3 **e.** 4 **f.** 1 **g.** 4 **h.** 3

2.4 Significant Figures in Calculations

- In multiplication or division, the final answer must have the same number of significant figures as in the measurement with the fewest significant figures.
- In addition or subtraction, the final answer must have the same number of decimal places as the measurement with the fewest decimal places.
- When evaluating a calculator answer, it is important to count the significant figures in the measurements and round off the calculator answer properly.
- Answers in chemical calculations rarely use all the numbers that appear in the calculator. Exact numbers are not included in the determination of the number of significant figures in an answer.

Study Note

1. To round off a number when the first number to be dropped is less than 5, keep the digits you need and drop all the digits that follow.

   Round off 42.8254 to 3 SFs → 42.8 (drop numbers 254)

2. To round off a number when the first number to be dropped is 5 or greater, keep the proper number of digits and increase the last retained digit by 1.

   Round off 8.4882 to 2 SFs → 8.5 (drop numbers 882, increase final digit by 1)

3. In large numbers, maintain the value of the answer by adding nonsignificant zeros as placeholders.

   Round off 356 835 to 3 SFs → 357 000
   (drop numbers 835, increase final digit by 1, add placeholder zeros)

◆ Learning Exercise 2.4A

Round off each of the following to give two significant figures:

a. 88.75 m ____________ **b.** 0.002 923 g ____________

c. 50.525 g ____________ **d.** 1.672 m ____________

e. 0.001 055 8 kg ____________ **f.** 82 080 mL ____________

Answers **a.** 89 m **b.** 0.0029 g **c.** 51 g
d. 1.7 m **e.** 0.0011 kg **f.** 82 000 mL

Study Note

1. An answer from multiplying and dividing has the same number of significant figures as the measurement that has the smallest number of significant figures.

   $1.5 \times 32.546 = 48.819 \rightarrow 49$ *Answer rounded off to two significant figures*
   2 SFs *5 SFs*

2. An answer from adding or subtracting has the same number of decimal places as the initial number with the fewest decimal places.

   $82.223 + 4.1 = 86.323 \rightarrow 86.3$ *Answer rounded off to one decimal place*
   3 decimal places *1 decimal place*

MasteringChemistry
Tutorial: Significant Figures in Calculations

◆ Learning Exercise 2.4B

Solve each problem and give the answer with the correct number of significant figures or decimal places:

a. $1.3 \times 71.5 =$ **b.** $\dfrac{8.00}{4.00} =$

c. $\dfrac{0.082 \times 25.4}{0.116 \times 3.4} =$ **d.** $\dfrac{3.05 \times 1.86}{118.5} =$

e. $\dfrac{376}{0.0073} =$ **f.** $38.520 - 11.4 =$

g. $4.2 + 8.15 =$ **h.** $102.56 + 8.325 - 0.8825 =$

Answers

a. 93	**b.** 2.00	**c.** 5.3	**d.** 0.0479
e. 52 000 (5.2×10^4)	**f.** 27.1	**g.** 12.4	**h.** 110.00

2.5 Prefixes and Equalities

- In the metric system, larger and smaller units use prefixes to change the size of the unit by factors of ten. For example, a prefix such as centi or milli preceding the unit meter gives a smaller length than a meter. A prefix such as kilo added to gram gives a unit that measures a mass that is 1000 times greater than a gram.
- An equality contains two units that measure the *same* length, volume, or mass.
- Some common metric equalities are: 1 m = 100 cm; 1 L = 1000 mL; 1 kg = 1000 g.
- Some useful metric–U.S. equalities are: 2.54 cm = 1 in.; 1 kg = 2.205 lb; 1 L = 1.057 qt
- Some metric (SI) prefixes are:

Prefix	Symbol	Numerical Value	Scientific Notation
Prefixes That Increase the Size of the Unit			
peta	P	1 000 000 000 000 000	10^{15}
tera	T	1 000 000 000 000	10^{12}
giga	G	1 000 000 000	10^{9}
mega	M	1 000 000	10^{6}
kilo	k	1 000	10^{3}
Prefixes That Decrease the Size of the Unit			
deci	d	0.1	10^{-1}
centi	c	0.01	10^{-2}
milli	m	0.001	10^{-3}
micro	μ	0.000 001	10^{-6}
nano	n	0.000 000 001	10^{-9}
pico	p	0.000 000 000 001	10^{-12}
femto	f	0.000 000 000 000 001	10^{-15}

MasteringChemistry

Self Study Activity: Metric System
Tutorial: SI Prefixes and Units

◆ Learning Exercise 2.5A

Match the items in column A with those from column B.

	A		B
1.	____ megameter	**a.**	nanometer
2.	____ 1000 meters	**b.**	decimeter
3.	____ 0.1 m	**c.**	10^{-6} m
4.	____ millimeter	**d.**	kilometer
5.	____ centimeter	**e.**	0.01 m
6.	____ 10^{-9} m	**f.**	1000 m
7.	____ micrometer	**g.**	10^{-3} m
8.	____ kilometer	**h.**	10^{6} m

Answers **1.** h **2.** d **3.** b **4.** g
5. e **6.** a **7.** c **8.** f

◆ Learning Exercise 2.5B

Place the following units in order from smallest to largest.

a. kilogram milligram gram ________________

b. centimeter kilometer millimeter ________________

c. dL mL L ________________

d. kg pg mg μg ________________

Answers **a.** milligram, gram, kilogram **b.** millimeter, centimeter, kilometer
c. mL, dL, L **d.** pg, μg, mg, kg

◆ Learning Exercise 2.5C

Complete each of the following metric relationships:

a. 1 L = ________ mL **b.** 1 L = ________ dL

c. 1 m = ________ cm **d.** 1 s = ________ ms

e. 1 kg = ________ g **f.** 1 cm = ________ mm

g. 1 mg = ________ μg **h.** 1 dL = ________ L

i. 1 m = ________ mm **j.** 1 cm = ________ m

Answers **a.** 1000 **b.** 10 **c.** 100 **d.** 1000 **e.** 1000
f. 10 **g.** 1000 **h.** 0.1 **i.** 1000 **j.** 0.01

2.6 Writing Conversion Factors

- Conversion factors are used in a chemical calculation to change from one unit to another. Each factor represents an equality that is expressed in the form of a fraction.
- Two forms of a conversion factor can be written for any equality. For example, the metric–U.S. equality 2.54 cm = 1 in. can be written as follows:

$$\frac{2.54\text{ cm}}{1\text{ in.}} \quad \text{and} \quad \frac{1\text{ in.}}{2.54\text{ cm}}$$

Study Note

Metric conversion factors are obtained from metric prefixes. For example, the metric equality 1 m = 100 cm is written as two factors:

$$\frac{1\text{ m}}{100\text{ cm}} \quad \text{and} \quad \frac{100\text{ cm}}{1\text{ m}}$$

When a unit is squared or cubed, the value of the conversion factor is also squared or cubed. For example, when the metric equality 1 m = 100 cm is squared, the relationship becomes

$$(1\text{ m})^2 = (100\text{ cm})^2$$

This equality gives the factors

$$\frac{(1\text{ m})^2}{(100\text{ cm})^2} \quad \text{and} \quad \frac{(100\text{ cm})^2}{(1\text{ m})^2}$$

◆ Learning Exercise 2.6A

Write two conversion factors for each of the following pairs of units:

a. millimeters and meters

b. kilogram and grams

c. kilograms and pounds

d. square inches and square centimeters

e. centimeters and meters

f. liters and quarts

g. deciliters and liters

h. millimeters cubed and centimeters cubed

Answers

a. $\frac{1000\text{ mm}}{1\text{ m}}$ and $\frac{1\text{ m}}{1000\text{ mm}}$

b. $\frac{1000\text{ g}}{1\text{ kg}}$ and $\frac{1\text{ kg}}{1000\text{ g}}$

c. $\frac{2.205\text{ lb}}{1\text{ kg}}$ and $\frac{1\text{ kg}}{2.205\text{ lb}}$

d. $\frac{(2.54\text{ cm})^2}{(1\text{ in.})^2}$ and $\frac{(1\text{ in.})^2}{(2.54\text{ cm})^2}$

e. $\frac{100\text{ cm}}{1\text{ m}}$ and $\frac{1\text{ m}}{100\text{ cm}}$

f. $\frac{1\text{ L}}{1.057\text{ qt}}$ and $\frac{1.057\text{ qt}}{1\text{ L}}$

g. $\frac{10\text{ dL}}{1\text{ L}}$ and $\frac{1\text{ L}}{10\text{ dL}}$

h. $\frac{(1\text{ cm})^3}{(10\text{ mm})^3}$ and $\frac{(10\text{ mm})^3}{(1\text{ cm})^3}$

Study Note

1. Sometimes, a statement within a problem gives an equality that is only true for that problem. Then conversion factors can be written that are true only for that problem. For example, a problem states that there are 50 mg of vitamin B in a tablet. The conversion factors are

$$\frac{1\text{ tablet}}{50\text{ mg vitamin B}} \quad \text{and} \quad \frac{50\text{ mg vitamin B}}{1\text{ tablet}}$$

2. If a problem gives a percentage (%), it can be stated as parts per 100 parts. For example, a candy bar contains 45% by mass chocolate. This percentage (%) equality can be written with factors using the same mass unit.

$$\frac{45\text{ g chocolate}}{100\text{ g candy bar}} \quad \text{and} \quad \frac{100\text{ g candy bar}}{45\text{ g chocolate}}$$

3. When a problem gives ppm or ppb, it can be stated as parts per million (ppm), which is mg/kg or parts per billion (ppb), which is μg/kg. For example, the level of nitrate in the Los Angeles water supply is 2.5 ppm. This ppm can be written as conversion factors using mg/kg.

$$\frac{2.5\text{ mg of nitrate}}{1\text{ kg of water}} \quad \text{and} \quad \frac{1\text{ kg of water}}{2.5\text{ mg of nitrate}}$$

◆ Learning Exercise 2.6B

Write two conversion factors for each of the following statements:

a. A cheese contains 55% fat by mass.

b. In the city, a car gets 14 miles to the gallon.

c. A 125-g steak contains 45 g of protein.

d. 18-karat pink gold contains 25% copper by mass.

e. A cadmium level of 1.8 ppm in food causes liver damage in rats.

f. A water sample contains 5.4 ppb of arsenic.

Answers

a. $\frac{55\text{ g fat}}{100\text{ g cheese}}$ and $\frac{100\text{ g cheese}}{55\text{ g fat}}$

b. $\frac{14\text{ mi}}{1\text{ gal}}$ and $\frac{1\text{ gal}}{14\text{ mi}}$

c. $\frac{45\text{ g protein}}{125\text{ g steak}}$ and $\frac{125\text{ g steak}}{45\text{ g protein}}$

d. $\frac{25\text{ g copper}}{100\text{ g pink gold}}$ and $\frac{100\text{ g pink gold}}{25\text{ g copper}}$

e. $\frac{1.8\text{ mg cadmium}}{1\text{ kg food}}$ and $\frac{1\text{ kg food}}{1.8\text{ mg cadmium}}$

f. $\frac{5.4\ \mu\text{g arsenic}}{1\text{ kg water}}$ and $\frac{1\text{ kg water}}{5.4\ \mu\text{g arsenic}}$

2.7 Problem Solving

- Conversion factors from metric, U.S. relationships, and percent can be used to change a quantity expressed in one unit to a quantity expressed in another unit.
- The process of solving a problem requires the change of the initial unit to one or more units until the final unit of the answer is obtained.

Guide to Problem Solving Using Conversion Factors

STEP 1 State the given unit and the needed unit.
STEP 2 Write a plan to convert the given unit to the final unit.
STEP 3 State the equalities and conversion factors needed to cancel units.
STEP 4 Set up problem to cancel units and calculate answer.

Example: How many liters is 2850 mL?

STEP 1: Given 2850 mL **Needed** liters

STEP 2: Plan milliliters $\xrightarrow{\text{metric factor}}$ liters

STEP 3: Equalities/Conversion Factors

1 L = 1000 mL

$\frac{1\text{ L}}{1000\text{ mL}}$ and $\frac{1000\text{ mL}}{1\text{ L}}$

STEP 4: Set Up Problem $2850\ \cancel{\text{mL}} \times \frac{1\text{ L}}{1000\ \cancel{\text{mL}}} = 2.85\text{ L}$

MasteringChemistry

Tutorial: Metric Conversions
Tutorial: Using Percentage as a Conversion Factors

◆ Learning Exercise 2.7A

Use metric conversion factors to solve each of the following problems:

a. 189 mL = _____ L

b. 2.7 cm = _____ mm

c. $0.274\ m^2 =$ ____ cm^2

d. 76 mg = ____ g

e. How many meters tall is a person whose height is 175 cm?

f. There are 285 mL in a cup of tea. How many liters is that?

g. An 18-karat ring contains 75.0% gold by mass. If the total mass of the ring is 13 500 mg, how many grams of gold does the ring contain?

h. You walked a distance of 1.5 km on the treadmill at the gym. How many meters did you walk?

Answers								
	a.	0.189 L	**b.**	27 mm	**c.**	$2740\ cm^2$	**d.**	0.076 g
	e.	1.75 m	**f.**	0.285 L	**g.**	10.1 g	**h.**	1500 m

◆ Learning Exercise 2.7B

Use metric–U.S. conversion factors to solve each of the following problems:

a. 18 in. = ________ cm

b. 4.0 qt = ________ L

c. 275 mL = ________ qt

d. 1300 mg = ________ lb

e. 150 lb = ________ kg

f. 840 g = ________ lb

g. 15 ft = ________ cm

h. 8.50 oz = ________ g

Answers				
	a. 46 cm	**b.** 3.8 L	**c.** 0.291 qt	**d.** 0.0029 lb
	e. 68 kg	**f.** 1.9 lb	**g.** 460 cm	**h.** 241 g

Study Note

1. For setups that require a series of conversion factors, it is helpful to write out the unit plan first. Work from the given unit to the needed unit. Then use a conversion factor for each unit change.

   Given unit → unit (1) → unit (2) = needed unit

2. To convert from unit to another, select conversion factors that cancel the given unit and provide a unit or the needed unit for the problem. Several factors may be needed to arrive at the needed unit.

$$\text{\sout{Given unit}} \times \frac{\text{\sout{unit (1)}}}{\text{\sout{given unit}}} \times \frac{\text{unit (2)}}{\text{\sout{unit (1)}}} = \text{needed unit (2)}$$

◆ Learning Exercise 2.7C

Use conversion factors to solve each of the following problems:

a. A piece of plastic tubing measures 120 mm. What is the length of the tubing in inches?

b. A statue weighs 245 lb. What is the mass of the statue in kilograms?

c. Your friend has a height of 6 ft 3 in. What is your friend's height in meters?

d. In a triple-bypass surgery, a patient requires 3.00 pt of whole blood. How many milliliters of blood were given?

e. Calculate the number of square meters in a rug with an area of 48 000 cm^2.

f. A mouthwash contains 22% alcohol by volume. How many milliliters of alcohol are in a 1.05-pt bottle of mouthwash?

g. An 18-karat gold bracelet has a mass of 2.0 oz. If 18-karat gold contains 75% pure gold, how many grams of pure gold are in the bracelet?

Answers	**a.**	4.7 in.	**b.**	111 kg	**c.**	1.9 m	**d.**	1420 mL
	e.	4.8 m^2	**f.**	110 mL	**g.**	43 g		

2.8 Density

- The density of a substance is the ratio of its mass to its volume, usually in units of g/mL or g/cm^3. For example, the density of sugar is 1.59 g/mL and silver is 10.5 g/mL.

$$\text{Density} = \frac{\text{mass of substance}}{\text{volume of substance}}$$

- The volume of 1 mL is equal to the volume of 1 cm^3.

Guide to Calculating Density

STEP 1 State the given and needed quantities.
STEP 2 Write the density expression.
STEP 3 Express mass in grams and volume in milliliters as g/mL.
STEP 4 Substitute mass and volume into density expression and solve.

Study Note

Density can be used as a factor to convert between the mass (g) and volume (mL) of a substance. The density of silver is 10.5 g/mL. What is the mass of 6.0 mL of silver?

$$6.0\ \cancel{\text{mL silver}} \times \frac{10.5\ \text{g silver}}{1\ \cancel{\text{mL silver}}} = 63\ \text{g of silver}$$

Density factor

What is the volume of 25 g of olive oil (D = 0.92 g/mL)?

$$25\ \cancel{\text{g olive oil}} \times \frac{1\ \text{mL olive oil}}{0.92\ \cancel{\text{g olive oil}}} = 27\ \text{mL of olive oil}$$

Density factor inverted

◆ Learning Exercise 2.8

Calculate the density or use density as a conversion factor to solve each of the following:

a. What is the density (g/mL) of glycerol if a 200.0-mL sample has a mass of 252 g?

b. A small solid has a mass of 5.5 oz. When placed in a graduated cylinder with a water level of 25.2 mL, the object causes the water level to rise to 43.8 mL. What is the density of the object, in g/mL?

c. A sugar solution has a density of 1.20 g/mL. What is the mass, in grams, of 0.250 L of the solution?

d. A piece of pure gold weighs 0.26 lb. If gold has a density of 19.3 mL, what is the volume, in milliliters, of the piece of gold?

e. A salt solution has a density of 1.15 g/mL and a volume of 425 mL. What is the mass, in grams, of the solution?

f. A 50.0-g sample of a glucose solution has a density of 1.28 g/mL. What is the volume, in liters, of the sample?

Answers			
	a. 1.26 g/mL	**b.** 8.4 g/mL	**c.** 300 g
	d. 6.1 mL	**e.** 489 g	**f.** 0.0391 L

Checklist for Chapter 2

You are ready to take the Practice Test for Chapter 2. Be sure that you have accomplished the following learning goals for this chapter. If you are not sure, review the section listed at the end of the goal. Then apply your new skills and understanding to the Practice Test.

After studying Chapter 2, I can successfully:

_____ Write the names and abbreviations for the metric (SI) units of measurement (2.1).

_____ Write large or small numbers using scientific notation (2.2).

_____ Identify a number as a measured number or an exact number (2.3).

_____ Count the number of significant figures in measured numbers (2.3).

_____ Report an answer with the correct number of significant figures (2.4).

_____ Write a metric equality from the numerical values of metric prefixes (2.5).

_____ Write two conversion factors for an equality (2.6).

_____ Use conversion factors to change from one unit to another unit (2.7).

_____ Calculate the density of a substance, or use density to calculate the mass or volume (2.8).

Practice Test for Chapter 2

1. Which of the following is a metric measurement of volume?
A. kilogram **B.** kilowatt **C.** kiloliter **D.** kilometer **E.** kiloquart

2. The measurement 24 000 g written in scientific notation is
A. 24 g **B.** 24×10^3 g **C.** 2.4×10^3 g **D.** 2.4×10^{-3} g **E.** 2.4×10^4 g

3. The measurement 0.005 m written in scientific notation is
A. 5 m **B.** 5×10^{-3} m **C.** 5×10^{-2} m **D.** 0.5×10^{-4} m **E.** 5×10^3 m

4. The measured number in the following is
A. 1 book **B.** 2 cars **C.** 4 flowers **D.** 5 rings **E.** 45 g

5. The number of significant figures in 105.4 m is
A. 1 **B.** 2 **C.** 3 **D.** 4 **E.** 5

6. The number of significant figures in 0.000 82 g is
A. 1 **B.** 2 **C.** 3 **D.** 4 **E.** 5

7. The calculator answer 5.7805 rounded to two significant figures is
A. 5 **B.** 5.7 **C.** 5.8 **D.** 5.78 **E.** 6.0

8. The calculator answer 3486.512 rounded to three significant figures is
A. 4000 **B.** 3500 **C.** 349 **D.** 3487 **E.** 3490

9. The reported answer for $16.0 \div 8.0$ is
A. 2 **B.** 2.0 **C.** 2.00 **D.** 0.2 **E.** 5.0

10. The reported answer for $58.5 + 9.158$ is
A. 67 **B.** 67.6 **C.** 67.7 **D.** 67.66 **E.** 67.658

11. The reported answer for $\frac{2.5 \times 3.12}{4.6}$ is
A. 0.54 **B.** 7.8 **C.** 0.85 **D.** 1.7 **E.** 1.69

12. Which of these prefixes has the largest value?
A. centi **B.** deci **C.** milli **D.** kilo **E.** micro

13. What is the decimal equivalent of the prefix *centi*?
A. 0.001 **B.** 0.01 **C.** 0.1 **D.** 10 **E.** 100

14. Which of the following is the smallest unit of measurement?
A. gram **B.** milligram **C.** kilogram **D.** decigram **E.** centigram

15. Which volume is the largest?
A. mL **B.** dL **C.** cm^3 **D.** L **E.** kL

16. Which of the following is a conversion factor?
A. 12 in. **B.** 3 ft **C.** 20 m **D.** $\frac{1000\text{ g}}{1\text{ kg}}$ **E.** 2 cm^3

17. Which is the correct conversion factor that relates milliliters to liters?
A. $\frac{1000\text{ mL}}{1\text{ L}}$ **B.** $\frac{100\text{ mL}}{1\text{ L}}$ **C.** $\frac{10\text{ mL}}{1\text{ L}}$ **D.** $\frac{0.01\text{ mL}}{1\text{ L}}$ **E.** $\frac{0.001\text{ mL}}{1\text{ L}}$

18. Which is the conversion factor for millimeters and centimeters?
A. $\frac{1\text{ mm}}{1\text{ cm}}$ **B.** $\frac{10\text{ mm}}{1\text{ cm}}$ **C.** $\frac{100\text{ cm}}{1\text{ mm}}$ **D.** $\frac{100\text{ mm}}{1\text{ cm}}$ **E.** $\frac{10\text{ cm}}{1\text{ mm}}$

19. Which is the conversion factor for ppm?
A. $\frac{1\text{ g}}{1\text{ kg}}$ **B.** $\frac{1\text{ dg}}{1\text{ kg}}$ **C.** $\frac{1\ \mu\text{g}}{1\text{ g}}$ **D.** $\frac{1\text{ mg}}{1\text{ kg}}$ **E.** $\frac{1\ \mu\text{g}}{1\text{ kg}}$

20. 294 mm is equal to
A. 2940 m **B.** 29.4 m **C.** 2.94 m **D.** 0.294 m **E.** 0.0294 m

21. The area of the face of a tennis racket measures 115 in.^{2}. What is that area, in square centimeters?
A. 742 cm^2 **B.** 17.8 cm^2 **C.** 45.3 cm^2 **D.** 292 cm^2 **E.** 115 cm^2

22. What is the volume of 65 mL, in liters?
A. 650 L **B.** 65 L **C.** 6.5 L **D.** 0.65 L **E.** 0.065 L

23. What is the mass, in kilograms, of a 22-lb turkey?
A. 10 kg **B.** 48 kg **C.** 10 000 kg **D.** 0.048 kg **E.** 22 000 kg

24. The number of milliliters in 2 dL of a liquid is
A. 20 mL **B.** 200 mL **C.** 2000 mL **D.** 20 000 mL **E.** 500 000 mL

25. A person who is 5 ft 4 in. tall would be
A. 64 m **B.** 25 m **C.** 14 m **D.** 1.6 m **E.** 1.3 m

26. How many ounces are in 1500 grams? (1 lb = 16 oz)
A. 94 oz **B.** 53 oz **C.** 24 000 oz **D.** 33 oz **E.** 3.3 oz

27. How many quarts of orange juice are in 255 mL of juice?
A. 0.255 qt **B.** 270 qt **C.** 236 qt **D.** 0.270 qt **E.** 0.400 qt

28. Your doctor places you on a 2200-Cal diet, with 18% of the Calories from fat. How many Calories are you allowed from fat?
A. 18 Cal **B.** 2200 Cal **C.** 1800 Cal **D.** 400 Cal **E.** 3960 Cal

29. How many milliliters of a salt solution with a density of 1.8 g/mL are needed to provide 400 g of salt solution?
A. 220 mL **B.** 22 mL **C.** 720 mL **D.** 400 mL **E.** 4.5 mL

30. Three liquids have densities of 1.15 g/mL, 0.79 g/mL, and 0.95 g/mL. When the liquids, which do not mix, are poured into a graduated cylinder, the liquid at the top is the one with a density of
A. 1.15 g/mL **B.** 1.00 g/mL **C.** 0.95 g/mL **D.** 0.79 g/mL **E.** 0.16 g/mL

31. A sample of oil has a mass of 65 g and a volume of 80.0 mL. What is the density of the oil?
A. 1.5 g/mL **B.** 1.4 g/mL **C.** 1.2 g/mL **D.** 0.90 g/mL **E.** 0.81 g/mL

32. What is the mass of a 10.0-mL sample of liquid with a density of 1.04 g/mL?
A. 104 g **B.** 10.4 g **C.** 1.04 g **D.** 1.40 g **E.** 9.62 g

33. Ethyl alcohol has a density of 0.785 g/mL. What is the mass of 0.250 L of the alcohol?
A. 196 g **B.** 158 g **C.** 3.95 g **D.** 0.253 g **E.** 0.160 g

34. 1.5 ft is the same length as
A. 46 cm **B.** 7.1 cm **C.** 18 cm **D.** 3.8 cm **E.** 0.59 cm

35. The level of lead in water was 4.6 ppb. What mass of lead is in 2.0 kg of water?
A. 9.2 μg **B.** 4.6 μg **C.** 2.3 μg **D.** 9.2 mg **E.** 4.6 mg

Answers to the Practice Test

1. C	**2.** E	**3.** B	**4.** E	**5.** D
6. B	**7.** C	**8.** E	**9.** B	**10.** C
11. D	**12.** D	**13.** B	**14.** B	**15.** E
16. D	**17.** A	**18.** B	**19.** D	**20.** D
21. A	**22.** E	**23.** A	**24.** B	**25.** D
26. B	**27.** D	**28.** D	**29.** A	**30.** D
31. E	**32.** B	**33.** A	**34.** A	**35.** A

3
Matter and Energy

Study Goals

- Classify an example of matter as a pure substance or a mixture.
- Describe some physical and chemical properties of matter.
- Identify the states of matter.
- Calculate temperature values in degrees Celsius, degrees Fahrenheit, and kelvins.
- Identify energy as potential or kinetic.
- Identify calorie and joule as units of energy and convert between them.
- Calculate the number of joules lost or gained by a specific quantity of a substance for a specific temperature change.
- Use the energy values to calculate the kilocalories (Cal) or kilojoules (kJ) in a food.

Think About It

1. What kinds of activities did you do today that used *kinetic* energy?

2. Why is the energy in your breakfast cereal called *potential* energy?

3. Why is the high specific heat of water important to our survival?

Key Terms

Match the following terms with the statements:

a.	potential energy	**b.**	joule	**c.**	matter
d.	chemical change	**e.**	mixture	**f.**	absolute zero

1. _____ the SI unit of energy

2. _____ anything that has mass and occupies space

3. _____ the lowest temperature possible, which is 0 on the Kelvin scale

4. _____ the physical combination of two or more substances that retain their identities

5. _____ stored energy

6. _____ a change in which an original substance is converted into a new substance with new properties

Answers			
	1. b	**2.** c	**3.** f
	4. e	**5.** a	**6.** d

3.1 Classification of Matter

- Matter is anything that has mass and occupies space.
- A pure substance, element or compound, has a definite composition.
- Elements are the simplest type of matter; compounds consist of a combination of two or more elements.
- Mixtures contain two or more substances that are physically, not chemically, combined.
- Mixtures are classified as homogeneous or heterogeneous.

MasteringChemistry
Tutorial: Classification of Matter

◆ Learning Exercise 3.1A

Identify each of the following as an element (E) or compound (C):

1. _____ iron

2. _____ carbon dioxide

3. _____ potassium iodide

4. _____ gold

5. _____ aluminum

6. _____ table salt (sodium chloride)

Answers			
	1. E	**2.** C	**3.** C
	4. E	**5.** E	**6.** C

◆ Learning Exercise 3.1B

Identify each of the following as a pure substance (P) or mixture (M):

1. _____ bananas and milk

2. _____ sulfur

3. _____ silver

4. _____ a bag of raisins and nuts

5. _____ water

6. _____ sand and water

Answers			
	1. M	**2.** P	**3.** P
	4. M	**5.** P	**6.** M

◆ Learning Exercise 3.1C

Identify each of the following mixtures as homogeneous (Ho) or heterogeneous (He):

1. _____ chocolate milk
2. _____ sand and water
3. _____ lemon drink
4. _____ a bag of raisins and nuts
5. _____ air
6. _____ vinegar

Answers **1.** Ho **2.** He **3.** Ho **4.** He **5.** Ho **6.** Ho

3.2 Properties of Matter

- Physical properties are those characteristics of a substance that can change without affecting the identity of the substance.
- Chemical properties are those characteristics of a substance that change when a new substance is produced.
- A substance undergoes a physical change when its shape, size, or state changes, but the type of substance itself does not change.
- Chemical properties are those characteristics of a substance that change when a new substance is produced.
- The states of matter are solid, liquid, and gas.
- Melting, freezing, boiling, and condensing are changes of state.

MasteringChemistry
Tutorial: Properties and Changes of Matter

◆ Learning Exercise 3.2A

State whether the following statements describe a gas (G), a liquid (L), or a solid (S):

1. _____ There are no attractions among the molecules.
2. _____ Particles are held close together in a definite pattern.
3. _____ The substance has a definite volume, but no definite shape.
4. _____ The particles are moving extremely fast.
5. _____ This substance has no definite shape and no definite volume.
6. _____ The particles are very far apart.
7. _____ This substance has a definite volume, but takes the shape of its container.
8. _____ The particles of this material bombard the sides of the container with great force.
9. _____ The particles in this substance are moving very, very slowly.
10. _____ This substance has a definite volume and a definite shape.

Answers **1.** G **2.** S **3.** L **4.** G **5.** G **6.** G **7.** L **8.** G **9.** S **10.** S

◆ Learning Exercise 3.2B

Classify each of the following as a physical (P) or chemical (C) property:

a. _____ Silver is shiny.

b. _____ Water fills a glass.

c. _____ Wood burns.

d. _____ Mercury is a very dense liquid.

e. _____ Helium is not reactive.

f. _____ Ice cubes float in water.

Answers	**a.** P	**b.** P	**c.** C
	d. P	**e.** C	**f.** P

◆ Learning Exercise 3.2C

Classify each of the following as a physical (P) or chemical change (C):

a. _____ Sodium melts at 98 °C.

b. _____ Iron forms rust in air and water.

c. _____ Water condenses on a cold window.

d. _____ Fireworks explode when ignited.

e. _____ Gasoline burns in a car engine.

f. _____ Paper is cut to make confetti.

Answers	**a.** P	**b.** C	**c.** P
	d. C	**e.** C	**f.** P

3.3 Temperature

- Temperature is measured in Celsius degrees (°C) or kelvins (K). In the United States, the Fahrenheit scale, (°F) or T_F, is still in use.
- The equation $T_F = 1.8T_C + 32$ is used to convert a Celsius temperature to a Fahrenheit temperature.
- When rearranged for T_C, the equation is used to convert from T_F to T_C.

$$T_C = \frac{(T_F - 32)}{1.8}$$

- The temperature on the Celsius scale is related to the Kelvin scale: $T_K = T_C + 273$.

MasteringChemistry
Tutorial: Temperature Conversions

◆ Learning Exercise 3.3

Calculate the temperature in each of the following problems:

a. To prepare yogurt, milk is warmed to 68 °C. What Fahrenheit temperature is needed to prepare the yogurt?

b. On a chilly day in Alaska, the temperature drops to −12 °C. What is that temperature on a Fahrenheit thermometer?

c. A patient has a temperature of 39.5 °C. What is that temperature in °F?

d. On a hot summer day, the temperature is 95 °F. What is the temperature on the Celsius scale?

e. A pizza is cooked at a temperature of 425 °F. What is the °C temperature?

f. A research experiment requires the use of liquid nitrogen to cool the reaction flask to −45 °C. What temperature will this be on the Kelvin scale?

Answers **a.** 154 °F **b.** 10 °F **c.** 103.1 °F **d.** 35 °C **e.** 218 °C **f.** 228 K

3.4 Energy

- Energy is the ability to do work.
- Potential energy is stored energy; kinetic energy is the energy of motion.
- The SI unit of energy is the joule (J), the metric unit is the calorie (cal). One calorie (cal) is equal to 4.184 joules (J).

$$\frac{4.184\ \text{J}}{1\ \text{cal}} \quad \text{and} \quad \frac{1\ \text{cal}}{4.184\ \text{J}}$$

MasteringChemistry

Tutorial: Heat
Tutorial: Energy Conversions

◆ Learning Exercise 3.4A

Match the words in column A with the descriptions in column B.

A	**B**
1. _____ kinetic energy	**a.** stored energy
2. _____ potential energy	**b.** the ability to do work
3. _____ energy	**c.** the energy of motion

Answers **1.** c **2.** a **3.** b

◆ Learning Exercise 3.4B

State whether each of the following statements describes potential (P) or kinetic (K) energy:

1. _____ a potted plant sitting on a ledge

2. _____ your breakfast cereal

3. _____ logs sitting in a fireplace

4. _____ a piece of candy

5. _____ an arrow shot from a bow

6. _____ a ski jumper at the top of the ski jump

7. _____ a jogger running

8. _____ a skydiver waiting to jump

9. _____ water flowing down a stream

10. _____ a bowling ball striking the pins

Answers **1.** P **2.** P **3.** P **4.** P **5.** K
6. P **7.** K **8.** P **9.** K **10.** K

◆ Learning Exercise 3.4C

Match the words in column A with the descriptions in column B.

A	B
1. _____ calorie	**a.** The SI unit of heat
2. _____ kilocalorie	**b.** The heat needed to raise 1 g of water by 1 °C
3. _____ joule	**c.** 1000 cal

Answers: **1.** b **2.** c **3.** a

◆ Learning Exercise 3.4D

Calculate an answer for each of the following conversions:

a. 58 000 cal to kcal

b. 3450 J to cal

c. 2.8 kJ to cal

d. 15 200 cal to kJ

Answers **a.** 58 kcal **b.** 825 cal **c.** 670 cal **d.** 63.6 kJ

3.5 Specific Heat

- Specific heat is the amount of energy required to raise the temperature of 1 g of a substance by 1 °C.
- The specific heat for liquid water is 4.184 J/g °C or 1.00 cal/g °C.
- The heat lost or gained can be calculated using the mass of the substance, temperature difference, and its specific heat: Heat (q) = mass × ΔT × SH
- A substance that loses heat has a negative sign for heat ($-q$); heat flows out.
- A substance that gains heat has a positive sign for heat ($+q$); heat flows in.

MasteringChemistry
Tutorial: Heat Capacity
Tutorial: Specific Heat Calculations

◆ Learning Exercise 3.5A

Calculate the specific heat for each of the following:

a. A 15.2-g sample of a metal that absorbs 231 J when its temperature rises from 84.5 °C to 125.3 °C.

b. A 31.8-g sample of a metal that absorbs 816 J when its temperature rises from 23.7 °C to 56.2 °C.

c. A 38.2-g sample of a metal that absorbs 125 J when its temperature rises from 62.1 °C to 68.4 °C.

Answers **a.** 0.372 J/g °C **b.** 0.790 J/g °C **c.** 0.52 J/g °C

Guide to Calculations Using Specific Heat
STEP 1 List the given and the needed data.
STEP 2 Calculate temperature change.
STEP 3 Write the equation for heat: Heat (joules) = mass (g) × ΔT × SH (J/g °C)
STEP 4 Substitute given values and solve, making sure units cancel.

◆ Learning Exercise 3.5B

Calculate the joules (J) and calories (cal) gained or released during the following:

1. heating 20.0 g of water from 22 °C to 77 °C

2. heating 10.0 g of water from 12.4 °C to 67.5 °C

3. cooling 0.450 kg of water from 80.0 °C to 35.0 °C

4. cooling 125 g of water from 72.0 °C to 45.0 °C

Answers	**1.** 4600 J; 1100 cal	**2.** 2310 J; 551 cal
	3. −84 700 J; −20 300 cal	**4.** −14 100 J; −3380 cal

Study Note

The mass of a substance is calculated by rearranging the heat equation:

$$\text{Mass} = \frac{\text{heat } (q)}{\Delta T \times SH}$$

The temperature change (ΔT) for a substance is calculated by rearranging the heat equation:

$$\Delta T = \frac{\text{heat } (q)}{\text{mass} \times SH}$$

◆ Learning Exercise 3.5C

a. Copper has a specific heat of 0.385 J/g °C. When 1250 J are added to a copper sample, its temperature rises from 24.6 °C to 61.3 °C. What is the mass, in grams, of the copper sample?

b. A sample of aluminum has a specific heat of 0.897 J/g °C. When 785 J are added to an aluminum sample, its temperature rises from 14 °C to 106 °C. What is the mass of the aluminum sample?

c. Silver has a specific heat of 0.235 J/g °C. What is the temperature change for a 15.2-g piece of silver at 10.0 °C when 426 J is added?

d. What is the final temperature of 15.0 g of water initially at 5.5 °C when 1140 J is added? (*Hint:* Determine the ΔT first.)

e. A 25.6-g sample of a metal heated to 100.0 °C is placed in 48.2 g of water initially at 14.5 °C. If the final temperature of the water (and metal) is 18.6 °C, what is the specific heat of the metal?

Answers		
	a. 88.5 g	**b.** 9.5 g
	c. 119 °C	**d.** ΔT = 18.2 °C; T_f = 23.7 °C
	e. 0.398 J/g °C	

3.6 Energy and Nutrition

- A nutritional calorie (Cal) is the same amount of energy as 1 kcal, or 1000 calories.
- When a substance is burned in a calorimeter, the water that surrounds the reaction chamber absorbs the heat given off. The heat absorbed by the water is calculated, and the caloric value (energy per gram) is determined for the substance.

MasteringChemistry

Tutorial: Nutritional Energy
Case Study: Calories from Hidden Sugar

Study Note

The caloric content of a food is the sum of calories from carbohydrate, fat, and protein. It is calculated by using the number of grams of each in a food and the caloric values of 17 kJ/g (4 kcal/g) for carbohydrate and protein, and 38 kJ/g (9 kcal/g) for fat.

◆ Learning Exercise 3.6A

Calculate the kilojoules (kJ) and kilocalories (kcal) for the following foods using the following data: (Round off the final answer to the nearest 10 kJ or kcal)

	Food	Carbohydrate	Fat	Protein	kJ	kcal
a.	green peas, cooked, 1 cup	19 g	1 g	9 g	______	______
b.	potato chips, 10 chips	10 g	8 g	1 g	______	______
c.	cream cheese, 8 oz	5 g	86 g	18 g	______	______
d.	lean hamburger, 3 oz	0	10 g	23 g	______	______
e.	banana, 1	26 g	0	1 g	______	______

Answers **a.** 510 kJ; 120 kcal **b.** 490 kJ; 120 kcal **c.** 3700 kJ; 870 kcal
d. 770 kJ; 180 kcal **e.** 460 kJ; 110 kcal

◆ Learning Exercise 3.6B

Use caloric values to solve each of the following:

1. How many kcal are in a single serving of pudding that contains 5 g of protein, 31 g of carbohydrate, and 5 g of fat (round kcal to the tens place)?

2. One serving of peanut butter (2 tbsp) has a caloric value of 190 kcal. If there are 8 g of protein and 10 g of carbohydrate, how many grams of fat are in one serving of peanut butter (round kcal to the tens place)?

3. A serving of breakfast cereal provides 220 kcal. In this serving, there are 8 g of protein and 6 g of fat. How many grams of carbohydrate are in the cereal (round kcal to the tens place)?

4. Complete the following table listing ingredients for a peanut butter sandwich (tbsp = tablespoon; tsp = teaspoon) (round kcal to the tens place):

	Protein	**Carbohydrate**	**Fat**	**kcal**
2 slices of bread	5 g	30 g	0 g	______
2 tbsp of peanut butter	8 g	10 g	13 g	______
2 tsp of jelly	0 g	10 g	0 g	______
1 tsp of margarine	0 g	0 g	7 g	______

Total kcal in sandwich ______

Answers

1. protein = 20 kcal; carbohydrate = 120 kcal; fat = 50 kcal; Total = 190 kcal

2. protein = 30 kcal; carbohydrate = 40 kcal Then 190 kcal − 30 kcal − 40 kcal = 120 kcal of fat
Converting 120 kcal of fat to g of fat: 120 kcal fat (1g fat/9 kcal fat) = 13 g of fat

3. protein = 30 kcal; fat = 50 kcal 80 kcal from protein and fat
220 kcal − 80 kcal = 140 kcal due to carbohydrate
140 kcal (1 g carbohydrate/4 kcal) = 35 g of carbohydrate

4. bread = 140 kcal; peanut butter = 190 kcal; jelly = 40 kcal; margarine = 60 kcal
total kcal in sandwich = 430 kcal (4.3×10^2 kcal)

Checklist for Chapter 3

You are ready to take the Practice Test for Chapter 3. Be sure you have accomplished the following learning goals for this chapter. If you are not sure, review the section listed at the end of the goal. Then apply your new skills and understanding to the Practice Test.

After studying Chapter 3, I can successfully:

_____ Identify matter as a pure substance or mixture (3.1).

_____ Identify a mixture as homogeneous or heterogeneous (3.1).

_____ Identify the physical state of a substance as a solid, liquid, or gas (3.2).

_____ Calculate a temperature in degrees Celsius or kelvins (3.3).

_____ Describe some forms of energy (3.4).

_____ Change a quantity in one energy unit to another energy unit (3.4).

_____ Calculate the specific heat of a substance (3.5).

_____ Given the mass of a sample, specific heat, and the temperature change, calculate the heat lost or gained (3.5).

_____ Using the energy values, calculate the kilocalories (Cal) or kilojoules (kJ) for a food sample (3.6).

Practice Test for Chapter 3

For questions 1 through 4, classify each of the following as a pure substance (P) or a mixture (M):

1. toothpaste _____

2. platinum _____

3. chromium _____

4. mouthwash _____

For questions 5 through 8, classify each of the following mixtures as homogeneous (Ho) or heterogeneous (He):

5. noodle soup _____

6. mineral water _____

7. chocolate chip cookie _____

8. mouthwash _____

9. Which of the following is a chemical property?
A. dynamite explodes **B.** a shiny metal **C.** a melting point of 110 °C
D. rain on a cool day **E.** breaking up cement

For questions 10 through 12, answer with solid (S), liquid (L), or gas (G):

10. _____ Has a definite volume, but takes the shape of a container.

11. _____ Does not have a definite shape or definite volume.

12. _____ Has a definite shape and a definite volume.

13. Which of the following is a chemical property of silver?
A. density of 10.5 g/mL **B.** shiny **C.** melts at 961 °C
D. good conductor of heat **E.** reacts to form tarnish

14. Which of the following is a physical property of silicon?
A. burns in chlorine **B.** has a black to gray color
C. reacts with nitric acid **D.** reacts with oxygen to form sand
E. used to form silicone

For problems 15 through 19, answer as physical change (P) or chemical change (C):

15. _____ butter melts in a hot pan

16. _____ iron forms rust with oxygen

17. _____ baking powder forms bubbles (CO_2) as a cake is baking

18. _____ water boils

19. _____ propane burns in a camp stove

20. 105 °F = _____°C
A. 73 °C **B.** 41°C **C.** 58 °C **D.** 90 °C **E.** 189 °C

21. The melting point of gold is 1064 °C. The Fahrenheit temperature needed to melt gold would be
A. 129 °F **B.** 623 °F **C.** 1031°F **D.** 1913 °F **E.** 1947 °F

22. The average daytime temperature on the planet Mercury is 683 K. What is this temperature on the Celsius scale?
A. 956 °C **B.** 715 °C **C.** 680 °C **D.** 410 °C **E.** 303 °C

23. Which of the following would be described as potential energy?
A. a car going around a racetrack **B.** a rabbit hopping
C. oil in an oil well **D.** a moving merry-go-round
E. a bouncing ball

24. Which of the following would be described as kinetic energy?
A. a car battery **B.** a can of tennis balls
C. gasoline in a car fuel tank **D.** a box of matches
E. a tennis ball crossing over the net

25. When 34 J are added to a 10.0-g piece of iron, the temperature rises by 7.6 °C. What is the specific heat of iron?
A. 2.2 J/g °C **B.** 0.22 J/g °C
C. 0.45 J/g °C **D.** 0.39 J/g °C
E. 2.6 J/g °C

26. The number of joules needed to raise the temperature of 5.0 g of water from 25 °C to 55 °C is
A. 5.0 J **B.** 36 J **C.** 30.0 J **D.** 335 J **E.** 630 J

27. The number of kilojoules (kJ) released when 15 g of water cools from 58 °C to 22 °C is
A. −0.13 kJ **B.** −0.54 kJ **C.** −2.3 kJ **D.** −63 kJ **E.** −150 kJ

For questions 28 to 30, consider a cup of milk that is 3.5% butterfat with a caloric value of 690 kJ. In the cup of milk, there are 9 g of fat, 12 g of carbohydrate, and some protein.

28. The number of kcal provided by the carbohydrate is
A. 4 kcal **B.** 9 kcal **C.** 36 kcal **D.** 48 kcal **E.** 81 kcal

29. The number of kJ provided by the fat is
A. 9.0 kJ **B.** 38 kJ **C.** 150 kJ **D.** 200 kJ **E.** 340 kJ

30. The number of kcal provided by the protein is
A. 4 kcal **B.** 9 kcal **C.** 36 kcal **D.** 48 kcal **E.** 81 kcal

Answers to the Practice Test

1. M	**2.** P	**3.** P	**4.** M	**5.** He
6. Ho	**7.** He	**8.** Ho	**9.** A	**10.** L
11. G	**12.** S	**13.** E	**14.** B	**15.** P
16. C	**17.** C	**18.** P	**19.** C	**20.** B
21. E	**22.** D	**23.** C	**24.** E	**25.** C
26. E	**27.** B	**28.** D	**29.** E	**30.** C

4

Atoms and Elements

Study Goals

- Given the name of an element, write its correct symbol; from the symbol, write its name.
- Use the periodic table to identify the group and the period of an element.
- Classify an element as a metal, nonmetal, or metalloid.
- Describe the subatomic particles, protons, neutrons, and electrons, in terms of their location in an atom, electrical charge, and relative mass.
- Describe Rutherford's gold foil experiment and explain how it led to the current model of the atom.
- Use the atomic number and mass number of an atom to determine the number of protons, neutrons, and electrons in the atom.
- Identify an isotope from the number of protons and neutrons.
- Calculate the atomic mass of an element from the masses of its naturally occurring isotopes and percent abundance.

Think About It

1. What are the subatomic particles that make up atoms?

2. How are the symbols of the elements related to their names?

3. What are some elements that are part of your vitamins?

Key Terms

Match each the following key terms with the correct definition:

a. element	**b.** atom	**c.** atomic number	**d.** mass number
e. isotope	**f.** group	**g.** halogen	**h.** representative element

1. _____ the number of protons and neutrons in the nucleus of an atom

2. _____ the smallest particle of an element

3. _____ a primary substance that cannot be broken down into simpler substances

4. _____ an atom of an element that has a different number of neutrons than another atom of the same element
5. _____ found in the first two and last six columns of the periodic table
6. _____ the number of protons in an atom
7. _____ an element found in Group 7A (17)
8. _____ vertical column on the periodic table in which elements have similar properties

Answers **1.** d **2.** b **3.** a **4.** e **5.** h **6.** c **7.** g **8.** f

4.1 Elements and Symbols

- Chemical symbols are one- or two-letter abbreviations for the names of the elements; only the first letter of a chemical symbol is a capital letter.
- Element names are derived from planets, mythology, and names of geographical features and famous people.

MasteringChemistry
Tutorial: Elements and Symbols in the Periodic Table

◆ Learning Exercise 4.1A

Write the symbol for each of the following elements:

1. carbon	_____	**2.** iron	_____	**3.** sodium	_____
4. phosphorus	_____	**5.** oxygen	_____	**6.** nitrogen	_____
7. iodine	_____	**8.** sulfur	_____	**9.** potassium	_____
10. lead	_____	**11.** calcium	_____	**12.** gold	_____
13. copper	_____	**14.** neon	_____	**15.** chlorine	_____

Answers					
	1. C	**2.** Fe	**3.** Na	**4.** P	**5.** O
	6. N	**7.** I	**8.** S	**9.** K	**10.** Pb
	11. Ca	**12.** Au	**13.** Cu	**14.** Ne	**15.** Cl

◆ Learning Exercise 4.1B

Write the name of the element represented by each of the following symbols:

1. Mg ____________________	**2.** H ____________________
3. Ag ____________________	**4.** F ____________________
5. Cr ____________________	**6.** Be ____________________
7. Si ____________________	**8.** Br ____________________
9. Zn ____________________	**10.** Al ____________________
11. Ba ____________________	**12.** Li ____________________

Answers				
	1. magnesium	**2.** hydrogen	**3.** silver	**4.** fluorine
	5. chromium	**6.** beryllium	**7.** silicon	**8.** bromine
	9. zinc	**10.** aluminum	**11.** barium	**12.** lithium

4.2 The Periodic Table

- The periodic table is an arrangement of the elements into vertical columns and horizontal rows.
- Each vertical column is a group of elements which have similar properties.
- A horizontal row of elements is called a period.
- On the periodic table, the metals are located on the left of the heavy zigzag line, the nonmetals are to the right, and metalloids are next to the zigzag line.
- Main group or representative elements are: 1A, 2A (1 and 2) and 3A to 8A (13 to 18). Transition elements are B-group elements (3–12).

Study Note

1. The periodic table consists of horizontal rows called *periods* and vertical columns called *groups*.
2. Elements in Group 1A (1) are the *alkali metals*. Elements in Group 2A (2) are the *alkaline earth metals*, and Group 7A (17) contains the *halogens*. Elements in Group 8A (18) are the *noble gases*.

◆ Learning Exercise 4.2A

Indicate whether the following elements are in a group (G), period (P), or neither (N):

a. Li, C, and O ____________ **b.** Br, Cl, and F ____________

c. Al, Si, and Cl ____________ **d.** C, N, and O ____________

e. Mg, Ca, and Ba ____________ **f.** C, S, and Br ____________

g. Li, Na, and K ____________ **h.** K, Ca, and Br ____________

Answers				
	a. P	**b.** G	**c.** P	**d.** P
	e. G	**f.** N	**g.** G	**h.** P

◆ Learning Exercise 4.2B

Complete the list of elements and symbols, group numbers, and period numbers in the following table:

Element and Symbol	Group Number	Period Number
	2A (2)	3
Silicon, Si		
	5A (15)	2
Aluminum, Al		
	4A (14)	5
	1A (1)	6

Answers

Element and Symbol	Group Number	Period Number
Magnesium, Mg	2A (2)	3
Silicon, Si	4A (14)	3
Nitrogen, N	5A (15)	2
Aluminum, Al	3A (13)	3
Tin, Sn	4A (14)	5
Cesium, Cs	1A (1)	6

◆ Learning Exercise 4.2C

Identify each of the following elements as a metal (M), a nonmetal (NM), or a metalloid (ML):

1. Cl ____ **2.** N ____ **3.** Fe ____ **4.** Si ____ **5.** Al ____
6. C ____ **7.** Ca ____ **8.** Zn ____ **9.** Sb ____ **10.** Mg ____

Answers **1.** NM **2.** NM **3.** M **4.** ML **5.** M
6. NM **7.** M **8.** M **9.** ML **10.** M

◆ Learning Exercise 4.2D

Match the names of the chemical groups with the elements K, Cl, He, Fe, Mg, Ne, Li, Cu, and Br.

1. halogens ________________

2. noble gases ________________

3. alkali metals ________________

4. alkaline earth metals ________________

5. transition elements ________________

Answers **1.** Cl, Br **2.** He, Ne **3.** K, Li **4.** Mg **5.** Fe, Cu

4.3 The Atom

- An atom is the smallest particle that retains the characteristics of an element.
- Atoms are composed of three subatomic particles. Protons have a positive charge (+), electrons carry a negative charge (–), and neutrons are electrically neutral.
- The protons and neutrons, with masses of about 1 amu, are found in the tiny, dense nucleus.
- The electrons are located outside the nucleus and have masses about 1/2000 that of a neutron or proton.

MasteringChemistry
Tutorial: The Anatomy of Atoms
Tutorial: Atomic Structure and Properties of Subatomic Particles

Dalton's Atomic Theory
1. All matter is made up of tiny particles called atoms.
2. All atoms of a given element are identical to one another and different from atoms of other elements.
3. Atoms of two or more different elements combine to form compounds. A particular compound is always made up of the same kinds of atoms and the same number of each kind of atom.
4. A chemical reaction involves the rearrangement, separation, or combination of atoms. Atoms are never created or destroyed during a chemical reaction.

◆ Learning Exercise 4.3A

True (T) or false (F): Each of the following statements is consistent with current atomic theory:

1. All matter is composed of atoms.

2. All atoms of an element are identical.

3. Atoms combine to form compounds.

4. Most of the mass of the atom is in the nucleus.

Answers **1.** T **2.** F **3.** T **4.** T

◆ Learning Exercise 4.3B

Match the following terms with the correct statements:

a. proton **b.** neutron **c.** electron **d.** nucleus

1. ____ is found in the nucleus of an atom

2. ____ has a 1− charge

3. ____ is found outside the nucleus

4. ____ has a mass of about 1 amu

5. ____ is the small, dense center of the atom

6. ____ is neutral

Answers **1.** a and b **2.** c **3.** c **4.** a and b **5.** d **6.** b

4.4 Atomic Number and Mass Number

- The *atomic number*, which can be found on the periodic table, is the number of protons in every atom of an element.
- In a neutral atom, the number of protons is equal to the number of electrons.
- The mass number is the total number of protons and neutrons in an atom.

MasteringChemistry
Tutorial: Atomic Number and Mass Number

◆ Learning Exercise 4.4A

Give the number of protons in each of the following neutral atoms:

a. an atom of carbon ____

b. an atom of the element with atomic number 15 ____

c. an atom with a mass number of 40 and atomic number 19 _____

d. an atom with 9 neutrons and a mass number of 19 _____

e. a neutral atom that has 18 electrons _____

Answers **a.** 6 **b.** 15 **c.** 19 **d.** 10 **e.** 18

Study Note

1. The *atomic number* is the number of protons in every atom of an element. In neutral atoms, the number of electrons equals the number of protons.
2. The *mass number* is the total number of neutrons and protons in the nucleus of an atom.
3. The number of neutrons is *mass number – atomic number.*

Example: How many protons and neutrons are in the nucleus of $^{109}_{47}Ag$?
Solution: The atomic number of Ag is 47. Thus, an atom of Ag has 47 protons.
Mass number – atomic number = number of neutrons: 109 − 47 = 62 neutrons

◆ Learning Exercise 4.4B

Determine the number of neutrons in each of the following atoms:

a. has a mass number of 42 and atomic number 20 __________

b. has a mass number of 10 and 5 protons __________

c. $^{30}_{14}Si$ __________

d. has a mass number of 9 and an atomic number of 4 __________

e. has a mass number of 22 and 10 protons __________

f. is a zinc atom with a mass number of 66 __________

Answers **a.** 22 **b.** 5 **c.** 16 **d.** 5 **e.** 12 **f.** 36

Study Note

In the atomic symbol for a particular atom, the mass number appears in the upper left corner and the atomic number in the lower left corner of the atomic symbol. Mass number → Atomic number → $^{32}_{16}S$

◆ Learning Exercise 4.4C

Complete the following table for neutral atoms:

Atomic Symbol	Atomic Number	Mass Number	Number of Protons	Number of Neutrons	Number of Electrons
	12			12	
			20	22	
		55		29	
	35			45	
		35	17		
$^{120}_{50}Sn$					

Answers

Atomic Symbol	Atomic Number	Mass Number	Number of Protons	Number of Neutrons	Number of Electrons
$^{24}_{12}Mg$	12	24	12	12	12
$^{42}_{20}Ca$	20	42	20	22	20
$^{55}_{26}Fe$	26	55	26	29	26
$^{80}_{35}Br$	35	80	35	45	35
$^{35}_{17}Cl$	17	35	17	18	17
$^{120}_{50}Sn$	50	120	50	70	50

4.5 Isotopes and Atomic Mass

- *Isotopes* are atoms that have the same number of protons, but different numbers of neutrons.
- The atomic mass of an element is the weighted average mass of all isotopes in a naturally occurring sample of that element.

MasteringChemistry

Self-Study Activity: Atoms and Isotopes
Tutorial: Atomic Mass Calculations

◆ Learning Exercise 4.5A

Identify the sets of atoms that are isotopes.

A. $^{20}_{10}X$ **B.** $^{20}_{11}X$ **C.** $^{21}_{11}X$ **D.** $^{19}_{10}X$ **E.** $^{19}_{9}X$

Answer Atoms A and D are isotopes (at. no. 10); atoms B and C are isotopes (at. no. 11).

◆ Learning Exercise 4.5B

Essay Copper has two naturally occurring isotopes, $^{63}_{29}Cu$ and $^{65}_{29}Cu$. If that is the case, why is the atomic mass of copper listed as 63.55 on the periodic table?

Answer Copper in nature consists of two isotopes that have different atomic masses. The atomic mass is the weighted average of the individual masses of the two isotopes, which takes into consideration their percent abundances in the sample. The atomic mass does not represent the mass of any individual atom.

◆ Learning Exercise 4.5C

In a sample of an element, 69.15% of the atoms have an atomic mass of 62.930 amu. If there is only one other isotope of that element in the sample and its atomic mass is 64.928 amu, what is the atomic mass of the element? Using the periodic table, write the symbol and name of the element.

Answer 100% − 69.15% = 30.85%

$$62.930 \text{ amu} \times \frac{69.15}{100} + 64.928 \text{ amu} \times \frac{30.85}{100} = 63.55 \text{ amu}$$

On the periodic table, the element with an atomic mass 63.55 amu is copper (Cu).

Checklist for Chapter 4

You are ready to take the Practice Test for Chapter 4. Be sure you have accomplished the following learning goals for this chapter. If you are not sure, review the section listed at the end of the goal. Then apply your new skills and understanding to the Practice Test.

After studying Chapter 4, I can successfully:

_____ Write the correct symbol or name for an element (4.1).

_____ Use the periodic table to identify the group and period of an element, and describe it as a metal, nonmetal, or metalloid and a representative or transition element (4.2).

_____ Use the periodic table to identify an element as a representative or transition element (4.2).

_____ State the electrical charge, mass, and location of the protons, neutrons, and electrons in an atom (4.3).

_____ Given the atomic number and mass number of an atom, state the number of protons, neutrons, and electrons (4.4).

_____ Identify an isotope (4.5).

_____ Calculate the atomic mass of an element from the naturally occurring isotopes and percent abundances (4.5).

Practice Test for Chapter 4

For questions 1 through 5, write the correct symbol for each of the elements listed:

1. potassium _____ **2.** phosphorus _____ **3.** calcium _____

4. carbon _____ **5.** sodium _____

For questions 6 through 10, write the correct element name for each of the symbols listed:

6. Fe ____________________ **7.** Cu ____________________

8. Cl ____________________ **9.** Pb ____________________

10. Ag ____________________

11. The elements C, N, and O are part of a
A. period **B.** group **C.** neither

12. The elements Li, Na, and K are part of a
A. period **B.** group **C.** neither

13. What is the classification of an atom with 15 protons and 17 neutrons?
A. metal **B.** nonmetal **C.** transition element **D.** noble gas **E.** halogen

14. What is the group number of the element with atomic number 3?
A. 1 **B.** 2 **C.** 3 **D.** 7 **E.** 8

For questions 15 through 18, consider an atom with 12 protons and 13 neutrons:

15. This atom has an atomic number of
A. 12 **B.** 13 **C.** 23 **D.** 24 **E.** 25

16. This atom has a mass number of
A. 12 **B.** 13 **C.** 23 **D.** 24 **E.** 25

17. This is an atom of
A. carbon **B.** sodium **C.** magnesium **D.** aluminum **E.** manganese

18. The number of electrons in this atom is
A. 12 **B.** 13 **C.** 23 **D.** 24 **E.** 25

For questions 19 through 22, consider an atom of calcium with a mass number of 42:

19. This atom of calcium has an atomic number of
A. 20 **B.** 22 **C.** 40 **D.** 41 **E.** 42

20. The number of protons in this atom of calcium is
A. 20 **B.** 22 **C.** 40 **D.** 41 **E.** 42

21. The number of neutrons in this atom of calcium is
A. 20 **B.** 22 **C.** 40 **D.** 41 **E.** 42

22. The number of electrons in this atom of calcium is
A. 20 **B.** 22 **C.** 40 **D.** 41 **E.** 42

23. A platinum atom, ${}^{195}_{78}\mathrm{Pt}$, has
A. $78\,p^+, 78\,e^-, 78\,n$ **B.** $195\,p^+, 195\,e^-, 195\,n$ **C.** $78\,p^+, 78\,e^-, 195\,n$
D. $78\,p^+, 78\,e^-, 117\,n$ **E.** $78\,p^+, 117\,e^-, 117\,n$

For questions 24 and 25, use the following list of atoms:

${}^{14}_{7}\mathrm{V}$ ${}^{16}_{8}\mathrm{W}$ ${}^{19}_{9}\mathrm{X}$ ${}^{16}_{7}\mathrm{Y}$ ${}^{18}_{8}\mathrm{Z}$

24. Which atom(s) is (are) isotopes of an atom with 8 protons and 9 neutrons?
A. W **B.** W, Z **C.** X, Y **D.** X **E.** Y

25. Which atom(s) are isotopes of an atom with 7 protons and 8 neutrons?
A. V **B.** W **C.** V, Y **D.** W, Z **E.** none

26. Which element would you expect to have properties most like oxygen?
A. nitrogen **B.** carbon **C.** chlorine **D.** argon **E.** sulfur

27. Which of the following is an isotope of nitrogen?
A. ${}^{14}_{8}\mathrm{N}$ **B.** ${}^{7}_{3}\mathrm{N}$ **C.** ${}^{10}_{5}\mathrm{N}$ **D.** ${}^{4}_{2}\mathrm{He}$ **E.** ${}^{15}_{7}\mathrm{N}$

Answers for the Practice Test

1. K	**2.** P	**3.** Ca	**4.** C	**5.** Na
6. iron	**7.** copper	**8.** chlorine	**9.** lead	**10.** silver
11. A	**12.** B	**13.** B	**14.** A	**15.** A
16. E	**17.** C	**18.** A	**19.** A	**20.** A
21. B	**22.** A	**23.** D	**24.** B	**25.** C
26. E	**27.** E			

5

Electronic Structure and Periodic Trends

Study Goals

- Identify electromagnetic radiation as energy that travels as waves at the speed of light.
- Compare the wavelength of radiation with its energy.
- Explain how atomic spectra correlate with the energy levels in atoms.
- Describe the sublevels and orbitals in atoms.
- Draw the orbital diagram and write the electron configurations for an element.
- Write the electron configuration for an atom using the sublevel blocks on the periodic table.
- Use the electron configurations of elements to explain periodic trends.

Think About It

1. How does the energy of microwaves compare with the energy of radio waves?

2. How much of the electromagnetic spectrum can we see?

3. What causes the different colors seen in fireworks?

Key Terms

Match each the following key terms with the correct definition:

a. photon	**b.** wavelength	**c.** electron configuration
d. ionization energy	**e.** electromagnetic radiation	**f.** orbital
g. valence electrons	**h.** *p* block	

1. _____ the distance between the peaks of two adjacent waves

2. _____ the electrons in the outermost energy level of an atom

3. _____ energy that travels as waves at the speed of light

4. _____ the smallest particle of light

5. _____ the organization of electrons within an atom arranged by orbitals of increasing energy

6. _____ elements in Groups 3A (13) to 8A (18)

7. _____ required to remove an electron from the outermost energy level of an atom

8. _____ the region within an atom where an electron of a certain energy is most likely to be found

Answers

1. b	**2.** g	**3.** e	**4.** a
5. c	**6.** h	**7.** d	**8.** f

5.1 Electromagnetic Radiation

- Electromagnetic radiation is energy that travels as waves in space at the speed of light.
- The wavelength (λ) is the distance between the peaks of adjacent waves.
- The frequency (ν) is the number of waves that pass a certain point in 1 second.
- The electromagnetic spectrum is all forms of electromagnetic radiation in order of decreasing wavelength.
- The speed of light (c) 3.00×10^8 m/s = wavelength $(\lambda) \times$ frequency (ν)

MasteringChemistry
Tutorial: Electromagnetic Radiation

◆ Learning Exercise 5.1A

For each of the following, calculate:

a. the frequency, in hertz, for a radio wave from an FM station that broadcasts at 105 MHz

b. the frequency, in kilohertz, from a police radar monitor that operates at 2.2×10^{10} Hz

Answers

a. $105\ \cancel{\text{MHz}} \times \dfrac{1 \times 10^6\ \text{Hz}}{1\ \cancel{\text{MHz}}} = 1.05 \times 10^8\ \text{Hz}$

b. $2.2 \times 10^{10}\ \cancel{\text{Hz}} \times \dfrac{1\ \text{kHz}}{1 \times 10^3\ \cancel{\text{Hz}}} = 2.2 \times 10^7\ \text{kHz}$

◆ Learning Exercise 5.1B

Using the electromagnetic spectrum, identify the type of radiation that would have the shorter wavelength.

a. microwave or infrared _______________

b. radio waves or X-rays _______________

c. gamma rays or radio waves _______________

d. ultraviolet or microwave _______________

Answers **a.** infrared **b.** X-rays **c.** gamma rays **d.** ultraviolet

5.2 Atomic Spectra and Energy Levels

- An atomic spectrum is a series of colored lines that correspond to photons of specific energies emitted by a heated element.
- A photon is a particle of electromagnetic radiation of a specified amount of energy called a quantum.
- In an atom, the energy levels indicated by the principal quantum number, n, contain electrons of similar energies.
- The principal quantum number, n, increases as the energy of the electrons increase.
- When electrons change energy levels, photons of specific energies are absorbed or emitted.

MasteringChemistry

Tutorial: Bohr's Shell Model
Tutorial: Energy Levels

◆ Learning Exercise 5.2

Identify the photon in each pair with the greater energy.

a. _____ yellow light or green light **b.** _____ $\lambda = 10^9$ nm or $\lambda = 10^5$ nm

c. _____ purple light or red light **d.** _____ $\lambda = 10^{-3}$ nm or $\lambda = 10^2$ nm

Answers **a.** green light **b.** $\lambda = 10^5$ nm **c.** purple light **d.** $\lambda = 10^{-3}$ nm

5.3 Sublevels and Orbitals

- Energy levels, indicated by the principal quantum numbers, contain electrons of similar energies.
- Within each energy level, electrons with identical energy are grouped in *sublevels*.
- An s sublevel can accommodate 2 electrons; a p sublevel can accommodate 6 electrons; a d sublevel can accommodate 10 electrons; and an f sublevel can accommodate 14 electrons.
- The energies of the sublevels increase in order: $s < p < d < f$.
- An orbital is a region in an atom where there is the greatest probability of finding an electron of certain energy.
- An s orbital is spherical, and p orbitals have two lobes along an axis. The d and f orbitals have more complex shapes.
- Each sublevel consists of a set of orbitals: an s sublevel consists of one orbital; a p sublevel consists of three orbitals; a d sublevel consists of five orbitals; and an f sublevel consists of seven orbitals.
- An orbital can contain a maximum of two electrons, which have opposite spins.

◆ Learning Exercise 5.3A

What is similar about the following orbitals?

a. 4*s* and 4*d*

b. 3*p* and 5*p*

Answers

a. The 4*s* and 4*d* orbitals are found in the same energy level with the same principal quantum number.
b. The 3*p* and 5*p* orbitals are the same type, which have the same shape.

◆ Learning Exercise 5.3B

State the maximum number of electrons for each of the following:

a. 3*p* sublevel _____ **b.** 3*d* sublevel _____

c. 2*s* orbital _____ **d.** energy level 4 _____

e. 1*s* sublevel _____ **f.** 4*p* orbital _____

g. 5*p* sublevel _____ **h.** 4*f* sublevel _____

Answers **a.** 6 **b.** 10 **c.** 2 **d.** 32
e. 2 **f.** 2 **g.** 6 **h.** 14

5.4 Drawing Orbital Diagrams and Writing Electron Configurations

- An orbital diagram shows the orbitals in an atom with squares that represent the orbitals and arrows that represent the electrons. In this example, the orbital diagram of helium consists of a square with two arrows.

- The electron configuration shows the number of electrons in each sublevel in order of increasing energy.
- In the abbreviated electron configuration, the electron configuration of the preceding noble gas is replaced by its symbol in square brackets.

MasteringChemistry
Tutorial: Electron Configurations

◆ Learning Exercise 5.4A

Draw the orbital diagram for each of the following elements:

a. beryllium _______________ **b.** carbon _______________

c. sodium _______________ **d.** nitrogen _______________

e. fluorine _______________ **f.** magnesium _______________

Answers

a. 1s [↑↓] 2s [↑↓]

b. 1s [↑↓] 2s [↑↓] 2p [↑][↑][ ]

c. 1s [↑↓] 2s [↑↓] 2p [↑↓][↑↓][↑↓] 3s [↑]

d. 1s [↑↓] 2s [↑↓] 2p [↑][↑][↑]

e. 1s [↑↓] 2s [↑↓] 2p [↑↓][↑↓][↑]

f. 1s [↑↓] 2s [↑↓] 2p [↑↓][↑↓][↑↓] 3s [↑↓]

◆ Learning Exercise 5.4B

Write the electron configuration for each of the following elements:

a. carbon ____________________

b. magnesium ____________________

c. iron ____________________

d. silicon ____________________

e. chlorine ____________________

f. phosphorus ____________________

Answers
a. $1s^22s^22p^2$ **b.** $1s^22s^22p^63s^2$ **c.** $1s^22s^22p^63s^23p^64s^23d^6$
d. $1s^22s^22p^63s^23p^2$ **e.** $1s^22s^22p^63s^23p^5$ **f.** $1s^22s^22p^63s^23p^3$

◆ Learning Exercise 5.4C

Write the abbreviated electron configuration for each of the following elements:

a. carbon ____________________

b. magnesium ____________________

c. iron ____________________

d. silicon ____________________

e. chlorine ____________________

f. phosphorus ____________________

Answers
a. $[He]2s^22p^2$ **b.** $[Ne]3s^2$ **c.** $[Ar]4s^23d^6$
d. $[Ne]3s^23p^2$ **e.** $[Ne]3s^23p^5$ **f.** $[Ne]3s^23p^3$

◆ Learning Exercise 5.4D

Name the element with an electron configuration ending with each of the following notations:

a. $3p^5$ __________ **b.** $2s^1$ __________

c. $3d^8$ __________ **d.** $4p^1$ __________

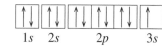

e. $5p^5$ ____________ **f.** $3p^2$ ____________

g. $1s^1$ ____________ **h.** $6s^2$ ____________

Answers
a. chlorine **b.** lithium **c.** nickel **d.** gallium
e. iodine **f.** silicon **g.** hydrogen **h.** barium

5.5 Electron Configurations and the Periodic Table

- The electron sublevels are organized as blocks on the periodic table. The *s* block corresponds to Groups 1A (1) and 2A (2), the *p* block extends from Group 3A (13) to 8A (18), the *d* block contains the transition elements, and the *f* block contains the lanthanides and actinides.

Guide to Writing Electron Configurations Using Sublevel Blocks

STEP 1 Locate the element on the periodic table.
STEP 2 Write the filled sublevels in order going across each period.
STEP 3 Complete the configuration by counting the number of electrons in the unfilled block.

◆ Learning Exercise 5.5A

Use the sublevel blocks on the periodic table to write the electron configuration for each of the following:

a. fluorine ____________________

b. sulfur ____________________

c. cadmium ____________________

d. strontium ____________________

Answers
a. $1s^22s^22p^5$ **b.** $1s^22s^22p^63s^23p^4$
c. $1s^22s^22p^63s^23p^64s^23d^{10}4p^65s^24d^{10}$ **d.** $1s^22s^22p^63s^23p^64s^23d^{10}4p^65s^2$

◆ Learning Exercise 5.5B

Use the sublevel blocks on the periodic table to write the abbreviated electron configuration for the each of the following elements:

a. iodine ____________ **b.** barium ____________

c. zinc ____________ **d.** tin ____________

e. cesium ____________ **f.** bromine ____________

Answers
a. $[Kr]5s^24d^{10}5p^5$ **b.** $[Xe]6s^2$ **c.** $[Ar]4s^23d^{10}$
d. $[Kr]5s^24d^{10}5p^2$ **e.** $[Xe]6s^1$ **f.** $[Ar]4s^23d^{10}4p^5$

◆ Learning Exercise 5.5C

Give the symbol of the element that has each of the following:

a. six $3d$ electrons ______ **b.** first to completely fill four s orbitals ______

c. $[Kr]5s^1$ ______ **d.** $[Ar]4s^2 3d^{10} 4p^5$ ______

e. two $6p$ electrons ______ **f.** first to have nine completely filled p orbitals ______

Answers **a.** Fe **b.** Ca **c.** Rb **d.** Br **e.** Pb **f.** Kr

5.6 Periodic Trends of the Elements

- The physical and chemical properties of elements change in a periodic manner going across each period and are repeated in each successive period.
- Representative elements in a group have similar behavior.
- The group number of a representative element gives the number of valence electrons.
- An electron-dot symbol is a convenient way to represent the valence electrons, which are shown as dots on the sides, top, or bottom of the symbol for the element. For beryllium, the electron-dot symbol is:

$$\dot{\text{Be}}\cdot$$

- The atomic radius of representative elements generally increases going down a group and decreases going from left to right across a period.
- The ionization energy generally decreases going down a group and increases going across a period.
- The sizes of the positive ions of metals are smaller than their neutral atoms, and the sizes of the negative ions of nonmetal atoms are larger than their neutral atoms.

MasteringChemistry

Tutorial: Ionization Energy
Tutorial: Periodic Trends
Tutorial: Patterns in the Periodic Table

◆ Learning Exercise 5.6A

State the number of valence electrons, the group number, and the electron-dot symbol for each of the following:

Element	Valence Electrons	Group Number	Electron-Dot Symbol
a. nitrogen	______	______	______
b. oxygen	______	______	______
c. magnesium	______	______	______
d. hydrogen	______	______	______
e. fluorine	______	______	______
f. aluminum	______	______	______

Answers

Element	**Valence Electrons**	**Group Number**	**Electron-Dot Symbol**
a. Nitrogen	$5\ e^-$	Group 5A (15)	$\cdot\ddot{\underset{\cdot}{N}}\cdot$
b. Oxygen	$6\ e^-$	Group 6A (16)	$\cdot\ddot{\underset{\cdot}{O}}:$
c. Magnesium	$2\ e^-$	Group 2A (2)	$\dot{Mg}\cdot$
d. Hydrogen	$1\ e^-$	Group 1A (1)	$H\cdot$
e. Fluorine	$7\ e^-$	Group 7A (17)	$\cdot\ddot{\underset{\cdot\cdot}{F}}:$
f. Aluminum	$3\ e^-$	Group 3A (13)	$\cdot\dot{Al}\cdot$

◆ Learning Exercise 5.6B

Indicate the element in each of the following pairs that has the larger atomic radius:

a. ____________ Mg or Ca **b.** ____________ Si or Cl

c. ____________ Sr or Rb **d.** ____________ Br or Cl

e. ____________ Li or Cs **f.** ____________ Li or N

g. ____________ N or P **h.** ____________ As or Ca

Answers **a.** Ca **b.** Si **c.** Rb **d.** Br
e. Cs **f.** Li **g.** P **h.** Ca

◆ Learning Exercise 5.6C

Indicate the element in each pair that has the lower ionization energy.

a. ____________ Mg or Na **b.** ____________ P or Cl

c. ____________ K or Rb **d.** ____________ Br or F

e. ____________ Li or O **f.** ____________ Sb or N

g. ____________ K or Br **h.** ____________ S or Na

Answers **a.** Na **b.** P **c.** Rb **d.** Br
e. Li **f.** Sb **g.** K **h.** Na

◆ Learning Exercise 5.6D

Which is larger in each of the following?

a. ____________ Mg or Mg^{2+} **b.** ____________ N or N^{3-}

c. ____________ Cl^- or Cl **d.** ____________ K^+ or K

Answers **a.** Mg **b.** N^{3-} **c.** Cl^- **d.** K

Checklist for Chapter 5

You are ready to take the Practice Test for Chapter 5. Be sure you have accomplished the following learning goals for this chapter. If you are not sure, review the section listed at the end of the goal. Then apply your new skills and understanding to the Practice Test.

After studying Chapter 5, I can successfully:

_____ Compare the wavelength of radiation with its energy (5.1).

_____ Compare frequency, wavelength, and energy of radiation on the electromagnetic spectrum (5.1).

_____ Explain how atomic spectra correlate with the energy levels in atoms (5.2).

_____ Describe the sublevels and orbitals in atoms (5.3).

_____ Write the orbital diagrams and electron configurations for hydrogen to argon (5.4).

_____ Use the sublevel blocks on the periodic table to write electron configurations (5.5).

_____ Explain the trends in valence electrons, group numbers, atomic size, and ionization energy going down a group and across a period (5.6).

_____ Compare the sizes of atoms and their ions (5.6).

Practice Test for Chapter 5

1. Which of the following has the longest wavelength?
A. microwave **B.** ultraviolet **C.** radio waves **D.** infrared **E.** visible

2. Which of the following has the highest frequency?
A. microwave **B.** ultraviolet **C.** radio waves **D.** infrared **E.** visible

3. What color of visible light has the highest frequency?
A. violet **B.** blue **C.** green **D.** yellow **E.** red

4. What color of visible light has the longest wavelength?
A. violet **B.** blue **C.** green **D.** yellow **E.** red

5. Arrange the following in order of increasing energy: infrared, ultraviolet, radio waves, X-rays.
A. X-rays, infrared, ultraviolet, radio waves
B. radio waves, ultraviolet, infrared, X-rays
C. ultraviolet, infrared, radio waves, X-rays
D. radio waves, X-rays, infrared, ultraviolet
E. radio waves, infrared, ultraviolet, X-rays

6. The wavelength is
A. the height to the peak of a wave
B. the length of a wave
C. the distance between two adjacent peaks of a wave
D. the number of waves that pass a point in 1 s
E. related directly to the frequency of a wave of electromagnetic radiation

7. A microwave has a wavelength of 3.0×10^{-2} m. This is equal to a wavelength of
A. 3.0×10^{-7} nm **B.** 3.0×10^{7} nm **C.** 3.0×10^{11} nm
D. 3.0×10^{8} nm **E.** 3.0×10^{17} nm

8. Atomic spectra
A. consist of lines of different colors
B. are different for each of the elements
C. occur when energy changes provide photons of certain wavelengths
D. indicate that electrons in atoms have specific energy levels
E. all of these

9. The maximum number of electrons that can be accommodated in the principal energy level $n = 4$ is
A. 2 **B.** 4 **C.** 8 **D.** 18 **E.** 32

10. The sublevels that make up the principal energy level $n = 3$ are
A. $1s2s3s$ **B.** $3s3p$ **C.** $3s3p3d$ **D.** $3s3p3d3f$ **E.** $3p3d3f$

11. A space within an atom that has a spherical shape is called a(n)
A. *d* orbital **B.** *p* orbital **C.** *s* orbital **D.** *f* orbital **E.** *g* orbital

12. The number of orbitals in a $4p$ sublevel is
A. one **B.** two **C.** three **D.** four **E.** five

13. The principal energy level that consists of one *s* orbital and three *p* orbitals has principal quantum number
A. $n = 1$ **B.** $n = 2$ **C.** $n = 3$ **D.** $n = 4$ **E.** $n = 5$

14. The maximum number of electrons that the $5d$ sublevel can accommodate is
A. 2 **B.** 3 **C.** 6 **D.** 10 **E.** 14

15. In the orbital diagram of nitrogen, the $2p$ sublevel would be

A.

↑↓		↑

B.

↑↑	↑	

C.

↑↓	↑↓	↑

D.

↑	↑	↑

E.

↑		↑↓

16. The electron configuraton for silicon is
A. $1s^22s^22p^63s^4$ **B.** $1s^22s^22p^63s^23p^2$ **C.** $1s^22s^22p^63s^23d^2$
D. $1s^22s^22p^63s^23p^4$ **E.** $1s^22s^22p^63p^4$

17. The electron configuration for oxygen is
A. $2s^22p^4$ **B.** $1s^22s^42p^4$ **C.** $1s^22s^6$
D. $1s^22s^22p^23s^2$ **E.** $1s^22s^22p^4$

18. The electron configuration for aluminum is
A. $1s^22s^22p^9$ **B.** $1s^22s^22p^63p^5$ **C.** $1s^22s^22p^63s^23p^1$
D. $1s^22s^22p^83p^1$ **E.** $1s^22s^22p^63p^3$

For questions 19 through 22, match the final notation in the electron configuration with the following elements:
A. As **B.** Rb **C.** Na **D.** N **E.** Xe

19. $4p^3$ _____ **20.** $5s^1$ _____

21. $3s^1$ _____ **22.** $5p^6$ _____

23. The element that has an electron configuration that ends with $5p^2$ is
A. strontium **B.** germanium **C.** titanium **D.** tin **E.** lead

24. The element that has an abbreviated electron configuration of $[Ar]4s^2 3d^6$ is
A. chromium **B.** iron **C.** krypton **D.** calcium **E.** zinc

25. The element that begins filling the 6*s* sublevel is
A. lithium **B.** rubidium **C.** cesium **D.** scandium **E.** barium

26. The element that has 5 electrons in its principal energy level $n = 4$ is
A. boron **B.** phosphorus **C.** vanadium **D.** bromine **E.** arsenic

27. The number of electrons in the 5*p* sublevel of iodine is
A. 2 **B.** 3 **C.** 5 **D.** 7 **E.** 15

28. The number of valence electrons in gallium is
A. 1 **B.** 2 **C.** 3 **D.** 13 **E.** 31

29. The atomic radius of oxygen is larger than that of
A. lithium **B.** sulfur **C.** fluorine **D.** boron **E.** argon

30. Of C, Si, Ge, Sn, and Pb, the element with the greatest ionization energy is
A. C **B.** Si **C.** Ge **D.** Sn **E.** Pb

31. The electron-dot symbol • X • would be correct for an element in
A. Group 1A (1) **B.** Group 2A (2) **C.** Group 4A (14) **D.** Group 6A (16) **E.** Group 7A (17)

32. The electron-dot symbol X • would be correct for
A. magnesium **B.** chlorine **C.** sulfur **D.** cesium **E.** nitrogen

33. Which of the following elements will produce an ion larger than the corresponding atom?
A. Cl **B.** Al **C.** K **D.** Ca **E.** Li

Answers for the Practice Test

1. C	**2.** B	**3.** A	**4.** E	**5.** E
6. C	**7.** B	**8.** E	**9.** E	**10.** C
11. C	**12.** C	**13.** B	**14.** D	**15.** D
16. B	**17.** E	**18.** C	**19.** A	**20.** B
21. C	**22.** E	**23.** D	**24.** B	**25.** C
26. E	**27.** C	**28.** C	**29.** C	**30.** A
31. B	**32.** D	**33.** A		

6
Inorganic and Organic Compounds: Names and Formulas

Study Goals

- Use the octet rule to determine the ionic charge of ions for representative elements.
- Use charge balance to write the formula of an ionic compound.
- Write the formulas and names for ionic compounds including those containing polyatomic ions.
- Write the formulas and names of covalent compounds including those of alkanes.
- From its properties, classify a compound as organic or inorganic.
- Identify the properties that are most characteristic of organic and inorganic compounds.
- Describe the type of bonding in organic compounds.
- Use the IUPAC system to name the alkanes with 1 to 10 carbon atoms.
- Draw the expanded and condensed structural formulas for alkanes.

Think About It

1. Why does an ion have a charge?

2. How does a compound differ from an element?

3. How is a covalent bond different from an ionic bond?

4. What two elements are found in all organic compounds?

Key Terms

Match each the following key terms with the correct definition:

a. covalent bond **b.** alkane **c.** ionic bond **d.** cation
e. ion **f.** octet **g.** hydrocarbon **h.** polyatomic ion

1. ______ an atom with a positive or negative charge
2. ______ an arrangement of eight valence electrons
3. ______ a hydrocarbon that contains only carbon–carbon single bonds
4. ______ a group of covalently bonded atoms that carries an overall charge
5. ______ the attraction between positively and negatively charged particles
6. ______ sharing of valence electrons by two atoms
7. ______ a positively charged ion with a noble gas configuration
8. ______ an organic compound that consist of only carbon and hydrogen atoms

Answers				
	1. e	**2.** f	**3.** b	**4.** h
	5. c	**6.** a	**7.** d	**8.** g

6.1 Octet Rule and Ions

- The stability of the noble gases is associated with an electron configuration of 8 electrons (s^2p^6), an octet, in their outer energy level. Helium is stable with 2 electrons in the outer energy level.

  He $1s^2$ Ne $1s^22s2p^6$

  Ar $1s^22s^22p^63s^23p^6$ Kr $1s^22s^22p^63s^23p^64s^23d^{10}4p^6$
- Atoms of elements other than the noble gases achieve stability by losing, gaining, or sharing valence electrons with other atoms in the formation of compounds.
- Metals of the representative elements in Groups 1A (1), 2A (2) and 3A (13) achieve a noble gas electron configuration by losing their valence electron(s) to form positively charged cations with a charge of 1+, 2+, or 3+.
- Nonmetals in Groups 5A (15), 6A (16) and 7A (17) gain valence electrons to achieve an octet forming negatively charged ions with a charge of 3–, 2–, or 1–.

MasteringChemistry

Tutorial: Octet Rule and Ions
Tutorial: Writing Electron-Dot Formulas

Study Note

When an atom loses or gains electrons, it acquires the electron configuration of the nearest noble gas. For example, sodium loses one electron, which gives the Na^+ ion the electron configuration of neon. Oxygen gains two electrons to give an oxide ion, O^{2-}, the electron configuration of neon.

◆ Learning Exercise 6.1A

The following elements lose electrons when they form ions. Indicate the group number, the number of electrons lost, and the ion (symbol and charge) for each of the following:

Element	Group Number	Electrons Lost	Ion Formed
Magnesium			
Sodium			
Calcium			
Potassium			
Aluminum			

Answers

Element	Group Number	Electrons Lost	Ion Formed
Magnesium	2A (2)	2	Mg^{2+}
Sodium	1A (1)	1	Na^{+}
Calcium	2A (2)	2	Ca^{2+}
Potassium	1A (1)	1	K^{+}
Aluminum	3A (13)	3	Al^{3+}

Study Note

The valence electrons are the electrons in the outermost energy level of an atom. For representative elements, the number of valence electrons is related to the group number.

◆ Learning Exercise 6.1B

The following elements gain electrons when they form ions. Indicate the group number, the number of electrons gained, and the ion (symbol and charge) for each of the following:

Element	Group Number	Electrons Gained	Ion Formed
Chlorine			
Oxygen			
Nitrogen			
Fluorine			
Sulfur			

Answers

Element	Group Number	Electrons Gained	Ion Formed
Chlorine	7A (17)	1	Cl^-
Oxygen	6A (16)	2	O^{2-}
Nitrogen	5A (15)	3	N^{3-}
Fluorine	7A (17)	1	F^-
Sulfur	6A (16)	2	S^{2-}

6.2 Ionic Compounds

- In the formulas of ionic compounds, the total positive charge is equal to the total negative charge. For example, the compound magnesium chloride, $MgCl_2$, contains Mg^{2+} and two Cl^-. The sum of the charges is zero: $(2+) + 2(1-) = 0$.
- When two or more ions are needed for charge balance, that number is indicated by subscripts in the formula.

MasteringChemistry
Tutorial: Ions
Tutorial: Ionic Compounds
Tutorial: Writing Ionic Formulas

For the following exercises, you may want to cut pieces of paper that represent typical positive and negative ions, as shown. To determine an ionic formula, place the positive and negative pieces together to complete a geometric shape. Write the number of positive ions and negative ions as the subscripts for the formula.

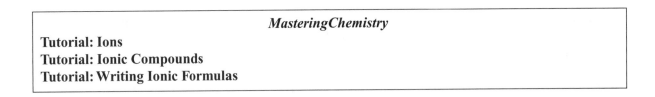

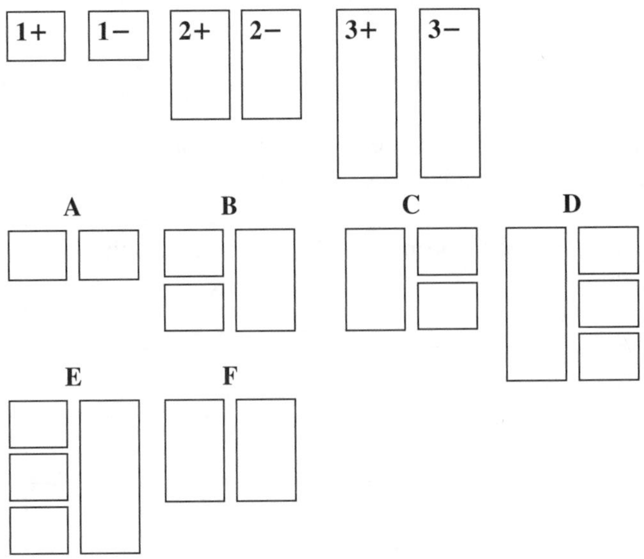

◆ Learning Exercise 6.2A

Give the letter (A, B, C, etc.) that matches the arrangement of ions in the following compounds:

Compound	Combination	Compound	Combination
1. $MgCl_2$	________	**2.** Na_2S	________
3. LiCl	________	**4.** CaO	________
5. K_3N	________	**6.** $AlBr_3$	________
7. MgS	________	**8.** $BaCl_2$	________

Answers **1.** C **2.** B **3.** A **4.** F
5. E **6.** D **7.** F **8.** C

Study Note

You can check that the ionic formula you write is electrically neutral by multiplying each of the ionic charges by their subscripts. When added together, their sum should equal zero. For example, the formula Na_2O gives $2(1+) + 1(2-) = (2+) + (2-) = 0$.

◆ Learning Exercise 6.2B

Write the correct formula of the ionic compound formed from each of the following pairs of ions:

1. Na^+ and Cl^- ____________ **2.** K^+ and S^{2-} ____________

3. Al^{3+} and O^{2-} ____________ **4.** Mg^{2+} and Cl^- ____________

5. Ca^{2+} and S^{2-} ____________ **6.** Al^{3+} and Cl^- ____________

7. Li^+ and N^{3-} ____________ **8.** Ba^{2+} and P^{3-} ____________

Answers **1.** NaCl **2.** K_2S **3.** Al_2O_3 **4.** $MgCl_2$
5. CaS **6.** $AlCl_3$ **7.** Li_3N **8.** Ba_3P_2

6.3 Naming and Writing Ionic Compounds

- In naming ionic compounds, the positive ion is named first, followed by the name of the negative ion. The name of a representative metal ion in Group 1A (1), 2A (2), or 3A (13) is the same as its elemental name. The name of a nonmetal ion is obtained by replacing the end of its element name with *ide*.
- Most transition elements form cations with two or more ionic charges. Then the ionic charge must be written as a Roman numeral in parentheses after the name of the metal. For example, the cations of iron, Fe^{2+} and Fe^{3+}, are named iron(II) and iron(III). The ions of copper are Cu^+, copper(I), and Cu^{2+}, copper(II).
- The only transition elements with fixed charges are zinc, Zn^{2+}; silver, Ag^+; and cadmium, Cd^{2+}.

MasteringChemistry

Tutorial: Writing Ionic Formulas

Guide to Naming Ionic Compounds with Metals that Form a Single Ion

STEP 1 Identify the cation and anion.
STEP 2 Name the cation by its element name.
STEP 3 Name the anion by using the first syllable of its element name followed by *ide.*
STEP 4 Write the name of the cation first and the anion second.

◆ Learning Exercise 6.3A

Write the ions and the correct formula for each of the following ionic compounds:

Formula	Ions		Name
1. Cs_2O	______	______	______________________
2. $BaBr_2$	______	______	______________________
3. Mg_3P_2	______	______	______________________
4. Na_2S	______	______	______________________

Answers
1. Cs^+, O^{2-}, cesium oxide
2. Ba^{2+}, Br^-, barium bromide
3. Mg^{2+}, P^{3-}, magnesium phosphide
4. Na^+, S^{2-}, sodium sulfide

◆ Learning Exercise 6.3B

Write the name of each of the following ions:

1. Cl^-	________	2. Fe^{2+}	________
3. Cu^+	________	4. Ag^+	________
5. O^{2-}	________	6. Ca^{2+}	________
7. S^{2-}	________	8. Al^{3+}	________
9. Fe^{3+}	________	10. Ba^{2+}	________
11. Cu^{2+}	________	12. N^{3-}	________

Answers
1. chloride
2. iron(II)
3. copper(I)
4. silver
5. oxide
6. calcium
7. sulfide
8. aluminum
9. iron(III)
10. barium
11. copper(II)
12. nitride

◆ Learning Exercise 6.3C

Write the symbol of each of the following ions:

Name	Symbol	Name	Symbol
1. iron(III) ion	________	2. cobalt(II) ion	________
3. zinc ion	________	4. lead(IV) ion	________
5. copper(I) ion	________	6. silver ion	________
7. potassium ion	________	8. nickel(III) ion	________

Answers
1. Fe^{3+}
2. Co^{2+}
3. Zn^{2+}
4. Pb^{4+}
5. Cu^+
6. Ag^+
7. K^+
8. Ni^{3+}

Guide to Naming Ionic Compounds with Variable Charge Metals

STEP 1 Determine the charge of the cation from the anion.
STEP 2 Name the cation by its element name and use a Roman numeral in parentheses for the charge.
STEP 3 Name the anion by using the first syllable of its element name followed by *ide*.
STEP 4 Write the name of the cation first and the name of the anion second.

◆ Learning Exercise 6.3D

Write the ions and a correct name for each of the following ionic compounds:

Formula	Ions		Name
1. $CrCl_2$	______	______	________________________
2. $SnBr_4$	______	______	________________________
3. Na_3P	______	______	________________________
4. Ni_2O_3	______	______	________________________
5. CuO	______	______	________________________
6. Mg_3N_2	______	______	________________________

Answers

1. Cr^{2+}, Cl^-, chromium(II) chloride
2. Sn^{4+}, Br^-, tin(IV) bromide
3. Na^+, P^{3-}, sodium phosphide
4. Ni^{3+}, O^{2-}, nickel(III) oxide
5. Cu^{2+}, O^{2-}, copper(II) oxide
6. Mg^{2+}, N^{3-}, magnesium nitride

Guide to Writing Formulas from the Name of An Ionic Compound

STEP 1 Identify the cation and anion.
STEP 2 Balance the charges.
STEP 3 Write the formula, cation first, using subscripts from charge balance.

◆ Learning Exercise 6.3E

Write the ions and the correct ionic formula for each of the following ionic compounds:

Compound	Positive Ion	Negative Ion	Formula of Compound
Aluminum sulfide			
Copper(II) chloride			
Magnesium oxide			
Gold(III) bromide			
Silver oxide			

Answers

Compound	Positive Ion	Negative Ion	Formula of Compound
Aluminum sulfide	Al^{3+}	S^{2-}	Al_2S_3
Copper(II) chloride	Cu^{2+}	Cl^-	$CuCl_2$
Magnesium oxide	Mg^{2+}	O^{2-}	MgO
Gold(III) bromide	Au^{3+}	Br^-	$AuBr_3$
Silver oxide	Ag^+	O^{2-}	Ag_2O

6.4 Polyatomic Ions

- A polyatomic ion is a group of nonmetal atoms that carries an electrical charge, usually negative, 1−, 2−, or 3−. The polyatomic ion NH_4^+ has a positive charge.
- Polyatomic ions cannot exist alone, but are combined with an ion of the opposite charge.
- Ionic compounds containing three elements (polyatomic ions) end with *ate* or *ite*.
- When there is more than one polyatomic ion in the formula for a compound, the entire polyatomic ion formula is enclosed in parentheses and the subscript written outside the parentheses.

MasteringChemistry

Tutorial: Polyatomic Ions

Study Note

By learning the most common polyatomic ions such as nitrate NO_3^-, carbonate CO_3^{2-}, sulfate SO_4^{2-}, and phosphate PO_4^{3-}, you can derive their related polyatomic ions. For example, the nitrite ion, NO_2^-, has one oxygen atom less than the nitrate ion: NO_3^-, nitrate, and NO_2^-, nitrite.

◆ Learning Exercise 6.4A

Write the polyatomic ion (symbol and charge) for each of the following:

1. Sulfate ion ______ **2.** Hydroxide ion ______

3. Carbonate ion ______ **4.** Sulfite ion ______

5. Ammonium ion ______ **6.** Phosphate ion ______

7. Nitrate ion ______ **8.** Nitrite ion ______

Answers **1.** SO_4^{2-} **2.** OH^- **3.** CO_3^{2-} **4.** SO_3^{2-}
5. NH_4^+ **6.** PO_4^{3-} **7.** NO_3^- **8.** NO_2^-

◆ Learning Exercise 6.4B

Write the formula of the ions, and the correct formula for each of the following compounds:

Compound	Positive Ion	Negative Ion	Formula
Sodium phosphate			
Iron(II) hydroxide			
Ammonium carbonate			
Silver bicarbonate			
Chromium(III) sulfate			
Lead(II) nitrate			
Potassium sulfite			
Barium phosphate			

Answers

Compound	Positive Ion	Negative Ion	Formula
Sodium phosphate	Na^+	PO_4^{3-}	Na_3PO_4
Iron(II) hydroxide	Fe^{2+}	OH^-	$Fe(OH)_2$
Ammonium carbonate	NH_4^+	CO_3^{2-}	$(NH_4)_2CO_3$
Silver bicarbonate	Ag^+	HCO_3^-	$AgHCO_3$
Chromium(III) sulfate	Cr^{3+}	SO_4^{2-}	$Cr_2(SO_4)_3$
Lead(II) nitrate	Pb^{2+}	NO_3^-	$Pb(NO_3)_2$
Potassium sulfite	K^+	SO_3^{2-}	K_2SO_3
Barium phosphate	Ba^{2+}	PO_4^{3-}	$Ba_3(PO_4)_2$

Guide to Naming Ionic Compounds with Polyatomic Ions

STEP 1 Identify the cation and polyatomic ion (anion).
STEP 2 Name the cation, using a Roman numeral in parentheses, if needed.
STEP 3 Name the polyatomic ion usually ending in *ite* or *ate*.
STEP 4 Write the name of the compound, cation first and the polyatomic ion second.

◆ Learning Exercise 6.4C

Write the ions and a correct name for each of the following ionic compounds:

Formula	Ions		Name
1. $Ba(NO_3)_2$	______	______	______________________________
2. $Fe_2(SO_4)_3$	______	______	______________________________
3. Na_3PO_3	______	______	______________________________
4. $Al(ClO_3)_3$	______	______	______________________________
5. $(NH_4)_2CO_3$	______	______	______________________________
6. $Cr(OH)_2$	______	______	______________________________

Answers
1. Ba^{2+}, NO_3^-, barium nitrate
2. Fe^{3+}, SO_4^{2-}, iron(III) sulfate
3. Na^+, PO_3^{3-}, sodium phosphite
4. Al^{3+}, $C1O_3^-$, aluminum chlorate
5. NH_4^+, CO_3^{2-}, ammonium carbonate
6. Cr^{2+}, OH^-, chromium(II) hydroxide

6.5 Covalent Compounds And Their Names

- In a covalent bond, atoms of nonmetals share electrons to achieve an octet. For example, oxygen with six valence electrons shares electrons with two hydrogen atoms to form the covalent compound water (H_2O). In water, oxygen has two lone pairs and two bonding pairs.

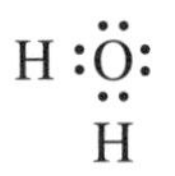

- Covalent compounds are composed of nonmetals bonded together to give discrete units called molecules.
- The formula of a covalent compound is written using the order of the symbols of the elements in the name, followed by subscripts indicated by prefixes.

MasteringChemistry
Self Study Activity: Covalent Bonds
Tutorial: Covalent Compounds and the Octet Rule
Tutorial: Naming Covalent Compounds
Tutorial: Atoms, Molecular Elements, Molecular Compounds, and Ionic Compounds
Tutorial: Covalent Molecules and the Octet Rule

Guide to Naming Covalent Compounds with Two Nonmetals

STEP 1 Name the first nonmetal by its element name.
STEP 2 Name the second nonmetal by using its first syllable of its element name followed by *ide*.
STEP 3 Add prefixes to indicate the number of atoms (subscripts).

◆ Learning Exercise 6.5A

Use the appropriate prefixes to name the following covalent compounds:

1. CS_2 ____________	**2.** CCl_4 ____________
3. CO ____________	**4.** SO_3 ____________
5. N_2O_4 ____________	**6.** PCl_3 ____________
7. P_4S_6 ____________	**8.** IF_7 ____________
9. ClO_2 ____________	**10.** S_2O ____________

Answers

1. carbon disulfide	**2.** carbon tetrachloride
3. carbon monoxide	**4.** sulfur trioxide
5. dinitrogen tetroxide	**6.** phosphorus trichloride
7. tetraphosphorus hexasulfide	**8.** iodine heptafluoride
9. chlorine dioxide	**10.** disulfur oxide

Guide to Writing Formulas for Covalent Compounds

STEP 1 Write the symbols in the order of the elements in the name.
STEP 2 Write any prefixes as subscripts.

◆ Learning Exercise 6.5B

Write the formula of each of the following covalent compounds:

1. dinitrogen oxide ____________	**2.** silicon tetrabromide ____________
3. nitrogen trichloride ____________	**4.** carbon dioxide ____________
5. sulfur hexafluoride ____________	**6.** oxygen difluoride ____________
7. phosphorus trifluoride ____________	**8.** phosphorus trihydride ____________
9. iodine trifluoride ____________	**10.** sulfur dioxide ____________

Answers

1. N_2O	**2.** $SiBr_4$	**3.** NCl_3	**4.** CO_2	**5.** SF_6
6. OF_2	**7.** PF_3	**8.** PH_3	**9.** IF_3	**10.** SO_2

6.6 Organic Compounds

- Organic compounds are compounds of carbon that typically have covalent bonds, low melting and boiling points, burn vigorously, are nonelectrolytes, and are soluble in nonpolar solvents.
- Alkanes are hydrocarbons that have only single bonds, C—C and C—H.
- Each carbon in an alkane has four bonds arranged so that the bonded atoms are in the corners of a tetrahedron.

> *MasteringChemistry*
>
> **Tutorial: Introduction to Organic Molecules**

◆ Learning Exercise 6.6A

Identify the following as typical of organic (O) or inorganic (I) compounds:

1. _____ have covalent bonds
2. _____ have low boiling points
3. _____ burn in air
4. _____ are soluble in water
5. _____ have high melting points
6. _____ are soluble in nonpolar solvents
7. _____ have ionic bonds
8. _____ form long chains
9. _____ contain carbon
10. _____ do not burn in air

Answers					
	1. O	**2.** O	**3.** O	**4.** I	**5.** I
	6. O	**7.** I	**8.** O	**9.** O	**10.** I

6.7 Names and Formulas of Hydrocarbons: Alkanes

- An expanded structural formula shows the bonds between all the atoms.
- A condensed structural formula depicts each carbon atom and its attached hydrogen atoms as a group.

Expanded Structural Formula	Condensed Structural Formula	Molecular Formula
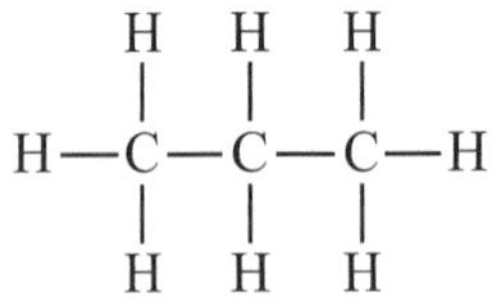	$CH_3—CH_2—CH_3$	C_3H_8

- In the IUPAC system used to name organic compounds, the alkanes containing one, two, three, or four carbon atoms connected in a row or a *continuous* chain are named: methane, ethane, propane, and butane, respectively, as shown:

Name	Number of Carbon Atoms	Condensed Structural Formula
Methane	1	CH_4
Ethane	2	$CH_3—CH_3$
Propane	3	$CH_3—CH_2—CH_3$
Butane	4	$CH_3—CH_2—CH_2—CH_3$
Pentane	5	$CH_3—CH_2—CH_2—CH_2—CH_3$

- Alkanes with five or more carbon atoms in a chain are named using the prefix *pent* (5), *hex* (6), *hept* (7), *oct* (8), *non* (9), or *dec* (10) and the suffix *ane*, which indicates single bonds (alkane).

> *MasteringChemistry*
>
> **Tutorial: IUPAC Naming of Alkanes**

◆ Learning Exercise 6.7A

Write the IUPAC name for each of the following:

1. $CH_3—CH_2—CH_3$ ____________________
2. $CH_3—CH_3$ ____________________
3. $CH_3—CH_2—CH_2—CH_3$ ____________________
4. $CH_3—CH_2—CH_2—CH_2—CH_2—CH_3$ ____________________

Answers **1.** propane **2.** ethane **3.** butane **4.** hexane

◆ Learning Exercise 6.7B

Draw the condensed structural formula for each of the following compounds:

1. hexane ______________________________
2. methane ______________________________
3. butane ______________________________
4. heptane ______________________________

Answers
1. $CH_3—CH_2—CH_2—CH_2—CH_2—CH_3$
2. CH_4
3. $CH_3—CH_2—CH_2—CH_3$
4. $CH_3—CH_2—CH_2—CH_2—CH_2—CH_2—CH_3$

Checklist for Chapter 6

You are ready to take the Practice Test for Chapter 6. Be sure you have accomplished the following learning goals for this chapter. If you are not sure, review the section listed at the end of the goal. Then apply your new skills and understanding to the Practice Test.

After studying Chapter 6, I can successfully:

_____ Illustrate the octet rule for the formation of ions (6.1).

_____ Write the formulas of ionic compounds containing the ions of metals and nonmetals of representative elements (6.2).

_____ Use charge balance to write an ionic formula (6.3).

_____ Write the name of an ionic compound (6.3).

_____ Write the formula of a compound containing a polyatomic ion (6.4).

_____ Write the names and formulas of covalent compounds (6.5).

_____ Identify properties as characteristic of organic or inorganic compounds (6.6).

_____ Use the IUPAC system to name alkanes with 1 to 10 carbon atoms (6.7).

_____ Draw the condensed structural formulas of alkanes with 1 to 10 carbon atoms (6.7).

Practice Test for Chapter 6

For questions 1 through 4, consider an atom of phosphorus.

1. Phosphorus is in group
   A. 2A (2) **B.** 3A (13) **C.** 5A (15) **D.** 7A (17) **E.** 8A (18)

2. How many valence electrons does an atom of phosphorus have?
   A. 2 **B.** 3 **C.** 5 **D.** 8 **E.** 15

3. To achieve an octet, a phosphorus atom will
   A. lose 1 electron **B.** lose 2 electrons **C.** lose 5 electrons
   D. gain 2 electrons **E.** gain 3 electrons

4. A phosphide ion has charge of
   A. 1+ **B.** 2+ **C.** 5+ **D.** 2− **E.** 3−

5. To achieve an octet, a calcium atom
   A. loses 1 electron **B.** loses 2 electrons **C.** loses 3 electrons
   D. gains 1 electron **E.** gains 2 electrons

6. To achieve an octet, a chlorine atom
   A. loses 1 electron **B.** loses 2 electrons **C.** loses 3 electrons
   D. gains 1 electron **E.** gains 2 electrons

7. Another name for a positive ion is
   A. anion **B.** cation **C.** proton **D.** positron **E.** sodium

8. The correct ionic charge for a calcium ion is
   A. 1+ **B.** 2+ **C.** 1− **D.** 2− **E.** 3−

9. The silver ion has a charge of
   A. 1+ **B.** 2+ **C.** 1− **D.** 2− **E.** 3−

10. The correct ionic charge for a phosphate ion is
    A. 1+ **B.** 2+ **C.** 1− **D.** 2− **E.** 3−

11. The correct ionic charge for fluoride is
    A. 1+ **B.** 2+ **C.** 1− **D.** 2− **E.** 3−

12. The correct ionic charge for a sulfate ion is
    A. 1+ **B.** 2+ **C.** 1− **D.** 2− **E.** 3−

13. When the elements magnesium and sulfur are mixed,
    A. an ionic compound forms
    B. a covalent compound forms
    C. no reaction occurs
    D. the two repel each other and will not combine
    E. none of the above

14. An ionic bond typically occurs between
    A. two different nonmetals
    B. two of the same type of nonmetals
    C. two noble gases
    D. two different metals
    E. a metal and a nonmetal

15. The formula for a compound between sodium and sulfur is
    A. SoS **B.** NaS **C.** Na_2S **D.** NaS_2 **E.** Na_2SO_4

16. The formula for a compound between aluminum and oxygen is
A. AlO **B.** Al_2O **C.** AlO_3 **D.** Al_2O_3 **E.** Al_3O_2

17. The formula for a compound between barium and sulfur is
A. BaS **B.** Ba_2S **C.** BaS_2 **D.** Ba_2S_2 **E.** $BaSO_4$

18. The correct formula for iron(III) chloride is
A. FeCl **B.** $FeCl_2$ **C.** Fe_2Cl **D.** Fe_3Cl **E.** $FeCl_3$

19. The correct formula for ammonium sulfate is
A. AmS **B.** $AmSO_4$ **C.** $(NH_4)_2S$ **D.** NH_4SO_4 **E.** $(NH_4)_2SO_4$

20. The correct formula for copper(II) chloride is
A. CoCl **B.** CuCl **C.** $CoCl_2$ **D.** $CuCl_2$ **E.** Cu_2Cl

21. The correct formula for lithium phosphate is
A. $LiPO_4$ **B.** Li_2PO_4 **C.** Li_3PO_4 **D.** $Li_2(PO_4)_3$ **E.** $Li_3(PO_4)_2$

22. The correct formula for silver oxide is
A. AgO **B.** Ag_2O **C.** AgO_2 **D.** Ag_3O_2 **E.** Ag_3O

23. The correct formula for magnesium carbonate is
A. $MgCO_3$ **B.** Mg_2CO_3 **C.** $Mg(CO_3)_2$ **D.** MgCO **E.** $Mg_2(CO_3)_3$

24. The correct formula for copper(I) sulfate is
A. $CuSO_3$ **B.** $CuSO_4$ **C.** Cu_2SO_3 **D.** $Cu(SO_4)_2$ **E.** Cu_2SO_4

25. The name of $Al_2(HPO_4)_3$ is
A. aluminum hydrogen phosphite
B. aluminum hydrogen phosphate
C. aluminum hydrogen phosphorus
D. aluminum hydrogen phosphorus oxide
E. trialuminum diphosphate

26. The name of CoS is
A. copper sulfide **B.** cobalt(II) sulfate **C.** cobalt(I) sulfide
D. cobalt sulfide **E.** cobalt(II) sulfide

27. The name of $MnCl_2$ is
A. magnesium chloride **B.** manganese(II) chlorine **C.** manganese(II) chloride
D. manganese chlorine **E.** manganese(III) chloride

28. The name of $ZnCO_3$ is
A. zinc(III) carbonate **B.** zinc(II) carbonate **C.** zinc bicarbonate
D. zinc carbon trioxide **E.** zinc carbonate

29. The name of Al_2O_3 is
A. aluminum oxide **B.** aluminum(II) oxide **C.** aluminum trioxide
D. dialuminum trioxide **E.** aluminum oxygenate

30. The name of $Cr_2(SO_3)_3$ is
A. chromium sulfite **B.** dichromium trisulfite **C.** chromium(III) sulfite
D. chromium(III) sulfate **E.** chromium sulfate

31. The name of PH_3 is
A. potassium trihydrogen
B. phosphorus trihydride
C. phosphorus trihydrogen
D. potassium trihydride
E. phosphorus trihydrogenide

32. The name of NCl_3 is
A. nitrogen chloride **B.** nitrogen trichloride **C.** trinitrogen chloride
D. nitrogen chlorine three **E.** nitrogen trichlorine

33. The name of CO is
A. carbon monoxide **B.** carbonic oxide **C.** carbon oxide
D. carbonious oxide **E.** carboxide

34. The formula for a compound between carbon and chlorine is
A. CCl **B.** CCl_2 **C.** C_4Cl **D.** CCl_4 **E.** C_4Cl_2

35.

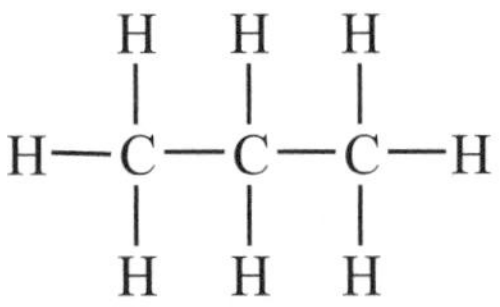

This is an example of a(n)
A. ball-and-stick model **B.** condensed structural formula **C.** space-filling model
D. tetrahedral model **E.** expanded structural formula

For questions 36 through 41, indicate whether the following characteristics are typical of organic (O) compounds or inorganic (I) compounds:

36. _____ higher melting points
37. _____ covalent bonds
38. _____ soluble in water
39. _____ ionic bonds
40. _____ burn easily in air
41. _____ soluble in nonpolar solvents

For questions 42 and 43, match the name of the hydrocarbon with each of the following:
A. ethane **B.** butane **C.** hexane **D.** heptane **E.** octane

42. $CH_3—CH_3$

43. $CH_3—CH_2—CH_2—CH_2—CH_2—CH_2—CH_2—CH_3$

44. $CH_3—CH_2—CH_2—CH_2—CH_2—CH_3$

45. $CH_3—CH_2—CH_2—CH_3$

46. $CH_3—CH_2—CH_2—CH_2—CH_2—CH_2—CH_3$

Answers to the Practice Test

1. C	**2.** C	**3.** E	**4.** E	**5.** B
6. D	**7.** B	**8.** B	**9.** A	**10.** E
11. C	**12.** D	**13.** A	**14.** E	**15.** C
16. D	**17.** A	**18.** E	**19.** E	**20.** D
21. C	**22.** B	**23.** A	**24.** E	**25.** B
26. E	**27.** C	**28.** E	**29.** A	**30.** C
31. B	**32.** B	**33.** A	**34.** D	**35.** E
36. I	**37.** O	**38.** I	**39.** I	**40.** O
41. O	**42.** A	**43.** E	**44.** C	**45.** B
46. D				

7

Chemical Quantities

Study Goals

- Use Avogadro's number to determine the number of particles in a given number of moles of an element or compound.
- Calculate the molar mass of a compound using its formula and the atomic masses on the periodic table.
- Use the molar mass to convert between the grams of a substance and the number of moles.
- Use the formula of a compound to calculate its percent composition.
- From percent composition, determine the empirical formula of a compound.
- Using molar mass and the empirical formula, determine the molecular formula.

Think About It

1. How is a mole analogous to a term for a collection such as one dozen?

2. How does the mole allow a conversion between mass and number of particles in a sample?

3. How are the atomic masses of elements in a compound used to calculate its percent composition?

4. What is the difference between the empirical formula and molecular formula of a compound?

Key Terms

Match the following terms with the statements below:

a. mole　　**b.** Avogadro's number　　**c.** empirical formula
d. molar mass　　**e.** molecular formula

1. _____ the formula that gives the actual number of atoms of each element in a compound

2. _____ the number of items in 1 mol that is equal to 6.022×10^{23}

3. _____ the smallest whole-number ratio of atoms in a formula

4. _____ the amount of a substance that contains 6.022×10^{23} particles of that substance

5. _____ the mass in grams of an element or compound that is equal numerically to its atomic mass or sum of atomic masses

Answers **1.** e **2.** b **3.** c **4.** a **5.** d

7.1 The Mole

- A mole of any compound contains Avogadro's number, 6.022×10^{23}, of particles.
  1 mol $= 6.022 \times 10^{23}$ particles (atoms, molecules, ions, formula units)
- Avogadro's number provides a conversion factor between moles and number of particles:

$$\frac{1 \text{ mol}}{6.022 \times 10^{23} \text{ particles}} \quad \text{and} \quad \frac{6.022 \times 10^{23} \text{ particles}}{1 \text{ mol}}$$

- The subscripts in a formula indicate the number of moles of each element in one mole of the compound.

Guide to Calculating the Atoms or Molecules of a Substance

STEP 1 Determine the given number of moles.
STEP 2 Write a plan to convert moles to atoms or molecules.
STEP 3 Use Avogadro's number to write conversion factors.
STEP 4 Set up problem to convert given moles to atoms or molecules.

MasteringChemistry

Tutorial: Using Avogadro's Number

◆ Learning Exercise 7.1A

Use Avogadro's number to calculate each of the following:

a. number of Ca atoms in 3.00 mol of Ca

b. number of Zn atoms in 0.250 mol of Zn

c. number of SO_2 molecules in 0.118 mol of SO_2

d. number of moles of Ag in 4.88×10^{23} atoms of Ag

e. number of moles of NH_3 (ammonia) in 7.52×10^{23} molecules of NH_3

Answers

a. 1.81×10^{24} atoms of Ca **b.** 1.51×10^{23} atoms of Zn
c. 7.11×10^{22} molecules of SO_2 **d.** 0.810 mol of Ag
e. 1.25 mol of NH_3

Study Note

The subscripts in the formula of a compound indicate the number of moles of each element in one mol of that compound. For example, consider the formula Mg_3N_2:

$$1 \text{ mol of } Mg_3N_2 = 3 \text{ mol of Mg atoms and } 2 \text{ mol of N atoms}$$

Some conversion factors for the moles of elements can be written as

$$\frac{3 \text{ mol Mg atoms}}{1 \text{ mol } Mg_3N_2} \text{ and } \frac{1 \text{ mol } Mg_3N_2}{3 \text{ mol Mg atoms}} \qquad \frac{2 \text{ mol N atoms}}{1 \text{ mol } Mg_3N_2} \text{ and } \frac{1 \text{ mol } Mg_3N_2}{2 \text{ mol N atoms}}$$

MasteringChemistry

Tutorial: Moles and the Chemical Formula

◆ Learning Exercise 7.1B

Consider the formula for vitamin C (ascorbic acid) $C_6H_8O_6$.

a. How many moles of carbon are in 2.0 mol of vitamin C?

b. How many moles of hydrogen are in 5.0 mol of vitamin C?

c. How many moles of oxygen are in 1.5 mol of vitamin C?

Answers

a. 12 mol of carbon **b.** 40. mol of hydrogen
c. $6 \times 1.5 = 9.0$ mol (2 SFs) mol of oxygen

◆ Learning Exercise 7.1C

For the compound ibuprofen ($C_{13}H_{18}O_2$) used in Advil and Motrin, determine each of the following:

a. moles of carbon atoms in 2.20 mol of ibuprofen

b. moles of hydrogen in 4.40×10^{24} molecules of ibuprofen

c. atoms of oxygen in 0.750 mol of ibuprofen

d. molecules of ibuprofen that contain 15 mol of hydrogen

e. molecules of ibuprofen that contain 2.84×10^{23} atoms of hydrogen

Answers
a. 28.6 mol of C
b. 132 mol of H
c. 9.03×10^{23} atoms of O
d. 5.0×10^{23} molecules of ibuprofen
e. 1.58×10^{22} molecules of ibuprofen

7.2 Molar Mass

- The molar mass (g/mol) of an element is numerically equal to its atomic mass in grams.
- The molar mass (g/mol) of a compound is determined by multiplying the molar mass of each element by its subscript in the formula and adding the results.

Guide to Calculating Molar Mass

STEP 1 Obtain the molar mass of each element.
STEP 2 Multiply each molar mass by the number of moles (subscript) in the formula.
STEP 3 Calculate the molar mass by adding the masses of the elements.

Example: What is the molar mass of silver nitrate, $AgNO_3$?

$$1\ \cancel{\text{mol Ag}} \times \frac{107.9\text{ g Ag}}{1\ \cancel{\text{mol Ag}}} = 107.9\text{ g of Ag}$$

$$1\ \cancel{\text{mol N}} \times \frac{14.01\text{ g N}}{1\ \cancel{\text{mol N}}} = 14.01\text{ g of N}$$

$$3\ \cancel{\text{mol O}} \times \frac{16.00\text{ g O}}{1\ \cancel{\text{mol O}}} = 48.00\text{ g of O}$$

Molar mass of $AgNO_3$ = (107.9 g + 14.01 g + 48.00 g) = 169.9 g

◆ Learning Exercise 7.2A

Determine the molar mass for each of the following:

a. K_2O

b. $AlCl_3$

c. $C_{13}H_{18}O_2$

d. C_4H_{10}

e. $Ca(NO_3)_2$

f. Mg_3N_2

g. $FeCO_3$

h. $(NH_4)_3PO_4$

Answers				
	a. 94.20 g/mol	**b.** 133.33 g/mol	**c.** 206.3 g/mol	**d.** 58.12 g/mol
	e. 164.10 g/mol	**f.** 100.95 g/mol	**g.** 115.86 g/mol	**h.** 149.10 g/mol

7.3 Calculations Using Molar Mass

- The molar mass is useful as a conversion factor to change a given quantity in moles to grams.
- The two conversion factors for the molar mass of NaOH (40.01 g/mol) have the following form:

$$\frac{40.01\text{ g NaOH}}{1\text{ mol NaOH}} \quad \text{and} \quad \frac{1\text{ mol NaOH}}{40.01\text{ g NaOH}}$$

Guide to Calculating the Moles (or Grams) of a Substance from Grams (or Moles)

STEP 1 Determine the given number of moles.
STEP 2 Write a plan to convert moles to grams (or grams to moles).
STEP 3 Determine the molar mass and write conversion factors.
STEP 4 Set up problem to convert given moles to grams (or grams to moles).

Example: What is the mass in grams of 0.254 mol of Na_2CO_3?

$$0.254\ \cancel{\text{mol } Na_2CO_3} \times \frac{105.99\text{ g } Na_2CO_3}{1\ \cancel{\text{mol } Na_2CO_3}} = 26.9\text{ g of } Na_2CO_3$$

Molar mass

◆ Learning Exercise 7.3A

Calculate the number of grams in each of the following:

a. 0.75 mol of S

b. 3.18 mol of K_2SO_4

c. 2.50 mol of NH_4Cl

d. 1.25×10^{24} molecules of O_2

e. 3.2×10^{23} Pb atoms

f. 4.08 mol of $FeCl_3$

g. 2.28 mol of PCl_3

h. 0.815 mol of $Mg(NO_3)_2$

Answers				
	a. 24 g	**b.** 554 g	**c.** 134 g	**d.** 66.4 g
	e. 110 g	**f.** 662 g	**g.** 313 g	**h.** 121 g

Study Note

When the grams of a substance are given, the molar mass is used to calculate the number of moles of substance present.

$$\cancel{\text{grams substance}} \times \frac{1 \text{ mol substance}}{\cancel{\text{grams substance}}} = \text{mol of substance}$$

Example: How many moles of NaOH are in 4.00 g of NaOH?

$$4.00 \cancel{\text{ g NaOH}} \times \frac{1 \text{ mol NaOH}}{40.01 \cancel{\text{ g NaOH}}} = 0.100 \text{ mol of NaOH}$$

Molar Mass (inverted)

MasteringChemistry

Tutorial: Conversions Involving Moles
Tutorial: Converting Between Grams and Moles
Self Study Activity: Stoichiometry

◆ Learning Exercise 7.3B

Calculate the number of moles in each of the following quantities:

a. 108 g of CH_4

b. 6.12 g of CO_2

c. 38.7 g of $CaBr_2$

d. 236 g of Cl_2

e. 128 g of $Mg(OH)_2$

f. 172 g of Al_2O_3

g. The methane used to heat a home has a formula of CH_4. If 725 g of methane are used in one month, how many moles of methane were used during this time?

h. A vitamin tablet contains 18 mg of iron. If there are 100 tablets in a bottle, how many moles of iron are contained in the vitamins in the bottle?

Answers

	a. 6.73 mol	**b.** 0.139 mol	**c.** 0.194 mol	**d.** 3.33 mol
	e. 2.19 mol	**f.** 1.69 mol	**g.** 45.2 mol	**h.** 0.032 mol

Guide to Converting Grams to Particles

STEP 1 Write a plan that converts grams to particles.
STEP 2 Write the conversion factors for molar mass and Avogadro's number.
STEP 3 Set up the problem to convert grams to moles to particles.
STEP 4 Set up problem to convert given moles to grams (or grams to moles).

Example: How many carbonate ions are in 26.92 g of Na_2CO_3?

$$26.92\ \cancel{g\ Na_2CO_3} \times \underbrace{\frac{1\ \cancel{mol\ Na_2CO_3}}{105.99\ \cancel{g\ Na_2CO_3}}}_{\textit{Molar mass}} \times \underbrace{\frac{6.022 \times 10^{23}\ CO_3^{2-}}{1\ \cancel{mol\ Na_2CO_3}}}_{\textit{Avogadro's number}} = 1.530 \times 10^{23}\ CO_3^{2-}\ \text{ions}$$

◆ Learning Exercise 7.4B

Calculate the number of particles in each of the following quantities:

a. molecules of C_3H_8 in 20.8 g of C_3H_8

b. molecules of CO_2 in 6.12 g of CO_2

c. bromine atoms in 38.7 g of PBr_3

d. chlorine atoms in 236 g of Cl_2

e. hydroxide ions in 128 g of $Mg(OH)_2$

f. oxide ions in 172 g of Al_2O_3

Answers

a. 2.84×10^{23} molecules of C_3H_8
b. 8.37×10^{22} molecules of CO_2
c. 2.58×10^{23} atoms of bromine
d. 4.01×10^{24} atoms of chlorine
e. 2.64×10^{24} hydroxide ions
f. 3.05×10^{24} oxide ions

7.4 Percent Composition and Empirical Formulas

- Percent composition is the percent by mass of each element in a compound.
- An empirical (simplest) formula gives the lowest whole-number ratio of atoms in a compound.

Guide to Calculating Percent Composition

STEP 1 Determine the total mass of each element in molar mass of a formula.
STEP 2 Divide the total mass of each element by the molar mass and multiply by 100%.

$$\frac{\text{Mass of element}}{\text{Molar mass of compound}} \times 100\% = \%\text{ by mass of element}$$

Example: What is the percent Na in NaOH?

$$\frac{22.99\text{ g Na}}{40.01\text{ g NaOH}} \times 100\% = 57.46\%\text{ Na}$$

Study Note

When given the percent composition of a substance, assume that you have a 100-g sample of that substance. Then the mass of each element can be expressed in grams equal to its percent composition.

MasteringChemistry

Tutorial: Mass Percent

◆ Learning Exercise 7.4A

Calculate the percent composition by mass for each of the following compounds:

a. C_3H_6O

b. $BaSO_4$

c. $Ca(NO_3)_2$

d. NaOCl

e. $(NH_4)_3PO_4$

Answers

a. C 62.04%, H 10.41%, O 27.55%
b. Ba 58.83%, S 13.74%, O 27.42%
c. Ca 24.42%, N 17.07%, O 58.50%
d. Na 30.88%, O 21.49%, Cl 47.62%
e. N 28.19%, H 8.11%, P 20.77%, O 42.92%

Guide to Calculating Empirical Formula

STEP 1 Calculate the moles of each element.
STEP 2 Divide by the smallest number of moles.
STEP 3 Use the lowest whole number ratio of moles as subscripts.

Example: What is the empirical formula of a compound containing 27.3% C and 72.7% O by mass? If we assume we have a 100 g-sample of the compound, there would be 27.3 g of C and 72.7 g of O.

1. Moles of each element

$$27.3\ \cancel{g\,C} \times \frac{1\ \text{mol C}}{12.01\ \cancel{g\,C}} = 2.27\ \text{mol of C} \qquad 72.7\ \cancel{g\,O} \times \frac{1\ \text{mol O}}{16.00\ \cancel{g\,O}} = 4.54\ \text{mol of O}$$

2. Divide by the smaller number of moles, which is 2.27

$$\frac{2.27\ \text{mol C}}{2.27} = 1.00\ \text{mol of C} \qquad \frac{4.54\ \text{mol O}}{2.27} = 2.00\ \text{mol of O}$$

3. Empirical formula $= C_{1.00}\,O_{2.00} = CO_2$

Study Note

When the number of moles is a decimal number that is very close (within 0.1) to a whole number, round off to the whole number. For example, 1.98 mol can be rounded off to 2 mol.

However, if a decimal value is obtained that is greater than 0.1 or less than 0.9, multiply the decimal value by a small integer to obtain a whole number. For example, 1.33 mol × 3 = 3.99, which rounds off to 4 mol. Then all other subscripts must be increased by multiplying by the same small integer.

◆ Learning Exercise 7.4B

Determine the empirical formula of each of the following compounds:

a. a compound that is 75.7% by mass tin (Sn) and 24.3% by mass fluorine (F)

b. a compound that is 20.2% by mass magnesium (Mg), 26.7% sulfur (S), and 53.2% oxygen (O)

c. a 4.58-g sample of a compound that contains 2.00 g of phosphorus (P) and the rest is oxygen (O)

d. a compound that contains 1.00 g of C, 0.168 g of H, and 0.890 g of O

Answers **a.** SnF_2 **b.** $MgSO_4$ **c.** P_2O_5 **d.** $C_3H_6O_2$

7.5 Molecular Formulas

- A molecular formula gives the actual number of atoms of each element in the compound.
- A molecular formula is related to the empirical formula by a small integer such as 1, 2, or 3.
- The molecular formula is obtained using the empirical formula and molar mass of the compound.

Study Note

When the empirical formula and molar mass of a compound are given, the molecular formula is obtained as follows:

1. Calculate the mass of the empirical formula.

2. Divide the molar mass by the mass of the empirical formula to obtain a small integer.

3. Multiply the empirical formula by the small integer to obtain the molecular formula.

Example: Lactic acid has the empirical formula CH_2O. If the molar mass of lactic acid is about 90 g, what is the molecular formula of lactic acid?

1. Mass of CH_2O $= 12.01$ g of C $+ 2(1.008)$ g of H $+ 16.00$ g of O $= 30.03$ g

2. $$\frac{\text{Molar mass}}{\text{Empirical formula mass}} = \frac{90\text{ g}}{30.03\text{ g}} = 3\text{ (small integer)}$$

3. Molecular formula $= 3 \times (CH_2O) = C_3H_6O_3$ Formula of lactic acid

MasteringChemistry
Tutorial: Empirical and Molecular Formulas

◆ Learning Exercise 7.5A

Determine the molecular formula of each compound from the empirical formula and approximate molar mass given:

a. NS, 184 g/mol

b. KCO_2, 166 g/mol

c. HIO_3, 176 g/mol

d. $NaPO_3$, 306 g/mol

e. $C_7H_8O_3$, 420 g/mol

f. C_3H_5O, 228 g/mol

Answers
a. N_4S_4 **b.** $K_2C_2O_4$ **c.** HIO_3
d. $Na_3P_3O_9$ **e.** $C_{21}H_{24}O_9$ **f.** $C_{12}H_{20}O_4$

◆ Learning Exercise 7.5B

Determine the molecular formula for each of the following:

a. pyrogallol with a composition of C 57.14%, H 4.80%, and O 38.06% and a molar mass of about 126 g

b. triazaborane with a composition of B 40.31%, H 7.51%, and N 52.18% and a molar mass of about 80 g

c. a substance used to tan leather with a composition of K 26.58%, Cr 35.36%, and O 38.07% and a molar mass of about 294 g

d. pyrocatechol, a topical anesthetic with a composition of C 65.44%, H 5.49%, and O 29.06% and a molar mass of about 110 g

Answers **a.** $C_6H_6O_3$ **b.** $B_3H_6N_3$ **c.** $K_2Cr_2O_7$ **d.** $C_6H_6O_2$

Checklist for Chapter 7

You are ready to take the Practice Test for Chapter 7. Be sure you have accomplished the following learning goals for this chapter. If you are not sure, review the section listed at the end of the goal. Then apply your new skills and understanding to the Practice Test.

After studying Chapter 7, I can successfully:

_____ Determine the number of particles in a mole of a substance (7.1).

_____ Calculate the molar mass given the formula of a substance (7.2).

_____ Convert the grams of a substance to moles, and moles to grams (7.3).

_____ Convert the number of grams of a substance to the number of particles, and number of particles to the number of grams (7.3).

_____ Calculate the percent composition of a compound (7.4).

_____ Determine the empirical formula for a compound (7.4).

_____ Determine the molecular formula for a compound (7.5).

Practice Test for Chapter 7

1. The number of carbon atoms in 2.0 mol of C is
A. 6.0 **B.** 12 **C.** 6.0×10^{23} **D.** 3.0×10^{23} **E.** 1.2×10^{24}

2. The number of molecules in 0.25 mol of NH_3 is
A. 0.25 **B.** 2.5×10^{23} **C.** 4.2×10^{-25} **D.** 1.5×10^{23} **E.** 2.43×10^{24}

3. The number of moles in 8.8×10^{24} molecules of SO_3 is
A. 0.068 **B.** 2.6 **C.** 1.5 **D.** 15 **E.** 53

4. The moles of oxygen (O) in 2.0 mol of $Al(OH)_3$ is
A. 1 **B.** 2 **C.** 3 **D.** 4 **E.** 6

5. The number of atoms of phosphorus in 1.00 mol of Be_3P_2 is
A. 3.01×10^{23} **B.** 6.02×10^{23} **C.** 1.20×10^{23} **D.** 9.03×10^{23} **E.** 1.20×10^{24}

6. What is the molar mass of Li_2SO_4?
A. 55.01 g **B.** 62.10 g **C.** 103.01 g **D.** 109.95 g **E.** 103.11 g

7. What is the molar mass of $NaNO_3$?
A. 34.00 g **B.** 37.00 g **C.** 53.00 g **D.** 75.00 g **E.** 85.00 g

8. The number of grams in 0.600 mol of Cl_2 is
A. 71.0 g **B.** 21.3 g **C.** 42.5 g **D.** 84.5 g **E.** 4.30 g

9. How many grams are in 4.00 mol of NH_3?
A. 4.00 g **B.** 17.0 g **C.** 34.0 g **D.** 68.1 g **E.** 0.240 g

10. How many grams are in 4.50 mol of N_2?
A. 6.22 g **B.** 28.0 g **C.** 56.0 g **D.** 112 g **E.** 126 g

11. How many moles is 8.00 g of NaOH?
A. 0.100 mol **B.** 0.200 mol **C.** 0.400 mol **D.** 2.00 mol **E.** 4.00 mol

12. The number of moles of aluminum in 54.0 g of Al is
A. 0.500 mol **B.** 1.00 mol **C.** 2.00 mol **D.** 3.00 mol **E.** 4.00 mol

13. The number of moles of water in 3.60 g of H_2O is
A. 0.0500 mol **B.** 0.100 mol **C.** 0.200 mol **D.** 0.300 mol **E.** 0.400 mol

14. What is the number of moles in 2.20 g of CO_2?
A. 2.00 mol **B.** 1.00 mol **C.** 0.200 mol **D.** 0.0500 mol **E.** 0.0100 mol

15. 0.200 g of H_2 = ___ mol of H_2
A. 0.100 mol **B.** 0.200 mol **C.** 0.400 mol **D.** 0.0400 mol **E.** 0.0100 mol

16. The percent by mass of carbon in C_3H_8 is
A. 18.29% **B.** 27.32% **C.** 59.84% **D.** 81.71% **E.** 92.24%

17. What is the empirical formula of a compound that is 27.3% C and 72.7% O?
A. CO **B.** CO_2 **C.** C_2O **D.** C_2O_3 **E.** CO_3

18. What is the empirical formula of a compound that has 3.88 g of Cl and 6.12 g of O?
A. ClO **B.** Cl_2O_7 **C.** Cl_3O **D.** Cl_2O_3 **E.** ClO_4

19. What is the molecular formula of a compound that has an empirical formula of CH_2 and molar mass of about 70 g?
A. CH_2 **B.** C_2H_4 **C.** C_3H_6 **D.** C_4H_8 **E.** C_5H_{10}

20. What is the molecular formula of a compound that has an empirical formula of P_2O_3 and molar mass of about 220 g?
A. P_2O_3 **B.** P_2O_5 **C.** P_4O_6 **D.** P_4O_8 **E.** P_4O_{10}

Answers to the Practice Test

1. E	**2.** D	**3.** D	**4.** E	**5.** E
6. D	**7.** E	**8.** C	**9.** D	**10.** E
11. B	**12.** C	**13.** C	**14.** D	**15.** A
16. E	**17.** B	**18.** B	**19.** E	**20.** C

8

Chemical Reactions of Inorganic and Organic Compounds

Study Goals

- Show that a balanced equation has an equal number of atoms of each element on the reactant side and the product side.
- Write a balanced equation for a chemical reaction when given the formulas of the reactants and products.
- Identify the functional groups in organic compounds.
- Classify an equation as a combination, decomposition, single replacement, or double replacement.
- Write equations for the combustion of organic compounds.
- Classify organic molecules according to their functional groups.
- Write equations for the combustion of organic compounds using condensed structural formulas.
- Write equations for the hydrogenation of alkenes and alkynes using condensed structural formulas.

Think About It

1. Why is the formation of rust on an iron post a chemical reaction?

2. How is a recipe like a chemical equation?

3. What are the different types of chemical reactions?

4. How is a recipe like a chemical equation?

5. A margarine is partially hydrogenated. What does that mean?

Key Terms

Match the following terms with the following statements:

a. chemical change	**b.** chemical equation	**c.** combination reaction
d. single replacement reaction	**e.** decomposition	**f.** functional group
g. reactant	**h.** alkene	**i.** ester

1. _____ the type of reaction in which an element replaces a different element in a reacting compound

2. _____ a change that alters the composition of a substance, producing a new substance with new properties

3. _____ the type of reaction in which a reactant breaks up into two or more products

4. _____ the type of reaction in which reactants combine to form a single product

5. _____ an atom or group of atoms that influences the chemical reactions of an organic compound

6. _____ a shorthand method of writing a chemical reaction with the formulas of the reactants written on the left side of an arrow and the formulas of the products on the right side

7. _____ an organic compound that contains a carboxyl group (—COO—) with an oxygen atom bonded to a carbon atom

8. _____ an initial substance that undergoes change in a chemical reaction

9. _____ a type of hydrocarbon that contains carbon to carbon double bonds (C═C).

Answers					
	1. d	**2.** a	**3.** e	**4.** c	**5.** f
	6. b	**7.** i	**8.** g	**9.** h	

8.1 Equations for Chemical Reactions

- A chemical change occurs when the atoms of the initial substances rearrange to form new substances.
- Chemical change is indicated by a change in properties of the reactants. For example, a rusting nail, souring milk, and a burning match are all chemical changes.
- When new substances form, a chemical reaction has taken place.
- A chemical equation shows the formulas of the reactants on the left side of the arrow and the formulas of the products on the right side.
- In a balanced equation, numbers called *coefficients*, which appear in front of the symbols or formulas, provide the same number of atoms for each kind of element on the reactant and product sides.
- Each formula in an equation is followed by an abbreviation, in parentheses, that gives the physical state of the substance: solid (*s*), liquid (*l*), or gas (*g*), and, if dissolved in water, an aqueous solution (*aq*).
- The Greek letter delta (Δ) over the arrow in an equation represents the application of heat to the reaction.

MasteringChemistry

Self-Study Activity: Chemical Reactions and Equations

Self-Study Activity: What is Chemistry?

◆ Learning Exercise 8.1A

Identify each of the following as a chemical (C) or a physical (P) change:

1. ____ tearing a piece of paper
2. ____ burning paper
3. ____ rusting iron
4. ____ digestion of food
5. ____ dissolving salt in water
6. ____ boiling water
7. ____ chewing gum
8. ____ removing tarnish with silver polish

Answers 1. P 2. C 3. C 4. C
5. P 6. P 7. P 8. C

◆ Learning Exercise 8.1B

State the number of atoms of each element on the reactant side and on the product side for each of the following balanced equations:

a. $CaCO_3(s) \longrightarrow CaO(g) + CO_2(g)$

Element	Atoms on Reactant Side	Atoms on Product Side
Ca		
C		
O		

b. $2Na(s) + 2H_2O(l) \longrightarrow 2NaOH(aq) + H_2(g)$

Element	Atoms on Reactant Side	Atoms on Product Side
Na		
H		
O		

c. $C_5H_{12}(g) + 8O_2(g) \longrightarrow 5CO_2(g) + 6H_2O(g)$

Element	Atoms on Reactant Side	Atoms on Product Side
C		
H		
O		

d. $2AgNO_3(aq) + K_2S(aq) \longrightarrow 2KNO_3(aq) + Ag_2S(s)$

Element	Atoms on Reactant Side	Atoms on Product Side
Ag		
N		
O		
K		
S		

e. $2Al(OH)_3(aq) + 3H_2SO_4(aq) \longrightarrow Al_2(SO_4)_3(s) + 6H_2O(l)$

Element	Atoms on Reactant Side	Atoms on Product Side
Al		
O		
H		
S		

Answers

a. $CaCO_3(s) \longrightarrow CaO(g) + CO_2(g)$

Element	Atoms on Reactant Side	Atoms on Product Side
Ca	1	1
C	1	1
O	3	3

b. $2Na(s) + 2H_2O(l) \longrightarrow 2NaOH(aq) + H_2(g)$

Element	Atoms on Reactant Side	Atoms on Product Side
Na	2	2
H	4	4
O	2	2

c. $C_5H_{12}(g) + 8O_2(g) \longrightarrow 5CO_2(g) + 6H_2O(g)$

Element	Atoms on Reactant Side	Atoms on Product Side
C	5	5
H	12	12
O	16	16

d. $2AgNO_3(aq) + K_2S(aq) \longrightarrow 2KNO_3(aq) + Ag_2S(s)$

Element	Atoms on Reactant Side	Atoms on Product Side
Ag	2	2
N	2	2
O	6	6
K	2	2
S	1	1

e. $2Al(OH)_3(aq) + 3H_2SO_4(aq) \longrightarrow Al_2(SO_4)_3(s) + 6H_2O(l)$

Element	Atoms on Reactant Side	Atoms on Product Side
Al	2	2
O	18	18
H	12	12
S	3	3

8.2 Balancing a Chemical Equation

- A chemical equation is balanced by placing numbers called coefficients in front of the symbols or formulas in the equation.

Study Note

Balance the following equation: $N_2(g) + H_2(g) \longrightarrow NH_3(g)$

1. Count the atoms of N and H on the reactant side and on the product side.

$$N_2(g) + H_2(g) \longrightarrow NH_3(g)$$

2N, 2H   1N, 3H

2. Balance the N atoms by placing a coefficient of 2 in front of NH_3. (This increases the H atoms, too.) Recheck the number of N atoms and the number of H atoms.

$$N_2(g) + H_2(g) \longrightarrow 2NH_3(g)$$

2N, 2H   2N, 6H

3. Balance the H atoms by placing a coefficient of 3 in front of H_2. Recheck the number of N atoms and the number of H atoms.

$$N_2(g) + 3H_2(g) \longrightarrow 2NH_3(g)$$

2N, 6H   2N, 6H *The equation is balanced.*

MasteringChemistry

Tutorial: Balancing Chemical Equations

Guide to Balancing a Chemical Equation

STEP 1 Write an equation using the correct formulas in the reactants and products.
STEP 2 Count the atoms of each element in reactants and products.
STEP 3 Use coefficients to balance each element.
STEP 4 Check the final equation for balance.

◆ Learning Exercise 8.2

Balance each of the following equations by placing appropriate coefficients in front of the formulas as needed:

a. ____ $MgO(s) \longrightarrow$ ____ $Mg(s) +$ ____ $O_2(g)$

b. ____ $Zn(s) +$ ____ $HCl(aq) \longrightarrow$ ____ $ZnCl_2(aq) +$ ____ $H_2(g)$

c. ____ $Al(s) +$ ____ $CuSO_4(aq) \longrightarrow$ ____ $Cu(s) +$ ____ $Al_2(SO_4)_3(aq)$

d. ____ $Al_2S_3(s) +$ ____ $H_2O(l) \longrightarrow$ ____ $Al(OH)_3(aq) +$ ____ $H_2S(aq)$

e. ____ $BaCl_2(aq) +$ ____ $Na_2SO_4(aq) \longrightarrow$ ____ $BaSO_4(s) +$ ____ $NaCl(aq)$

f. ____ $CO(g) +$ ____ $Fe_2O_3(s) \longrightarrow$ ____ $Fe(s) +$ ____ $CO_2(g)$

g. ____ $K(s) +$ ____ $H_2O(l) \longrightarrow$ ____ $KOH(aq) +$ ____ $H_2(g)$

h. ____ $Fe(OH)_3(s) \xrightarrow{\Delta}$ ____ $Fe_2O_3(s) +$ ____ $H_2O(l)$

Answers

a. $2MgO(s) \longrightarrow 2Mg(s) + O_2(g)$

b. $Zn(s) + 2HCl(aq) \longrightarrow ZnCl_2(aq) + H_2(g)$

c. $2Al(s) + 3CuSO_4(aq) \longrightarrow 3Cu(s) + Al_2(SO_4)_3(aq)$

d. $Al_2S_3(s) + 6H_2O(l) \longrightarrow 2Al(OH)_3(aq) + 3H_2S(aq)$

e. $BaCl_2(aq) + Na_2SO_4(aq) \longrightarrow BaSO_4(s) + 2NaCl(aq)$

f. $3CO(g) + Fe_2O_3(s) \longrightarrow 2Fe(s) + 3CO_2(g)$

g. $2K(s) + 2H_2O(l) \longrightarrow 2KOH(aq) + H_2(g)$

h. $2Fe(OH)_3(s) \xrightarrow{\Delta} Fe_2O_3(s) + 3H_2O(l)$

8.3 Reaction Types

- Reactions are classified as: combination, decomposition, single replacement, and double replacement.
- In a *combination* reaction, two or more reactants form one product: $A + B \longrightarrow AB$
- In a *decomposition* reaction, a reactant splits into two or more simpler products: $AB \longrightarrow A + B$
- In a *single replacement* reaction, an uncombined element takes the place of an element in a compound:

$$A + BC \longrightarrow AC + B$$

- In a *double replacement* reaction, the positive ions in the reactants switch places:

$$AB + CD \longrightarrow AD + CB$$

MasteringChemistry

Tutorial: Chemical Reactions and Equations
Tutorial: Classifying Chemical Reactions by What Atoms Do

◆ Learning Exercise 8.3A

Match each of the following reactions with the type of reaction:

a. combination **b.** decomposition **c.** single replacement **d.** double replacement

1. ____ $N_2(g) + 3H_2(g) \longrightarrow 2NH_3(g)$

2. ____ $BaCl_2(aq) + K_2CO_3(aq) \longrightarrow BaCO_3(s) + 2KCl(aq)$

3. ____ $2H_2O_2(aq) \longrightarrow 2H_2O(l) + O_2(g)$

4. ____ $CuO(s) + H_2(g) \longrightarrow Cu(s) + H_2O(l)$

5. ____ $N_2(g) + 2O_2(g) \longrightarrow 2NO_2(g)$

6. ____ $2NaHCO_3(aq) \longrightarrow Na_2CO_3(aq) + CO_2(g) + H_2O(l)$

7. ____ $PbCO_3(s) \longrightarrow PbO(s) + CO_2(g)$

8. ____ $2Al(s) + Fe_2O_3(s) \longrightarrow 2Fe(s) + Al_2O_3(s)$

Answers **1.** a **2.** d **3.** b **4.** c
5. a **6.** b **7.** b **8.** c

◆ Learning Exercise 8.3B

1. Tarnish, Ag_2S, is removed from silver objects in a reaction with aluminum. The unbalanced chemical equation is

$$Al(s) + Ag_2S(s) \longrightarrow Ag(s) + Al_2S_3(s)$$

a. What is the balanced equation?

b. What type of reaction takes place?

2. Predict the products of each of the following reactions:

a. Combination: $Ca(s) + Br_2(l) \longrightarrow$

b. Double replacement: $NaI(aq) + AgNO_3(aq) \longrightarrow$

Answers **1a.** $2Al(s) + 3Ag_2S(s) \longrightarrow 6Ag(s) + Al_2S_3(s)$
b. single replacement reaction
2a. $Ca(s) + Br_2(l) \longrightarrow CaBr_2(s)$
b. $NaI(aq) + AgNO_3(aq) \longrightarrow NaNO_3(aq) + AgI(s)$

8.4 Functional Groups

- Organic compounds are classified by *functional groups*, which are atoms or groups of atoms where specific chemical reactions occur.

- Alkenes are hydrocarbons that contain one or more double bonds $(C=C)$; alkynes contain a triple bond $(C\equiv C)$.

- Alcohols contain a hydroxyl $(-OH)$ group; ethers have an oxygen atom $(-O-)$ between carbon groups.

- Aldehydes contain a carbonyl group $(C=O)$ bonded to at least one H atom; ketones contain the carbonyl group bonded to two other carbon groups.

- Carboxylic acids contain a carboxyl group $(-COOH)$, which is a combination of a carbonyl group $(C=O)$ and a hydroxyl group $(-OH)$. Esters contain the carboxyl groups, in which an oxygen is attached to another carbon atom.

- Amines are derived from ammonia, NH_3, in which carbon groups replace one or more of the H atoms. Amides replace the $-OH$ group of a carboxylic acid with a nitrogen group.

- Amino acids contain two functional groups, an amino group and a carboxyl group, attached to a central carbon atom.

- A peptide bond is an amide bond between the carboxyl group of one amino acid and the amino group of the second amino acid.

$$\overset{+}{H_3N}-\underset{\underset{CH_3}{|}}{CH}-\underset{\overset{\parallel}{O}}{C}-\underset{\underset{H}{|}}{N}-\underset{\underset{CH_2-OH}{|}}{CH}-COO^-$$

peptide bond

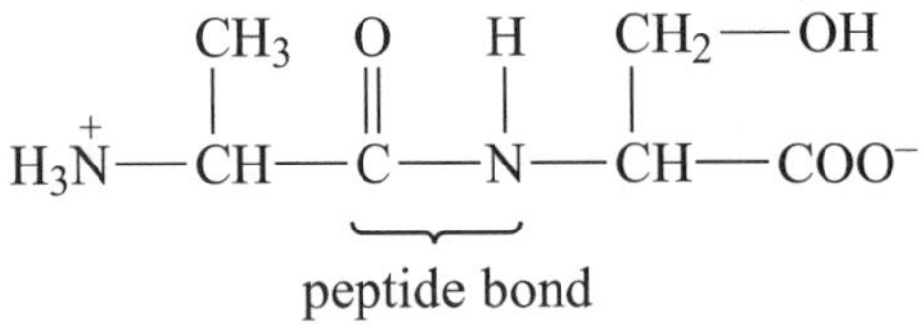

MasteringChemistry

Tutorial: Drawing Organic Compounds with Functional Groups
Self-Study Activity: Functional Groups
Tutorial: Identifying Functional Groups
Self-Study Activity: Aldehydes and Ketones
Tutorial: Carboxylic Acids
Case Study: Death by Chocolate

◆ Learning Check 8.4A

Classify the organic compounds shown according to the following functional groups:

a. alkane **b.** alkene **c.** alcohol **d.** ether **e.** aldehyde **f.** carboxylic acid

1. ____ $CH_3-CH_2-CH=CH_2$ 2. ____ $CH_3-CH_2-CH_3$

3. ____ $CH_3-CH_2-\overset{\overset{O}{\parallel}}{C}-H$ 4. ____ $CH_3-CH_2-CH_2-OH$

5. ____ $CH_3-CH_2-O-CH_2-CH_3$ 6. ____ $CH_3-CH_2-CH_2-CH_3$

7. ____ $CH_3-CH_2-CH_2-COOH$

Answers **1.** b **2.** a **3.** e **4.** c **5.** d **6.** a **7.** f

◆ Learning Exercise 8.4B

Classify the following compounds according to their functional groups:

a. alcohol **b.** aldehyde **c.** ketone **d.** ether **e.** amine

1. ____ $CH_3-CH_2-CH_2-\overset{O}{\overset{\|}{C}}-H$ **2.** ____ $CH_3-CH_2-CH_2-NH_2$

3. ____ $CH_3-CH_2-\overset{O}{\overset{\|}{C}}-CH_2-CH_3$ **4.** ____ $CH_3-CH_2-O-CH_3$

5. ____ $CH_3-\overset{O}{\overset{\|}{C}}-CH_2-CH_3$ **6.** ____ $CH_3-\overset{O}{\overset{\|}{C}}-H$

7. ____ $CH_3-CH_2-\overset{NH_2}{\overset{|}{C}H}-CH_3$ **8.** ____ $CH_3-CH_2-\overset{OH}{\overset{|}{C}H}-CH_3$

Answers				
	1. b	**2.** e	**3.** c	**4.** d
	5. c	**6.** b	**7.** e	**8.** a

◆ Learning Exercise 8.4C

Using the appropriate side chain, complete the condensed structural formula of each of the following amino acids:

Glycine

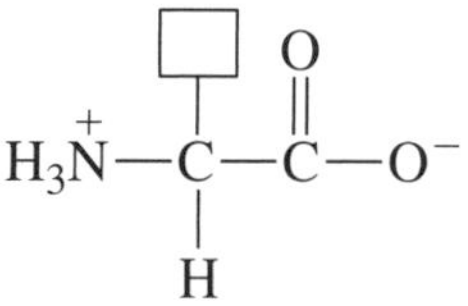

Alanine

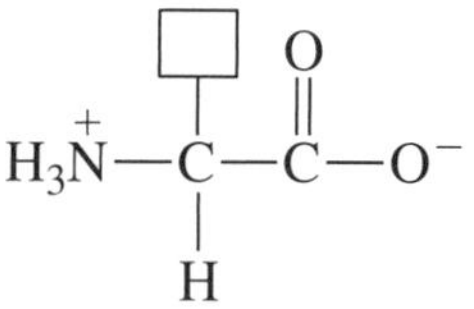

Serine

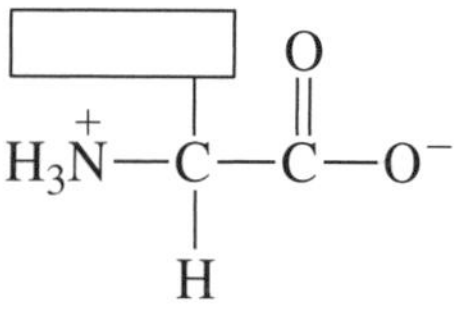

Answers

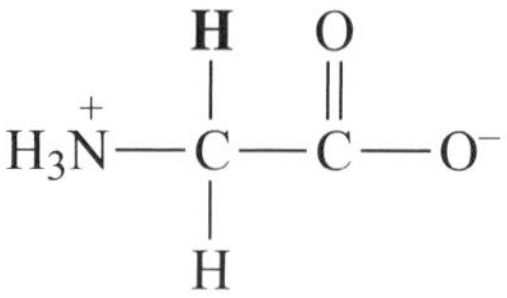

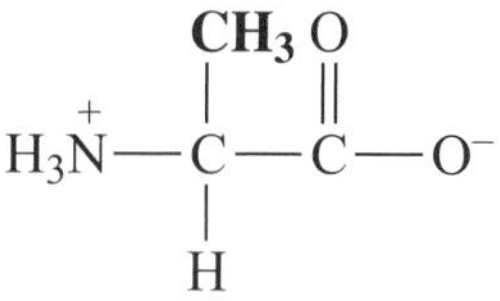

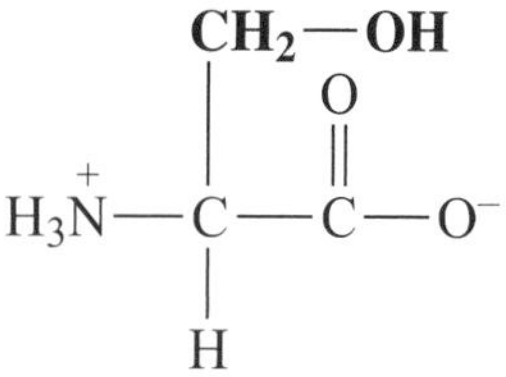

◆ Learning Exercise 8.4D

Draw the structural formulas of the following dipeptides:

1. Ser—Gly

2. Cys — Val

Answers

1.
$$HO-CH_2 \quad O \quad H$$
$$H_3\overset{+}{N}-CH-C-N-CH_2-COO^-$$

2.
$$HS-CH_2 \quad O \quad H \quad \overset{\displaystyle CH_3}{\overset{|}{CH}}-CH_3$$
$$H_3\overset{+}{N}-CH-C-N-CH-COO^-$$

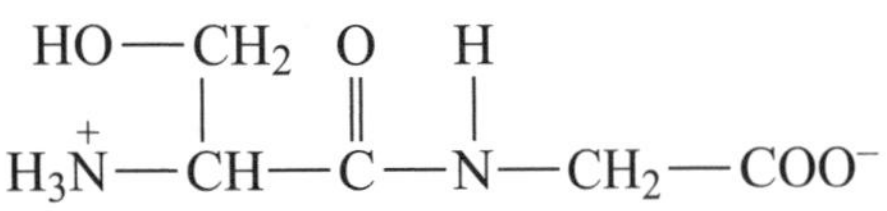

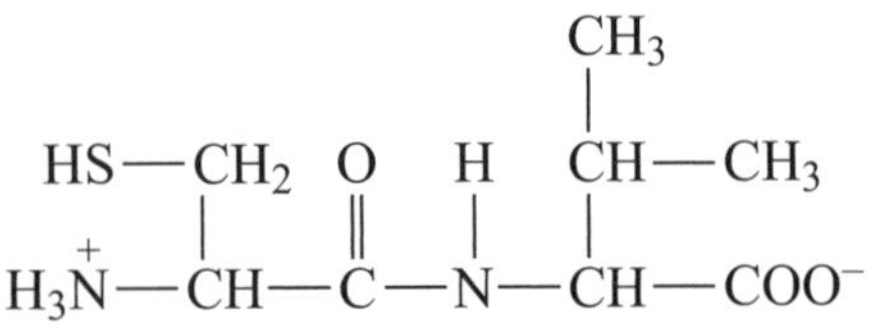

8.5 Reactions of Organic Compounds

- Organic compounds undergo a *combustion reaction* when they react with oxygen to produce carbon dioxide, water, and energy usually as heat.

$$CH_4(g) + O_2(g) \xrightarrow{\Delta} CO_2(g) + 2H_2O(g)$$

- Hydrogenation adds hydrogen atoms to the double bond of an alkene or the triple bond of an alkyne to yield an alkane.

$$CH_2{=}CH_2(g) + H_2(g) \xrightarrow{Pt} CH_3-CH_3(g)$$

$$CH{\equiv}CH(g) + 2H_2(g) \xrightarrow{Pt} CH_3-CH_3(g)$$

MasteringChemistry

Tutorial: Addition Reactions
Tutorial: Writing Balanced Equations for the Combustion of Alkanes

◆ Learning Exercise 8.5A

Write a balanced equation for the complete combustion of each of the following:

1. propane

2. hexane

3. pentane

Answers

1. $C_3H_8(g) + 5O_2(g) \xrightarrow{\Delta} 3CO_2(g) + 4H_2O(g)$ + heat

2. $2C_6H_{14}(g) + 19O_2(g) \xrightarrow{\Delta} 12CO_2(g) + 14H_2O(g)$ + heat

3. $C_5H_{12}(g) + 8O_2(g) \xrightarrow{\Delta} 5CO_2(g) + 6H_2O(g)$ + heat

◆ Learning Exercise 8.5B

Octane, C_8H_{18}, a compound in gasoline, burns in oxygen to produce carbon dioxide and water.

a. What is the balanced equation for the reaction?

b. What type of reaction takes place?

Answers

a. $2C_8H_{18}(g) + 25O_2(g) \xrightarrow{\Delta} 16CO_2(g) + 18H_2O(g)$

b. combustion

◆ Learning Exercise 8.5C

Write the products for each of the following reactions:

1. $CH_3—CH═CH_2 + H_2 \xrightarrow{Pt}$

2. $CH_3—CH═CH—CH_2—CH_3 + H_2 \xrightarrow{Pt}$

3. $CH_3—CH≡CH—CH_3 + 2H_2 \xrightarrow{Pt}$

Answers
1. $CH_3—CH_2—CH_3$
2. $CH_3—CH_2—CH_2—CH_2—CH_3$
3. $CH_3—CH_2—CH_2—CH_3$

Checklist for Chapter 8

You are ready to take the Practice Test for Chapter 8. Be sure you have accomplished the following learning goals for this chapter. If you are not sure, review the section listed at the end of the goal. Then apply your new skills and understanding to the Practice Test.

After studying Chapter 8, I can successfully:

_____ Identify a chemical and physical change (8.1).

_____ State a chemical equation in words and calculate the total atoms of each element in the reactants and products (8.1).

_____ Write a balanced equation for a chemical reaction from the formulas of the reactants and products (8.2).

_____ Identify a reaction as a combination, decomposition, single replacement, or double replacement (8.3).

_____ Identify the functional groups in organic compounds (8.4).

_____ Draw the structure for an amino acid (8.4).

_____ Describe a peptide bond and draw the structure for a peptide (8.4).

_____ Balance equations for the combustion reaction of organic compounds (8.5).

_____ Draw the condensed structural formula for the product of a hydrogenation reaction of an alkene or alkyne (8.5).

Practice Test for Chapter 8

For questions 1 through 5, indicate whether each change is a (P) physical change or a (C) chemical change:

1. _____ a melting ice cube

2. _____ breaking glass

3. _____ bleaching a stain

4. _____ a burning candle

5. _____ milk turning sour

For questions 6 through 10, balance and select the correct coefficient for the component in the equation written in boldface type.

A. 1 **B.** 2 **C.** 3 **D.** 4 **E.** 5

6. _____ $Sn(s) + \mathbf{Cl_2}(g) \longrightarrow SnCl_4(s)$

7. _____ $Al(s) + H_2O(l) \longrightarrow Al_2O_3(s) + \mathbf{H_2}(g)$

8. _____ $C_3H_8(g) + \mathbf{O_2}(g) \longrightarrow CO_2(g) + H_2O(g)$

9. _____ $\mathbf{NH_3}(g) + O_2(g) \longrightarrow N_2(g) + H_2O(g)$

10. _____ $N_2O(g) \longrightarrow N_2(g) + \mathbf{O_2}(g)$

For questions 11 through 16, classify each reaction as one of the following:

A. combination **B.** decomposition **C.** single replacement **D.** double replacement

11. _____ $S(s) + O_2(g) \longrightarrow SO_2(g)$

12. _____ $Fe_2O_3(s) + 3C(s) \longrightarrow 2Fe(s) + 3CO(g)$

13. _____ $CaCO_3(s) \longrightarrow CaO(s) + CO_2(g)$

14. _____ $Mg(s) + 2AgNO_3(aq) \longrightarrow Mg(NO_3)_2(aq) + 2Ag(s)$

15. _____ $Na_2S(aq) + Pb(NO_3)_2(aq) \longrightarrow PbS(s) + 2NaNO_3(aq)$

16. _____ $2KClO_3(s) \longrightarrow 2KCl(s) + 3O_2(g)$

For questions 17 through 22, classify each of the following compounds according to its functional group:

A. alkane **B.** alkene **C.** alcohol **D.** aldehyde
E. ketone **F.** ether **G.** amine

17. ____ $CH_3-CH_2-\overset{O}{\overset{\|}{C}}-CH_3$

18. ____ $CH_3-CH_2-CH_2-OH$

19. ____ $CH_3-CH=\overset{CH_3}{\overset{|}{C}}-CH_3$

20. ____ $CH_3-CH_2-O-CH_3$

21. ____ $CH_3-CH_2-NH-CH_3$

22. ____ $CH_3-\overset{O}{\overset{\|}{C}}-H$

In questions 23 through 26, classify each of the following compounds according to its functional group:

A. alcohol **B.** aldehyde **C.** carboxylic acid **D.** ester **E.** ketone

23. ____ $CH_3-CH_2-\overset{O}{\overset{\|}{C}}-O-CH_3$

24. ____ $CH_3-CH_2-\overset{O}{\overset{\|}{C}}-H$

25. ____ $CH_3-\overset{O}{\overset{\|}{C}}-CH_2-CH_3$

26. ____ $CH_3-CH_2-\overset{O}{\overset{\|}{C}}-OH$

27. The correctly balanced equation for the complete combustion of ethane is:

A. $C_2H_6(g) + O_2(g) \xrightarrow{\Delta} 2CO(g) + 3H_2O(g) + \text{heat}$
B. $C_2H_6(g) + O_2(g) \xrightarrow{\Delta} CO_2(g) + H_2O(g) + \text{heat}$
C. $C_2H_6(g) + 2O_2(g) \xrightarrow{\Delta} 2CO_2(g) + 3H_2O(g) + \text{heat}$
D. $2C_2H_6(g) + 7O_2(g) \xrightarrow{\Delta} 4CO_2(g) + 6H_2O(g) + \text{heat}$
E. $2C_2H_6(g) + 4O_2(g) \xrightarrow{\Delta} 4CO_2(g) + 6H_2O(g) + \text{heat}$

28. Hydrogenation of $CH_3-CH=CH_2$ gives:

A. $3CO_2(g) + 6H_2(g)$
B. $CH_3-CH_2-CH_3$
C. $CH_2=CH-CH_3$
D. $CH_3-CH_2-CH_2-CH_3$
E. no reaction

29. Hydrogenation of $CH_3—CH_2—CH═CH_2$ gives:
A. $3CO_2(g) + 6H_2(g)$
B. $CH_3—CH_2—CH_3$
C. $CH_2═CH—CH_3$
D. $CH_3—CH_2—CH_2—CH_3$
E. no reaction

30. The sequence Ala—Gly
A. is a dipeptide
B. has one peptide bond
C. alanine has a free $—\overset{+}{N}H_3$
D. glycine has a free $—COO^-$
E. all of these

Answers to the Practice Test

1. P	**2.** P	**3.** C	**4.** C	**5.** C
6. B	**7.** C	**8.** E	**9.** D	**10.** A
11. A	**12.** C	**13.** B	**14.** C	**15.** D
16. B	**17.** E	**18.** C	**19.** B	**20.** F
21. G	**22.** D	**23.** D	**24.** B	**25.** E
26. C	**27.** D	**28.** B	**29.** D	**30.** E

9

Chemical Quantities in Reactions

Study Goals

- Given a quantity in moles of a reactant or product, use a mole–mole factor from the balanced equation to calculate the number of moles of another substance in the reaction.
- Given the mass in grams of a substance in a reaction, use the appropriate conversion factors and molar masses to calculate the mass in grams of another substance in the reaction.
- Identify a limiting reactant when given the quantities of two reactants.
- Calculate the amount of product formed from a limiting reactant.
- Given the actual quantity of product, determine the percent yield for the reaction.
- Given the heat of reaction (enthalpy change), describe a reaction as an exothermic reaction or endothermic reaction.
- Given the heat of reaction, ΔH, calculate the loss or gain of heat for an exothermic or endothermic reaction.
- Describe the role of ATP in providing energy for the cells of the body.

Think About It

1. How does an equation tell us the reacting ratios of the substances in the reaction?

2. What is a limiting reactant?

3. Why does a hot pack get hot, and a cold pack cold?

4. What is the difference between theoretical yield and actual yield?

5. What is the role of ATP in reactions that occur in the body?

Key Terms

Match the following terms with the following statements:

a. limiting reactant **b.** ATP **c.** exothermic reaction
d. theoretical yield **e.** percent yield **f.** mole–mole factor

1. _____ the relationship between the number of moles of two substances in an equation

2. _____ a high-energy compound in the cells of the body that provides energy for energy-requiring reactions

3. _____ the maximum amount of product that can be produced for a given amount of reactant

4. _____ the reacting substance that is used up during a chemical reaction

5. _____ the ratio of the amount of product actually produced compared with the theoretical yield

6. _____ a reaction in which the energy of the reactants is greater than that of the products

Answers **1.** f **2.** b **3.** d **4.** a **5.** e **6.** c

9.1 Mole Relationships in Chemical Equations

- The law of conservation of mass states that mass is not lost or gained during a chemical reaction.
- The coefficients in a balanced chemical equation describe the moles of reactants and products in the reactions.
- Using the coefficients, mole–mole conversion factors are written for any two substances in the equation.
- For the reaction of oxygen-forming ozone, $3O_2(g) \longrightarrow 2O_3(g)$, the mole–mole conversion factors are the following:

$$\frac{3 \text{ mol } O_2}{2 \text{ mol } O_3} \text{ and } \frac{2 \text{ mol } O_3}{3 \text{ mol } O_2}$$

MasteringChemistry
Tutorial: Stoichiometry
Tutorial: Moles of Reactants and Products

◆ Learning Exercise 9.1A

Write all of the possible conversion factors for the following equation:

$$N_2(g) + O_2(g) \longrightarrow 2NO(g)$$

Answers

For N_2 and O_2:

$$\frac{1 \text{ mol } N_2}{1 \text{ mol } O_2} \text{ and } \frac{1 \text{ mol } O_2}{1 \text{ mol } N_2}$$

For N_2 and NO:

$$\frac{1 \text{ mol } N_2}{2 \text{ mol NO}} \text{ and } \frac{2 \text{ mol NO}}{1 \text{ mol } N_2}$$

For NO and O_2:

$$\frac{1 \text{ mol } O_2}{2 \text{ mol NO}} \text{ and } \frac{2 \text{ mol NO}}{1 \text{ mol } O_2}$$

Guide to Using Mole–Mole Factors

STEP 1 Write the given and needed number of moles.
STEP 2 Write a plan to convert the given to the needed moles.
STEP 3 Use coefficients to write relationships and mole–mole factors.
STEP 4 Set up problem using the mole–mole factor that cancels given moles.

Study Note

The appropriate mole–mole factor is used to change the number of moles of the given to moles of the needed.

Example: Using the equation, $N_2(g) + O_2(g) \longrightarrow 2NO(g)$, calculate the moles of NO obtained from 3.0 mol of N_2.

$$3.0 \cancel{\text{mol } N_2} \times \frac{2 \text{ mol NO}}{1 \cancel{\text{mol } N_2}} = 6.0 \text{ mol of NO}$$

◆ Learning Exercise 9.1B

Use this equation to answer the following questions:

$$C_3H_8(g) + 5O_2(g) \xrightarrow{\Delta} 3CO_2(g) + 4H_2O(g)$$

a. How many moles of O_2 are needed to react with 2.00 mol of C_3H_8?

b. How many moles of CO_2 are produced when 4.00 mol of O_2 reacts?

c. How many moles of C_3H_8 react with 3.00 mol of O_2?

d. How many moles of H_2O are produced from 0.500 mol of C_3H_8?

Answers **a.** 10.0 mol of O_2 **b.** 2.40 mol of CO_2
c. 0.600 mol of C_3H_8 **d.** 2.00 mol of H_2O

9.2 Mass Calculations for Reactions

- The number of grams or moles of one substance in an equation is converted to grams or moles of another substance in an equation using appropriate molar masses and mole–mole factors.

Guide to Calculating the Masses of Reactants and Products for a Chemical Reaction

STEP 1 Use molar mass to convert grams of given to moles (if necessary).
STEP 2 Write a mole–mole factor from the coefficients in the equation.
STEP 3 Convert moles of given to moles of needed substance using the mole–mole factor.
STEP 4 Convert moles of needed substance to grams using the molar mass.

Study Note

How many grams of NO can be produced from 8.00 g of O_2?

Equation: $N_2(g) + O_2(g) \rightarrow 2NO(g)$

Solution: g of O_2 $\longrightarrow$ moles of O_2 $\longrightarrow$ moles of NO $\longrightarrow$ g of NO

$$8.00\ \cancel{g\ O_2} \times \frac{1\ \cancel{mol\ O_2}}{32.00\ \cancel{g\ O_2}} \times \frac{2\ \cancel{mol\ NO}}{1\ \cancel{mol\ O_2}} \times \frac{26.01\ g\ NO}{1\ \cancel{mol\ NO}} = 13.0\ g\ of\ NO$$

MasteringChemistry
Tutorial: Masses of Reactants and Products

◆ Learning Exercise 9.2

Consider the equation for the following questions:

$$2C_2H_6(g) + 7O_2(g) \longrightarrow 4CO_2(g) + 6H_2O(g)$$

a. How many grams of oxygen (O_2) are needed to react with 4.00 mol of C_2H_6?

b. How many grams of C_2H_6 are needed to react with 115 g of O_2?

c. How many grams of C_2H_6 react if 2.00 g of CO_2 is produced?

d. How many grams of CO_2 are produced when 2.00 mol of C_2H_6 reacts with sufficient oxygen?

e. How many grams of water are produced when 82.5 g of O_2 reacts with sufficient C_2H_6?

f. How many molecules of H_2O are produced when 25.0 g of O_2 reacts?

Answers

a. 448 g of O_2	**b.** 30.9 g of C_2H_6	**c.** 0.683 g of C_2H_6
d. 176 g of CO_2	**e.** 39.8 g of H_2O	**f.** 4.03×10^{23} molecules of H_2O

9.3 Limiting Reactants

- In a limiting reactant problem, the availability of one of the reactants limits the amount of product.
- The reactant that is used up is the limiting reactant; the reactant that remains is the excess reactant.
- The limiting reactant produces the smaller number of moles or grams of product.

MasteringChemistry

Tutorial: What Will Run Out First?

Guide to Calculating Product from a Limiting Reactant

STEP 1 Use molar mass to convert the grams of each reactant to moles.
STEP 2 Write a mole–mole factor using the coefficients in the equation.
STEP 3 Calculate moles of product from each reactant and determine the limiting reactant.
STEP 4 Determine the moles of product or calculate grams of product using molar mass.

Study Note

The amount of product possible from a mixture of two reactants is determined by calculating the moles of product each will produce. The limiting reactant produces the smaller amount of product. In the reaction, $S(l) + 3F_2(g) \longrightarrow SF_6(g)$, how many grams of SF_6 are possible when 5.00 mol of S is mixed with 12.0 mol of F_2?

To determine the limiting reactant, we calculate the number of moles that each reactant can produce and determine which reactant produces the smaller number of moles.

$$5.00\ \cancel{\text{mol S}} \times \frac{1\text{ mol SF}_6}{1\ \cancel{\text{mol S}}} = 5.00\text{ mol of SF}_6$$

$$12.0\ \cancel{\text{mol F}_2} \times \frac{1\text{ mol SF}_6}{3\ \cancel{\text{mol F}_2}} = 4.00\text{ mol of SF}_6 \text{ limiting reactant (smaller number of moles)}$$

From the limiting reactant, we convert the number of moles of product to grams.

$$4.00\ \cancel{\text{mol SF}_6} \times \frac{146.07\text{ g SF}_6}{1\ \cancel{\text{mol SF}_6}} = 584\text{ g of SF}_6$$

◆ Learning Exercise 9.3

a. How many grams of Co_2S_3 can be produced from the reaction of 2.20 mol of Co and 3.60 mol of S?

$$2Co(s) + 3S(s) \longrightarrow Co_2S_3(s)$$

b. How many grams of NO_2 can be produced from the reaction of 32.0 g of NO and 24.0 g of O_2?

$$2NO(g) + O_2(g) \longrightarrow 2NO_2(g)$$

Answers
a. 236 g of Co_2S_3 (Co is the limiting reactant)
b. 49.2 g of NO_2 (NO is the limiting reactant)

9.4 Percent Yield

- In most reactions, some product is lost due to incomplete reactions or side reactions. The amount of product collected and measured is called the actual yield.
- Theoretical yield is the maximum amount of product calculated for a given amount of a reactant.
- Percent yield is the ratio of the actual amount (yield) of product obtained to the theoretical yield.

$$\text{Percent yield} = \frac{\text{actual yield}}{\text{theoretical yield}} \times 100\%$$

Study Note

The percent yield is the ratio of the actual yield obtained to the theoretical yield, which is calculated for a given amount of starting reactant. If we calculate that the reaction of 35.5 g of N_2 can theoretically produce 43.1 g of NH_3, but the actual yield is 26.0 g of NH_3, the percent yield is:

$$\frac{26.0 \text{ g } NH_3 \text{ (actual)}}{43.1 \text{ g } NH_3 \text{ (theoretical)}} \times 100\% = 60.3\% \qquad \text{Percent yield}$$

Guide to Calculations for Percent Yield

STEP 1 Write the given and needed quantities.
STEP 2 Write a plan to calculate the theoretical yield and the percent yield.
STEP 3 Write the molar mass for the reactant and the mole–mole factor from the balanced equation.
STEP 4 Solve for the percent yield ratio by dividing the actual yield (given) by the theoretical yield × 100%.

◆ Learning Exercise 9.4

Consider the following reaction: $2H_2S(g) + 3O_2(g) \longrightarrow 2SO_2(g) + 2H_2O(g)$

a. If 60.0 g of H_2S reacts with oxygen to produce 45.5 g of SO_2, what is the percent yield of SO_2?

b. If 25.0 g of O_2 reacts with H_2S to produce 18.6 g of SO_2, what is the percent yield of SO_2?

Consider the reaction $2C_2H_6(g) + 7O_2(g) \longrightarrow 4CO_2(g) + 6H_2O(g)$
Ethane

c. If 125 g of C_2H_6 reacting with sufficient oxygen produces 175 g of CO_2, what is the percent yield of CO_2?

d. When 35.0 g of O_2 reacts with sufficient ethane to produce 12.5 g of H_2O, what is the percent yield of water?

Answers **a.** 40.3% **b.** 55.7% **c.** 47.8% **d.** 74.0%

9.5 Energy in Chemical Reactions

- The heat of reaction is the energy released or absorbed when reactions occur.
- The heat of reaction or enthalpy change, ΔH, is the energy difference between the reactants and the products.
- In exothermic reactions, the energy of the reactants is greater than that of the products, which means that energy is released. The heat of reaction is designated by a $-\Delta H$ or by writing the heat of reaction as a product.
- In endothermic reactions, the energy of the reactants is less than that of the products, which means that energy is absorbed. The heat of reaction is designated by a $+\Delta H$ or by writing the heat of reaction as a reactant.
- The value of ΔH refers to the heat change for the number of moles of each substance in the balanced chemical equation.
- Heat conversion factors can be written for each substance: heat/mol of substance and mol of substance/heat.

MasteringChemistry

Tutorial: Heat of Reaction

◆ Learning Exercise 9.5A

Indicate whether each of the following is an endothermic or exothermic reaction:

1. $2H_2(g) + O_2(g) \longrightarrow 2H_2O(g) + 582\ \text{kJ}$ ______________
2. $C_2H_4(g) \longrightarrow H_2(g) + C_2H_2(g) \quad \Delta H = +176\ \text{kJ}$ ______________
3. $2C(s) + O_2(g) \longrightarrow 2CO(g) \quad \Delta H = -220\ \text{kJ}$ ______________
4. $C_6H_{12}O_6(s) + 6O_2(g) \longrightarrow 6CO_2(g) + 6H_2O(l) + 1350\ \text{kcal}$ ______________
   glucose
5. $C_2H_5OH(l) + 21\ \text{kcal} \longrightarrow C_2H_4(g) + H_2O(g)$ ______________

Answers **1.** exothermic **2.** endothermic **3.** exothermic
4. exothermic **5.** endothermic

Guide to Calculations Using Heat of Reaction (ΔH)

STEP 1 List given and needed data for the equation.
STEP 2 Write a plan using heat of reaction and any molar mass needed.
STEP 3 Write the conversion factors including heat of reaction.
STEP 4 Set up the problem.

◆ Learning Exercise 9.5B

Consider the following reaction:

$$C_6H_{12}O_6(s) + 6O_2(g) \longrightarrow 6CO_2(g) + 6H_2O(l) \quad \Delta H = -1350\ \text{kcal}$$

How much heat, in kcal and kJ, is produced when 50.0 g of CO_2 is formed?

Answer 256 kcal; 1070 kJ

◆ Learning Exercise 9.5C

Calculate the kilojoules in each of the following reactions and specify if energy is absorbed or released:

1. $C_2H_4(g) \longrightarrow H_2(g) + C_2H_2(g) \quad \Delta H = +176\ \text{kJ}$

   a. 3.50 mol of C_2H_2 is produced

b. 75.0 g of C_2H_4 reacts

2. $2H_2(g) + O_2(g) \longrightarrow 2H_2O(g)$ $\quad\quad \Delta H = -582\ \text{kJ}$

a. 0.820 mol of H_2 reacts

b. 2.25 g of H_2O is produced

Answers				
	1a.	616 kJ is absorbed	**b.**	471 kJ is absorbed
	2a.	239 kJ is released	**b.**	36.3 kJ is released

9.6 Energy in the Body

- ATP is a high-energy compound that can be broken down to provide energy to do work in the cells.
- ATP is produced when large molecules are broken down in the body. ATP is required for the synthesis of large molecules.
- The breakdown of ATP is linked with many reactions in the cell that require energy.

MasteringChemistry

Self Study Activity: ATP
Tutorial: ATP: Energy Rich

◆ Learning Exercise 9.6A

Complete the following statements for ATP:

The ATP molecule is composed of (1) ____________, (2) ____________, and three (3) ____________.

When ATP breaks down, (4) ____________ is released. ATP is called a (5) ____________ compound.

The phosphate group, called inorganic phosphate, is abbreviated as (6) ______________.

The equation for the breakdown of ATP can be written as (7) ______________________________.

The energy from ATP is linked to cellular reactions that are (8) ____________________________.

Answers
1. adenine **2.** ribose **3.** phosphate groups
4. energy **5.** high-energy **6.** P_i
7. $ATP \longrightarrow ADP + P_i + \text{energy}$ (7.3 kcal/mol, 31 kJ/mol)
8. energy requiring

◆ Learning Exercise 9.6B

The reaction of fructose and a phosphate (P_i) to give fructose-6-phosphate requires 16 kJ of energy:

$$\text{Fructose} + P_i + 16\text{ kJ} \longrightarrow \text{fructose-6-phosphate}$$

The breakdown of ATP provides the energy to drive this reaction.

a. Write the equations and the combined equation for the energy-requiring reaction and the energy-producing reactions including the energy change, in kJ, for the combined reaction.

Answers

$$\text{Fructose} + P_i + 16\text{ kJ} \longrightarrow \text{fructose-6-phosphate}$$
$$\text{ATP} \longrightarrow \text{ADP} + P_i + 31\text{ kJ}$$
$$\text{Fructose} + \text{ATP} \longrightarrow \text{fructose-6-phosphate} + \text{ADP} + 15\text{ kJ}$$

Checklist for Chapter 9

You are ready to take the Practice Test for Chapter 9. Be sure you have accomplished the following learning goals for this chapter. If you are not sure, review the section listed at the end of the goal. Then apply your new skills and understanding to the Practice Test.

After studying Chapter 9, I can successfully:

_____ Use mole–mole factors for the mole relationships in an equation to calculate the moles of another substance in an equation for a chemical reaction (9.1).

_____ Calculate the mass of a substance in an equation using mole factors and molar masses (9.2).

_____ Given the mass of reactants, find the limiting reactant and calculate the amount of product formed (9.3).

_____ Calculate the percent yield for a reaction given the actual yield of a product (9.4).

_____ Given the heat of reaction, describe a reaction as exothermic or endothermic (9.5).

_____ Calculate the heat, in kcal or kilojoules, released or absorbed by a chemical reaction (9.5).

_____ Describe the role of ATP in "driving" reactions in the cells of the body (9.6).

_____ Calculate the energy change when ATP is used in an energy-requiring reaction (9.6).

Practice Test for Chapter 9

For questions 1 to 7, use the reaction: $C_2H_5OH(g) + 3O_2(g) \longrightarrow 2CO_2(g) + 3H_2O(g)$
Ethanol

1. How many grams of oxygen are needed to react with 1.00 mol of ethanol?
   A. 8.00 g **B.** 16.0 g **C.** 32.0 g **D.** 64.0 g **E.** 96.0 g
2. How many moles of water are produced when 12.0 mol of oxygen reacts?
   A. 3.00 mol **B.** 6.00 mol **C.** 8.00 mol **D.** 12.0 mol **E.** 36.0 mol
3. How many grams of carbon dioxide are produced when 92.0 g of ethanol reacts?
   A. 22.0 g **B.** 44.0 g **C.** 88.0 g **D.** 92.0 g **E.** 176 g

4. How many moles of oxygen would be needed to produce 44.0 g of CO_2?
A. 0.670 mol **B.** 1.00 mol **C.** 1.50 mol **D.** 2.00 mol **E.** 3.00 mol

5. How many grams of water will be produced if 23.0 g of ethanol reacts?
A. 54.0 g **B.** 27.0 g **C.** 18.0 g **D.** 9.00 g **E.** 6.00 g

6. How many grams of water will be produced if 3.00 mol of ethanol and 5.00 mol of O_2 react?
A. 3.60 g **B.** 18.0 g **C.** 54.1 g **D.** 90.1 g **E.** 162 g

7. How many grams of water will be produced if 5.00 mol of ethanol and 12.0 mol of O_2 react?
A. 270 g **B.** 216 g **C.** 54.1 g **D.** 36.0 g **E.** 18.0 g

For questions 8 and 9, use the reaction: $C_3H_8(g) + 5O_2(g) \longrightarrow 3CO_2(g) + 4H_2O(g)$

8. If 50.0 g of C_3H_8 and 150 g of O_2 react, the limiting reactant is:
A. C_3H_8 **B.** O_2 **C.** both **D.** neither **E.** CO_2

9. If 50.0 g of C_3H_8 and 150 g of O_2 react, the mass of CO_2 formed is:
A. 150 g **B.** 206 g **C.** 200 g **D.** 124 g **E.** 132 g

10. In an experiment using the equation below, 36.0 g of NO is produced from 20.0 g of N_2. What is the percent yield? $N_2(g) + O_2(g) \longrightarrow 2NO(g)$
A. 42.9% **B.** 55.6% **C.** 71.4% **D.** 84.1% **E.** 93.4%

For questions 11 and 12, use the following equation: $N_2(g) + 3H_2(g) \longrightarrow 2NH_3(g)$

11. When 25.0 g of N_2 reacts, what is the theoretical yield of NH_3?
A. 7.60 g **B.** 15.2 g **C.** 28.02 g **D.** 30.4 g **E.** 60.8 g

12. When 25.0 g of N_2 reacts, the actual yield of NH_3 is 20.8 g. What is the percent yield?
A. 68.4% **B.** 20.8% **C.** 25.0% **D.** 82.5% **E.** 100%

For questions 13 through 16, use the following equation and heat of reaction:
$C_2H_6O(l) + 3O_2(g) \longrightarrow 2CO_2(g) + 3H_2O(l) \quad \Delta H = -1420\text{ kJ}$

13. The reaction is
A. thermic **B.** endothermic **C.** exothermic **D.** nonthermic **E.** subthermic

14. How much heat, in kilojoules, is released when 0.500 mol of C_2H_6O reacts?
A. 710 kJ **B.** 1420 kJ **C.** 2130 kJ **D.** 2840 kJ **E.** 4260 kJ

15. How many grams of CO_2 are produced if 355 kJ of heat are released?
A. 11.0 g **B.** 22.0 g **C.** 33.0 g **D.** 44.0 g **E.** 88.0 g

16. How much heat, in kilojoules, is released when 16.0 g of O_2 reacts?
A. 4260 kJ **B.** 1420 kJ **C.** 710 kJ **D.** 237 kJ **E.** 118 kJ

17. ATP
A. contains adenosine, ribose, and one phosphate group
B. contains adenosine, ribose, and two phosphate groups
C. contains adenosine, ribose, and three phosphate groups
D. requires energy to break down
E. is a low-energy compound

18. The energy change when ATP is used to drive the following reaction is:
Glucose + P_i + 14 kJ $\longrightarrow$ glucose-6-phosphate
A. 17 kJ **B.** 31 kJ **C.** 45 kJ **D.** 70. kJ **E.** 14 kJ

Answers to the Practice Test

1. E	**2.** D	**3.** E	**4.** C	**5.** B
6. D	**7.** B	**8.** B	**9.** D	**10.** D
11. D	**12.** A	**13.** C	**14.** A	**15.** B
16. D	**17.** C	**18.** A		

10

Structures of Solids and Liquids

Study Goals

- Draw the electron-dot formulas for covalent compounds or polyatomic ions with multiple bonds.
- Draw resonance structures when two or more electron-dot formulas are possible.
- Use the valence-shell electron-pair repulsion (VSEPR) theory to determine the three-dimensional structure and bond angles of a molecule or polyatomic ion.
- Use electronegativity values to identify polar covalent, nonpolar covalent, and ionic bonds.
- Classify a molecule as polar or nonpolar.
- Describe the attractive forces between ions, polar covalent molecules, and nonpolar covalent molecules.
- Describe the changes of state between solids, liquids, and gases.
- Determine the energy lost or gained during a change of state at the melting or boiling point.
- Draw a heating or cooling curve for a substance when heated or cooled.

Think About It

1. How is a covalent bond different from an ionic bond?

2. When is an electron-dot formula written with multiple bonds?

3. How does the VSEPR theory predict the bond angles in a molecule of a covalent compound?

4. How do you determine if a molecule is polar or nonpolar?

Key Terms

Match each the following key terms with the correct definition:

a. polar covalent bond
b. multiple bond
c. VSEPR theory
d. nonpolar covalent bond
e. resonance structures
f. dipole–dipole attraction
g. deposition
h. polarity
i. dispersion forces

1. _____ equal sharing of valence electrons by two atoms
2. _____ occurs when two or more electron-dot formulas can be written for the same compound
3. _____ the unequal attraction for shared electrons in a covalent bond
4. _____ the atoms in a molecule are arranged to minimize repulsion between electrons
5. _____ the sharing of two or three pairs of electrons by two atoms
6. _____ interaction between the positive end of a polar molecule and the negative end of another
7. _____ results from momentary polarization of nonpolar molecules
8. _____ measures unequal sharing of electrons
9. _____ change of a gas directly to a solid

Answers	**1.** d	**2.** e	**3.** a	**4.** c	**5.** b
	6. f	**7.** I	**8.** h	**9.** g	

10.1 Electron-Dot Formulas

- In a covalent bond, atoms of nonmetals share electrons to achieve an octet.
- For example, oxygen with six valence electrons shares electrons with two hydrogen atoms to form the covalent compound water (H_2O).

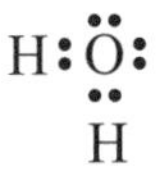

- In a double bond, two pairs of electrons are shared between the same two atoms to complete octets.
- In a triple bond, three pairs of electrons are shared to complete octets.
- Bonding pairs of electrons are shared between atoms. Nonbonding electrons are also called lone pairs.
- When two or more electron-dot formulas can be written for the same compound, the resulting formulas are called *resonance structures*.

MasteringChemistry
Tutorial: Covalent Lewis (Electron) Dot Structures
Tutorial: Writing Electron-Dot Structures
Tutorial: Lewis (Electron) Dot Structures and Resonance

◆ Learning Exercise 10.1A

List the number of bonds typically formed by the following atoms in covalent compounds:

a. N ________	**b.** S ________	**c.** P ________	**d.** C ________
e. Cl ________	**f.** O ________	**g.** H ________	**h.** F ________

Answers	**a.** 3	**b.** 2	**c.** 3	**d.** 4
	e. 1	**f.** 2	**g.** 1	**h.** 1

Guide to Drawing Electron-Dot Formulas

STEP 1 Determine the arrangement of atoms.
STEP 2 Determine the total number of valence electrons.
STEP 3 Attach each bonded atom to the central atom with a pair of electrons.
STEP 4 Place the remaining electrons using single or multiple bonds to complete octets (two for H, six for B).

Study Note

An electron-dot formula for SCl_2 can be drawn as follows:

STEP 1 Determine the arrangement of atoms.

Cl S Cl

STEP 2 Determine the total number of valence electrons.

1 S and 2 Cl $= 1(6\,e^-) + 2(7\,e^-) = 20$ valence electrons

STEP 3 Attach each bonded atom to the central atom with a pair of electrons (single bond).

Cl—S—Cl $20\,e^- - 4\,e^- = 16$ remaining valence electrons (8 pairs)

STEP 4 Place the remaining electrons using single or multiple bonds to complete octets (two for H, six for B).

$:\!\ddot{C}l\!—\!\ddot{S}\!—\!\ddot{C}l\!:$

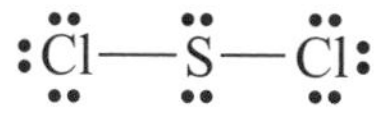

◆ Learning Exercise 10.1B

Draw the electron-dot formulas for each of the following covalent compounds:

H_2 NCl_3 HCl

Cl_2 H_2S CCl_4

Answers H:H $:\!\ddot{C}l\!:\!\ddot{N}\!:\!\ddot{C}l\!:$ $H\!:\!\ddot{C}l\!:$ $:\!\ddot{C}l\!:\!\ddot{C}l\!:$ $H\!:\!\ddot{S}\!:$ $:\!\ddot{C}l\!:\!\ddot{C}\!:\!\ddot{C}l\!:$
 $:\!\ddot{C}l\!:$ H $:\!\ddot{C}l\!:$

Study Note

An electron-dot formula for CO_2 can be drawn with double bonds as follows:

STEP 1 Determine the arrangement of atoms.

O C O

STEP 2 Determine the total number of valence electrons.

$$1\text{ C} \quad \text{and} \quad 2\text{ O} = 1(4\,e^-) + 2(6\,e^-) = 16 \text{ valence electrons}$$

STEP 3 Attach each bonded atom to the central atom with a pair of electrons (single bond).

$$\text{O—C—O} \qquad 16\,e^- - 4\,e^- = 12 \text{ remaining valence electrons (6 pairs)}$$

STEP 4 Place the remaining electrons using single or multiple bonds to complete octets.

$$:\underset{..}{\ddot{O}}—C—\underset{..}{\ddot{O}}:$$

When octets cannot be completed with available valence electrons, one or more lone pairs are shared between the central atom and the bonded atoms.

$$:\ddot{O}=C=\ddot{O}:$$

◆ Learning Exercise 10.1C

Each of the following compounds contains a multiple bond. Draw the electron-dot formulas.

a. CS_2

b. HCN

c. H_2CCH_2

d. HONO

Answers

a. $:\ddot{S}::C::\ddot{S}:$

b. $H:C:::N:$

c. $H:\overset{H}{\ddot{C}}::\overset{H}{\ddot{C}}:H$

d. $H:\underset{..}{\ddot{O}}:\dot{\dot{N}}::\ddot{O}:$

Study Note

When a molecule or polyatomic ion contains multiple bonds, it may be possible to draw more than one electron-dot formula. Draw two resonance structures for selenium dioxide, SeO_2.

Solution

STEP 1 Determine the arrangement of atoms. In SeO_2, the Se atom is the central atom.

$$O \quad Se \quad O$$

STEP 2 Determine the total number of valence electrons. We can use the group numbers to determine the valence electrons.

1 Se atom \times 6 e^-	= 6 valence e^-
2 O atoms \times 6 e^-	= 12 valence e^-
Total valence electrons	= 18 valence e^-

STEP 3 Attach each bonded atom to the central atom with a pair of electrons. Four electrons are used to form two single bonds to connect the Se atom and each O atom. We will use a single line to represent a pair of bonding electrons.

$$O-Se-O$$

STEP 4 Place the remaining electrons using single or multiple bonds to complete octets. The remaining 14 electrons are drawn as lone pairs to complete the octets of the O atoms, but not the Se atom.

$$:\ddot{O}-\ddot{S}e-\ddot{O}:$$

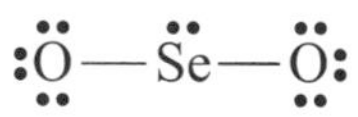

To complete the octet for Se, a lone pair from one of the O atoms is shared with Se to form a double bond. Because the lone pair that is shared can come from either O atom, two resonance structures can be drawn.

$$:\ddot{O}-\ddot{S}e=\ddot{O}: \longleftrightarrow :\ddot{O}=\ddot{S}e-\ddot{O}:$$

◆ Learning Exercise 10.1D

Draw two or more resonance structures for each of the following:

a. SO_2

b. CO_3^{2-}

c. NCO^-

d. NO_3^-

Answers

a. $:\ddot{O}-\ddot{S}=\ddot{O}: \leftrightarrow :\ddot{O}=\ddot{S}-\ddot{O}:$

b. $\left[:\ddot{O}-\overset{\ddot{O}:}{\underset{}{C}}=\ddot{O}:\right]^{2-} \leftrightarrow \left[:\ddot{O}-\overset{:O:}{\underset{}{C}}-\ddot{O}:\right]^{2-} \leftrightarrow \left[:\ddot{O}=\overset{\ddot{O}:}{\underset{}{C}}-\ddot{O}:\right]^{2-}$

c. $\left[:\ddot{N}-C\equiv O:\right]^- \leftrightarrow \left[:\ddot{N}=C=\ddot{O}:\right]^- \leftrightarrow \left[:N\equiv C-\ddot{O}:\right]^-$

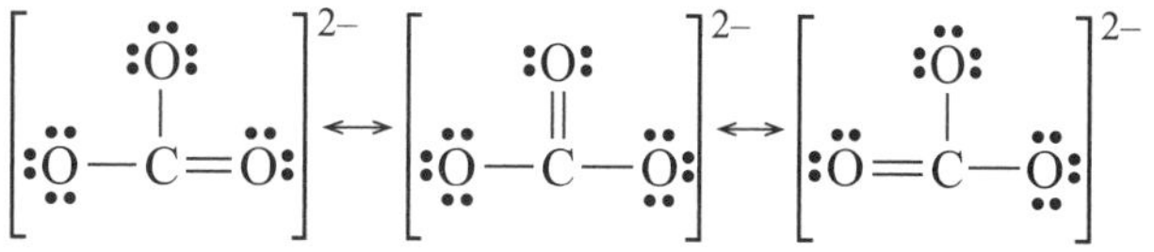

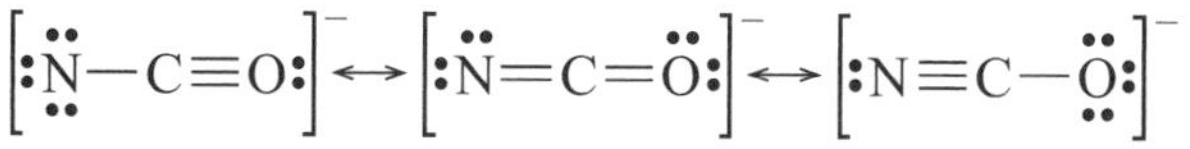

$$
\text{d.} \quad \left[\begin{array}{c} :\ddot{O}: \\ | \\ :\ddot{O}=N-\ddot{O}: \\ \cdot\cdot \end{array} \right]^{-} \longleftrightarrow \left[\begin{array}{c} :O: \\ \| \\ :O-N-\ddot{O}: \\ \cdot\cdot \quad \cdot\cdot \end{array} \right]^{-} \longleftrightarrow \left[\begin{array}{c} :\ddot{O}: \\ | \\ :\ddot{O}-N=\ddot{O}: \\ \cdot\cdot \end{array} \right]^{-}
$$

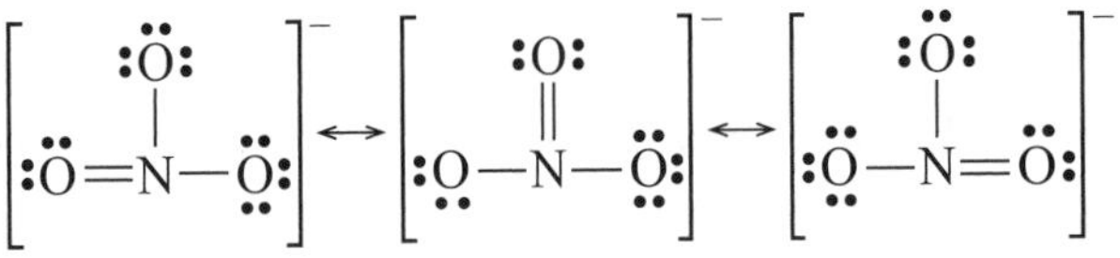

10.2 Shapes of Molecules and Ions (VSEPR Theory)

- The VSEPR theory predicts the geometry of a molecule by placing the electron groups around a central atom as far apart as possible.

- The atoms bonded to the central atom determine the three-dimensional geometry of a molecule or a polyatomic ion.

- A central atom with two electron groups has a linear geometry (180°); three electron groups give a trigonal planar geometry (120°); and four electron pairs give a tetrahedral geometry (109.5°).

- A linear molecule has a central atom bonded to two atoms and no lone pairs.

- A trigonal planar molecule has a central atom bonded to three atoms and no lone pairs. A bent molecule with a bond angle of 120° has a central atom bonded to two atoms and one lone pair.

- A tetrahedral molecule has a central atom bonded to four atoms and no lone pairs. In a trigonal pyramidal molecule, a central atom is bonded to three atoms and one lone pair. In a bent molecule with a bond angle of 109.5°, a central atom is bonded to two atoms and two lone pairs.

MasteringChemistry

Tutorial: Shapes of Molecules

◆ Learning Exercise 10.2A

Match the shape of a molecule with the following descriptions of the electron groups around the central atoms and the number of bonded atoms:

a. linear **b.** trigonal planar **c.** tetrahedral
d. trigonal pyramidal **e.** bent (120°) **f.** bent (109.5°)

1. three electron groups with three bonded atoms _____

2. two electron groups with two bonded atoms _____

3. four electron groups with three bonded atoms _____

4. three electron groups with two bonded atoms _____

5. four electron groups with four bonded atoms _____

6. four electron groups with two bonded atoms _____

Answers **1.** b **2.** a **3.** d **4.** e **5.** c **6.** f

Guide to Predicting Molecular Shape (VSEPR Theory)

STEP 1 Draw the electron-dot formula.
STEP 2 Arrange the electron groups around the central atom to minimize repulsion.
STEP 3 Use the atoms bonded to the central atom to determine the molecular shape.

◆ Learning Exercise 10.2B

For each of the following, draw the electron-dot formula, state the number of electron groups and bonded atoms, and predict the shape and angles of each of the following molecules or ions:

Molecule or Ion	Electron-dot Formula	Number of Electron Groups	Number of Bonded Atoms	Shape and Angle
CH_4				
PCl_3				
SO_3				
H_2S				
NO_2^-				

Answers

Molecule or Ion	Electron-dot Formula	Number of Electron Groups	Number of Bonded Atoms	Shape and Angle
CH_4	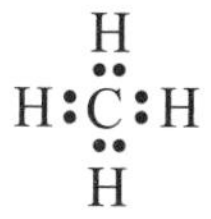	4	4	Tetrahedral, 109.5°
PCl_3	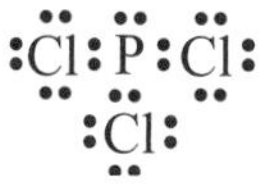	4	3	Trigonal pyramidal, 109.5°
SO_3	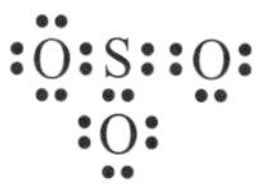	3	3	Trigonal planar, 120°
H_2S		4	2	Bent, 109.5°
NO_2^-		3	2	Bent, 120°

10.3 Electronegativity and Polarity

- Electronegativity values indicate the ability of an atom to attract electrons in a chemical bond. In general, metals have low electronegativity values and nonmetals have high values.
- When atoms sharing electrons have the same or similar electronegativity values, electrons are shared equally and the bond is *nonpolar covalent.*
- Electrons are shared unequally in *polar covalent* bonds because they are attracted to the more electronegative atom.
- An electronegativity difference of 0 to 0.4 indicates a nonpolar covalent bond, whereas a difference of 0.5 to 1.7 indicates a polar covalent bond.
- An electronegativity difference of 1.8 or greater indicates a bond that is ionic.
- A polar bond with its charge separation is called a dipole; the positive end is marked as δ^+ and the negative end as δ^-.
- Nonpolar molecules can have polar bonds when the dipoles are in a symmetric arrangement and the bond dipoles cancel each other.
- In polar molecules, the dipoles do not cancel each other.

MasteringChemistry

Self Study Activity: Electronegativity
Self Study Activity: Polar Attraction
Self Study Activity: Bonds and Bond Polarity
Tutorial: Distinguishing Polar and Nonpolar Molecules

◆ Learning Exercise 10.3A

Using the electronegativity values, determine the following:

1. the electronegativity difference for each pair
2. the type of bonding as (I) ionic, (PC) polar covalent, or (NC) nonpolar covalent

	Elements	Electronegativity Difference	Bonding		Elements	Electronegativity Difference	Bonding
a.	H and O	______	______	b.	N and S	______	______
c.	Al and O	______	______	d.	Li and F	______	______
e.	H and Cl	______	______	f.	Cl and Cl	______	______
g.	S and F	______	______	h.	H and C	______	______

Answers **a.** 1.4, PC **b.** 0.5, PC **c.** 2.0, I **d.** 3.0, I
e. 0.9, PC **f.** 0, NP **g.** 1.5, PC **h.** 0.4, NP

◆ Learning Exercise 10.3B

Write the symbols δ^+ and δ^- over the atoms in polar bonds.

1. H—O **2.** N—H **3.** C—Cl

4. O—F **5.** N—F **6.** P—Cl

Answers

1. $\overset{\delta^+}{H}—\overset{\delta^-}{O}$ **2.** $\overset{\delta^+}{N}—\overset{\delta^-}{H}$ **3.** $\overset{\delta^+}{C}—\overset{\delta^-}{Cl}$

4. $\overset{\delta^+}{O}—\overset{\delta^-}{F}$ **5.** $\overset{\delta^+}{N}—\overset{\delta^-}{F}$ **6.** $\overset{\delta^+}{P}—\overset{\delta^-}{Cl}$

◆ Learning Exercise 10.4C

Indicate the dipoles in each of the following molecules and determine whether the molecule is polar or nonpolar:

1. CF_4 **2.** HCl

3. NH_3 **4.** OF_2

Answers

1. CF_4

F
↑
F←C→F
↓
F

dipoles cancel, nonpolar

2. HCl

$\overset{\delta^+}{H}\rightarrow\overset{\delta^-}{Cl}$

dipole does not cancel, polar

3. NH_3

H→N̈←H
↑
H

dipoles do not cancel, polar

4. OF_2

:Ö→F
↓
F

dipoles do not cancel, polar

10.4 Attractive Forces In Compounds

- The interactions between particles in solids and liquids determine their melting and boiling points.
- Ionic solids have high melting points due to strong ionic interactions between positive and negative ions.
- In polar substances, dipole–dipole attractions occur between the positive end of one molecule and the negative end of another.
- Hydrogen bonding, a particularly strong type of dipole–dipole attraction, occurs between partially positive hydrogen atoms and the strongly electronegative atoms F, O, or N.
- Dispersion forces occur when temporary dipoles form within the nonpolar molecules, causing weak attractions to other nonpolar molecules.

MasteringChemistry
Tutorial: Intermolecular Forces
Tutorial: Forces Between Molecules
Self Study Activity: Polar Attraction

◆ Learning Exercise 10.4A

Indicate the major type of interactive force that occurs in each of the following substances:

a. ionic **b.** dipole–dipole attractions **c.** hydrogen bond **d.** dispersion forces

1. ______ KCl **2.** ______ NF_3 **3.** ______ SBr_2 **4.** ______ Cl_2

5. ______ HF **6.** ______ H_2O **7.** ______ C_4H_{10} **8.** ______ Na_2O

Answers **1.** a **2.** b **3.** b **4.** d
5. c **6.** c **7.** d **8.** a

◆ Learning Exercise 10.4B

Identify the substance that would have the higher boiling point in each pair and give a reason:

1. NaCl or HCl ______ **2.** Br_2 or HBr ______ **3.** H_2O or H_2S ______

4. C_2H_6 or C_8H_{18} ______ **5.** $MgCl_2$ or OCl_2 ______ **6.** NH_3 or PH_3 ______

Answers

1. NaCl (ionic) **2.** HBr (dipole–dipole) attractions

3. H_2O (hydrogen bonding) **4.** C_8H_{18} (dispersion forces of larger molecule)

5. $MgCl_2$ (ionic) **6.** NH_3 (hydrogen bonding)

10.5 Changes of State

- The three states of matter are solid, liquid, and gas.
- Physical properties such as shape, state, or color can change without affecting the identity of a substance.

- A substance melts and freezes at its melting (freezing) point. During the process of melting or freezing, the temperature remains constant.
- The *heat of fusion* is the energy required to change 1 g of solid to liquid. For water to freeze at 0 °C, the heat of fusion is 80 cal/g or 334 J/g. This is also the amount of heat lost when 1 g of water freezes at 0 °C.
- A substance boils and condenses at its boiling point. During the process of boiling or condensing, the temperature remains constant.
- Evaporation is a surface phenomenon while boiling occurs throughout the liquid.
- When water boils at 100 °C, 2260 J or 540 cal, the *heat of vaporization*, is required to change 1 g of liquid to gas (steam); it is also the amount of heat released when 1 g of water vapor condenses at 100 °C.
- Sublimation is the change of state from a solid directly to a gas. Deposition is the opposite process.
- A heating or cooling curve illustrates the changes in temperature and state as heat is added to or removed from a substance.
- The heat given off or gained when a substance undergoes temperature changes, as well as changes of state, is the total of two or more energy calculations.

MasteringChemistry

Tutorial: Heat of Vaporization and Heat of Fusion
Tutorial: Heat, Energy, and Changes of State
Tutorial: Energetics of Phase Changes

◆ Learning Exercise 10.5A

State whether each of the following statements describes a gas (G), a liquid (L), or a solid (S):

1. ________ There are no attractions among the molecules in this substance.
2. ________ The particles in this substance are held close together in a definite pattern.
3. ________ This substance has a definite volume, but no definite shape.
4. ________ The particles in this substance are moving extremely fast.
5. ________ This substance has no definite shape and no definite volume.
6. ________ The particles in this substance are very far apart.
7. ________ This substance has its own volume, but takes the shape of its container.
8. ________ The particles of this substance hit the sides of the container with great force.
9. ________ The particles in this substance are moving very, very slowly.
10. ________ This substance has a definite volume and a definite shape.

Answers **1.** G **2.** S **3.** L **4.** G **5.** G
6. G **7.** L **8.** G **9.** S **10.** S

◆ Learning Exercise 10.5B

Identify each of the following as

1. melting **2.** freezing **3.** sublimation **4.** condensation

a. ________ A liquid changes to a solid.

b. ________ Ice forms on the surface of a lake in winter.

c. ________ Dry ice in an ice cream cart changes to a gas.

d. ________ Butter in a hot pan turns to liquid.

e. ________ A gas changes to a liquid.

Answers **a.** 2 **b.** 2 **c.** 3 **d.** 1 **e.** 4

Study Note

The amount of heat needed or released during melting or freezing can be calculated using the heat of fusion: Heat = mass × heat of fusion

◆ Learning Exercise 10.5C

Calculate each of the following when a substance melts or freezes:

a. How many joules of heat are needed to melt 24.0 g of ice at 0 °C?

b. How much heat, in kilojoules, is released when 325 g of water freezes at 0 °C?

c. How many joules of heat are required to melt 125 g of ice at 0 °C?

d. How many grams of ice would melt when 1680 J of heat were absorbed?

Answers **a.** 8020 J **b.** 109 kJ **c.** 41 800 J **d.** 5.03 g

◆ Learning Exercise 10.5D

Calculate each of the following when a substance boils or condenses:

a. How many joules are needed to completely change 5.00 g of water to vapor at 100 °C?

b. How many kilojoules are released when 515 g of steam at 100 °C condenses to form liquid water at 100 °C?

c. How many grams of water can be converted to steam at 100 °C when 155 kJ of energy is absorbed?

d. How many grams of steam condense if 24 800 J is released at 100 °C?

Answers **a.** 11 300 J **b.** 1160 kJ **c.** 68.6 g **d.** 11.0 g

◆ Learning Exercise 10.5E

On each heating or cooling curve, indicate the portion that corresponds to a solid, liquid, or gas and the changes in state.

1. Draw a heating curve for water that begins at −20 °C and ends at 120 °C. Water has a melting point of 0 °C and a boiling point of 100 °C.

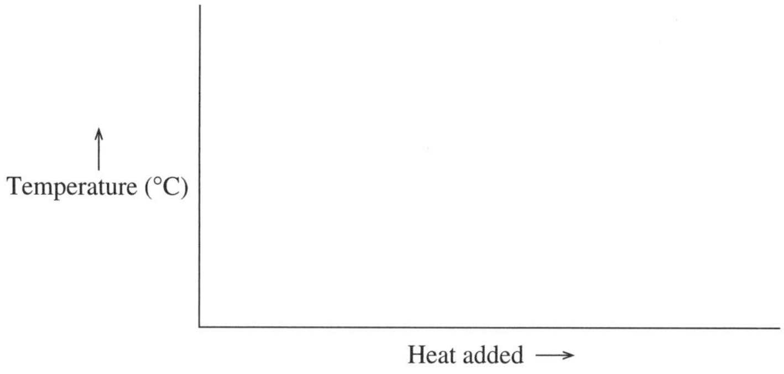

2. Draw a heating curve for bromine from −25 °C to 75 °C. Bromine has a melting point of −7 °C and a boiling point of 59 °C.

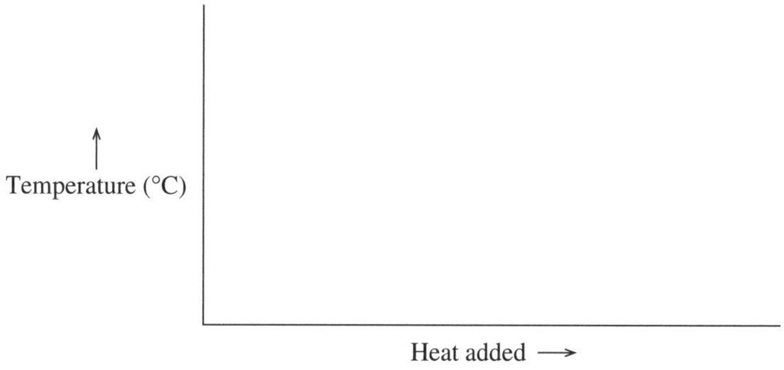

3. Draw a cooling curve for sodium from 1000 °C to 0 °C. Sodium has a freezing point of 98 °C and a boiling (condensation) point of 883 °C.

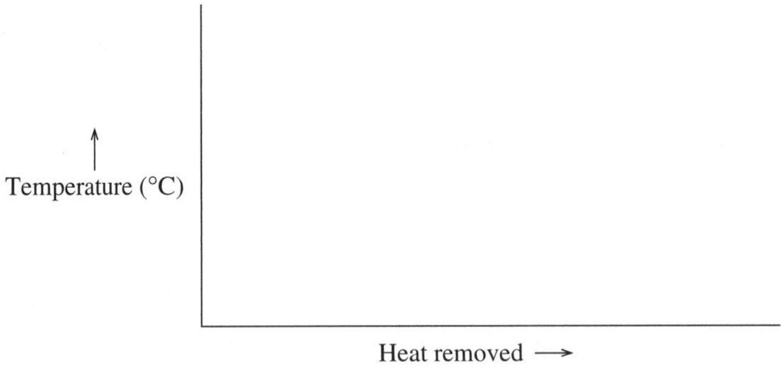

Answers

1.

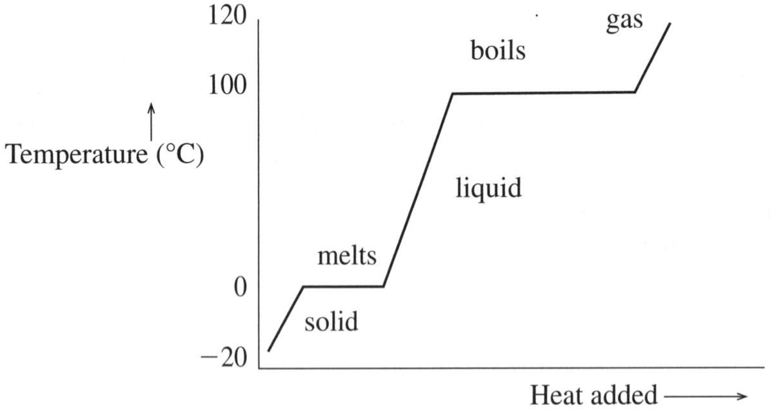

2.

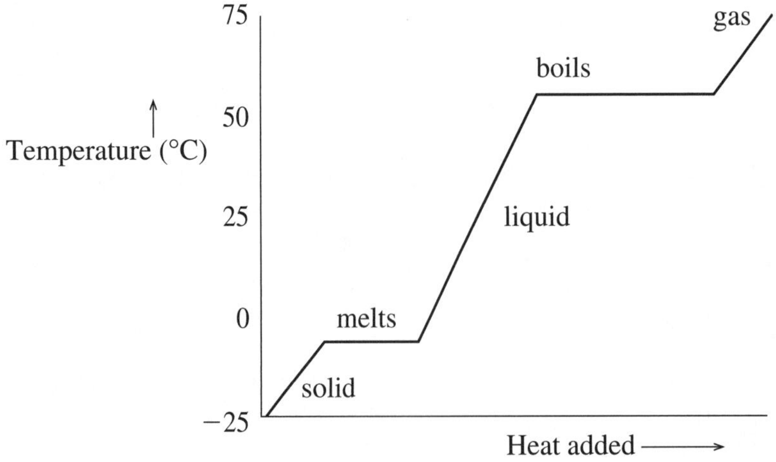

3.

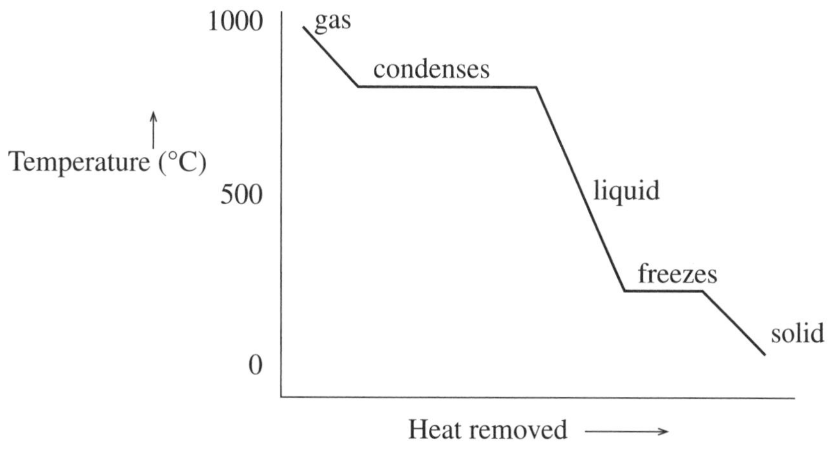

◆ Learning Check 10.5F

a. How many kilojoules are released when 35.0 g of water at 65.0 °C is cooled to 0 °C and frozen? (*Hint:* Two steps are needed.)

b. How many kilojoules are needed to melt 15.0 g of ice at 0 °C, heat the water to 100 °C, and convert the water to gas at 100 °C? (*Hint:* Three steps are needed.)

c. Calculate the number of kilojoules released when 25.0 g of steam at 100 °C condenses, then cools, and finally freezes at 0 °C. (*Hint:* Three steps are needed.)

Answers **a.** 21.2 kJ **b.** 45.2 kJ **c.** 75.3 kJ

Checklist for Chapter 10

You are ready to take the Practice Test for Chapter 10. Be sure you have accomplished the following learning goals for this chapter. If you are not sure, review the section listed at the end of the goal. Then apply your new skills and understanding to the Practice Test.

After studying Chapter 10, I can successfully:

_____ Use electron configurations to show the formation of ions (10.1).

_____ Draw an electron-dot formula of a covalent compound or polyatomic ion with multiple bonds (10.1).

_____ Draw resonance structures for an electron-dot formula when two or more electron-dot formulas are possible (10.1).

_____ Predict the shape and bond angles for a molecule or polyatomic ion (10.2).

_____ Determine the electronegativity difference for two atoms in a compound (10.3).

_____ Classify a bond as nonpolar covalent, polar covalent, or ionic (10.3).

_____ Classify a molecule as polar or nonpolar (10.3).

_____ Identify the attractive forces between particles in a liquid or solid (10.4).

_____ Identify the physical state of a substance as a solid, liquid, or gas (10.5).

_____ Calculate the heat change when a specific amount of a substance changes temperature (10.5).

_____ Calculate the heat change for a change of state for a specific amount of a substance (10.5).

_____ Draw heating and cooling curves using the melting and boiling points of a substance (10.5).

Practice Test for Chapter 10

1. A polar covalent bond typically occurs between
  A. two different nonmetals
  B. two of the same type of nonmetals
  C. two noble gases
  D. two different metals
  E. a metal and a nonmetal

For questions 2 through 7, indicate the type of bonding expected between the following elements:

I. ionic **N.** nonpolar covalent **P.** polar covalent

2. _______ silicon and oxygen **3.** _______ barium and chlorine

4. _______ aluminum and fluorine **5.** _______ chlorine and chlorine

6. _______ sulfur and oxygen **7.** _______ nitrogen and oxygen

For questions 8 through 13, match each the following shapes with one of the descriptions:

A. linear **B.** trigonal planar **C.** tetrahedral
D. trigonal pyramidal **E.** bent (120°) **F.** bent (109.5°)

8. has a central atom with three electron groups and three bonded atoms _______

9. has a central atom with two electron groups and two bonded atoms _______

10. has a central atom with four electron groups and three bonded atoms _______

11. has a central atom with three electron groups and two bonded atoms _______

12. has a central atom with four electron groups and four bonded atoms _______

13. has a central atom with four electron groups and two bonded atoms _______

For questions 14 through 18, determine the shape and bond angles for each of the following molecules:

A. linear, 180° **B.** trigonal planar, 120° **C.** bent, 120°
D. tetrahedral, 109.5° **E.** trigonal pyramidal, 109.5° **F.** bent, 109.5°

14. PCl_3 **15.** CBr_4 **16.** H_2S **17.** BCl_3 **18.** $BeBr_2$

19. Which of the following describes a liquid?
A. a substance that has no definite shape and no definite volume
B. a substance with particles that are far apart
C. a substance with a definite shape and a definite volume
D. a substance containing particles that are moving very fast
E. a substance that has a definite volume, but takes the shape of its container

20. Which of the following describes a solid?
A. a substance that has no definite shape and no definite volume
B. a substance with particles that are far apart
C. a substance with a definite shape and a definite volume
D. a substance containing particles that are moving very fast
E. a substance that has a definite volume, but takes the shape of its container

For questions 21 through 25, match items A to E with one the statements:

A. energy **B.** evaporation **C.** heat of fusion
D. heat of vaporization **E.** boiling

21. ________ the energy required to convert a gram of solid to liquid

22. ________ the heat needed to boil a liquid

23. ________ the conversion of liquid molecules to gas at the surface of a liquid

24. ________ the ability to do work

25. ________ the formation of a gas within the liquid as well as on the surface

26. Ice cools down a drink because
A. the ice is warmer than the drink and heat flows out of the ice cubes
B. heat is absorbed from the drink to melt the ice cubes
C. the heat of fusion of the ice is higher than the heat of fusion for water
D. both A and B
E. none of the above

27. A can full of steam is tightly stoppered. As the can cools
A. nothing will happen
B. the steam will blow up the can
C. the steam condenses and the can collapses
D. the steam condenses and the water formed will expand and explode the can
E. none of the above

28. The number of joules needed to convert 15.0 g of ice to liquid at 0 °C is
A. 73 J **B.** 334 J **C.** 2260 J **D.** 4670 J **E.** 5010 J

29. The number of joules released when 2.0 g of water at 50 °C are cooled and frozen at 0 °C is
A. 418 J **B.** 670 J **C.** 1100 J **D.** 2100 J **E.** 3400 J

30. What is the total number of kilojoules required to convert 25 g of ice at 0 °C to gas at 100 °C?
A. 8.4 kJ **B.** 19 kJ **C.** 59 kJ **D.** 67 kJ **E.** 75 kJ

31. The number of kcal needed to convert 10 g of ice at 0 °C to steam at 100 °C is
A. 0.8 kcal **B.** 1.8 kcal **C.** 6.2 kcal **D.** 6.4 kcal **E.** 7.2 kcal

For questions 32 through 35, consider the heating curve below for p-toluidine. Answer the following questions when heat is added to p-toluidine at −20 °C, where toluidine is below its melting point.

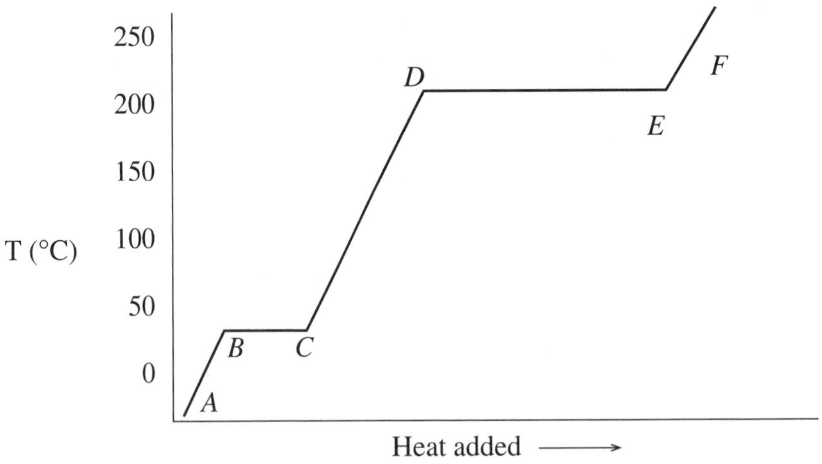

32. On the heating curve, segment *BC* indicates
A. solid **B.** melting **C.** liquid **D.** boiling **E.** gas

33. On the heating curve, segment *CD* shows toluidine as
A. solid **B.** melting **C.** liquid **D.** boiling **E.** gas

34. The boiling point of toluidine would be
A. 20 °C **B.** 45 °C **C.** 100 °C **D.** 200 °C **E.** 250 °C

35. On the heating curve, segment *EF* shows toluidine as
A. solid **B.** melting **C.** liquid **D.** boiling **E.** gas

Answers to the Practice Test

1. A	**2.** P	**3.** I	**4.** I	**5.** N
6. P	**7.** P	**8.** B	**9.** A	**10.** D
11. E	**12.** C	**13.** F	**14.** E	**15.** D
16. F	**17.** B	**18.** A	**19.** E	**20.** C
21. C	**22.** D	**23.** B	**24.** A	**25.** E
26. B	**27.** C	**28.** E	**29.** C	**30.** E
31. E	**32.** B	**33.** C	**34.** D	**35.** E

11
Gases

Study Goals

- Use the kinetic molecular theory of gases to describe the properties of gases.
- Describe the units of measurement used for pressure, and change from one unit to another.
- Use the pressure–volume relationship (Boyle's law) to determine the new pressure or volume of a certain amount of gas at a constant temperature.
- Use the volume–temperature relationship (Charles's law) to determine the new temperature or volume of a certain amount of gas at a constant pressure.
- Use the pressure–temperature relationship (Gay–Lussac's law) to determine the new temperature or pressure of a certain amount of gas at a constant volume.
- Use the combined gas law to find the new pressure, volume, or temperature of a gas when changes in two of these properties are given.
- Describe the relationship between the amount of a gas and its volume (Avogadro's law), and use this relationship in calculations.
- Use the ideal gas law to solve for P, V, T, or n of a gas when given three of the four values in the ideal gas equation.
- Use the ideal gas law to calculate density and molar mass.
- Use partial pressures to calculate the total pressure of a mixture of gases.
- Determine the mass or volume of a gas in a chemical reaction under given conditions.

Think About It

1. How does a barometer work?

2. Why must the pressure of the inhaled gas mixture increase when a person is scuba diving?

3. Why are airplanes pressurized?

4. How do we picture a gas on the molecular level?

Key Terms

Match each of the following key terms with the correct definition:

a. kinetic molecular theory	**b.** pressure	**c.** Boyle's law
d. Charles's law	**e.** STP	**f.** partial pressure

1. _____ the volume of a gas varies directly with the Kelvin temperature when pressure and amount of gas remain constant

2. _____ a force exerted by gas particles that collide with the sides of a container

3. _____ the pressure exerted by the individual gases in a gas mixture

4. _____ the volume of a gas varies inversely with the pressure of a gas when temperature and amount of gas are constant

5. _____ a model that explains the behavior of gaseous particles

6. _____ condition of a gas at 1.00 atm and 0 °C

Answers **1.** d **2.** b **3.** f **4.** c **5.** a **6.** e

11.1 Properties of Gases

- In a gas, particles are far apart and moving so fast that they are not attracted to each other.
- A gas is described by the physical properties of pressure (P), volume (V), temperature (T), and amount in moles (n).

> *MasteringChemistry*
> **Self Study Activity: Properties of Gases**

◆ Learning Exercise 11.1

Answer true (T) or false (F) for each of the following:

a. _______ Gases are composed of small particles.

b. _______ Gas molecules are usually close together.

c. _______ Gas molecules move rapidly because they are strongly attracted.

d. _______ Distances between gas molecules are large.

e. _______ Gas molecules travel in straight lines until they collide.

f. _______ Pressure is the force of gas particles striking the walls of a container.

g. _______ Kinetic energy decreases with increasing temperature.

Answers **a.** T **b.** F **c.** F **d.** T
e. T **f.** T **g.** F

11.2 Gas Pressure

- A gas exerts pressure, which is the force that gas particles exert on the walls of a container.
- Units of gas pressure include torr, mmHg, Pascal (Pa), kilopascal (kPa), $lb/in.^2$ (psi), and atm.
- Gas pressure can be converted from one unit to another using conversion factors.

MasteringChemistry
Case Study: Scuba Diving and Blood Gases

◆ Learning Exercise 11.2

Complete the following:

a. 1.50 atm = _____ mmHg

b. 150 mmHg = _____ atm

c. 0.725 atm = _____ kPa

d. 1520 mmHg = _____ atm

e. 30.5 psi = _____ mmHg

f. During the weather report on TV, the pressure is given as 29.4 in. (of mercury). What is this pressure in mmHg and in atm?

Answers			
	a. 1140 mmHg	**b.** 0.20 atm	**c.** 73.2 kPa
	d. 2.00 atm	**e.** 1580 mmHg	**f.** 747 mmHg; 0.983 atm

11.3 Pressure and Volume (Boyle's Law)

- According to Boyle's law, pressure increases if volume decreases and pressure decreases if volume increases.
- If two properties change in opposite directions, the properties have an *inverse relationship*.
- The volume (V) of a gas changes inversely with the pressure (P) of the gas when T and n are held constant: $P_1V_1 = P_2V_2$. The subscripts 1 and 2 represent initial and final conditions.

Guide to Using the Gas Laws
STEP 1 Organize the data in a table of initial and final conditions.
STEP 2 Rearrange the gas law to solve for the unknown quantity.
STEP 3 Substitute values into the gas law equation to solve for the unknown.

MasteringChemistry
Tutorial: Pressure and Volume

◆ Learning Exercise 11.3A

Complete each of the following with *increases* or *decreases:*

1. Gas pressure increases (T and n constant) when volume ________________.
2. Gas volume increases at constant T and n when pressure ________________.

Answers **1.** decreases **2.** decreases

◆ Learning Exercise 11.3B

Calculate the unknown quantity in each of the following gas problems using Boyle's law:

a. A sample of 4.0 L of helium gas has a pressure of 800 mmHg. What is the new pressure, in mmHg, when the volume is reduced to 1.0 L (n and T constant)?

b. A gas occupies a volume of 360 mL at 750 mmHg. What volume, in mL, does it occupy at a pressure of (1) 1500 mmHg? (2) 375 mmHg (n and T constant)?

c. A gas sample at a pressure of 5.00 atm has a volume of 3.00 L. If the gas pressure is changed to 760 mmHg, what volume, in liters, will the gas occupy (n and T constant)?

d. A sample of 250 mL of nitrogen is initially at a pressure of 2.50 atm. If the pressure changes to 825 mmHg, what is the new volume, in mL (n and T constant)?

Answers **a.** 3200 mmHg **b.** (1) 180 mL (2) 720 mL
c. 15.0 L **d.** 576 mL

11.4 Temperature and Volume (Charles's Law)

- According to Charles's law, temperature increases if the volume of the gas increases and temperature decreases if volume decreases.
- In a direct relationship, two properties increase or decrease together.
- The volume (V) of a gas is directly related to its Kelvin temperature (T) when there is no change in the pressure or moles of the gas:

$$\frac{V_1}{T_1} = \frac{V_2}{T_2}$$ The subscripts 1 and 2 represent initial and final conditions.

◆ Learning Exercise 11.4A

Complete each of the following with *increases* or *decreases:*

a. When the temperature of a gas increases at constant pressure and amount, its volume ___________.

b. When the volume of a gas decreases at constant pressure and amount, its temperature ___________.

Answers: **a.** increases **b.** decreases

◆ Learning Exercise 11.4B

Solve each of the following gas law problems using Charles's law:

a. A balloon has a volume of 2.5 L at a temperature of 0 °C. What is the new volume of the balloon when the temperature rises to 120 °C and the pressure and moles remain constant?

b. Consider a balloon filled with helium to a volume of 6600 L at a temperature of 223 °C. To what temperature, in K and °C, must the gas be cooled to decrease the volume to 4100 L (P and n constant)?

c. A sample of 750 mL of neon is heated from 120 °C to 350 °C. If pressure and amount remain constant, what is the new volume, in mL?

d. What is the final temperature, in K and °C, of 350 mL of oxygen gas at 22 °C if its volume decreases to 0.100 L (P and n constant)?

Answers **a.** 3.6 L **b.** 308 K; 35 °C **c.** 1200 mL **d.** 84 K; −189 °C

11.5 Temperature and Pressure (Gay-Lussac's Law)

- According to Gay-Lussac's law, at constant volume and amount, the gas pressure increases when the temperature of the gas increases, and the pressure decreases when the temperature decreases.
- The pressure (P) of a gas is directly related to its Kelvin temperature (T), when n and V are constant:

$$\frac{P_1}{T_1} = \frac{P_2}{T_2}$$ The subscripts 1 and 2 represent initial and final conditions.

Vapor pressure is the pressure of the gas that forms when a liquid evaporates. At the boiling point of a liquid, the vapor pressure equals the atmospheric pressure.

MasteringChemistry
Tutorial: Temperature and Pressure
Tutorial: Vapor Pressure and Boiling Point

◆ Learning Exercise 11.5A

Solve the following gas law problems using Gay-Lussac's law:

a. A sample of helium gas has a pressure of 860 mmHg at a temperature of 225 K. At what pressure (mmHg) will the helium sample reach a temperature of 675 K (V and n constant)?

b. A balloon contains a gas with a pressure of 580 mmHg and a temperature of 227 °C. What is the new pressure (mmHg) of the gas when the temperature drops to 27 °C (V and n constant)?

c. A spray can contains a gas with a pressure of 3.0 atm at a temperature of 17 °C. What is the pressure (atm) inside the container if the temperature rises to 110 °C (V and n constant)?

d. A gas has a pressure of 1200 mmHg at 300 °C. What will the temperature (°C) be when the pressure falls to 1.10 atm (V and n constant)?

Answers **a.** 2580 mmHg **b.** 348 mmHg **c.** 4.0 atm **d.** 126 °C

11.6 The Combined Gas Law

- At constant amount of a gas, the gas laws can be combined into a relationship of pressure (P), volume (V), and temperature (T):

$$\frac{P_1V_1}{T_1} = \frac{P_2V_2}{T_2}$$ Subscripts 1 and 2 represent initial and final conditions.

MasteringChemistry
Tutorial: The Combined Gas Law

◆ Learning Exercise 11.6

Solve each of the following using the combined gas law (n is constant):

a. A 5.0-L sample of nitrogen gas has a pressure of 1200 mmHg at 220 K. What is the pressure, in mmHg, of the sample when the volume increases to 20.0 L at 440 K?

b. A 25.0-mL bubble forms at the ocean depths where the pressure is 10.0 atm and the temperature is 5.0 °C. What is the volume of that bubble at the ocean surface where the pressure is 760.0 mmHg and the temperature is 25 °C?

c. A 35.0-mL sample of argon gas has a pressure of 1.0 atm and a temperature of 15 °C. What is the final volume, in millimeters, if the pressure goes to 2.0 atm and the temperature to 45 °C?

d. A weather balloon with a volume of 315 L is launched at the Earth's surface where the temperature is 12 °C and the pressure is 0.930 atm. What is the volume, in liters, of the balloon in the upper atmosphere where the pressure is 116 mmHg and the temperature is −35 °C?

e. A 10.0-L sample of gas is emitted from a volcano with a pressure of 1.20 atm and a temperature of 150.0 °C. What is the volume, in liters, of the gas when its pressure is 0.900 atm and the temperature is −40.0 °C?

Answers	**a.** 600 mmHg	**b.** 268 mL	**c.** 19 mL
	d. 1.60×10^3 L	**e.** 7.34 L	

11.7 Volume and Moles (Avogadro's Law)

- If the number of moles of gas increases, the volume increases; if the number of moles of gas decreases, the volume decreases.
- Avogadro's law states that equal volumes of gases at the same temperature and pressure contain the same number of moles. The volume (V) of a gas is directly related to the number of moles of the gas when the pressure and temperature of the gas do not change:

$$\frac{V_1}{n_1} = \frac{V_2}{n_2}$$ Subscripts 1 and 2 represent initial and final conditions.

- At STP conditions, standard pressure (1 atm) and temperature (0 °C), one mole of a gas occupies a volume of 22.4 L, the *molar volume*.

Study Note

At STP, the molar volume factor 22.4 L/mol converts between moles of a gas and volume.

Example: What volume, in liters, would 2.00 mol of N_2 occupy at STP?

Solution: $2.00\ \cancel{\text{mol } N_2} \times \dfrac{22.4\text{ L}}{1\ \cancel{\text{mol } N_2}} = 44.8\text{ L}$

MasteringChemistry

Tutorial: Volume and Moles

◆ Learning Exercise 11.7

Use Avogadro's law to solve the following gas problems.

a. A gas containing 0.50 mol of helium has a volume of 4.00 L. What is the new volume, in liters, when 1.0 mol of nitrogen is added to the container (*P* and *T* constant)?

b. A balloon containing 1.00 mol oxygen has a volume of 15 L. What is the new volume, in liters, of the balloon when 2.00 mol of helium is added (*P* and *T* constant)?

c. What is the volume, in liters, occupied by 210 g of nitrogen (N_2) at STP (*P* and *T* constant)?

d. What is the volume, in liters, of 6.40 g of O_2 at STP?

Answers **a.** 12 L **b.** 45 L **c.** 168 L **d.** 4.48 L

11.8 The Ideal Gas Law

- The ideal gas law $PV = nRT$ gives the relationship among the four variables: pressure, volume, moles, and temperature. When any three variables are given, the fourth can be calculated.
- R is the universal gas constant whose value depends on the unit used to measure P:

$$0.0821\ \text{L} \cdot \text{atm/mol} \cdot \text{K} \quad \text{or} \quad 62.4\ \text{L} \cdot \text{mmHg/mol} \cdot \text{K}$$

- The ideal gas law can be used to determine the mass or volume of a gas in a chemical reaction.
- The molar mass and density of a gas can be determined using the ideal gas law.

Guide to Using the Ideal Gas Law

STEP 1 Organize the data given for the gas.
STEP 2 Solve the ideal gas law to solve for the unknown quantity.
STEP 3 Substitute gas data and calculate unknown value.

MasteringChemistry
Tutorial: Introduction to the Ideal Gas Law
Self Study Activity: The Ideal Gas Law

Study Note

Identify the three known variables for the ideal gas law, and arrange the equation to solve for the unknown variable.

Example: Solve the ideal gas law for P. $PV = nRT \quad P = \dfrac{nRT}{V}$

◆ Learning Exercise 11.8A

Use the ideal gas law to solve for the unknown variable in each of the following:

a. What volume, in liters, is occupied by 0.250 mol of nitrogen gas (N_2) at 0 °C and 1.50 atm?

b. What is the temperature, in °C, of 0.500 mol of helium that occupies a volume of 15.0 L at a pressure of 1250 mmHg?

c. What is the pressure, in atm, of 1.50 mol of neon in a 5.00-L steel container at a temperature of 125 °C?

d. What is the pressure, in atm, of 8.0 g of oxygen (O_2) that has a volume of 245 mL at a temperature of 22 °C?

Answers **a.** 3.74 L **b.** 328 °C **c.** 9.80 atm **d.** 25 atm

◆ Learning Exercise 11.8B

Use gas laws to determine the molar mass of a gas. (Hint: Determine the moles (n) of gas and then calculate the mass/mole ratio to obtain molar mass.)

a. A gas has a mass of 0.650 g and a volume of 560 mL at STP. What is the molar mass of the gas?

b. A sample of gas with a mass of 0.412 g has a volume of 273 mL and a pressure of 746 mmHg at 25 °C. What is the molar mass of the gas?

Answers **a.** 26.0 g/mol **b.** 37.6 g/mol

11.9 Gas Laws and Chemical Reactions

- The ideal gas law or molar volume at STP is used to determine the number of moles (and therefore, mass using the molar mass) or volume of a gas in a chemical reaction.
- Mole-mole factors are then used to calculate moles of other substances in a reaction.

MasteringChemistry
Tutorial: The Ideal Gas Law and Stoichiometry

◆ Learning Exercise 11.9

Use gas laws to determine the quantity of a reactant or product in each of the following chemical reactions:

a. How many liters of hydrogen gas at STP are produced when 12.5 g of magnesium react?

$$Mg(s) + 2HCl(aq) \longrightarrow MgCl_2(aq) + H_2(g)$$

b. How many grams of KNO_3 must decompose to produce 35.8 L of O_2 at 28 °C and 745 mmHg?

$$2KNO_3(s) \longrightarrow 2KNO_2(s) + O_2(g)$$

c. At a temperature of 325 °C and a pressure of 1.20 atm, how many liters of CO_2 can be produced when 50.0 g of propane undergoes combustion?

$$C_3H_8(g) + 5O_2(g) \longrightarrow 3CO_2(g) + 4H_2O(g)$$

Answers **a.** 11.5 L of H_2 **b.** 287 g of KNO_3 **c.** 139 L of CO_2

11.10 Partial Pressures (Dalton's Law)

- In a mixture of two or more gases, the total pressure is the sum of the partial pressures (subscripts 1, 2, 3 . . . represent the partial pressures of the individual gases).

$$P_{total} = P_1 + P_2 + P_3 + \cdots$$

- The partial pressure of a gas in a mixture is the pressure it would exert if it were the only gas in the container.

Guide to Solving for Partial Pressure

STEP 1 Write the equation for the sum of partial pressures.
STEP 2 Solve for the unknown quantity.
STEP 3 Substitute known pressures and calculate the unknown.

MasteringChemistry

Tutorial: Mixture of Gases

◆ Learning Exercise 11.10A

Use Dalton's law to solve the following problems about gas mixtures:

a. What is the pressure, in mmHg, of a sample of gases that contains oxygen at 0.500 atm, nitrogen (N_2) at 132 torr, and helium at 224 mmHg?

b. What is the pressure (atm) of a gas sample that contains helium at 285 mmHg and oxygen (O_2) at 1.20 atm?

c. A gas sample containing nitrogen (N_2) and oxygen (O_2) has a pressure of 1500 mmHg. If the partial pressure of the nitrogen is 0.900 atm, what is the partial pressure, in mmHg, of the oxygen gas in the mixture?

Answers **a.** 736 mmHg **b.** 1.58 atm **c.** 816 mmHg

◆ Learning Exercise 11.10B

Fill in the blanks by writing I (*increases*) or D (*decreases*) for a gas in a closed container.

	Pressure	Volume	Moles	Temperature
a.	______	Increases	Constant	Constant
b.	Increases	Constant	______	Constant
c.	Constant	Decreases	______	Constant
d.	______	Constant	Constant	Increases
e.	Constant	______	Constant	Decreases
f.	______	Constant	Increases	Constant

Answers

a. D **b.** I **c.** D
d. I **e.** D **f.** I

◆ Learning Exercise 11.10C

Complete the following table for typical blood gas values for partial pressures:

Gas	Alveoli	Oxygenated blood	Deoxygenated blood	Tissues
O_2	______	______	______	______
CO_2	______	______	______	______

Answers

O_2	100 mmHg	100 mmHg	40 mmHg or less	30 mmHg or less
CO_2	40 mmHg	40 mmHg	46 mmHg	50 mmHg or greater

Checklist for Chapter 11

You are ready to take the Practice Test for Chapter 11. Be sure you have accomplished the following learning goals for this chapter. If you are not sure, review the section listed at the end of the goal. Then apply your new skills and understanding to the Practice Test.

After studying Chapter 11, I can successfully:

_____ Describe the kinetic molecular theory of gases (11.1).

_____ Change the units of pressure from one to another (11.2).

_____ Use the pressure–volume relationship (Boyle's law) to determine the new pressure or volume of a fixed amount of gas at constant temperature (11.3).

_____ Use the temperature–volume relationship (Charles's law) to determine the new temperature or volume of a fixed amount of gas at a constant pressure (11.4).

_____ Use the temperature–pressure relationship (Gay-Lussac's law) to determine the new temperature or pressure of a certain amount of gas at a constant volume (11.5).

_____ Use the combined gas law to find the new pressure, volume, or temperature of a certain amount of gas when changes in two of these properties are given (11.6).

_____ Describe the relationship between the amount of a gas and its volume (Avogadro's law) at constant P and T, and use this relationship in calculations (11.7).

_____ Use the ideal gas law to solve for pressure, volume, temperature, amount of a gas, molar mass, or density (11.8).

_____ Use the gas laws to calculate the quantity of a reactant or product in a chemical reaction (11.9).

_____ Calculate the total pressure of a gas mixture from the partial pressures (11.10).

Practice Test for Chapter 11

For questions 1 through 5, answer using T (true) or F (false):

1. _____ A gas does not have its own volume or shape.

2. _____ The molecules of a gas are moving extremely fast.

3. _____ The collisions of gas molecules with the walls of their container create pressure.

4. _____ Gas molecules are close together and move in straight lines.

5. _____ Gas molecules have no attractions between them.

6. When a gas is heated in a closed metal container, the
A. pressure increases **B.** pressure decreases **C.** volume increases
D. volume decreases **E.** number of molecules increases

7. The pressure of a gas will increase when
A. the volume increases
B. the temperature decreases
C. more molecules of gas are added
D. molecules of gas are removed
E. none of these

8. If the temperature of a gas is increased,
A. the pressure will decrease
B. the volume will increase
C. the volume will decrease
D. the number of molecules will increase
E. none of these

9. The relationship that the volume of a gas is inversely related to its pressure at constant temperature is known as
A. Boyle's law **B.** Charles's law **C.** Gay-Lussac's law
D. Dalton's law **E.** Avogadro's law

10. What is the pressure (atm) of a gas with a pressure of 1200 mmHg?
A. 0.63 atm **B.** 0.79 atm **C.** 1.2 atm
D. 1.6 atm **E.** 2.0 atm

11. A 6.00-L sample of oxygen has a pressure of 660 mmHg. When the volume is reduced to 2.00 L at constant temperature and moles, it will have a new pressure of
A. 1980 mmHg **B.** 1320 mmHg **C.** 330 mmHg
D. 220 mmHg **E.** 110 mmHg

12. A sample of nitrogen gas at 110 K has a pressure of 1.0 atm. When the temperature is increased to 360 K at constant moles and volume, the new pressure will be
A. 0.50 atm **B.** 1.0 atm **C.** 1.5 atm
D. 3.3 atm **E.** 4.0 atm

13. If two gases have the same volume, temperature, and pressure, they also have the same
A. density **B.** number of molecules **C.** molar mass
D. speed **E.** size molecules

14. A gas sample with a volume of 4.0 L has a pressure of 750 mmHg and a temperature of 77 °C. What is its new volume at 277 °C and 250 mmHg (*n* constant)?
A. 7.6 L **B.** 19 L **C.** 2.1 L
D. 0.00056 L **E.** 3.3 L

15. If the temperature and amount of a gas does not change, but its volume doubles, its pressure will
A. double
B. triple
C. decrease to one-half the original pressure
D. decrease to one-fourth the original pressure
E. not change

16. A sample of oxygen with a pressure of 400 mmHg contains 2.0 mol of gas and has a volume of 4.0 L. What will the new pressure be when the volume expands to 5.0 L and 3.0 mol of helium gas is added while temperature is constant?
A. 160 mmHg **B.** 250 mmHg **C.** 800 mmHg
D. 1000 mmHg **E.** 1560 mmHg

17. A sample of 2.00 mol of gas initially at STP is converted to a volume of 5.0 L and a temperature of 27 °C. What is its new pressure in atm?
A. 0.12 atm **B.** 5.5 atm **C.** 7.5 atm
D. 9.8 atm **E.** 10. atm

18. The conditions for standard temperature and pressure (STP) are
A. 0 K, 1 atm **B.** 0 °C, 10 atm **C.** 25 °C, 1 atm
D. 273 K, 1 atm **E.** 273 K, 0.5 atm

19. The volume occupied by 1.50 mol of CH_4 at STP is
A. 44.10 L **B.** 33.6 L **C.** 22.4 L
D. 11.2 L **E.** 5.60 L

20. How many grams of oxygen gas (O_2) are present in 44.1 L of oxygen at STP?
A. 10.0 g **B.** 16.0 g **C.** 32.0 g
D. 410.0 g **E.** 63.0 g

21. What is the volume, in liters, of 0.50 mol of nitrogen gas (N_2) at 25 °C and 2.0 atm?
A. 0.51 L **B.** 1.0 L **C.** 4.2 L
D. 6.1 L **E.** 24 L

22. A gas mixture contains helium with a partial pressure of 0.100 atm, oxygen with a partial pressure of 445 mmHg, and nitrogen with a partial pressure of 235 mmHg. What is the total pressure, in atm, for the gas mixture?
A. 0.995 atm **B.** 1.39 atm **C.** 1.69 atm
D. 2.00 atm **E.** 10.0 atm

23. A mixture of oxygen and nitrogen has a total pressure of 1040 mmHg. If the oxygen has a partial pressure of 510 mmHg, what is the partial pressure of the nitrogen?
A. 240 mmHg **B.** 530 mmHg **C.** 775 mmHg
D. 1040 mmHg **E.** 1350 mmHg

24. 3.00 mol of He in a steel container has a pressure of 12.0 atm. What is the new pressure after 4.00 mol of He is added (*T* is constant)?

A. 5.14 atm **B.** 16.0 atm **C.** 28.0 atm
D. 32.0 atm **E.** 45.0 atm

25. 2.50 g of $KClO_3$ decomposes by the equation: $2KClO_3(s) \longrightarrow 2KCl(s) + 3O_2(g)$. What volume of O_2 is produced at 25 °C and a pressure of 750 mmHg?

A. 0.00998 L **B.** 0.759 L **C.** 0.0636 L
D. 0.506 L **E.** 1.32 L

26. A gas sample with a mass of 0.440 g has a volume of 0.224 L at STP. The gas has a molar mass of

A. 19.6 g/mol **B.** 22.4 g/mol **C.** 44.0 g/mol
D. 44.8 g/mol **E.** 50.9 g/mol

27. The exchange of gases between the alveoli, blood, and tissues of the body is a result of

A. pressure gradients **B.** different molecular masses
C. shapes of molecules **D.** altitude
E. all of these

28. Oxygen moves into the tissues from the blood because its partial pressure

A. in arterial blood is higher than in the tissues
B. in venous blood is higher than in the tissues
C. in arterial blood is lower than in the tissues
D. in venous blood is lower than in the tissues
E. is equal in the blood and in the tissues

Answers to the Practice Test

1. T	**2.** T	**3.** T	**4.** F	**5.** T
6. A	**7.** C	**8.** B	**9.** A	**10.** D
11. A	**12.** D	**13.** B	**14.** B	**15.** C
16. C	**17.** D	**18.** D	**19.** B	**20.** E
21. D	**22.** A	**23.** B	**24.** C	**25.** B
26. C	**27.** A	**28.** A		

12
Solutions

Study Goals

- Identify the solute and solvent in a solution.
- Describe the formation of a solution.
- Define solubility, distinguish between a saturated and unsaturated solution, and determine whether a salt will dissolve in water.
- Identify solutes as electrolytes or nonelectrolytes and distinguish between strong electrolytes, weak electrolytes, and nonelectrolytes in solution.
- Calculate the mass percent (m/m) and volume percent (v/v) concentrations of a solution.
- Describe how to prepare a dilute solution.
- Calculate the molarity of a solution; use molarity as a conversion factor to determine moles of solute or volume of solution.
- Calculate the quantities of reactants or products in reactions involving solutions based on molarity and volume.
- Identify a mixture as a solution, a colloid, or a suspension.
- Describe how number of particles affects the freezing point, boiling point, and osmotic pressure of a solution.
- Describe osmosis and isotonic, hypotonic, and hypertonic solutions.

Think About It

1. Why is tea called a solution, but the water used to make tea is a pure substance?

2. Sugar dissolves in water, but oil does not. Why?

3. Why is salt used in the preparation of ice cream?

Key Terms

Match each of the following key terms with the correct definitions:

a. solution **b.** percent concentration (m/m) **c.** molarity
d. weak electrolyte **e.** strong electrolyte **f.** solubility
g. net ionic equation

1. _____ a substance that dissociates into ions when it dissolves in water

2. _____ the grams of solute dissolved in 100 g of solution

3. _____ the number of moles of solute in 1 L of solution

4. _____ a solute that forms a few ions, but mostly molecules, in water

5. _____ a mixture of at least two components called a solute and a solvent

6. _____ shows only the reactants undergoing a change

7. _____ the maximum amount of a solution that can dissolve under given conditions

Answers **1.** e **2.** b **3.** c **4.** d **5.** a **6.** g **7.** f

12.1 Solutions

- A solution is a homogeneous mixture that forms when a solute dissolves in a solvent. The solvent is present in a greater amount.
- Solvents and solutes can be solids, liquids, or gases. The solution is the same physical state as the solvent.
- A polar solute is soluble in a polar solvent; a nonpolar solute is soluble in a nonpolar solvent.
- Water molecules form hydrogen bonds because the partial positive charge of the hydrogen in one water molecule is attracted to the partial negative charge of oxygen in another water molecule.
- An ionic solute dissolves in water, a polar solvent, because the polar water molecules attract and pull the positive and negative ions into solution. In solution, water molecules surround the ions in a process called hydration.

MasteringChemistry
Self Study Activity: Hydrogen Bonding

◆ Learning Exercise 12.1A

Water is polar, and hexane is nonpolar. In which solvent is each of the following soluble?

a. bromine, Br_2, nonpolar _______ **b.** HCl, polar _______

c. cholesterol, nonpolar _______ **d.** vitamin D, nonpolar _______

e. vitamin C, polar _______

Answers **a.** hexane **b.** water **c.** hexane **d.** hexane **e.** water

◆ Learning Exercise 12.1B

Indicate the solute and solvent in each of the following:

	Solute	Solvent
a. 10 g of KCl dissolved in 100 g of water	________	________
b. soda water: $CO_2(g)$ dissolved in water	________	________
c. an alloy composed of 80% Zn and 20% Cu	________	________
d. a mixture of O_2 (200 mmHg) and He (500 mmHg)	________	________
e. a solution of 40 mL of CCl_4 and 2 mL of Br_2	________	________

Answers **a.** KCl; water **b.** CO_2; water **c.** Cu; Zn
d. O_2; He **e.** Br_2; CCl_4

12.2 Electrolytes and Nonelectrolytes

- Electrolytes conduct an electrical current because they produce ions in aqueous solutions.
- Strong electrolytes are completely ionized, whereas weak electrolytes are slightly ionized.
- Nonelectrolytes do not form ions in solution but dissolve as molecules.

◆ Learning Exercise 12.2A

Write an equation for the formation of an aqueous solution of each of the following strong electrolytes:

a. LiCl

b. $Mg(NO_3)_2$

c. Na_3PO_4

d. K_2SO_4

e. $MgCl_2$

Answers

a. $LiCl(s) \xrightarrow{H_2O} Li^+(aq) + Cl^-(aq)$

b. $Mg(NO_3)_2(s) \xrightarrow{H_2O} Mg^{2+}(aq) + 2NO_3^-(aq)$

c. $Na_3PO_4(s) \xrightarrow{H_2O} 3Na^+(aq) + PO_4^{3-}(aq)$

d. $K_2SO_4(s) \xrightarrow{H_2O} 2K^+(aq) + SO_4^{2-}(aq)$

e. $MgCl_2(s) \xrightarrow{H_2O} Mg^{2+}(aq) + 2Cl^-(aq)$

◆ Learning Exercise 12.2B

Indicate whether an aqueous solution of each of the following contains ions only, molecules only, or mostly molecules with some ions. Write an equation for the formation of the solution.

a. glucose, $C_6H_{12}O_6$, a nonelectrolyte

b. NaOH, a strong electrolyte

c. KCl, a strong electrolyte

d. HF, a weak electrolyte

Answers

a. $C_6H_{12}O_6(s) \xrightarrow{H_2O} C_6H_{12}O_6(aq)$	molecules only
b. $NaOH(s) \xrightarrow{H_2O} Na^+(aq) + OH^-(aq)$	ions only
c. $KCl(s) \xrightarrow{H_2O} K^+(aq) + Cl^-(aq)$	ions only
d. $HF(aq) + H_2O(l) \rightleftharpoons H_3O^+(aq) + F^-(aq)$	mostly molecules and a few ions

12.3 Solubility

- The amount of solute that dissolves depends on the nature of the solute and solvent.
- Solubility describes the maximum amount of a solute that dissolves in exactly 100 g solvent at a given temperature.
- A saturated solution contains the maximum amount of dissolved solute at a certain temperature whereas an unsaturated solution contains less than this amount.
- An increase in temperature increases the solubility of most solids, but decreases the solubility of gases in water.
- Henry's law states that the solubility of gases in liquids is directly related to the partial pressure of that gas over the liquid.
- The solubility rules describe the kinds of ionic combinations that are soluble and insoluble in water. If a salt contains Li^+, Na^+, K^+, NO_3^-, or NH_4^+ ion, it is soluble in water. Most halides and sulfates are soluble.

MasteringChemistry

Self Study Activity: Solubility
Tutorial: Solubility
Case Study: Kidney Stones and Saturated Solutions

◆ Learning Exercise 12.3A

Identify each of the following as a saturated solution (S) or an unsaturated solution (U):

1. A sugar cube dissolves when added to a cup of coffee. ______
2. A KCl crystal added to a KCl solution does not change in size. ______
3. A layer of sugar forms in the bottom of a glass of iced tea. ______
4. The rate of crystal formation is equal to the rate of dissolving. ______
5. Upon heating, all the sugar in a solution dissolves. ______

Answers **1.** U **2.** S **3.** S **4.** S **5.** U

◆ Learning Exercise 12.3B

Use the solubility of $NaNO_3$ to answer each of the following problems:

Temperature (°C)	**Solubility (g of $NaNO_3$/100 g of H_2O)**
40	100. g
60	120. g
80	150. g
100	200. g

a. How many grams of $NaNO_3$ will dissolve in 100. g water at 40 °C?

b. How many grams of $NaNO_3$ will dissolve in 300. g of water at 60 °C?

c. A solution is prepared using 200. g of water and 350. g of $NaNO_3$ at 80 °C. Will any solute remain undissolved: If so how much?

d. Will 150. g of $NaNO_3$ all dissolve when added to 100. g of water at 100 °C?

Answers **a.** 100. g **b.** 360. g
c. 300. g of $NaNO_3$ will dissolve leaving 50. g of $NaNO_3$ that will not dissolve.
d. Yes, all 150. g of $NaNO_3$ will dissolve.

◆ Learning Exercise 12.3C

Predict whether the following salts are soluble (S) or insoluble (I) in water:

1. ______ NaCl	**2.** ______ $AgNO_3$	**3.** ______ $PbCl_2$
4. ______ Ag_2S	**5.** ______ $BaSO_4$	**6.** ______ Na_2CO_3
7. ______ K_2S	**8.** ______ $MgCl_2$	**9.** ______ BaS

Answers **1.** S **2.** S **3.** I **4.** I **5.** I
6. S **7.** S **8.** S **9.** I

◆ Learning Exercise 12.3D

Predict whether an insoluble salt forms in the following mixtures of soluble salts. If so, write the formula of the solid.

1. $NaCl(aq) + Pb(NO_3)_2(aq) \longrightarrow$

2. $BaCl_2(aq) + Na_2SO_4(aq) \longrightarrow$

3. $K_3PO_4(aq) + NaNO_3(aq) \longrightarrow$

4. $Na_2S(aq) + AgNO_3(aq) \longrightarrow$

Answers **1.** Yes; $PbCl_2$ **2.** Yes; $BaSO_4$ **3.** None **4.** Yes; Ag_2S

Guide to Writing Net Ionic Equations for Formation of an Insoluble Salt

STEP 1 Write the ions of the reactants.
STEP 2 Write the combinations of ions and determine if any are insoluble.
STEP 3 Write the ionic equation including any solid.
STEP 4 Write the net ionic equation.

Study Note

Write the ionic and net ionic equations for the reaction that occurs when solutions of $AgNO_3(aq)$ and $Na_2SO_4(aq)$ are mixed.

Solution:

STEP 1 Write the ions of the reactants.

$$AgNO_3(aq) = Ag^+(aq) + NO_3^-(aq) \qquad Na_2SO_4(aq) = 2Na^+ + SO_4^{2-}(aq)$$

STEP 2 Write the combinations of ions and determine if any are insoluble.

$$Na^+(aq) + NO_3^-(aq) \longrightarrow NaNO_3(aq) \quad \text{a soluble salt}$$
$$2Ag^+(aq) + SO_4^{2-}(aq) \longrightarrow AgSO_4(s) \quad \text{an insoluble salt}$$

STEP 3 Write the ionic equation including any solid. By breaking up all the ionic formulas into ions, we can write

$$Na^+(aq) + NO_3^-(aq) + 2Ag^+(aq) + SO_4^{2-}(aq) \longrightarrow Na^+(aq) + NO_3^-(aq) + Ag_2SO_4(s)$$

STEP 4 Write the net ionic equation. Remove the spectator ions, which appear on both sides of the ionic equation, to write the net ionic equation.

$$Na^+(aq) + NO_3^-(aq) + \mathbf{2Ag^+(aq)} + \mathbf{SO_4^{2-}(aq)} \longrightarrow Na^+(aq) + NO_3^-(aq) + \mathbf{Ag_2SO_4(s)}$$
$$2Ag^+(aq) + SO_4^{2-}(aq) \longrightarrow Ag_2SO_4(s) \quad \text{net ionic equation}$$

◆ Learning Exercise 12.3E

For the mixtures that form insoluble salts in Learning Exercise 12.3D, write the ionic equation and the net ionic equation.

1.

2.

3.

4.

Answers

1. $2Na^+(aq) + 2Cl^-(aq) + Pb^{2+}(aq) + 2NO_3^-(aq) \longrightarrow 2Na^+(aq) + PbCl_2(s) + 2NO_3^-(aq)$ (ionic)
   $2Cl^-(aq) + Pb^{2+}(aq) \longrightarrow PbCl_2(s)$ (net ionic equation)
2. $Ba^{2+}(aq) + 2Cl^-(aq) + 2Na^+(aq) + SO_4^{2-}(aq) \longrightarrow 2Na^+(aq) + BaSO_4(s) + 2Cl^-(aq)$ (ionic)
   $Ba^{2+}(aq) + SO_4^{2-}(aq) \longrightarrow BaSO_4(s)$ (net ionic equation)
3. No insoluble salt forms.
4. $2Na^+(aq) + S^{2-}(aq) + 2Ag^+(aq) + 2NO_3^-(aq) \longrightarrow 2Na^+(aq) + Ag_2S(s) + 2NO_3^-(aq)$ (ionic)
   $S^{2-}(aq) + 2Ag^+(aq) \longrightarrow Ag_2S(s)$ (net ionic equation)

12.4 Percent Concentration

- The concentration of a solution is the relationship between the amount of solute (g or mL) and the amount (g or mL) of solution.
- A mass percent (m/m) expresses the ratio of the mass of solute to the mass of solution multiplied by 100%.

$$\text{Percent (m/m)} = \frac{\text{grams of solute}}{\text{grams of solution}} \times 100\%$$

- A volume percent (v/v) expresses the ratio of the volume of the solute to the volume of the solution multiplied by 100%.

Guide to Calculating Solution Concentration

STEP 1 Determine quantities of solute and solution.
STEP 2 Write the % concentration expression.
STEP 3 Substitute solute and solution quantities into the expression.

Study Note

Calculate a mass percent concentration (m/m) as

$$\text{Mass percent} = \frac{\text{grams of solute}}{\text{grams of solution}} \times 100\%$$

What is the percent (m/m) when 2.4 g of $NaHCO_3$ dissolves in 120 g of solution?

Solution: $\frac{2.4 \text{ g } NaHCO_3}{120 \text{ g solution}} \times 100\% = 2.0\% \text{ (m/m)}$

MasteringChemistry

Tutorial: Calculating Percent Concentration
Tutorial: Percent Concentration as a Conversion Factor

◆ Learning Exercise 12.4A

Determine the percent concentration (m/m or v/v) for each of the following solutions:

a. mass percent (m/m) for 18.0 g of NaCl in 90.0 g of solution

b. mass percent (m/m) for 35.2 g of KCl in 425 g of KCl solution

c. mass percent (m/m) for 1.0 g of KOH in 25 g of KOH solution

d. volume percent (v/v) for 18 mL of alcohol in 350 mL of mouthwash solution

Answers **a.** 20.0% **b.** 8.28% **c.** 4.0% **d.** 5.1%

Guide to Using Concentration to Calculate Mass or Volume

STEP 1 State the given and needed quantities.
STEP 2 Write a plan to calculate mass or volume.
STEP 3 Write equalities and conversion factors.
STEP 4 Set up problem to calculate mass or volume.

Study Note

How many grams of KI are needed to prepare 225 g of a 4.0% (m/m) KI solution?

Solution:

STEP 1 State the given and needed quantities.

Given	225 g of a 4.0% (m/m) KI solution
Need	grams of KI

STEP 2 Write a plan to calculate mass or volume.

$$\text{grams of KI solution} \longrightarrow \text{grams of KI}$$

STEP 3 Write equalities and conversion factors.

$$100\text{ g of KI solution} = 4.0\text{ g of KI}$$

$$\frac{4.0\text{ g KI}}{100\text{ g KI solution}} \quad \text{and} \quad \frac{100\text{ g KI solution}}{4.0\text{ g KI}}$$

STEP 4 Set up problem to calculate mass or volume.

$$225\ \cancel{\text{g solution}} \times \frac{4.0\text{ g KI}}{100\ \cancel{\text{g solution}}} = 9.0\text{ g of KI}$$

◆ Learning Exercise 12.4B

Calculate the number of grams of solute needed to prepare each of the following solutions:

a. How many grams of glucose are needed to prepare 480 g of a 5.0% (m/m) solution?

b. How many grams of lidocaine hydrochloride are needed to prepare 50.0 g of a 2.0% (m/m) solution?

c. How many grams of KCl are needed to prepare 1250 g of a 4.00% (m/m) KCl solution?

d. How many grams of NaCl are needed to prepare 75.6 g of a 1.50% (m/m) solution?

Answers **a.** 24 g **b.** 1.0 g **c.** 50.0 g of KCl **d.** 1.13 g of NaCl

◆ Learning Exercise 12.4C

Use percent concentration factors to calculate the mass of each solution that contains the amount of solute stated in each problem.

a. 8.75 g of NaCl from a 3.00% (m/m) NaCl solution

b. 24 g of glucose from a 5.0% (m/m) glucose solution

c. 10.5 g of KCl from a 2.5% (m/m) KCl solution

d. 75.0 g of NaOH from a 25.0% (m/m) NaOH solution

Answers **a.** 292 g **b.** 480 g **c.** 420 g **d.** 300 g

12.5 Molarity and Dilution

- Molarity is a concentration term that describes the number of moles of solute dissolved in 1 L (1000 mL) of solution.

$$\frac{\text{moles solute}}{\text{L solution}} \quad \text{and} \quad \frac{\text{L solution}}{\text{moles solute}}$$

- Molarity can be used as a conversion factor to convert the moles of a solute to a volume of exactly 1 L of solution and the volume of exactly 1 L of solution to moles of the solute.

$$\frac{\text{moles solute}}{\text{1 L solution}} \quad \text{and} \quad \frac{\text{1 L solution}}{\text{moles solute}}$$

- *Dilution* is the process of mixing a solution with solvent to obtain a lower concentration.
- For dilutions, solve for the unknown value in the expression $M_1V_1 = M_2V_2$ (subscript 1 represents the concentrated solution and subscript 2 represents the diluted solution.

Guide to Calculating Molarity

STEP 1 State the given and needed quantities.
STEP 2 Write a plan to calculate molarity.
STEP 3 Write equalities and conversion factors needed.
STEP 4 Set up problem to calculate molarity.

MasteringChemistry
Tutorial: Dilution

◆ Learning Exercise 12.5A

Calculate the molarity of the following solutions:

a. 1.45 mol of HCl in 0.250 L of HCl solution

b. 10.0 mol of glucose ($C_6H_{12}O_6$) in 2.50 L of glucose solution

c. 80.0 g of NaOH in 1.60 L of NaOH solution (*Hint:* Find moles of NaOH.)

d. 38.8 g of NaBr in 175 mL of NaBr solution

Answers **a.** 5.80 M HCl solution **b.** 4.00 M glucose solution
c. 1.25 M NaOH solution **d.** 2.15 M NaBr solution

Study Note

Molarity is used as a conversion factor to convert between the amount of solute and the volume of solution. How many grams of NaOH are in 0.250 L of a 5.00 M NaOH solution?

Solution:

STEP 1 State the given and needed quantities.

Given 0.250 L of a 5.00 M NaOH solution
Need grams of NaOH

STEP 2 Write a plan to calculate mass or volume.

liters of solution ⟶ moles of solution ⟶ grams of solute

STEP 3 Write equalities and conversion factors. The concentration 5.00 M can be expressed as conversion factors.

$$1 \text{ L of NaOH solution} = 5.00 \text{ mol of NaOH}$$

$$\frac{5.00 \text{ mol NaOH}}{1 \text{ L NaOH solution}} \quad \text{and} \quad \frac{1 \text{ L NaOH solution}}{5.00 \text{ mol NaOH}}$$

STEP 4 Set up problem to calculate mass or volume.

$$0.250 \cancel{\text{L solution}} \times \frac{5.00 \cancel{\text{mol NaOH}}}{1 \cancel{\text{L solution}}} \times \frac{40.00 \text{ g NaOH}}{1 \cancel{\text{mol NaOH}}} = 50.0 \text{ g of NaOH}$$

◆ Learning Exercise 12.5B

Calculate the quantity of solute in each of the following solutions:

a. How many moles of HCl are in 1.50 L of a 6.00 M HCl solution?

b. How many moles of KOH are in 0.750 L of a 10.0 M KOH solution?

c. How many grams of NaOH are needed to prepare 0.500 L of a 4.40 M NaOH solution? (*Hint:* Find moles of NaOH.)

d. How many grams of NaCl are in 285 mL of a 1.75 M NaCl solution?

Answers **a.** 9.00 mol of HCl **b.** 7.50 mol of KOH **c.** 88.0 g of NaOH **d.** 29.2 g of NaCl

◆ Learning Exercise 12.5C

Calculate the milliliters needed of each solution to obtain the given quantity of solute:

a. 1.50 mol of $Mg(OH)_2$ from a 2.00 M $Mg(OH)_2$ solution

b. 0.150 mol of glucose from a 2.20 M glucose solution

c. 18.5 g of KI from a 3.00 M KI solution

d. 18.0 g of NaOH from a 6.00 M NaOH solution

Answers **a.** 750 mL **b.** 68.2 mL **c.** 37.1 mL **d.** 75.0 mL

Guide to Calculating Dilution Quantities

STEP 1 Prepare a table of the initial and diluted volumes and concentrations.
STEP 2 Solve the dilution expression for the unknown quantity.
STEP 3 Set up problem by substituting known quantities in the dilution expression.

Study Note

What is the final concentration after 150. mL of a 2.00 M NaCl solution is diluted to a volume of 400. mL?

Solution:

STEP 1 Prepare a table of the initial and diluted volumes and concentrations.

$$M_1 = 2.00\text{ M} \quad V_1 = 150.\text{ mL} \qquad M_2 = ? \quad V_2 = 400.\text{ mL}$$

STEP 2 Solve the dilution expression for the unknown quantity.

$$M_1V_1 = M_2V_2 \qquad M_2 = \frac{M_1V_1}{V_2}$$

STEP 3 Set up problem by placing known quantities in the dilution expression.

$$M_2 = \frac{(2.00\text{ M})(150.\text{ mL})}{400.\text{ mL}} = 0.750\text{ M}$$

◆ Learning Exercise 12.5D

Solve each of the following dilution problems (assume the volumes add):

a. What is the final concentration after 100. mL of a 5.0 M KCl solution is diluted with water to give a final volume of 200. mL?

b. What is the final concentration of the diluted solution if 5.0 mL of a 1.5 M KCl solution is diluted to 25 mL?

c. What is the final concentration after 250 mL of an 8.0 M NaOH solution is diluted with 750 mL of water?

d. 160. mL of water is added to 40. mL of a 1.0 M NaCl solution. What is the final concentration?

e. What volume of 6.0 M HCl is needed to prepare 300. mL of a 1.0 M HCl solution? How much water must be added?

Answers **a.** 2.5 M **b.** 0.30 M **c.** 2.7 M **d.** 0.20 M
e. $V_1 = 50.$ mL; add 250. mL of water

12.6 Solutions in Chemical Reactions

- The volume and molarity of a solution are used to determine the moles or liters of a substance required or produced in a chemical reaction.
- The balanced equation can be used to convert moles of one substance to moles of another substance.
- Molar mass is used to convert moles to grams or grams to moles.
- The number of moles of a solute and its molarity are used to determine the volume of a solution in a chemical reaction.

Guide to Calculations Involving Solutions in Chemical Reactions

STEP 1 State the given and needed quantities.
STEP 2 Write a plan to calculate needed quantity or concentration.
STEP 3 Write equalities and conversion factors including mole–mole and concentration factors.
STEP 4 Set up problem to calculate needed quantity or concentration.

◆ Learning Exercise 12.6A

For the reaction, $2AgNO_3(aq) + H_2SO_4(aq) \longrightarrow Ag_2SO_4(s) + 2H_2O(l)$

a. How many milliliters of a 1.5 M $AgNO_3$ solution will react with 40.0 mL of a 1.0 M H_2SO_4 solution?

b. How many grams of Ag_2SO_4 will be produced?

Answers

a. $$40.0\ \cancel{\text{mL } H_2SO_4} \times \frac{1\ \cancel{\text{L } H_2SO_4}}{1000\ \cancel{\text{mL } H_2SO_4}} \times \frac{1.0\ \cancel{\text{mol } H_2SO_4}}{1\ \cancel{\text{L } H_2SO_4}} \times \frac{2\ \cancel{\text{mol } AgNO_3}}{1\ \cancel{\text{mol } H_2SO_4}} \times \frac{1000\text{ mL } AgNO_3}{1.5\ \cancel{\text{mol } AgNO_3}} = 53\text{ mL of } AgNO_3 \text{ solution}$$

b. $$40.0\ \cancel{\text{mL } H_2SO_4} \times \frac{1\ \cancel{\text{L } H_2SO_4}}{1000\ \cancel{\text{mL } H_2SO_4}} \times \frac{1.0\ \cancel{\text{mol } H_2SO_4}}{1\ \cancel{\text{L } H_2SO_4}} \times \frac{1\ \cancel{\text{mol } Ag_2SO_4}}{1\ \cancel{\text{mol } H_2SO_4}} \times \frac{311.9\text{ g } Ag_2SO_4}{1\ \cancel{\text{mol } Ag_2SO_4}} = 12\text{ g of } Ag_2SO_4$$

◆ Learning Exercise 12.6B

For the following reaction, calculate the number of milliliters of a 1.80 M KOH solution that react with 18.5 mL of a 2.20 M HCl solution.

$$HNO_3(aq) + KOH(aq) \longrightarrow KNO_3(aq) + H_2O(l)$$

Answer

$$18.5\ \cancel{\text{mL HCl}} \times \frac{1\ \cancel{\text{L HCl}}}{1000\ \cancel{\text{mL HCl}}} \times \frac{2.20\ \cancel{\text{mol HCl}}}{1\ \cancel{\text{L HCl}}} \times \frac{1\ \cancel{\text{mol KOH}}}{1\ \cancel{\text{mol HCl}}} \times \frac{1000\ \text{mL KOH}}{1.80\ \cancel{\text{mol KOH}}} = 22.6\ \text{mL of KOH solution}$$

12.7 Properties of Solutions

- Colloids contain particles that do not settle out and pass through filters but not through semipermeable membranes.
- Suspensions are composed of large particles that settle out of solution.
- The freezing point of a solution is lower than that of the solvent and boiling point is higher and depend only on the number of particles of solute in the solution.
- The freezing point and boiling point of a solution depends on the molal (m) concentration of the solution.

$$\text{Molality } (m) = \frac{\text{moles of solute}}{\text{kilogram of solvent}}$$

- The freezing point and freezing point depression (ΔT_f) of a solution can be calculated based on the molality of the solution and the freezing point depression constant (K_f) for the solvent.

$$\Delta T_f = mK_f$$

- The boiling point and boiling point elevation (ΔT_b) of a solution can be calculated based on the molality of the solution and the boiling point elevation constant (K_b) for the solvent

$$\Delta T_b = mK_b$$

- In the process of osmosis, water (solvent) moves through a semipermeable membrane from the solution that has a lower solute concentration to a solution where the solute concentration is higher.
- Osmotic pressure is the pressure that prevents the flow of water into a more concentrated solution.
- Particles in a solution lower the freezing point and elevate the boiling point of the solvent.
- Isotonic solutions have osmotic pressures equal to that of body fluids. A hypotonic solution has a lower osmotic pressure than body fluids; a hypertonic solution has a higher osmotic pressure.
- A red blood cell maintains its volume in an isotonic solution, but it swells (hemolysis) in a hypotonic solution and shrinks (crenation) in a hypertonic solution.
- In dialysis, water and small solute particles can pass through a dialyzing membrane, while larger particles such as blood cells are retained.

MasteringChemistry
Self Study Activity: Diffusion
Self Study Activity: Osmosis
Tutorial: Osmosis
Tutorial: Dialysis

◆ Learning Exercise 12.7A

Identify each of the following as a solution, colloid, or suspension:

1. ________ contains single atoms, ions, or small molecules

2. ________ settles out with gravity

3. ________ retained by filters

4. ________ cannot diffuse through a cellular membrane

5. ________ aggregates of atoms, molecules, or ions larger in size than solution particles

6. ________ contains large particles that are visible

Answers **1.** solution **2.** suspension **3.** suspension
4. colloid **5.** colloid **6.** suspension

◆ Learning Exercise 12.7B

Fill in the blanks:

In osmosis, the direction of solvent flow is from the (1) [higher/lower] solute concentration to the (2) [higher/lower] solute concentration. A semipermeable membrane separates 5% (m/v) and 10% (m/v) sucrose solutions. The (3) ________% (m/v) solution has the greater osmotic pressure. Water will move from the (4) ________% (m/v) solution into the (5) ________% (m/v) solution. The compartment that contains the (6) ________% (m/v) solution increases in volume.

Answers **(1)** lower **(2)** higher **(3)** 10 **(4)** 5 **(5)** 10 **(6)** 10

◆ Learning Exercise 12.7C

A semipermeable membrane separates a 2% starch solution from a 10% starch solutions. Complete each of the following with 2%, 6% or 10%:

a. Water will flow from the ________ starch solution to the ________ starch solution.

b. The volume of the ________ starch solution will increase and the volume of the ________ starch solution will decrease.

Answers **a.** 2%, 10% **b.** 10%, 2%

Study Note

Calculate the freezing point lowering and freezing point of a solution containing 125 grams of KCl in 500. g of water.

Solution:

STEP 1 Given 125 g of KCl; 500. g of water (solvent) = 0.500 kg of water
Need freezing point lowering ΔT_f and freezing point of KCl solution

STEP 2 Plan

$$\text{grams of KCl} \longrightarrow \text{moles of KCl} \times \frac{2 \text{ mol particles}}{1 \text{ mol KCl}} = \text{moles of particles}$$

$$m = \frac{\text{moles of particles}}{\text{kilogram of water}} \quad \text{and} \quad m \times \text{freezing point factor} = \Delta T_f$$

Final temperature = 0.0 °C − ΔT_f

STEP 3 Equalities/Conversion Factors

$$1 \text{ mol of KCl} = 74.55 \text{ g of KCl} \qquad 1\ m = 1.86\ °\text{C}$$

$$\frac{1 \text{ mol KCl}}{74.55 \text{ g KCl}} \text{ and } \frac{74.55 \text{ g KCl}}{1 \text{ mol KCl}} \qquad \frac{1\ m}{1.86\ °\text{C}} \text{ and } \frac{1.86\ °\text{C}}{1\ m}$$

STEP 4 Set Up Problem We use our plan and conversion factors to calculate the molality of the KCl solution and the new freezing point.

$$\text{Moles of particles} = 125 \cancel{\text{g KCl}} \times \frac{1 \cancel{\text{mol KCl}}}{74.55 \cancel{\text{g KCl}}} \times \frac{2 \text{ mol particles}}{1 \cancel{\text{mol KCl}}} = 3.35 \text{ mol of particles (ions)}$$

$$\text{Molality } (m) = \frac{3.35 \text{ mol particles}}{0.500 \text{ kg water}} = 6.70\ m$$

$$\Delta T_f = 6.70 \cancel{m} \times \frac{1.86\ °\text{C}}{1 \cancel{m}} = 12.5\ °\text{C}$$

Freezing point = 0.0 °C − 12.5 °C = −12.5 °C

◆ Learning Exercise 12.7D

One mole of particles lowers the freezing point of 1000 g of water by 1.86 °C and raises the boiling point by 0.52 °C. For the following solutes each in 1000 g of water, indicate:

(1) the number of moles of particles

(2) the freezing point change and freezing point of the solution

(3) the boiling point change and boiling point of the solution

Solute	Number of Moles of Particles	Freezing Point Change and Freezing Point	Boiling Point Change and Boiling Point
a. 1.00 mol of fructose (nonelectrolyte)			
b. 1.50 mol of KCl (strong electrolyte)			
c. 0.650 mol of $Ca(NO_3)_2$ (strong electrolyte)			

Answers

Solute	Number of Moles of Particles	Freezing Point Change and Freezing Point	Boiling Point Change and Boiling Point
a. 1.00 mol fructose (nonelectrolyte)	1.00 mol	Decreases 1.86 °C; fp = −1.86 °C	Increases 0.52 °C; bp = 100.52 °C
b. 1.50 mol of KCl (strong electrolyte)	3.00 mol	Decreases 5.58 °C; fp = −5.58 °C	Increases 1.6 °C; bp = 101.6 °C
c. 0.650 mol of $Ca(NO_3)_2$ (strong electrolyte)	1.95 mol	Decreases 3.63 °C; fp = −3.63 °C	Increases 1.0 °C; bp = 101.0 °C

◆ Learning Exercise 12.7E

Fill in the blanks:

A (1) ________% (m/v) NaCl solution and a (2) ________% (m/v) glucose solution are isotonic to the body fluids. A red blood cell placed in these solutions does not change in volume because these solutions are (3) ________ tonic. When a red blood cell is placed in water, it undergoes (4) ________ because water is (5) ________ tonic. A 20% (m/v) glucose solution will cause a red blood cell to undergo (6) ________ because the 20% (m/v) glucose solution is (7) ________ tonic.

Answers **(1)** 0.9 **(2)** 5 **(3)** iso **(4)** hemolysis **(5)** hypo **(6)** crenation **(7)** hyper

◆ Learning Exercise 12.7F

Indicate whether the following solutions are

a. hypotonic **b.** hypertonic **c.** isotonic

1. _____ 5% (m/v) glucose **2.** _____ 3% (m/v) NaCl **3.** _____ 2% (m/v) glucose

4. _____ water **5.** _____ 0.9% (m/v) NaCl **6.** _____ 10% (m/v) glucose

Answers **1.** c **2.** b **3.** a **4.** a **5.** c **6.** b

◆ Learning Exercise 12.7G

Indicate whether the following solutions will cause a red blood cell to undergo

a. crenation **b.** hemolysis **c.** no change (stays the same)

1. _____ 10% (m/v) NaCl **2.** _____ 1% (m/v) glucose **3.** _____ 5% (m/v) glucose

4. _____ 0.5% (m/v) NaCl **5.** _____ 10% (m/v) glucose **6.** _____ water

Answers **1.** a **2.** b **3.** c **4.** b **5.** a **6.** b

◆ Learning Exercise 12.7H

A dialysis bag contains starch, glucose, NaCl, protein, and urea.

a. When the dialysis bag is placed in water, what components would you expect to dialyze through the bag? Why?

b. Which components will stay inside the dialysis bag? Why?

Answers

a. Glucose, NaCl, and urea are solution particles. Solution particles will pass through semipermeable membranes.

b. Starch and protein are colloidal particles; colloids are retained by semipermeable membranes.

Checklist for Chapter 12

You are ready to take the Practice Test for Chapter 12. Be sure you have accomplished the following learning goals for this chapter. If you are not sure, review the section listed at the end of the goal. Then apply your new skills and understanding to the Practice Test.

After studying Chapter 12, I can successfully:

_____ Identify the solute and solvent in a solution and describe the process of dissolving an ionic solute in water (12.1).

_____ Identify the components in solutions of electrolytes and nonelectrolytes (12.2).

_____ Identify a saturated and an unsaturated solution (12.3).

_____ Identify a salt as soluble or insoluble (12.3).

_____ Write a chemical equation (or ionic or net ionic equation) to show the formation of an insoluble salt (12.3).

_____ Describe the effects of temperature and nature of the solute on its solubility in a solvent (12.3).

_____ Determine the solubility of an ionic compound (salt) in water (12.3).

_____ Calculate the percent concentration, m/m and v/v, of a solute in a solution, and use percent concentration to calculate the amount of solute or solution (12.4).

_____ Calculate the molarity of a solution (12.5).

_____ Use molarity as a conversion factor to calculate the moles (or grams) of a solute or the volume of the solution (12.5).

_____ Calculate the new concentration or new volume when a solution is diluted (12.5).

_____ Use the volume and molarity of a solution to calculate the grams or moles of a substance produced in a chemical reaction (12.6).

_____ Calculate the molality of a solution (12.7).

_____ Use the concentration of particles in a solution to calculate the freezing point or boiling point of the solution (12.7).

_____ Identify a solution as isotonic, hypotonic, or hypertonic (12.7).

_____ Identify a mixture as a solution, a colloid, or a suspension (12.7).

_____ Explain the processes of osmosis and dialysis (12.7).

Practice Test for Chapter 12

For questions 1 through 4, indicate if each of the following is more soluble in (W) water, a polar solvent, or (B) benzene, a nonpolar solvent:

1. $I_2(s)$, nonpolar

2. $NaBr(s)$, polar

3. $KI(s)$, polar

4. C_6H_{12}, nonpolar

5. When dissolved in water, $Ca(NO_3)_2(s)$ dissociates into
A. $Ca^{2+}(aq) + (NO_3)_2^{2-}(aq)$
B. $Ca^{+}(aq) + NO_3^{-}(aq)$
C. $Ca^{2+}(aq) + 2NO_3^{-}(aq)$
D. $Ca^{2+}(aq) + 2N^{5+}(aq) + 2O_3^{6-}(aq)$
E. $CaNO_3^{+}(aq) + NO_3^{-}(aq)$

6. Ethanol, $CH_3—CH_2—OH$, is a nonelectrolyte. When placed in water it
A. dissociates completely. **B.** dissociates partially. **C.** does not dissociate.
D. makes the solution acidic. **E.** makes the solution basic.

7. The solubility of NH_4Cl is 46 g in 100 g of water at 40 °C. How much NH_4Cl can dissolve in 500 g of water at 40 °C?
A. 9.2 g **B.** 46 g **C.** 100 g **D.** 184 g **E.** 230 g

For questions 8 through 11, indicate if each of the following are soluble (S) or insoluble (I) in water:

8. NaCl **9.** AgCl **10.** $BaSO_4$ **11.** FeO

12. A solution containing 1.20 g of sucrose in 50.0 g of solution has a percent (m/m) concentration of
A. 0.600% **B.** 1.20% **C.** 2.40% **D.** 30.0% **E.** 41.6%

13. The amount of lactose in 250 g of a 3.0% (m/m) lactose solution for infant formula is
A. 0.15 g **B.** 1.2 g **C.** 6.0 g **D.** 7.5 g **E.** 30 g

14. The mass of solution needed to obtain 0.40 g of glucose from a 5.0% (m/m) glucose solution is
A. 1.0 g **B.** 2.0 g **C.** 4.0 g **D.** 5.0 g **E.** 8.0 g

15. The amount of NaCl needed to prepare 50.0 g of a 4.00% (m/m) NaCl solution is
A. 20.0 g **B.** 15.0 g **C.** 10.0 g **D.** 4.00 g **E.** 2.00 g

16. A solution containing 6.0 g of NaCl in 1500 g of solution has a mass percent concentration of
A. 0.40% (m/m) **B.** 0.25% (m/m) **C.** 4.0% (m/m) **D.** 0.90% (m/m) **E.** 2.5% (m/m)

17. The moles of KOH needed to prepare 2400 mL of a 2.0 M KOH solution is
A. 1.2 mol **B.** 2.4 mol **C.** 4.8 mol **D.** 12 mol **E.** 48 mol

18. The grams of NaOH needed to prepare 7.5 mL of a 5.0 M NaOH is
A. 1.5 g **B.** 3.8 g **C.** 6.7 g **D.** 15 g **E.** 38 g

For questions 19 through 21, consider a 20.0-g sample of a solution that contains 2.0 g of NaOH.

19. The percent (m/m) concentration of the solution is
A. 1.0% **B.** 4.0% **C.** 5% **D.** 10% **E.** 20%

20. The moles of NaOH in the sample is
A. 0.050 mol **B.** 0.40 mol **C.** 1.0 mol **D.** 2.5 mol **E.** 4.0 mol

21. If the solution has a volume of 0.025 L, what is the molarity of the sample?
A. 0.10 M **B.** 0.5 M **C.** 1.0 M **D.** 1.5 M **E.** 2.0 M

22. Which of the following is soluble in water?
A. AgCl **B.** $BaCO_3$ **C.** K_2SO_4 **D.** PbS **E.** MgO

23. The insoluble salt that forms when a solution of NaCl mixes with a $Pb(NO_3)_2$ solution is
A. Na_2Pb **B.** $ClNO_3$ **C.** $NaNO_3$ **D.** $PbCl_2$ **E.** none

24. Which of the following salts is insoluble in water?
A. $CuCl_2$ **B.** $Pb(NO_3)_2$ **C.** K_2CO_3 **D.** $(NH_4)SO_4$ **E.** $CaCO_3$

25. A 20.-mL sample of 5.0 M HCl is diluted with water to give 100. mL of solution. The final concentration of the HCl solution is
A. 10 M **B.** 5.0 M **C.** 2.0 M **D.** 1.0 M **E.** 0.50 M

26. Water is added to 200. mL of a 4.00 M KNO_3 solution to give 400. mL of solution. The final concentration of the diluted solution is
A. 1.00 M **B.** 2.00 M **C.** 4.00 M **D.** 0.500 M **E.** 0.100 M

27. 5.0 mL of a 2.0 M KOH solution is diluted with water to give 50.0 mL of solution. The final concentration of the KOH solution is
A. 1.5 M **B.** 1.0 M **C.** 20 M **D.** 0.20 M **E.** 0.50 M

28. What mass of Ag_2SO_4 is formed when 25.0 mL of a 0.111 M $AgNO_3$ solution reacts with excess H_2SO_4? $2AgNO_3(aq) + H_2SO_4(aq) \longrightarrow Ag_2SO_4(s) + 2H_2O(l)$
A. 433 g **B.** 1.74 g **C.** 0.866 g **D.** 0.433 g **E.** 2.78 g

For questions 31 through 35, indicate whether each statement describes a
A. solution **B.** colloid **C.** suspension

29. _____ contains single atoms, ions, or small molecules of solute

30. _____ settles out upon standing

31. _____ can be separated by filtering

32. _____ can be separated by semipermeable membranes

33. _____ passes through semipermeable membranes

34. Two solutions that have identical osmotic pressures are
A. hypotonic **B.** hypertonic **C.** isotonic **D.** isotopic **E.** hyperactive

35. In osmosis, the net flow of water is
A. between solutions of equal concentrations
B. from higher solute concentration to lower solute concentration
C. from lower solute concentration to higher solute concentration
D. from a colloid to a solution of equal concentration
E. from lower solvent concentration to higher solvent concentration

36. A red blood cell undergoes hemolysis when placed in a solution that is
A. isotonic **B.** hypotonic **C.** hypertonic **D.** colloidal **E.** semitonic

37. A solution that has the same osmotic pressure as body fluids is
A. 0.1% (m/v) NaCl **B.** 0.9 % (m/v) NaCl **C.** 5% (m/v) NaCl
D. 10% (m/v) glucose **E.** 15% (m/v) glucose

38. If a solution contains 0.50 mole of $CaCl_2$ in 1000 g of water, what is the freezing point of the solution?
A. 0 °C **B.** 2.8 °C **C.** −2.8 °C **D.** 0.93 °C **E.** −0.93 °C

Answers to the Practice Test

1. B	**2.** W	**3.** W	**4.** B	**5.** C
6. C	**7.** E	**8.** S	**9.** I	**10.** I
11. I	**12.** C	**13.** D	**14.** E	**15.** E
16. A	**17.** C	**18.** A	**19.** D	**20.** A
21. E	**22.** C	**23.** D	**24.** E	**25.** D
26. B	**27.** D	**28.** D	**29.** A	**30.** C
31. C	**32.** B	**33.** A	**34.** C	**35.** C
36. B	**37.** B	**38.** C		

13
Chemical Equilibrium

Study Goals

- Describe how temperature, concentration, and catalysts affect the rate of a reaction.
- Use the concept of reversible reactions to explain chemical equilibrium.
- Calculate the equilibrium constant for a reversible reaction using the concentrations of reactants and products at equilibrium.
- Use an equilibrium constant to predict the extent of reaction and to calculate equilibrium concentrations.
- Use Le Châtelier's principle to describe the changes made in equilibrium concentrations when reaction conditions change.
- Calculate the solubility product for a saturated solution; use the solubility product to calculate the solubility.

Think About It

1. Why does a high temperature cook food faster than a low temperature?

2. Why do automobile engines use a catalytic converter?

3. What does the size of an equilibrium constant tell you about the relative concentrations of the reactants and products?

4. What happens to a system at equilibrium when the conditions are altered?

Key Terms

Match each of the following key terms with the correct definition:

a. activation energy　**b.** equilibrium　**c.** catalyst
d. equilibrium constant expression　**e.** collision theory　**f.** heterogeneous equilibrium

1. _____ a substance that lowers the activation energy and increases the rate of reaction

2. _____ equilibrium components are present in at least two different states

3. _____ the ratio of the concentrations of products to reactants with each component raised to an exponent equal to its coefficient

4. _____ the energy that is required in a collision to break the bonds in the reactants

5. _____ a reaction requires that reactants collide to form products

6. _____ the condition in which the rate of the forward reaction is equal to the rate of the reverse reaction

Answers **1.** c **2.** f **3.** d **4.** a **5.** e **6.** b

13.1 Rates of Reaction

- The rate of a reaction is the speed at which reactants are consumed or products are formed.
- Collision theory is used to explain reaction rates and the factors that affect them.
- The rate of a reaction depends on temperature, concentration, and the presence of a catalyst.
- At higher temperatures, reaction rates increase because reactants move faster, collide more often, and produce more collisions with the required energy of activation. The opposite occurs at lower temperatures.
- Increasing the concentrations of reactants increases the rate of a reaction because collisions occur more often.
- The reaction rate is increased by the addition of a catalyst since a catalyst provides an alternate reaction pathway with a lower energy of activation.

MasteringChemistry
Self Study Activity: Factors That Affect Rate
Tutorial: Activation Energy and Catalysis
Tutorial: Factors That Affect Rate

◆ Learning Exercise 13.1A

Indicate the effect of each of the following on the rate of a chemical reaction:

increase (I) decrease (D) no effect (N)

1. _____ adding a catalyst

2. _____ running the reaction at a lower temperature

3. _____ doubling the concentrations of the reactants

4. _____ removing a catalyst

5. _____ running the experiment in a different laboratory

6. _____ increasing the temperature

7. _____ using a container with a different shape

8. _____ using lower concentrations of reactants

Answers **1.** I **2.** D **3.** I **4.** D
5. N **6.** I **7.** N **8.** D

◆ Learning Exercise 13.1B

For the reaction, $NO_2(g) + CO(g) \longrightarrow NO(g) + CO_2(g)$, indicate the effect of each of the following conditions as:

increase (I) decrease (D) no effect (N)

1. _____ adding CO
2. _____ running the experiment on Wednesday
3. _____ removing NO_2
4. _____ adding a catalyst
5. _____ adding NO_2

Answers **1.** I **2.** N **3.** D **4.** I **5.** I

13.2 Chemical Equilibrium

- A reversible reaction proceeds in both the forward and reverse directions.
- Chemical equilibrium is achieved when the rate of the forward reaction becomes equal to the rate of the reverse reaction.
- In a system at equilibrium, there is no change in the concentrations of reactants and products.
- At equilibrium, the concentrations of reactants can be greater than, less than, or in some cases, equal to the concentrations of products.

MasteringChemistry

Self Study Activity: Equilibrium
Tutorial: Chemical Equilibrium

◆ Learning Exercise 13.2

Indicate if each of the following indicates a system at equilibrium (E) or not at equilibrium (NE):

a. _____ The rate of the forward reaction is faster than the rate of the reverse reaction.

b. _____ There is no change in the concentrations of reactants and products.

c. _____ The rate of the forward reaction is equal to the rate of the reverse reaction.

d. _____ The concentrations of reactants are decreasing.

e. _____ The concentrations of products are increasing.

Answers **a.** NE **b.** E **c.** E **d.** NE **e.** NE

13.3 Equilibrium Constants

- The equilibrium constant expression for a system at equilibrium is the ratio of the concentrations of the products to the concentrations of the reactants with the concentration of each substance raised to a power equal to its coefficient in the equation.
- For the general equation, $aA + bB \rightleftarrows cC + dD$, the equilibrium constant is written:

$$K_c = \frac{[C]^c[D]^d}{[A]^a[B]^b}$$

MasteringChemistry

Tutorial: Equilibrium Constant

Guide to Writing the K_c Expression

STEP 1 Write the balanced equilibrium equation.
STEP 2 Write the products in brackets as the numerator and reactants in brackets as the denominator. Do not include pure solids or liquids.
STEP 3 Write the coefficient of each substance as an exponent.

◆ Learning Exercise 13.3A

Write the expression for the equilibrium constant (K_c) for each of the following reactions:

a. $2SO_3(g) \rightleftharpoons 2SO_2(g) + O_2(g)$ **b.** $2NO(g) + Br_2(g) \rightleftharpoons 2NOBr(g)$

c. $N_2(g) + 3H_2(g) \rightleftharpoons 2NH_3(g)$ **d.** $2NO_2(g) \rightleftharpoons N_2O_4(g)$

Answers **a.** $K_c = \dfrac{[SO_2]^2[O_2]}{[SO_3]^2}$ **b.** $K_c = \dfrac{[NOBr]^2}{[NO]^2[Br_2]}$

c. $K_c = \dfrac{[NH_3]^2}{[N_2][H_2]^3}$ **d.** $K_c = \dfrac{[N_2O_4]}{[NO_2]^2}$

◆ Learning Exercise 13.3B

Write the equilibrium constant expression (K_c) for each of the following heterogeneous systems at equilibrium:

a. $H_2(g) + S(l) \rightleftharpoons H_2S(g)$ **b.** $H_2O(g) + C(s) \rightleftharpoons H_2(g) + CO(g)$

c. $2PbS(s) + 3O_2(g) \rightleftharpoons 2PbO(s) + 2SO_2(g)$

d. $SiH_4(g) + 2O_2(g) \rightleftharpoons SiO_2(s) + 2H_2O(g)$

Answers

a. $K_c = \dfrac{[H_2S]}{[H_2]}$ **b.** $K_c = \dfrac{[H_2][CO]}{[H_2O]}$

c. $K_c = \dfrac{[SO_2]^2}{[O_2]^3}$ **d.** $K_c = \dfrac{[H_2O]^2}{[SiH_4][O_2]^2}$

Guide to Calculating the K_c Value

STEP 1 Write the K_c expression for the equilibrium.
STEP 2 Substitute equilibrium (molar) concentrations and calculate K_c.

◆ Learning Exercise 13.3C

Calculate the K_c value for each of the following equilibrium concentrations:

a. $H_2(g) + I_2(g) \rightleftharpoons 2HI(g)$

$[H_2] = 0.28$ M $[I_2] = 0.28$ M $[HI] = 2.0$ M

b. $2NO_2(g) \rightleftharpoons N_2(g) + 2O_2(g)$

$[NO_2] = 0.60$ M $[N_2] = 0.010$ M $[O_2] = 0.020$ M

c. $N_2(g) + 3H_2(g) \rightleftharpoons 2NH_3(g)$

$[N_2] = 0.50$ M $[H_2] = 0.20$ M $[NH_3] = 0.80$ M

Answers

a. $K_c = \dfrac{[HI]^2}{[H_2][I_2]} = \dfrac{[2.0]^2}{[0.28][0.28]} = 51$

b. $K_c = \dfrac{[N_2][O_2]^2}{[NO_2]^2} = \dfrac{[0.010][0.020]^2}{[0.60]^2} = 1.1 \times 10^{-5}$

c. $K_c = \dfrac{[NH_3]^2}{[N_2][H_2]^3} = \dfrac{[0.80]^2}{[0.50][0.20]^3} = 1.6 \times 10^{2}$

13.4 Using Equilibrium Constants

- A large K_c results when a reaction favors products and the concentration of the products is greater than that of the reactants.
- A small K_c results when a reaction favors reactants and the concentration of the products is less than that of the reactants.
- The concentration of a component in an equilibrium mixture is calculated from the K_c and the concentrations of all other components.

MasteringChemistry
Tutorial: Calculations Using the Equilibrium Constant

Guide to Using K_c Value

STEP 1 Write the K_c expression for the equilibrium equation.
STEP 2 Solve the K_c expression for the unknown concentration.
STEP 3 Substitute the known values into the rearranged K_c expression.
STEP 4 Check answer by using the calculated concentrations in the K_c expression.

◆ Learning Exercise 13.4A

Consider the reaction $2NOBr(g) \rightleftharpoons 2NO(g) + Br_2(g)$.

a. Write the equilibrium constant expression for the reaction at equilibrium.

b. If the equilibrium constant is 2×10^3, does the equilibrium mixture contain mostly reactants, mostly products, or about equal amounts of both reactants and products? Explain.

Answers **a.** $K_c = \dfrac{[NO]^2[Br_2]}{[NOBr]^2}$

b. A large $K_c(>1)$ means the equilibrium mixture contains mostly products.

◆ Learning Exercise 13.4B

Consider the reaction $2HI(g) \rightleftharpoons H_2(g) + I_2(g)$.

a. Write the expression for the equilibrium constant for the reaction.

b. If the equilibrium constant is 1.6×10^{-2}, does the equilibrium mixture contain mostly reactants, mostly products, or about equal amounts of both reactants and products? Explain.

Answers **a.** $K_c = \dfrac{[H_2][I_2]}{[HI]^2}$

b. A small $K_c < 1$ means that the equilibrium mixture contains mostly reactants.

◆ Learning Exercise 13.4C

Calculate the concentration of the indicated component for each of the following equilibrium systems:

a. $PCl_5(g) \rightleftharpoons PCl_3(g) + Cl_2(g)$ $K_c = 1.2 \times 10^{-2}$

$[PCl_5] = 2.50$ M $[PCl_3] = 0.50$ M $[Cl_2] = ?$

b. $CO(g) + H_2O(g) \rightleftharpoons CO_2(g) + H_2(g)$ $K_c = 1.6$

$[CO] = 1.0$ M $[H_2O] = 0.80$ M $[CO_2] = ?$ $[H_2] = 1.2$ M

Answers

a. $K_c = \dfrac{[PCl_3][Cl_2]}{[PCl_5]}$ $[Cl_2] = \dfrac{K_c[PCl_5]}{[PCl_3]} = \dfrac{1.2 \times 10^{-2}[2.50]}{[0.50]} = 6.0 \times 10^{-2}$ M

b. $K_c = \dfrac{[CO_2][H_2]}{[CO][H_2O]}$ $[CO_2] = \dfrac{K_c[CO][H_2O]}{[H_2]} = \dfrac{1.6[1.0][0.80]}{[1.2]} = 1.1$ M

13.5 Changing Equilibrium Conditions: Le Châtelier's Principle

- Le Châtelier's principle states when a stress (change in conditions) is placed on a reaction at equilibrium, the equilibrium will shift in a direction (forward or reverse) that relieves the stress.
- A change in the concentration of a reactant or product or the temperature of any reaction will cause the reaction to shift to relieve the stress.
- A change in the volume of a reaction involving gases will cause a shift if the reaction involves different number of moles of gaseous reactants and products.
- The addition of a catalyst does not affect the position of an equilibrium reaction.

MasteringChemistry
Tutorial: Le Châtelier's Principle

◆ Learning Exercise 13.5A

Identify the effect of each of the following on the reaction at equilibrium:

$$N_2(g) + O_2(g) + 180\text{ kJ} \rightleftharpoons 2NO(g)$$

a. shift toward products **b.** shift toward reactants **c.** no change

1. ____ adding $O_2(g)$ **2.** ____ removing $N_2(g)$

3. ____ removing NO(*g*) **4.** ____ adding heat

5. ____ reducing the container volume **6.** ____ increasing the container volume

7. ____ adding a platinum catalyst

Answers **1.** a **2.** b **3.** a **4.** a **5.** c **6.** c **7.** c

◆ Learning Exercise 13.5B

Identify the effect of the each of the following on the equilibrium of the reaction:

$$2NOBr(g) \rightleftharpoons 2NO(g) + Br_2(g) + 340\ \text{kJ}$$

a. shift toward products **b.** shift toward reactants **c.** no change

1. _____ adding $NO(g)$

2. _____ removing $Br_2(g)$

3. _____ removing $NOBr(g)$

4. _____ adding a catalyst

5. _____ lowering the temperature

6. _____ increasing the volume of the container

Answers **1.** b **2.** a **3.** b **4.** c **5.** a **6.** a

13.6 Equilibrium in Saturated Solutions

- The solubility of a slightly soluble salt in an aqueous solution is represented by an equilibrium expression called the **solubility product constant, (K_{sp})**.
- The K_{sp} can be used to calculate the solubility of the salt or the solubility can be used to calculate the K_{sp}.
- The solubility of a salt is affected by the addition of a common ion to the equilibrium.
- As in other heterogeneous equilibria, the concentration of the solid is constant and not included in the K_{sp} expression. For example, the K_{sp} of AgBr is: $K_{sp} = [Ag^+][Br^-]$

MasteringChemistry
Tutorial: Solubility-Product Constant Expression

◆ Learning Exercise 13.6A

Write the equilibrium equation and the K_{sp} expression for each of the following slightly soluble salts:

a. BaF_2

b. NiS

c. Ag_2CO_3

d. Ag_3PO_4

Answers

a. $BaF_2(s) \rightleftharpoons Ba^{2+}(aq) + 2F^-(aq)$ $K_{sp} = [Ba^{2+}][F^-]^2$

b. $NiS(s) \rightleftharpoons Ni^{2+}(aq) + S^{2-}(aq)$ $K_{sp} = [Ni^{2+}][S^{2-}]$

c. $Ag_2CO_3(s) \rightleftharpoons 2Ag^+(aq) + CO_3^{2-}(aq)$ $K_{sp} = [Ag^+]^2[CO_3^{2-}]$

d. $Ag_3PO_4(s) \rightleftharpoons 3Ag^+(aq) + PO_4^{3-}(aq)$ $K_{sp} = [Ag^+]^3[PO_4^{3-}]$

Guide to Calculating K_{sp}

STEP 1 Write the equilibrium equation for the dissociation of the ionic compound.
STEP 2 Write the solubility product expression (K_{sp}).
STEP 3 Substitute the molarity of each ion into the K_{sp} and calculate.

◆ Learning Exercise 13.6B

Calculate the K_{sp} value for each of the following slightly soluble salts:

a. CdS with $[Cd^{2+}] = [S^{2-}] = 3 \times 10^{-14}$ M

b. $AgIO_3$ with $[Ag^+] = [IO_3^-] = 1.7 \times 10^{-4}$ M

c. SrF_2 with $[Sr^{2+}] = 8.5 \times 10^{-4}$ and $[F^-] = 1.7 \times 10^{-3}$

d. Ag_2SO_3 with $[Ag^+] = 3.2 \times 10^{-5}$ and $[SO_3^{2-}] = 1.6 \times 10^{-5}$

Answers **a.** $K_{sp} = 9 \times 10^{-28}$ **b.** $K_{sp} = 2.9 \times 10^{-8}$
c. $K_{sp} = 2.5 \times 10^{-9}$ **d.** $K_{sp} = 1.6 \times 10^{-14}$

Guide to Calculating Molar Solubility from K_{sp}

STEP 1 Write the equilibrium equation for the dissociation of the slightly soluble salt.
STEP 2 Write the solubility product expression (K_{sp}).
STEP 3 Substitute S for the molarity of each ion into the K_{sp} expression.
STEP 4 Calculate the molar solubility (S).

◆ Learning Exercise 13.6C

Calculate the molar solubility (S) of each of the following slightly soluble salts:

a. $CaSO_4$ $K_{sp} = 9 \times 10^{-6}$

b. $ZnCO_3$ $K_{sp} = 1.4 \times 10^{-11}$

c. FeS $K_{sp} = 8 \times 10^{-19}$

d. CuI $K_{sp} = 1.3 \times 10^{-12}$

Answers **a.** $CaSO_4$ $S = 3 \times 10^{-3}$ M **b.** $ZnCO_3$ $S = 3.7 \times 10^{-6}$ M
c. FeS $S = 9 \times 10^{-10}$ M **d.** CuI $S = 1.1 \times 10^{-6}$ M

◆ Learning Exercise 13.6D

What is the $[Ca^{2+}]$ in each of the following solutions, given that for $CaSO_4$ the $K_{sp} = 9 \times 10^{-6}$:

a. A solution with $[SO_4^{2-}] = 1.0 \times 10^{-1}$ M

b. A solution with $[SO_4^{2-}] = 2.7 \times 10^{-2}$ M

Answers **a.** $[Ca^{2+}] = 9 \times 10^{-5}$ M **b.** $[Ca^{2+}] = 3 \times 10^{-4}$ M

Checklist for Chapter 13

You are ready to take the Practice Test for Chapter 13. Be sure you have accomplished the following learning goals for this chapter. If you are not sure, review the section listed at the end of the goal. Then apply your new skills and understanding to the Practice Test.

After studying Chapter 13, I can successfully:

_____ Describe the factors that increase or decrease the rate of a reaction (13.1).

_____ Write the forward and reverse reactions of a reversible reaction (13.2).

_____ Explain how equilibrium occurs when the rate of a forward reaction is equal to the rate of a reverse reaction (13.2).

_____ Write the equilibrium constant expression for a reaction system at equilibrium (13.3).

_____ Calculate the equilibrium constant from the equilibrium concentrations (13.3).

_____ Use the equilibrium constant to determine whether an equilibrium favors the reactants or products (13.4).

_____ Use the equilibrium constant expression to determine the equilibrium concentration of a component in the reaction (13.4).

_____ Use Le Châtelier's principle to describe the shift in a system at equilibrium when stress is applied to the system (13.5).

_____ Calculate the solubility product for a saturated salt solution (13.6).

_____ Use the solubility product expression to calculate the solubility of a slightly soluble salt (13.6).

_____ Use the solubility product expression to calculate the solubility of a slightly soluble salt in the presence of a common ion (13.6).

Practice Test for Chapter 13

1. The number of molecular collisions increases when
A. more reactants are added.
B. products are removed.
C. the energy of collision is below the energy of activation.
D. the reaction temperature is lowered.
E. the reacting molecules have an incorrect orientation upon impact.

2. The energy of activation is lowered when
A. more reactants are added.
B. products are removed.
C. a catalyst is used.
D. the reaction temperature is lowered.
E. the reaction temperature is raised.

3. Food deteriorates more slowly in a refrigerator because
A. more reactants are added.
B. products are removed.
C. the energy of activation is higher.
D. fewer collisions have the energy of activation.
E. collisions have the wrong orientation upon impact.

4. A reaction reaches equilibrium when
A. the rate of the forward reaction is faster than the rate of the reverse reaction.
B. the rate of the reverse reaction is faster than the rate of the forward reaction.
C. the concentrations of reactants and products are changing.
D. fewer collisions have the energy of activation.
E. the rate of the forward reaction is equal to the rate of the reverse reaction.

5. The equilibrium constant expression for the following reaction is
$2NOCl(g) \rightleftharpoons 2NO(g) + Cl_2(g)$

A. $\frac{[NO][Cl_2]}{[NOCl]}$ **B.** $\frac{[NOCl_2]^2}{[NO]^2[Cl_2]}$ **C.** $\frac{[NOCl_2]}{[NO][Cl_2]}$

D. $\frac{[NO^2][Cl_2]}{[NOCl]}$ **E.** $\frac{[NO]^2[Cl_2]}{[NOCl]^2}$

6. The equilibrium constant expression for the following reaction is
$MgO(s) \rightleftharpoons CO_2(g) + MgCO_3(s)$

A. $[CO_2]$ **B.** $\frac{[CO_2][MgCO_3]}{[MgO]}$ **C.** $\frac{[MgO]}{[CO_2][MgCO_3]}$

D. $\frac{1}{[CO_2]}$ **E.** $\frac{[CO_2]}{[MgO]}$

7. The equilibrium constant expression for the following reaction is
$2PbS(s) + 3O_2(g) \rightleftharpoons 2PbO(s) + 2SO_2(g)$

A. $\frac{[PbO][SO_2]}{[PbS][O_2]}$ **B.** $\frac{[PbO]^2[SO_2]^2}{[PbS]^2[O_2]^3}$ **C.** $\frac{[SO_2]^2}{[O_2]^3}$

D. $\frac{[SO_2]}{[O_2]}$ **E.** $\frac{[O_2]^3}{[SO_2]^2}$

8. The equilibrium equation that has the following equilibrium constant expression is
$\frac{[H_2S]^2}{[H_2]^2[S_2]}$

A. $H_2S(g) \rightleftharpoons H_2(g) + S_2(g)$
B. $2H_2S(g) \rightleftharpoons H_2(g) + S_2(g)$
C. $2H_2(g) + S_2(g) \rightleftharpoons 2H_2S(g)$
D. $2H_2S(g) \rightleftharpoons 2H_2(g)$
E. $2H_2(g) \rightleftharpoons 2H_2S(g)$

9. The value of the equilibrium constant for the following equilibrium is
$COBr_2(g) \rightleftharpoons CO(g) + Br_2(g)$
$[COBr_2] = 0.93$ M $[CO] = [Br_2] = 0.42$ M

A. 0.19 **B.** 0.39 **C.** 0.42
D. 2.2 **E.** 5.3

10. The value of the equilibrium constant for the following equilibrium is
$2NO(g) + O_2(g) \rightleftharpoons 2NO_2(g)$ $[NO] = 2.7$ M $[O_2] = 1.0$ $[NO_2] = 3.0$ M

A. 0.81 **B.** 1.1 **C.** 1.2
D. 8.1 **E.** 9.0

11. Calculate the $[PCl_5]$ for the decomposition of PCl_5 that has a $K_c = 0.050$.
$PCl_5(g) \rightleftharpoons PCl_3(g) + Cl_2(g)$ $[PCl_3] = [Cl_2] = 0.20$ M

A. 0.01 M **B.** 0.050 M **C.** 0.20 M
D. 0.40 M **E.** 0.80 M

12. The reaction that has a much greater concentration of products at equilibrium has a K_c value of

A. 1.6×10^{-15} **B.** 2×10^{-11} **C.** 1.2×10^{-5}
D. 3×10^{-3} **E.** 1.4×10^{5}

13. The reaction that has a much greater concentration of reactants at equilibrium has a K_c value of
A. 1.1×10^{-11} **B.** 2×10^{-2} **C.** 1.2×10^{2}
D. 2×10^{4} **E.** 1.3×10^{12}

14. The reaction that has about the same concentration of reactant and products at equilibrium has a K_c value of
A. 1.4×10^{-12} **B.** 2×10^{-8} **C.** 1.2
D. 3×10^{2} **E.** 1.3×10^{7}

For questions 15–19, indicate how each of the following affects the equilibrium of the reaction shown:
$PCl_5(g) + \text{heat} \rightleftharpoons PCl_3(g) + Cl_2(g)$

A. shifts toward products **B.** shifts toward reactants **C.** no change

15. add more Cl_2 **16.** cool the reaction **17.** remove some PCl_3

18. add more PCl_5 **19.** remove some PCl_5

For questions 20–24, indicate whether the equilibrium will

A. shift toward products **B.** shift toward reactants **C.** no change

$2NO(g) + O_2(g) \rightleftharpoons 2NO_2(g) + \text{heat}$

20. add more NO **21.** increase temperature **22.** add a catalyst

23. add some O_2 **24.** remove NO_2

25. The solubility product expression for the insoluble salt $Ca_3(PO_4)_2$ is
A. $[Ca^{2+}][PO_4^{3-}]$ **B.** $[Ca^{2+}]_3[PO_4^{3-}]_2$ **C.** $[Ca^{2+}]^3[PO_4^{3-}]^2$

D. $[Ca^{2+}]^2[PO_4^{3-}]^3$ **E.** $\dfrac{[Ca^{2+}][PO_4^{3-}]}{[Ca_3(PO_4^{3-})_2]}$

26. The K_{sp} of CuI when a saturated solution has $[Cu^+] = 1 \times 10^{-6}$ M and $[I^-] = 1 \times 10^{-6}$ M is
A. 1×10^{-12} **B.** 1×10^{-6} **C.** 1
D. 2×10^{-6} **E.** 2×10^{-12}

27. What is the molar solubility of $SrCO_3$ if it has a K_{sp} of 5.6×10^{-10}?
A. 2.8×10^{-10} M **B.** 2.8×10^{-5} M **C.** 5.6×10^{-10} M
D. 2.3×10^{-20} M **E.** 2.4×10^{-5} M

28. What is the molar solubility of $SrCO_3$ in a solution with $[CO_3^{2-}] = 0.1$ M? $SrCO_3$ has a K_{sp} of 5.6×10^{-10}.
A. 0.1 M **B.** 2.4×10^{-5} M **C.** 5.6×10^{-9} M
D. 5.6×10^{-10} M **E.** 5.6×10^{-11} M

Answers to the Practice Test

1. A	**2.** C	**3.** D	**4.** E	**5.** E
6. A	**7.** C	**8.** C	**9.** A	**10.** C
11. E	**12.** E	**13.** A	**14.** C	**15.** B
16. B	**17.** A	**18.** A	**19.** B	**20.** A
21. B	**22.** C	**23.** A	**24.** A	**25.** C
26. A	**27.** E	**28.** C		

14
Acids and Bases

Study Goals

- Describe the characteristics of Arrhenius, Brønsted–Lowry, and organic acids and bases.
- Identify conjugate acid–base pairs in Brønsted–Lowry acids and bases.
- Write equations for the dissociation of strong and weak acids and identify the direction of the reaction.
- Write the expression for the dissociation constants of a weak acid or weak base.
- Use the ion product of water to calculate and convert among $[H_3O^+]$, $[OH^-]$, pH, and pOH.
- Write balanced equations for reactions of acids with metals, carbonates, and bases.
- Calculate the volume or concentration of an acid or base from titration data.
- Predict if a salt solution will be acidic, basic, and neutral.
- Describe the function of a buffer.
- Calculate the pH of a buffer solution.

Think About It

1. Why do a lemon, grapefruit, and vinegar taste sour?

2. What do antacids do? What are some bases listed on the labels of antacids?

3. Why are some aspirin products buffered?

Key Terms

Match each of the following key terms with the correct definition:

a. acid	**b.** base	**c.** pH	**d.** neutralization
e. buffer	**f.** K_a	**g.** endpoint	**h.** dissociation

1. _____ a substance that forms hydroxide ions (OH^-) in water and/or accepts protons (H^+)
2. _____ a reaction between an acid and a base to form a salt and water
3. _____ a substance that forms hydrogen ions (H^+) in water
4. _____ a mixture of a weak acid (or base) and its salt that maintains the pH of a solution

5. ____ a measure of the acidity (H_3O^+) of a solution

6. ____ separation of a substance into ions in water

7. ____ product of the molar concentrations of the ions from a weak acid divided by the molar concentration of that weak acid

8. ____ the point in a titration where the indicator changes color

Answers **1.** b **2.** d **3.** a **4.** e **5.** c **6.** h **7.** f **8.** g

14.1 Acids and Bases

- In water, an Arrhenius acid produces hydrogen ions (H^+) and an Arrhenius base produces hydroxide ions (OH^-).
- The names of inorganic acids are based on the anion present.
- Arrhenius bases are named as hydroxides.
- According to the Brønsted–Lowry theory, acids are proton (H^+) donors and bases are proton acceptors.
- Protons form hydronium ions, H_3O^+, in water when they bond to polar water molecules.
- In the IUPAC system, a carboxylic acid is named by replacing the *e* ending of the alkane name with *oic acid*. Simple acids usually are named by the common naming system using the prefixes form- (1C), acet- (2C), propion- (4C), butyr- (4C), followed by *ic acid*: HCOOH methanoic acid (formic acid), CH_3—COOH ethanoic acid (acetic acid), CH_3—CH_2—COOH propanoic acid (propionic acid), CH_3—CH_2—CH_2—COOH butanoic acid (butyric acid).

MasteringChemistry

Tutorial: Acid and Base Formulas
Tutorial: Naming Acids and Bases
Tutorial: Definitions of Acids and Bases

◆ Learning Exercise 14.1A

Indicate if each of the following characteristics describe an (A) acid or (B) base:

1. ____ turns blue litmus red
2. ____ tastes sour
3. ____ contains more OH^- than H_3O^+
4. ____ neutralizes bases
5. ____ tastes bitter
6. ____ turns red litmus blue
7. ____ contains more H_3O^+ than OH^-
8. ____ neutralizes acids

Answers **1.** A **2.** A **3.** B **4.** A **5.** B **6.** B **7.** A **8.** B

◆ Learning Exercise 14.1B

Fill in the blanks with the formula or name of an acid or base.

1. HCl ____________
2. ________ sodium hydroxide
3. ________ sulfurous acid
4. ________ nitric acid

5. $Ca(OH)_2$ ________________ **6.** H_2CO_3 ________________

7. $Al(OH)_3$ ________________ **8.** ________ potassium hydroxide

9. $HClO_4$ ________________ **10.** H_3PO_3 ________________

Answers

1. hydrochloric acid **2.** NaOH
3. H_2SO_3 **4.** HNO_3
5. calcium hydroxide **6.** carbonic acid
7. aluminum hydroxide **8.** KOH
9. perchloric acid **10.** phosphorous acid

◆ Learning Exercise 14.1C

Give the IUPAC and common names or condensed structural formula for each of the following:

1. ethanoic acid ________________

2. $\mathrm{CH_3-CH_2-\overset{\overset{\displaystyle O}{\|}}{C}-OH}$ ________________

3. butanoic acid ________________

4. formic acid ________________

5. $\mathrm{CH_3-CH_2-CH_2-CH_2-COOH}$ ________________

Answers

1. $\mathrm{CH_3-\overset{\overset{\displaystyle O}{\|}}{C}-OH}$
2. propanoic acid; propionic acid
3. $\mathrm{CH_3-CH_2-CH_2-\overset{\overset{\displaystyle O}{\|}}{C}-OH}$
4. $\mathrm{H-\overset{\overset{\displaystyle O}{\|}}{C}-OH}$
5. pentanoic acid

14.2 Brønsted–Lowry Acids and Bases

- According to the Brønsted–Lowry theory, acids are proton (H^+) donors, and bases are proton acceptors.
- Conjugate acid–base pairs are molecules or ions linked by the loss and gain of one H^+.
- Every proton transfer reaction involves two acid–base conjugate pairs.

MasteringChemistry
Tutorial: Identifying Conjugate Acid–Base Pairs

Study Note

Identify the conjugate acid–base pairs in the following equation:

$$HCl(aq) + H_2O(l) \longrightarrow H_3O^+(aq) + Cl^-(aq)$$

Solution: HCl (proton donor) and Cl^- (proton acceptor)
H_2O (proton acceptor) and H_3O^+ (proton donor)

◆ Learning Exercise 14.2A

Complete the following conjugate acid–base pairs:

	Conjugate Acid	Conjugate Base
1.	H_2O	________
2.	HSO_4^-	________
3.	________	F^-
4.	________	CO_3^{2-}
5.	HNO_3	________
6.	NH_4^+	________
7.	________	HS^-
8.	________	$H_2PO_4^-$

Answers **1.** OH^- **2.** SO_4^{2-} **3.** HF **4.** HCO_3^-
5. NO_3^- **6.** NH_3 **7.** H_2S **8.** H_3PO_4

◆ Learning Exercise 14.2B

Identify the conjugate acid–base pairs in each of the following reactions:

1. $HF(aq) + H_2O(l) \rightleftharpoons H_3O^+(aq) + F^-(aq)$

2. $NH_4^+(aq) + SO_4^{2-}(aq) \rightleftharpoons NH_3(aq) + HSO_4^-(aq)$

3. $NH_3(aq) + H_2O(l) \rightleftharpoons NH_4^+(aq) + OH^-(aq)$

4. $HNO_3(aq) + OH^-(aq) \rightleftharpoons H_2O(l) + NO_3^-(aq)$

Answers **1.** HF/F^- and H_2O/H_3O^+ **2.** NH_4^+/NH_3 and SO_4^{2-}/HSO_4^-
3. NH_3/NH_4^+ and H_2O/OH^- **4.** HNO_3/NO_3^- and OH^-/H_2O

◆ Learning Exercise 14.2C

Write an equation with conjugate acid–base pairs starting with each of the following reactants:

1. HBr (acid) and CO_3^{2-} (base)

2. HSO_4^- (acid) and OH^- (base)

3. NH_4^+ (acid) and H_2O (base)

4. HCl (acid) and SO_4^{2-} (base)

Answers

1. $HBr(aq) + CO_3^{2-}(aq) \rightleftharpoons HCO_3^-(aq) + Br^-(aq)$
2. $HSO_4^-(aq) + OH^-(aq) \rightleftharpoons H_2O(l) + SO_4^{2-}(aq)$
3. $NH_4^+(aq) + H_2O(l) \rightleftharpoons H_3O^+(aq) + NH_3(aq)$
4. $HCl(aq) + SO_4^{2-}(aq) \rightleftharpoons Cl^-(aq) + HSO_4^-(aq)$

14.3 Strengths of Acids and Bases

- In aqueous solution, a strong acid donates nearly all its protons to water, whereas a weak acid donates only a small percentage of protons to water.
- Most hydroxides of Groups 1A (1) and 2A (2) are strong bases, which dissociate nearly completely in water.
- In an aqueous ammonia solution, a weak base, NH_3, accepts only a small percentage of protons to form the conjugate acid, NH_4^+.
- As the strength of an acid decreases, the strength of the conjugate base increases. By comparing relative strengths, the direction of an acid-base reaction can be predicted.

MasteringChemistry
Tutorial: Using Dissociation Constants
Tutorial: Properties of Acids and Bases

Study Note

Only six common acids are strong acids; other acids are weak acids.

HCl	HNO_3
HBr	H_2SO_4 (first H)
HI	$HClO_4$

Example: Is H_2S a strong or a weak acid?

Solution: H_2S is a weak acid because it is not one of the six strong acids.

◆ Learning Exercise 14.3A

Identify each of the following as a strong or weak acid or base:

1. HNO_3 ________ 2. H_2CO_3 ________ 3. $H_2PO_4^-$ ________
4. NH_3 ________ 5. LiOH ________ 6. H_3BO_3 ________
7. $Ca(OH)_2$ ________ 8. H_2SO_4 ________

Answers **1.** strong acid **2.** weak acid **3.** weak acid **4.** weak base
5. strong base **6.** weak acid **7.** strong base **8.** strong acid

◆ Learning Exercise 14.3B

Using Table 14.4 (in textbook), identify the stronger acid in each of the following pairs of acids:

1. HCl or H_2CO_3 ________ 2. HNO_2 or HCN ________
3. H_2S or HBr ________ 4. H_2SO_4 or HSO_4^- ________
5. HF or H_3PO_4 ________

Answers **1.** HCl **2.** HNO_2 **3.** HBr **4.** H_2SO_4 **5.** H_3PO_4

Study Note

In an acid–base reaction, the relative strengths of the two acids or two bases indicate whether the equilibrium position favors the reactants or products.

Example: Does the following reaction favor the reactants or products?

$$NO_3^-(aq) + H_2O(l) \rightleftharpoons HNO_3(aq) + OH^-(aq)$$

Solution: Because the reactants contain the weaker base and acid, NO_3^- and H_2O, the equilibrium favors the reactants. This is represented with a long arrow to the left.

$$NO_3^-(aq) + H_2O(l) \leftrightharpoons HNO_3(aq) + OH^-(aq)$$

◆ Learning Exercise 14.3C

Indicate whether each of the following reactions favors the reactants or the products:

1. $HNO_3(aq) + H_2O(l) \rightleftharpoons H_3O^+(aq) + NO_3^-(aq)$
2. $I^-(aq) + H_3O^+(aq) \rightleftharpoons H_2O(l) + HI(aq)$
3. $NH_3(aq) + H_2O(l) \rightleftharpoons NH_4^+(aq) + OH^-(aq)$
4. $HCl(aq) + CO_3^{2-}(aq) \rightleftharpoons Cl^-(aq) + HCO_3^-(aq)$

Answers **1.** products **2.** reactants **3.** reactants **4.** products

14.4 Dissociation Constants

- An acid or base dissociation constant is the equilibrium constant when the concentration of water is considered a constant.
- For an acid dissociation constant, as with other equilibrium expressions, the molar concentrations of the products are divided by the molar concentrations of the reactants.

$$\mathbf{HA}(aq) + H_2O(l) \rightleftharpoons \mathbf{H_3O^+}(aq) + \mathbf{A^-}(aq)$$

$$K_a = K_c[H_2O] = \frac{[\mathbf{H_3O^+}][\mathbf{A^-}]}{[\mathbf{HA}]}$$

- An acid or a base with a large dissociation constant is more fully dissociated than an acid or base with a small dissociation constant.
- Dissociation constants greater than one ($K > 1$) favor the products, whereas constants smaller than one ($K < 1$) favor the reactants.

◆ Learning Exercise 14.4A

Write the acid dissociation constant expression for the ionization of the following weak acids:

1. HCN

2. HNO_2

3. H_2CO_3 (first ionization only)

4. H_2S (first ionization only)

Answers

1. $K_a = \frac{[H_3O^+][CN^-]}{[HCN]}$

2. $K_a = \frac{[H_3O^+][NO_2^-]}{[HNO_2]}$

3. $K_a = \frac{[H_3O^+][HCO_3^-]}{[H_2CO_3]}$

4. $K_a = \frac{[H_3O^+][HS^-]}{[H_2S]}$

◆ Learning Exercise 14.4B

For each of the following pairs of acid dissociation constants, indicate the constant of the weaker acid.

1. 5.2×10^{-5} or 3.8×10^{-3} ______________

2. 3.0×10^{8} or 1.6×10^{-10} ______________

3. 4.5×10^{-2} or 7.2×10^{-6} ______________

Answers **1.** 5.2×10^{-5} **2.** 1.6×10^{-10} **3.** 7.2×10^{-6}

◆ Learning Exercise 14.4C

Indicate whether the equilibrium favors the reactants or products for each of the following dissociation constants:

1. 5.2×10^{-5} ______________

2. 3.0×10^{8} ______________

3. 4.5×10^{-5} ______________

4. 7.2×10^{15} ______________

Answers **1.** reactants **2.** products **3.** reactants **4.** products

14.5 Ionization of Water

- In pure water, a few water molecules transfer protons (H^+) to other water molecules, forming two conjugate acid-base pairs. This reaction produces small, but equal, amounts of each ion such that $[H_3O^+]$ and $[OH^-]$ each $= 1.0 \times 10^{-7}$ mol/L.

$$H_2O(l) + H_2O(l) \rightleftarrows H_3O^+(aq) + OH^-(aq)$$

- K_w, the ion product constant of water, $[H_3O^+][OH^-] = [1.0 \times 10^{-7}][1.0 \times 10^{-7}] = 1.0 \times 10^{-14}$, applies to all aqueous solutions.
- In acidic solutions, the $[H_3O^+]$ is greater than the $[OH^-]$. In basic solutions, the $[OH^-]$ is greater than the $[H_3O^+]$.

MasteringChemistry
Tutorial: Ionization of Water

Guide to Calculating $[H_3O^+]$ and $[OH^-]$ in Aqueous Solutions

STEP 1 Write the K_w for water.
STEP 2 Solve the K_w for the unknown $[H_3O^+]$ or $[OH^-]$.
STEP 3 Substitute the known $[H_3O^+]$ or $[OH^-]$ and calculate.

Study Note

Example: What is the $[H_3O^+]$ in a solution that has $[OH^-] = 2.0 \times 10^{-9}$ M?

Solution: $K_w = [H_3O^+][OH^-] = 1.0 \times 10^{-14}$

Rearrange the K_w expression for $[H_3O^+]$, substitute $[OH^-] = 2.0 \times 10^{-9}$ M, and calculate.

$$[H_3O^+] = \frac{1.0 \times 10^{-14}}{[2.0 \times 10^{-9}]} = 5.0 \times 10^{-6}\text{ M}$$

◆ Learning Exercise 14.5A

Use the K_w to calculate the $[H_3O^+]$ when the $[OH^-]$ has each of the following values:

a. $[OH^-] = 1.0 \times 10^{-10}$ M $[H_3O^+] =$ ____________

b. $[OH^-] = 2.0 \times 10^{-5}$ M $[H_3O^+] =$ ____________

c. $[OH^-] = 4.5 \times 10^{-7}$ M $[H_3O^+] =$ ____________

d. $[OH^-] = 8.0 \times 10^{-4}$ M $[H_3O^+] =$ ____________

e. $[OH^-] = 5.5 \times 10^{-8}$ M $[H_3O^+] =$ ____________

Answers **a.** 1.0×10^{-4} M **b.** 5.0×10^{-10} M **c.** 2.2×10^{-8} M
d. 1.3×10^{-11} M **e.** 1.8×10^{-7} M

◆ Learning Exercise 14.5B

Use the K_w to determine the $[OH^-]$ when the $[H_3O^+]$ has each of the following values:

a. $[H_3O^+] = 1.0 \times 10^{-3}$ M $[OH^-] =$ ____________

b. $[H_3O^+] = 3.0 \times 10^{-10}$ M $[OH^-] =$ ________________

c. $[H_3O^+] = 4.0 \times 10^{-6}$ M $[OH^-] =$ ________________

d. $[H_3O^+] = 2.8 \times 10^{-13}$ M $[OH^-] =$ ________________

e. $[H_3O^+] = 8.6 \times 10^{-7}$ M $[OH^-] =$ ________________

Answers **a.** 1.0×10^{-11} M **b.** 3.3×10^{-5} M **c.** 2.5×10^{-9} M
d. 3.6×10^{-2} M **e.** 1.2×10^{-8} M

14.6 The pH Scale

- The pH scale is a range of numbers from 0 to 14 related to the $[H_3O^+]$ of the solution.
- A neutral solution has a pH of 7.0. In acidic solutions, the pH is below 7.0, and in basic solutions, the pH is above 7.0.
- Mathematically, pH is the negative logarithm of the hydronium ion concentration: $pH = -\log[H_3O^+]$.
- The pOH is similar to pH; $pOH = -\log[OH^-]$. The pH plus the pOH are equal to 14.00.

MasteringChemistry
Case Study: Hyperventilation and Blood pH
Self Study Activity: The pH Scale
Tutorial: Logarithms
Tutorial: The pH Scale

◆ Learning Exercise 14.6A

State whether the following pH values are acidic, basic, or neutral:

1. ________ plasma, pH = 7.4 **2.** ________ soft drink, pH = 2.8

3. ________ maple syrup, pH = 6.8 **4.** ________ beans, pH = 5.0

5. ________ tomatoes, pH = 4.2 **6.** ________ lemon juice, pH = 2.2

7. ________ saliva, pH = 7.0 **8.** ________ eggs, pH = 7.8

9. ________ lye, pH = 12.4 **10.** ________ strawberries, pH = 3.0

Answers **1.** basic **2.** acidic **3.** acidic **4.** acidic **5.** acidic
6. acidic **7.** neutral **8.** basic **9.** basic **10.** acidic

Guide to Calculating pH of an Aqueous Solution

STEP 1 Enter the $[H_3O^+]$ value.
STEP 2 Press the *log* key and change the sign.
STEP 3 Adjust the number of digits on the *right* of the decimal point to equal the SFs in the coefficient.

◆ Learning Exercise 14.6B

Calculate the pH of each of the following solutions:

a. $[H_3O^+] = 1.0 \times 10^{-8}$ M ______ b. $[OH^-] = 1 \times 10^{-12}$ M ______

c. $[H_3O^+] = 1 \times 10^{-3}$ M ______ d. $[OH^-] = 1 \times 10^{-10}$ M ______

e. $[H_3O^+] = 3.4 \times 10^{-8}$ M ______ f. $[OH^-] = 7.8 \times 10^{-2}$ M ______

Answers **a.** 8.00 **b.** 2.0 **c.** 3.0 **d.** 4.0 **e.** 7.47 **f.** 12.89

◆ Learning Exercise 14.6C

Calculate the pH of each of the following solutions:

a. $[H_3O^+] = 5.0 \times 10^{-3}$ M b. $[OH^-] = 4.0 \times 10^{-6}$ M

c. $[H_3O^+] = 7.5 \times 10^{-8}$ M d. $[OH^-] = 2.5 \times 10^{-10}$ M

Answers **a.** 2.30 **b.** 8.60 **c.** 7.12 **d.** 4.40

◆ Learning Exercise 14.6D

Complete the following table:

	$[H_3O^+]$	$[OH^-]$	pH
a.	______	1×10^{-12} M	______
b.	______	______	8.0
c.	5.0×10^{-11} M	______	______
d.	______	______	7.80
e.	______	______	4.25
f.	2.0×10^{-10} M	______	______

Answers

	$[H_3O^+]$	$[OH^-]$	pH
a.	1×10^{-2} M	1×10^{-12} M	2.0
b.	1×10^{-8} M	1×10^{-6} M	8.0
c.	5.0×10^{-11} M	2.0×10^{-4} M	10.3
d.	1.6×10^{-8} M	6.3×10^{-7} M	7.80
e.	5.6×10^{-5} M	1.8×10^{-10} M	4.25
f.	2.0×10^{-10} M	5.0×10^{-5} M	9.70

◆ Learning Exercise 14.6E

Calculate the pOH of each of the following solutions:

a. $[OH^-] = 1 \times 10^{-4}$ M ______ b. $[H_3O^+] = 1.0 \times 10^{-2}$ M ______

c. pH = 8.15 ______ d. $[OH^-] = 4.0 \times 10^{-10}$ M ______

e. $[H_3O^+] = 3.4 \times 10^{-8}$ M ______ f. pH = 2.22 ______

Answers **a.** 4.0 **b.** 12.00 **c.** 5.85 **d.** 9.40 **e.** 6.53 **f.** 11.78

14.7 Reactions of Acids and Bases

- Acids react with *active* metals to yield hydrogen gas, H_2, and the salt of the metal.
- Acids react with carbonates and bicarbonates to yield CO_2, H_2O, and the salt of the metal.
- Acids neutralize bases in a reaction that produces water and a salt.
- The net ionic equation for any strong acid–strong base neutralization is $H^+(aq) + OH^-(aq) \longrightarrow H_2O(l)$.
- In a balanced neutralization equation, an equal number of moles of H^+ and OH^- must react.
- The concentration of an acid solution can be determined by titration.

> ***MasteringChemistry***
> **Self Study Activity: Reactions of Acids and Bases**

◆ Learning Exercise 14.7A

Complete and balance each of the following reactions of acids:

1. ____ $Zn(s)$ + ____ $HCl(aq) \longrightarrow$ ____ $ZnCl_2(aq)$ + ____

2. ____ $HCl(aq)$ + ____ $Li_2CO_3(aq) \longrightarrow$ ____ + ____ + ____

3. ____ $HCl(aq)$ + ____ $NaHCO_3(aq) \longrightarrow$ ____ $CO_2(g)$ + ____ $H_2O(l)$ + ____ $NaCl(aq)$

4. ____ $Al(s)$ + ____ $H_2SO_4(aq) \longrightarrow$ ____ $Al_2(SO_4)_3(aq)$ + ____

Answers

1. $Zn(s) + 2HCl(aq) \longrightarrow ZnCl_2(aq) + H_2(g)$
2. $2HCl(aq) + Li_2CO_3(aq) \longrightarrow CO_2(g) + H_2O(l) + 2LiCl(aq)$
3. $HCl(aq) + NaHCO_3(aq) \longrightarrow CO_2(g) + H_2O(l) + NaCl(aq)$
4. $2Al(s) + 3H_2SO_4(aq) \longrightarrow Al_2(SO_4)_3(aq) + 3H_2(g)$

> **Guide to Balancing an Equation for Neutralization**
>
> **STEP 1** Write the the reactants and products.
> **STEP 2** Balance the H^+ in the acid with the OH^- in the base.
> **STEP 3** Balance the H_2O with the H^+ and the OH^-.
> **STEP 4** Write of the salt from the remaining ions.

◆ Learning Exercise 14.7B

Balance each of the following neutralization reactions:

1. ______ $NaOH(aq)$ + ______ $H_2SO_4(aq) \longrightarrow$ ______ $Na_2SO_4(aq)$ + ______ $H_2O(l)$

2. ______ $Mg(OH)_2(aq)$ + ______ $HCl(aq) \longrightarrow$ ______ $MgCl_2(aq)$ + ______ $H_2O(l)$

3. ______ $Al(OH)_3(aq)$ + ______ $HNO_3(aq) \longrightarrow$ ______ $Al(NO_3)_3(aq)$ + ______ $H_2O(l)$

4. ______ $Ca(OH)_2(aq)$ + ______ $H_3PO_4(aq) \longrightarrow$ ______ $Ca_3(PO_4)_2(s)$ + ______ $H_2O(l)$

Answers

1. $2NaOH(aq) + H_2SO_4(aq) \longrightarrow Na_2SO_4(aq) + 2H_2O(l)$
2. $Mg(OH)_2(aq) + 2HCl(aq) \longrightarrow MgCl_2(aq) + 2H_2O(l)$
3. $Al(OH)_3(aq) + 3HNO_3(aq) \longrightarrow Al(NO_3)_3(aq) + 3H_2O(l)$
4. $3Ca(OH)_2(aq) + 2H_3PO_4(aq) \longrightarrow Ca_3(PO_4)_2(s) + 6H_2O(l)$

◆ Learning Exercise 14.7C

Complete each of the following neutralization reactions and then balance:

a. ______ $KOH(aq)$ + ______ $H_3PO_4(aq) \longrightarrow$ ______ + ______ $H_2O(l)$

b. ______ $NaOH(aq)$ + ______ $\longrightarrow$ ______ $Na_2SO_4(aq)$ + ______

c. ______ + ______ $\longrightarrow$ ______ $AlCl_3(aq)$ + ______

d. ______ + ______ $\longrightarrow$ ______ $Fe_2(SO_4)_3(aq)$ + ______

Answers

a. $3KOH(aq) + H_3PO_4(aq) \longrightarrow K_3PO_4(aq) + 3H_2O(l)$
b. $2NaOH(aq) + H_2SO_4(aq) \longrightarrow Na_2SO_4(aq) + 2H_2O(l)$
c. $Al(OH)_3(s) + 3HCl(aq) \longrightarrow AlCl_3(aq) + 3H_2O(l)$
d. $2Fe(OH)_3(s) + 3H_2SO_4(aq) \longrightarrow Fe_2(SO_4)_3(aq) + 6H_2O(l)$

14.8 Acid–Base Titration

- In a laboratory procedure called a titration, an acid sample is neutralized with a known amount of a base.
- From the volume and molarity of the base, the concentration of the acid is calculated.
- A titration is used to determine the volume or concentration of an acid or a base from the laboratory data.

Guide to Calculations for an Acid–Base Titration

STEP 1 State the given and needed quantities and concentrations.
STEP 2 Write a plan to calculate molarity or volume.
STEP 3 State equalities and conversion factors including concentrations.
STEP 4 Set up problem to calculate needed quantity.

MasteringChemistry

Tutorial: Acid–Base Titrations

◆ Learning Exercise 14.8

Solve the following problems using the titration data given:

a. A 5.00-mL sample of HCl is placed in a flask. In the titration, 15.0 mL of a 0.200 M NaOH solution is required for neutralization. What is the molarity of the HCl in the sample?

$$HCl(aq) + NaOH(aq) \longrightarrow NaCl(aq) + H_2O(l)$$

b. How many milliliters of a 0.200 M NaOH solution are required to neutralize completely 8.50 mL of a 0.500 M H_2SO_4 solution?

$$H_2SO_4(aq) + 2NaOH(aq) \longrightarrow Na_2SO_4(aq) + 2H_2O(l)$$

c. A 10.0-mL sample of H_3PO_4 solution is placed in a flask. If titration requires 42.0 mL of a 0.100 M NaOH solution to reach the endpoint, what is the molarity of the H_3PO_4 solution?

$$H_3PO_4(aq) + 3NaOH(aq) \longrightarrow Na_3PO_4(aq) + 3H_2O(l)$$

d. A 24.6-mL sample of HCl solution reacts with 33.0 mL of a 0.222 M NaOH solution. What is the molarity of the HCl solution (see reaction in part **a**)?

e. A 15.7-mL sample of H_2SO_4 reacts with 27.7 mL of a 0.187 M NaOH solution. What is the molarity of the H_2SO_4 solution (see reaction in part **b**)?

Answers **a.** 0.600 M HCl solution **b.** 42.5 mL **c.** 0.140 M H_3PO_4 solution
d. 0.298 M HCl solution **e.** 0.165 M H_2SO_4 solution

14.9 Acid–Base Properties of Salt Solutions

- Salts of strong acids and strong bases produce neutral aqueous solutions.
- Salts of weak acids and strong bases form basic aqueous solutions because the conjugate base of the weak acid attracts a proton from water to form OH^- in the solution.
- Salts of strong acids and weak bases form acidic aqueous solutions because the conjugate acid of the weak base donates a proton to water, which produces H_3O^+ in the solution.

MasteringChemistry

Tutorial: Salts of Weak Acids and Bases

Study Note

Use the cation and anion of a salt to determine the acidity of its aqueous solution.

Example: Will the salt Na_2CO_3 form an acidic, basic, or neutral aqueous solution?

Solution: A salt of a strong base (NaOH) and a weak acid (HCO_3^-) will remove protons (H^+) from water to produce a basic solution: $CO_3^{2-}(aq) + H_2O(l) \longrightarrow HCO_3^-(aq) + OH^-(aq)$

◆ Learning Exercise 14.9A

Identify solutions of each of the following salts as acidic, basic, or neutral:

1. NaBr ______ **2.** KNO_2 ______ **3.** NH_4Cl ______

4. Li_2SO_4 ______ **5.** KF ______

Answers **1.** neutral **2.** basic **3.** acidic **4.** basic **5.** basic

◆ Learning Exercise 14.9B

Determine if each of the following salts dissolved in water forms a solution that is acidic, basic, or neutral. If acidic or basic, write an equation for the reaction:

	Acidic, Basic, or Neutral	Equation
1. NaCN	________	________
2. LiBr	________	________
3. NH_4NO_3	________	________
4. Na_2S	________	________
5. $BaCl_2$	________	________

Answers

1.	NaCN	basic	$CN^-(aq) + H_2O(l) \longrightarrow HCN(aq) + OH^-(aq)$
2.	LiBr	neutral	
3.	NH_4NO_3	acidic	$NH_4^+(aq) + H_2O(l) \longrightarrow H_3O^+(aq) + NH_3(aq)$
4.	Na_2S	basic	$S^{2-}(aq) + H_2O(l) \longrightarrow HS^-(aq) + OH^-(aq)$
5.	$BaCl_2$	neutral	

14.10 Buffers

- A buffer solution resists a change in pH when small amounts of acid or base are added.
- A buffer contains either (1) a weak acid and its salt or (2) a weak base and its salt.
- The weak acid reacts with added OH^-, and the anion of the salt reacts with added H_3O^+, which maintains the pH of the solution.
- The weak base reacts with added H_3O^+, and the cation of the salt reacts with added OH^-, which maintains the pH of the solution.
- The pH of a buffer can be calculated by rearranging and solving the K_a for $[H_3O^+]$, and then converting to pH.

MasteringChemistry

Self Study Activity: pH and Buffers
Tutorial: Buffer Solutions
Tutorial: Preparing Buffer Solutions
Tutorial: Calculating the pH of a Buffer

◆ Learning Exercise 14.10A

State whether each of the following represents a buffer system and explain why:

a. HCl + NaCl

b. K_2SO_4

c. H_2CO_3

d. $H_2CO_3 + NaHCO_3$

Answers

a. No. A strong acid and its salt do not make a buffer.
b. No. A salt alone cannot act as a buffer.
c. No. A weak acid alone cannot act as a buffer.
d. Yes. A weak acid and its salt act as a buffer system.

Guide to Calculating pH of a Buffer

STEP 1 Write the K_a or K_b expression.
STEP 2 Rearrange the K_a or K_b for $[H_3O^+]$ or $[OH^-]$.
STEP 3 Substitute in the [HA] and $[A^-]$.
STEP 4 Use $[H_3O^+]$ to calculate pH.

Study Note

Rearranging the K_a to solve for $[H_3O^+]$ gives the following expression:

$$[H_3O^+] = K_a \times \frac{[HA]}{[A^-]}$$

Using the $[H_3O^+]$, write the equation to calculate the pH of the buffer.

$$pH = -\log[H_3O^+]$$

◆ Learning Exercise 14.10B

The K_a for acetic acid CH_3COOH is 1.8×10^{-5}.

a. What is the pH of a buffer that contains 1.0 M CH_3COOH and 1.0 M $NaCH_3COO$?

b. What is the pH of a buffer made from 0.20 M CH_3COOH and 0.10 M $NaCH_3COO$?

c. What is the pH of a buffer made from 0.10 M CH_3COOH and 1.0 M $NaCH_3COO$?

Answers

a. $[H_3O^+] = 1.8 \times 10^{-5} \times \frac{[1.0\ \cancel{M}]}{[1.0\ \cancel{M}]} = 1.8 \times 10^{-5}$ pH = 4.74

b. $[H_3O^+] = 1.8 \times 10^{-5} \times \frac{[0.20\ \cancel{M}]}{[0.10\ \cancel{M}]} = 3.6 \times 10^{-5}$ pH = 4.44

c. $[H_3O^+] = 1.8 \times 10^{-5} \times \frac{[0.10\ \cancel{M}]}{[1.0\ \cancel{M}]} = 1.8\ 10^{-6}$ pH = 5.74

Checklist for Chapter 14

You are ready to take the Practice Test for Chapter 14. Be sure you have accomplished the following learning goals for this chapter. If you are not sure, review the section listed at the end of the goal. Then apply your new skills and understanding to the Practice Test.

After studying Chapter 14, I can successfully:

_______ Describe the properties of Arrhenius acids and bases and write their names (14.1).

_______ Write the IUPAC and common names and draw structural formulas of carboxylic acids (14.1).

_______ Describe the Brønsted–Lowry concept of acids and bases (14.2).

_______ Write conjugate acid–base pairs for an acid–base reaction (14.2).

_______ Write equations for the ionization of strong and weak acids and bases (14.3).

_______ Write a dissociation constant for a weak acid or base (14.4).

_______ Use the ion product constant of water to calculate $[H_3O^+]$ and $[OH^-]$ (14.5).

_______ Calculate pH or pOH from the $[H_3O^+]$ or $[OH^-]$ of a solution (14.6).

_______ Write a balanced equation for the reactions of acids with metals, carbonates, and bases (14.7).

_______ Calculate the molarity or volume of an acid or base from titration information (14.8).

_______ Determine if a salt solution will be acidic, basic, or neutral (14.9).

_______ Describe the role of buffers in maintaining the pH of a solution, and calculate the pH of a buffer solution (14.10).

Practice Test for Chapter 14

1. An acid is a compound that when placed in water yields this characteristic ion:
A. H_3O^+ **B.** OH^- **C.** Na^+ **D.** Cl^- **E.** CO_3^{2-}

2. $MgCl_2$ would be classified as a(n)
A. acid **B.** base **C.** salt **D.** buffer **E.** nonelectrolyte

3. $Mg(OH)_2$ would be classified as a
A. weak acid **B.** strong base **C.** salt **D.** buffer **E.** nonelectrolyte

4. In the K_w expression for pure H_2O, the $[H_3O^+]$ has the value
A. 1.0×10^{-7} M **B.** 1.0×10^{-1} M **C.** 1.0×10^{-14} M
D. 1.0×10^{-6} M **E.** 1.0×10^{12} M

5. Of the following pH values, which is the most acidic?
A. 8.0 **B.** 5.5 **C.** 1.5 **D.** 3.2 **E.** 9.0

6. Of the following pH values, which is the most basic?
A. 10.0 **B.** 4.0 **C.** 2.2 **D.** 11.5 **E.** 9.0

For questions 7 through 9, consider a solution with $[H_3O^+] = 1 \times 10^{-11}$ M.

7. The pH of the solution is
A. 1.0 **B.** 2.0 **C.** 3.0 **D.** 11.0 **E.** 14.0

8. The hydroxide ion concentration is
A. 1×10^{-1} M **B.** 1×10^{-3} M **C.** 1×10^{-4} M
D. 1×10^{-7} M **E.** 1×10^{-11} M

9. The solution is
A. acidic **B.** basic **C.** neutral **D.** a buffer **E.** neutralized

For questions 10 through 13, consider a solution with $[OH^-] = 1 \times 10^{-5}$ M.

10. The hydrogen ion concentration of the solution is
A. 1×10^{-5} M **B.** 1×10^{-7} M **C.** 1×10^{-9} M
D. 1×10^{-10} M **E.** 1×10^{-14} M

11. The pH of the solution is
A. 2.0 **B.** 5.0 **C.** 9.0 **D.** 11 **E.** 14

12. The pOH of the solution is
A. 2.0 **B.** 5.0 **C.** 9.0 **D.** 11.0 **E.** 14

13. The solution is
A. acidic **B.** basic **C.** neutral **D.** a buffer **E.** neutralized

14. Acetic acid is a weak acid because
A. it forms a dilute acid solution
B. it is isotonic
C. it is less than 50% ionized in water
D. it is a nonpolar molecule
E. it can form a buffer

15. A weak base when added to water
A. makes the solution slightly basic
B. does not affect the pH
C. dissociates completely
D. does not dissociate
E. makes the solution slightly acidic

16. Which is an equation for neutralization of an acid and a base?
A. $CaCO_3(s) \longrightarrow CaO(s) + CO_2(g)$
B. $Na_2SO_4(s) \longrightarrow 2Na^+(aq) + SO_4^{2-}(aq)$
C. $H_2SO_4(aq) + 2NaOH(aq) \longrightarrow Na_2SO_4(aq) + 2H_2O(l)$
D. $Na_2O(s) + SO_3(g) \longrightarrow Na_2SO_4(aq)$
E. $H_2CO_3(aq) \longrightarrow CO_2(g) + H_2O(l)$

17. What is the name given to components in the body that keep blood pH within its normal 7.35 to 7.45 range?
A. nutrients **B.** buffers **C.** metabolites **D.** regufluids **E.** neutralizers

18. A buffer system
A. maintains a pH of 7.0
B. contains a weak base
C. contains a salt
D. contains a strong acid and its salt
E. maintains the pH of a solution

19. Which of the following would act as a buffer system?
A. HCl
B. Na_2CO_3
C. $NaOH + NaNO_3$
D. NH_4OH
E. $NaHCO_3 + H_2CO_3$

20. What is the molarity of a 10.0-mL sample of HCl that is neutralized by 15.0 mL of a 2.0 M NaOH solution?
A. 0.50 M HCl **B.** 1.0 M HCl **C.** 1.5 M HCl **D.** 2.0 M HCl **E.** 3.0 M HCl

21. In a titration, 6.0 mol NaOH will completely neutralize ______ mol of H_2SO_4.
A. 1.0 **B.** 2.0 **C.** 3.0 **D.** 6.0 **E.** 11

22. Which of the following pairs is a conjugate acid–base pair?
A. HCl/HNO_3 **B.** HNO_2/NO_2^- **C.** NaOH/KOH
D. HSO_4^-/HCO_3^- **E.** Cl^-/F^-

23. The conjugate base of HSO_4^- is
A. SO_4^{2-} **B.** H_2SO_4 **C.** HS^- **D.** H_2S **E.** SO_3^{2-}

24. In which of the following reactions does H_2O act as an acid?

A. $H_3PO_4(aq) + H_2O(l) \longrightarrow H_3O^+(aq) + H_2PO_4^-(aq)$

B. $H_2SO_4(aq) + H_2O(l) \longrightarrow H_3O^+(aq) + HSO_4^-(aq)$

C. $H_2O(l) + HS^-(aq) \longrightarrow H_3O^+(aq) + S^{2-}(aq)$

D. $NaOH(aq) + HCl(aq) \longrightarrow NaCl(aq) + H_2O(l)$

E. $NH_3(g) + H_2O(l) \longrightarrow NH_4^+(aq) + OH^-(aq)$

25. Which of the following acids has the smallest K_a value?

A. HNO_3 **B.** H_2SO_4 **C.** HCl **D.** H_2CO_3 **E.** HBr

26. Using the following K_a values, identify the strongest acid in the group:

A. 7.5×10^{-3}

B. 1.8×10^{-5}

C. 4.5×10^{8}

D. 4.9×10^{-10}

E. 3.2×10^{4}

27. Which of the following salts produces an acidic aqueous solution?

A. KCl **B.** NH_4Cl **C.** Na_2SO_4 **D.** K_2CO_3 **E.** $NaNO_2$

28. A buffer is made with 1.0 M HF and 1.0 M NaF. If HF has a K_a of 7.2×10^{-4} what is the pH of the buffer?

A. 3.14 **B.** 4.00 **C.** 4.14 **D.** 4.72 **E.** 7.20

29. Which of the following would turn litmus blue?

A. HCl **B.** NH_4Cl **C.** Na_2SO_4 **D.** KOH **E.** $NaNO_3$

30. What is the name of $HClO_3$?

A. hydrochloric acid **B.** chloric acid **C.** chlorous acid

D. chloric trioxide acid **E.** perchloric acid

31. What is the name of NH_4OH?

A. ammonium oxide

B. nitrogen tetrahydride hydroxide

C. ammonium hydroxide

D. perammonium hydroxide

E. amine oxide hydride

32. What is the name of $CH_3—CH_2—COOH$?

A. methyl peroxidic acid

B. propanoic acid

C. acetic acid

D. methyl acetic acid

E. butanoic acid

Answers to the Practice Test

1. A	**2.** C	**3.** B	**4.** A	**5.** C
6. D	**7.** D	**8.** B	**9.** B	**10.** C
11. C	**12.** B	**13.** B	**14.** C	**15.** A
16. C	**17.** B	**18.** E	**19.** E	**20.** E
21. C	**22.** B	**23.** A	**24.** E	**25.** D
26. C	**27.** B	**28.** A	**29.** D	**30.** B
31. C	**32.** B			

15

Oxidation and Reduction

Study Goals

- Identify the atoms that are oxidized and reduced in an oxidation–reduction reaction.
- Assign an oxidation number to the atoms in a compound to determine the oxidized and reduced components.
- Use oxidation numbers to identify what is oxidized, what is reduced, the reducing agent, and the oxidizing agent.
- Balance oxidation–reduction equations using the oxidation number or the half-reaction method.
- Classify alcohols as primary, secondary, or tertiary.
- Draw the condensed structural formulas of the products from the oxidation of alcohols.
- Write the half-reactions that occur at the anode and cathode of a voltaic cell; write the shorthand cell notation.
- Describe the half-cell reactions and the overall reactions that occur in electrolytic cells.
- Use the activity series to determine if a reaction is spontaneous or nonspontaneous and if it requires energy to proceed.

Think About It

1. What change occurs in iron metal (Fe) when it is oxidized to form rust?

2. How does the oxidation number change when an atom is reduced?

3. How does a battery provide electricity to run a flashlight or watch?

4. How can corrosion be prevented?

5. How can a nonspontaneous oxidation–reduction reaction be used in electroplating?

Key Terms

Match each of the following key terms with the correct definition:

a. cathode **b.** oxidation **c.** reduction **d.** oxidizing agent
e. reducing agent **f.** voltaic cell **g.** secondary alcohol

1. _____ the reactant that is reduced

2. _____ the gain of electrons by a substance

3. _____ a cell that uses an oxidation–reduction reaction to produce electrical energy

4. _____ the loss of electrons by a substance

5. _____ the reactant that is oxidized

6. _____ the electrode in an electrochemical cell where reduction takes place

7. _____ the carbon atom with —OH group is bonded to two other carbon atoms

Answers **1.** d **2.** c **3.** f **4.** b **5.** e **6.** a **7.** g

15.1 Oxidation and Reduction

- In an oxidation–reduction reaction, electrons are transferred from one reactant to another.
- In an oxidation, electrons are lost. In a reduction, there is a gain of electrons.
- In any oxidation–reduction reaction, the number of electrons lost must be equal to the number of electrons gained.
- A reducing agent (oxidized) provides the electrons, and an oxidizing agent (reduced) accepts electrons.

Study Note

Oxidation **i**s the **l**oss of electrons (OIL). **R**eduction **i**s the **g**ain of electrons (RIG).

A balanced oxidation–reduction equation contains an oxidation of one reactant and a reduction of the other reactant. $Cu(s) + 2Ag^{+}(aq) \longrightarrow Cu^{2+}(aq) + 2Ag(s)$

Oxidation: $Cu(s) \longrightarrow Cu^{2+}(aq) + 2\,e^{-}$

Reduction: $2Ag^{+}(aq) + 2\,e^{-} \longrightarrow 2Ag(s)$

In a balanced oxidation–reduction equation, the loss of electrons must be equal to the gain of electrons.

◆ Learning Exercise 15.1A

Identify each of the following as oxidation (O) or reduction (R):

1. _______ loss of two electrons **2.** _______ $Zn(s) \longrightarrow Zn^{2+}(aq) + 2\,e^{-}$

3. _______ $Cu^{2+}(aq) + e^{-} \longrightarrow Cu^{+}(aq)$ **4.** _______ gain of electrons

Answers **1.** O **2.** O **3.** R **4.** R

◆ Learning Exercise 15.1B

For each of the following, indicate whether the underlined element is *oxidized* or *reduced*:

a. $4\underline{Al}(s) + 3O_2(g) \longrightarrow 2Al_2O_3(s)$ Al is ________

b. $\underline{Fe}^{3+}(aq) + e^- \longrightarrow Fe^{2+}(aq)$ Fe^{3+} is ________

c. $\underline{Cu}O(s) + H_2(g) \longrightarrow Cu(s) + H_2O(l)$ Cu^{2+} is ________

d. $2\underline{Cl}^-(aq) \longrightarrow Cl_2(g) + 2\,e^-$ Cl^- is ________

e. $2H\underline{Br}(aq) + Cl_2(g) \longrightarrow 2HCl(aq) + Br_2(g)$ Br^- is ________

f. $2\underline{Na}(s) + Cl_2(g) \longrightarrow 2NaCl(aq)$ Na is ________

g. $\underline{Cu}Cl_2(aq) + Zn(s) \longrightarrow ZnCl_2(aq) + Cu(s)$ Cu^{2+} is ________

Answers

a. Al^0 is oxidized to Al^{3+}; loss of electrons.

b. Fe^{3+} is reduced to Fe^{2+}; gain of electrons.

c. Cu^{2+} is reduced to Cu^0; gain of electrons.

d. Cl^- is oxidized to Cl^0; loss of electrons.

e. Br^- is oxidized to Br^0; loss of electrons.

f. Na^0 is oxidized to Na^+; loss of electrons.

g. Cu^{2+} is reduced to Cu^0; gain of electrons.

◆ Learning Exercise 15.1C

Determine whether each of the following is an oxidation–reduction reaction. If the reaction is an oxidation–reduction, identify the element that is oxidized and the element that is reduced.

a. $2Na(s) + Cl_2(g) \longrightarrow 2NaCl(aq)$

b. $Zn(s) + 2H^+(aq) \longrightarrow Zn^{2+}(aq) + H_2(g)$

c. $H_2S(aq) + I_2(s) \longrightarrow 2HI(aq) + S(s)$

d. $BaCl_2(aq) + 2AgNO_3(aq) \longrightarrow 2AgCl(s) + Ba(NO_3)_2(aq)$

e. $2Mg(s) + O_2(g) \longrightarrow 2MgO(s)$

f. $B_2O_3(s) + 3Mg(s) \longrightarrow 3MgO(s) + 2B(s)$

Answers

a. oxidation: $Na \longrightarrow Na^+ + 1\,e^-$ reduction: $Cl_2 + 2\,e^- \longrightarrow 2Cl^-$
b. oxidation: $Zn \longrightarrow Zn^{2+} + 2\,e^-$ reduction: $2H^+ + 2\,e^- \longrightarrow H_2$
c. oxidation: $S^{2-} \longrightarrow S + 2\,e^-$ reduction: $I_2 + 2\,e^- \longrightarrow 2I^-$
d. not oxidation–reduction (double replacement reaction)
e. oxidation: $Mg \longrightarrow Mg^{2+} + 2\,e^-$ reduction: $O_2 + 4\,e^- \longrightarrow 2O^{2-}$
f. oxidation: $Mg \longrightarrow Mg^{2+} + 2\,e^-$ reduction: $B^{3+} + 3\,e^- \longrightarrow B$

15.2 Oxidation Numbers

- An oxidation number is assigned to atoms to determine which substance is oxidized and which is reduced. These numbers do not necessarily represent actual charges.
- Oxidation numbers are assigned based on a set of rules.
- An increase in oxidation number indicates an oxidation; a decrease in oxidation number indicates a reduction.
- An oxidation–reduction equation is balanced by identifying the oxidized and reduced atoms from the changes in electrons and equalizing the changes in oxidation number.
- The oxidation number method is typically used to balance equations written in molecular form.

Table 15.1 Rules for Assigning Oxidation Numbers

1. The sum of the oxidation numbers in a molecule is equal to zero (0), or for a polyatomic ion the sum of the oxidation numbers is equal to its charge.
2. The oxidation number of an element (monatomic or diatomic) is zero (0).
3. The oxidation number of a monatomic ion is equal to its charge.
4. In compounds, the oxidation number of Group 1A (1) elements is +1 and of Group 2A (2) elements is +2.
5. In compounds, the oxidation number of fluorine is always −1. Other elements in Group 7A (17) also are −1 except when combined with oxygen or fluorine.
6. In compounds, the oxidation number of oxygen is usually −2 except in OF_2.
7. In compounds with nonmetals, the oxidation number of hydrogen is +1; in compounds with metals, the oxidation number of hydrogen is −1.

MasteringChemistry

Tutorial: Assigning Oxidation States

Guide to Balancing Equations Using Oxidation Numbers

STEP 1 Assign oxidation numbers to all elements.
STEP 2 Identify the oxidized and reduced elements from the changes in oxidation numbers.
STEP 3 Multiply the changes in oxidation numbers by small integrers to equalize increase and decrease.
STEP 4 Balance the remaining elements by inspection.

Study Note

Balance the following equation using the oxidation number method:

$$Cr(s) + HCl(aq) \longrightarrow CrCl_3(aq) + H_2(g)$$

1. Assign oxidation numbers to all elements.

$$\underset{0}{Cr(s)} + \underset{+1-1}{HCl(aq)} \longrightarrow \underset{+3-1}{CrCl_3(aq)} + \underset{0}{H_2(g)}$$

2. Identify the oxidized and reduced elements from the changes in oxidation numbers.

Oxidation of Cr

$$\underset{0}{Cr(s)} + \underset{+1-1}{2HCl(aq)} \longrightarrow \underset{+3-1}{CrCl_3(aq)} + \underset{0}{H_2(g)}$$

Reduction of H

3. Multiply the changes by small integers to equalize the increase and decrease in oxidation numbers. The Cr oxidation loses $3\,e^-(\times 2 = 6\,e^-)$. The 2H reduction is multiplied by $3(= 6\,e^-)$.

$$2Cr(s) + 6HCl(aq) \longrightarrow 2CrCl_3(aq) + 3H_2(g)$$

◆ Learning Exercise 15.2A

Assign oxidation numbers to the elements in each of the following:

a. PbO **b.** Fe **c.** NO_2 **d.** $CuCl_2$

e. H_2CO_3 **f.** HNO_3 **g.** K_3PO_3 **h.** $Cr_2O_7^{2-}$

Answers

a. PbO
+2−2

b. Fe
0

c. NO_2
+4−2
$N + 2(-2) = 0$
$N = +4$

d. $CuCl_2$
+2−1
$Cu + 2(-1) = 0$
$Cu = +2$

e. H_2CO_3
+1+4−2
$2(+1) + C + 3(-2) = 0$
$C = -2 + 6 = +4$

f. HNO_3
+1 +5 −2
$+1 + N + 3(-2) = 0$
$N = -1 + 6 = +5$

g. K_3PO_3
+1+3−2
$3(+1) + P + 3(-2)$
$P = -3 + 6 = +3$

h. $Cr_2O_7^{2-}$
+6 −2
$2Cr + 7(-2) = -2$
$2Cr = -2 + 14 = +12$
$Cr = +6$

◆ Learning Exercise 15.2B

Complete each of the following statements about oxidation–reduction reactions by using the term *loses* or *gains*:

a. In reduction, an element ______ electrons.

b. The oxidizing agent ______ electrons.

c. The element that ______ electrons is oxidized.

d. The reducing agent is the element that ______ electrons.

e. The oxidation number increases when an element ______ electrons.

f. The oxidizing agent is the substance that ______ electrons.

g. The oxidation number decreases when an element ______ electrons.

h. In oxidation, an element ______ electrons.

Answers				
	a. gains	**b.** gains	**c.** loses	**d.** loses
	e. loses	**f.** gains	**g.** gains	**h.** loses

Study Note

The oxidizing agent gains electrons from the oxidation. The reducing agent provides electrons for the reduction.

◆ Learning Exercise 15.2C

For each of the following, assign oxidation numbers, identify the elements in the reactants that are oxidized and reduced, and identify the reactant that is the oxidizing agent and the reactant that is the reducing agent.

a. $PbO(s) + CO(g) \longrightarrow Pb(s) + CO_2(g)$

substance oxidized ______ substance reduced ______

reducing agent ______ oxidizing agent ______

b. $Fe_2O_3(s) + 3C(s) \longrightarrow 2Fe(s) + 3CO(g)$

substance oxidized ______ substance reduced ______

reducing agent ______ oxidizing agent ______

c. $Cu(s) + 4HNO_3(aq) \longrightarrow Cu(NO_3)_2(aq) + 2NO_2(g) + 2H_2O(l)$

substance oxidized ______ substance reduced ______

reducing agent ______ oxidizing agent ______

Answers

a. $PbO(s) + CO(g) \longrightarrow Pb(s) + CO_2(g)$
+2−2 +2−2 0 +4−2

element oxidized C(+2 ⟶ +4) element reduced Pb(+2 ⟶ 0)
reducing agent CO oxidizing agent PbO

b. $Fe_2O_3(s) + 3C(s) \longrightarrow 2Fe(s) + 3CO(g)$
+3 −2 0 0 +2 −2

element oxidized C(0 ⟶ +2) element reduced Fe(+3 ⟶ 0)
reducing agent C oxidizing agent Fe_2O_3

c. $Cu(s) + 4HNO_3(aq) \longrightarrow Cu(NO_3)_2(aq) + 2NO_2(g) + 2H_2O(l)$
0 +1+5−2 +2 +5−2 +4−2 +1 −2

element oxidized Cu(0 ⟶ +2) element reduced N(+5 ⟶ +4)
reducing agent Cu oxidizing agent HNO_3

◆ Learning Exercise 15.2D

Use oxidation numbers to balance each of the following equations by placing appropriate coefficients in front of the formulas:

a. ____ $MgO(s) \longrightarrow$ ____ $Mg(s) +$ ____ $O_2(g)$

b. ____ $Zn(s) +$ ____ $HCl(aq) \longrightarrow$ ____ $ZnCl_2(aq) +$ ____ $H_2(g)$

c. ____ $Al(s) +$ ____ $CuSO_4(aq) \longrightarrow$ ____ $Cu(s) +$ ____ $Al_2(SO_4)_3(aq)$

d. ____ $CO(g) +$ ____ $Fe_2O_3(s) \longrightarrow$ ____ $Fe(s) +$ ____ $CO_2(g)$

e. ____ $K(s) +$ ____ $H_2O(l) \longrightarrow$ ____ $K_2O(aq) +$ ____ $H_2(g)$

Answers

a. $2MgO(s) \longrightarrow 2Mg(s) + O_2(g)$ — O loses 2 e^-; Mg gains 2 e^-
+2 −2 0 0

b. $Zn(s) + 2HCl(aq) \longrightarrow ZnCl_2(aq) + H_2(g)$ — Zn loses 2 e^-; 2H gains 2 e^-
0 +1 −1 +2 −1 0

c. $2Al(s) + 3CuSO_4(aq) \longrightarrow 3Cu(s) + Al_2(SO_4)_3(aq)$ — Al loses 3 e^- (× 2); Cu gains 2 e^- (× 3)
0 +2 +6−2 0 +3 +6−2

d. $3CO(g) + Fe_2O_3(s) \longrightarrow 2Fe(s) + 3CO_2(g)$ — C loses 2 e^- (× 3); Fe gains 3 e^- (× 2)
+2 −2 +3 −2 0 +4 −2

e. $2K(s) + H_2O(l) \longrightarrow K_2O(aq) + H_2(g)$ — K loses 1 e^- (× 2); 2H gains 1 e^- (× 2)
0 +1−2 +1−2 0

15.3 Balancing Oxidation–Reduction Equations Using Half-Reactions

- The half-reaction method of balancing oxidation–reduction equations is typically used to balance oxidation–reduction equations that are written in the ionic form.
- Using the half-reaction method for balancing redox equations, the elements are balanced, O is balanced by adding H_2O, and H is balanced by adding H^+. Then charge is balanced by adding electrons, and each half-reaction is multiplied by an integer to equalize electron loss and gain, and added together.

MasteringChemistry

Tutorial: Balancing Redox Reactions

Guide to Balancing Redox Equations Using Half-Reactions

STEP 1 Balance elements other than H and O in each half-reaction. In acidic solution, add H_2O to the side that needs O and add H^+ to the side that needs H.

STEP 2 Balance each half-reaction for charge by adding electrons to the side with more positive charge.

STEP 3 Multiply each half-reaction by factors that equalize the loss and gain of electrons.

STEP 4 Add half-reactions, cancel electrons, and combine H_2O and H^+. Check balance of atoms and charge.

Study Note

Using the Half-Reaction Method to Balance Oxidation–Reduction Equations

Consider the following unbalanced oxidation–reduction equation:

$$I^-(aq) + SO_4^{2-}(aq) \longrightarrow H_2S(g) + I_2(s)$$

1. **Balance elements other than H and O in each half-reaction.**

$$2I^-(aq) \longrightarrow I_2(s) \qquad SO_4^{2-}(aq) \longrightarrow H_2S(g)$$

In acidic solution, add H_2O to the side that needs O and add H^+ to the side that needs H.

$$SO_4^{2-}(aq) \longrightarrow H_2S(g) + \mathbf{4H_2O}(l) \qquad SO_4^{2-}(aq) + \mathbf{10H^+}(aq) \longrightarrow H_2S(g) + 4H_2O(l)$$

2. **Balance each half-reaction for charge by adding electrons to the side with more positive charge.**

$$2I^-(aq) \longrightarrow I_2(s) + \mathbf{2\,e^-} \qquad SO_4^{2-}(aq) + 10H^+(aq) + \mathbf{8\,e^-} \longrightarrow H_2S(g) + 4H_2O(l)$$

3. **Multiply each half-reaction by factors that equalize the loss and gain of electrons.**

$$[2I^-(aq) \longrightarrow I_2(s) + 2\,e^-] \times 4 \quad (= 8\,e^- \text{ loss})$$

$$8I^-(aq) \longrightarrow 4I_2(g) + \mathbf{8\,e^-} \qquad SO_4^{2-}(aq) + 10H^+(aq) + \mathbf{8\,e^-} \longrightarrow H_2S(g) + 4H_2O(l)$$

4. **Add half-reactions, cancel electrons, and combine H_2O and H^+. Check balance of atoms and charge.**

$$8I^-(aq) + SO_4^{2-}(aq) + 10H^+(aq) + \cancel{8e^-} \longrightarrow 4I_2(s) + H_2S(g) + 4H_2O(l) + \cancel{8e^-}$$

$$8I^-(aq) + SO_4^{2-}(aq) + 10H^+(aq) \longrightarrow 4I_2(s) + H_2S(g) + 4H_2O(l) \text{ balanced}$$

◆ Learning Exercise 15.3B

Use half-reactions to balance the following oxidation–reduction equations:

a. $NO_3^-(aq) + S(s) \longrightarrow NO_2(g) + H_2SO_4(aq)$

b. $MnO_4^-(aq) + Cl^-(aq) \longrightarrow Mn^{2+}(aq) + Cl_2(g)$

c. $Cr_2O_7^{2-}(aq) + Fe^{2+}(aq) \longrightarrow Cr^{3+}(aq) + Fe^{3+}(aq)$

d. $NO_2(g) + ClO^-(aq) \longrightarrow NO_3^-(aq) + Cl^-(aq)$

e. $BrO_3^-(aq) + MnO_2(s) \longrightarrow Br^-(aq) + MnO_4^-(aq)$

Answers

a. $6NO_3^-(aq) + S(s) + 6H^+(aq) \longrightarrow 6NO_2(g) + H_2SO_4(aq) + 2H_2O(l)$

b. $2MnO_4^-(aq) + 10Cl^-(aq) + 16H^+(aq) \longrightarrow 2Mn^{2+}(aq) + 5Cl_2(g) + 8H_2O(l)$

c. $Cr_2O_7^{2-}(aq) + 6Fe^{2+}(aq) + 14H^+(aq) \longrightarrow 2Cr^{3+}(aq) + 6Fe^{3+}(aq) + 7H_2O(l)$

d. $2NO_2(g) + ClO^-(aq) + H_2O(l) \longrightarrow 2NO_3^-(aq) + Cl^-(aq) + 2H^+(aq)$

e. $BrO_3^-(aq) + 2MnO_2(s) + H_2O(l) \longrightarrow Br^-(aq) + 2MnO_4^-(aq) + 2H^+(aq)$

◆ Learning Exercise 15.3C

1. Oxalic acid, $H_2C_2O_4$, present in plants such as spinach, reacts with permanganate, MnO_4^-,

$H_2C_2O_4(aq) + MnO_4^-(aq) \longrightarrow Mn^{2+}(aq) + CO_2(g)$

a. Balance the equation in acidic solution.

b. If 25.0 mL of a 0.0200 M MnO_4^- solution (from $KMnO_4$) is required to react with 5.00 mL of a $H_2C_2O_4$ solution, what is the molarity of the $C_2O_4^{2-}$ solution?

2. Iodine reacts with a thiosulfate, $S_2O_3^{2-}$, solution: $I_2(s) + S_2O_3^{2-}(aq) \longrightarrow I^-(aq) + S_4O_6^{2-}(aq)$

a. Balance the equation in acidic solution.

b. If 37.6 mL of 0.250 M $S_2O_3^{2-}$ solution is required to completely react with I_2, how many grams of I_2 are present in the sample?

Answers

1. **a.** $5H_2C_2O_4(aq) + 6H^+(aq) + 2MnO_4^-(aq) \longrightarrow 2Mn^{2+}(aq) + 10CO_2(g) + 8H_2O(l)$

b. $$0.0250\ \cancel{L} \times \frac{0.0200\ \cancel{mol\ MnO_4^-}}{1\ \cancel{L}} \times \frac{5\ mol\ H_2C_2O_4}{2\ \cancel{mol\ MnO_4^-}} \times \frac{1}{0.00500\ L} = 0.250\ M$$

2. **a.** $I_2(s) + 2S_2O_3^{2-}(aq) \longrightarrow 2I^-(aq) + S_4O_6^{2-}(aq)$

b. $$0.0376\ \cancel{L} \times \frac{0.0250\ \cancel{mol\ S_2O_3^{2-}}}{1\ \cancel{L}} \times \frac{1\ \cancel{mol\ I_2}}{2\ \cancel{mol\ S_2O_3^{2-}}} \times \frac{253.8\ g\ I_2}{1\ \cancel{mol\ I_2}} = 1.19\ g\ of\ I_2$$

15.4 Oxidation of Alcohols

- For organic and biochemical compounds, oxidation typically involves the addition of oxygen or the loss of hydrogen. Reduction can be the loss of oxygen or the gain of hydrogen.
- In organic chemistry, oxidation occurs when there is an increase in the number of carbon to oxygen bonds and reduction when there is a decrease in the number of carbon to oxygen bonds.
- Alcohols are classified as primary, secondary, or tertiary. A primary alcohol has one carbon group attached to the carbon bonded to the hydroxyl group (—OH). Methanol, CH_3OH, is considered a primary alcohol. A secondary alcohol has two carbon groups attached to the carbon atom bonded to the hydroxyl group (—OH). A tertiary alcohol has three carbon groups attached to the carbon atom bonded to the hydroxyl group (—OH).
- Primary (1°) alcohols oxidize to aldehydes, and secondary (2°) alcohols oxidize to ketones, but tertiary (3°) alcohols do not oxidize. To review aldehydes and ketones, see Chapter 8, Section 8.4.

Study Note

Example: Identify the following as primary, secondary, or tertiary alcohols.

Solution: Determine the number of carbon groups attached to the carbon atom attached to the hydroxyl ($-OH$) group.

$$CH_3-CH_2-OH \qquad CH_3-\overset{\displaystyle CH_3}{\underset{\displaystyle |}{C}}H-OH \qquad CH_3-\overset{\displaystyle CH_3}{\underset{\displaystyle \underset{CH_3}{|}}{\overset{|}{C}}}-OH$$

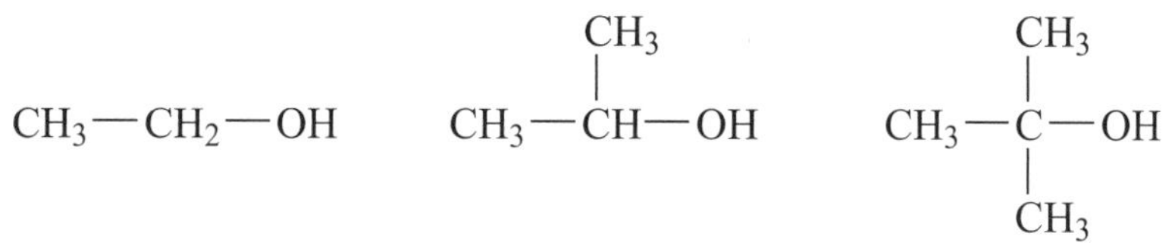

Primary (1°) Secondary (2°) Tertiary (3°)

◆ Learning Exercise 15.4A

Match the following terms with the statements shown:

a. primary alcohol **b.** secondary alcohol **c.** tertiary alcohol
d. aldehyde **e.** ketone

1. _____ an organic compound with one alkyl group bonded to the carbon with the $-OH$ group

2. _____ an alcohol that oxidizes to a ketone

3. _____ the oxidation product of a secondary alcohol

4. _____ an organic compound with three alkyl groups bonded to the carbon with the $-OH$ group

5. _____ the oxidation product of a primary alcohol

Answers **1.** a **2.** b **3.** e **4.** c **5.** d

◆ Learning Exercise 15.4B

Classify each of the following alcohols as primary (1°), secondary (2°), or tertiary (3°):

1. CH_3-CH_2-OH _____

2. $CH_3-CH_2-\overset{\displaystyle OH}{\underset{\displaystyle |}{C}}H-CH_3$ _____

3. $CH_3-\overset{\displaystyle OH}{\underset{\displaystyle \underset{CH_3}{|}}{\overset{|}{C}}}-CH_2-CH_3$ _____

4. $CH_3-\overset{\displaystyle CH_3}{\underset{\displaystyle \underset{CH_3}{|}}{\overset{|}{C}}}-CH_2-CH_2-OH$ _____

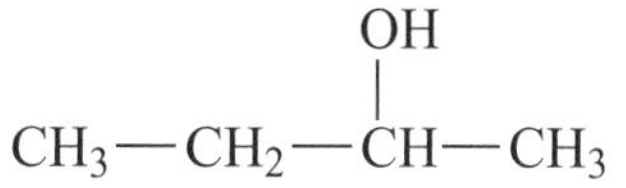

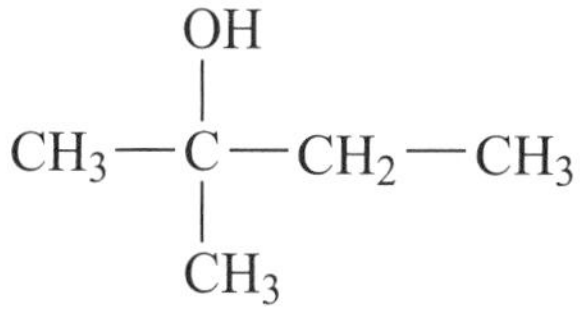

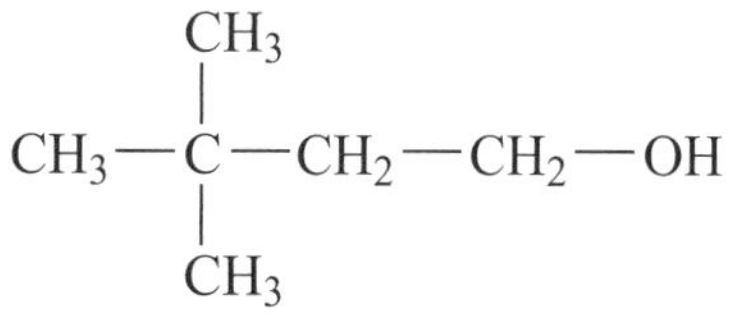

Answers **1.** primary (1°) **2.** secondary (2°) **3.** tertiary (3°) **4.** primary (1°)

◆ **Learning Exercise 15.3C**

Draw the condensed structural formulas of the products expected in the oxidation reaction of each of the following reactants:

1. $CH_3—CH_2—CH_2—CH_2—OH \xrightarrow{[O]}$

2. $CH_3—\overset{\overset{\displaystyle OH}{|}}{CH}—CH_3 \xrightarrow{[O]}$

3. $CH_3—CH_2—\overset{\overset{\displaystyle CH_3}{|}}{CH}—CH_2—OH \xrightarrow{[O]}$

4. $CH_3—\overset{\overset{\displaystyle OH}{|}}{\underset{\underset{\displaystyle CH_3}{|}}{C}}—CH_2—CH_3 \xrightarrow{[O]}$

Answers

1. $CH_3—CH_2—CH_2—\overset{\overset{\displaystyle O}{\|}}{C}—H$

2. $CH_3—\overset{\overset{\displaystyle O}{\|}}{C}—CH_3$

3. $CH_3—CH_2—\overset{\overset{\displaystyle CH_3}{|}}{CH}—\overset{\overset{\displaystyle O}{\|}}{C}—H$

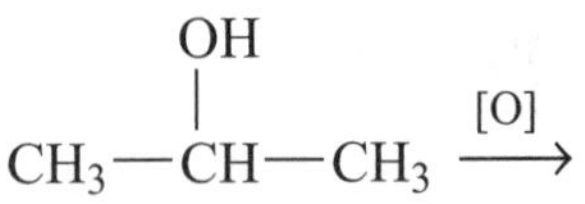

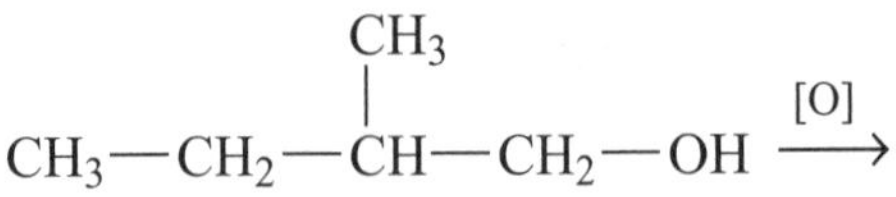

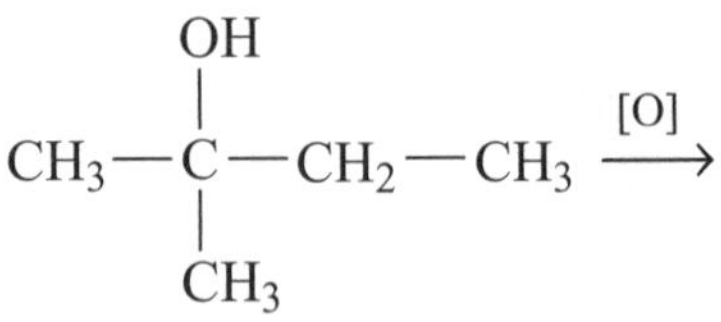

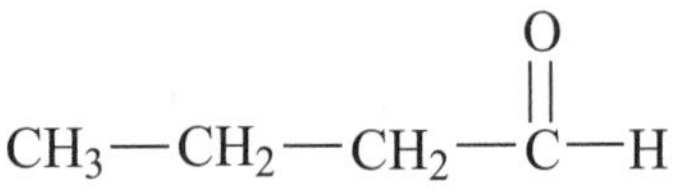

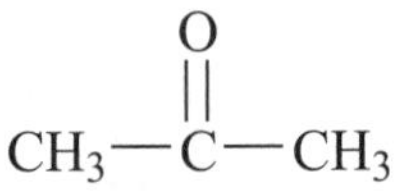

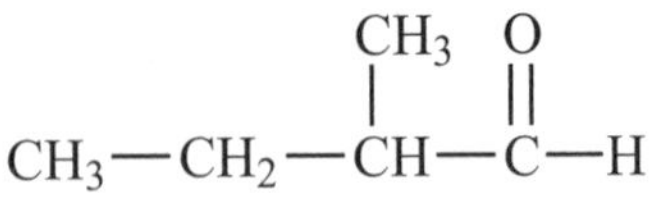

4. No reaction occurs with a tertiary (3°) alcohol.

15.5 Electrical Energy from Oxidation–Reduction Reactions

- In an electrochemical cell, the half-reactions of an oxidation–reduction reaction are physically separated, so the electrons flow through an external circuit.

- In a voltaic cell, the flow of electrons through an external circuit generates electrical energy.

- Oxidation takes place at the anode; reduction takes place at the cathode.

- Electrons flow from the anode to the cathode.

- Corrosion is an oxidation–reduction process and can be controlled using principles of electrochemical cells.

- Batteries are examples of electrochemical cells that provide electrical energy.

Study Note

We can diagram the cell using a shorthand notation as follows:

$$Zn(s) \mid Zn^{2+}(aq) \parallel Cu^{2+}(aq) \mid Cu(s)$$

The components of the oxidation half-cell (anode) are written on the left side in this shorthand notation, and the components of the reduction half-cell (cathode) are written on the right. A single vertical line separates the solid Zn anode from the ionic Zn^{2+} solution and the Cu^{2+} solution from the Cu cathode.

◆ Learning Exercise 15.5A

For each of the following voltaic cells, diagram the cell, write the cell notation, write the half-reactions at the anode and cathode and the balanced equation, and indicate the direction of electron flow.

a. Mg anode in a solution containing Mg^{2+}, and a Ag cathode in a solution containing Ag^+

b. Cr cathode in a solution containing Cr^{3+}, and an Al anode in a solution containing Al^{3+}

c. Fe anode in a solution containing Fe^{2+}, and a Ni cathode in a solution containing Ni^{2+}

Answers

a.

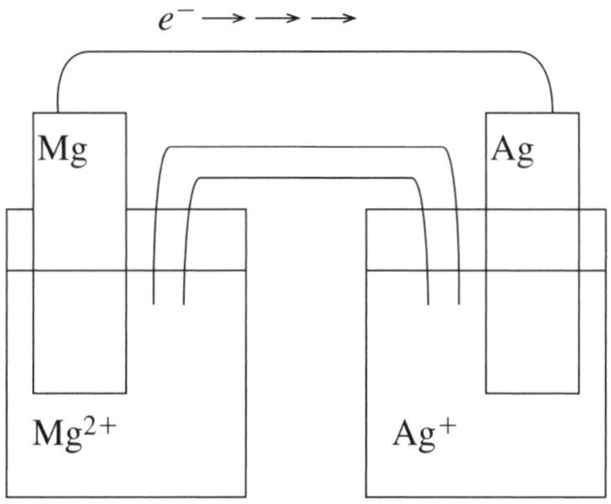

$Mg(s)\,|\,Mg^{2+}(aq)\,\|\,Ag^{+}(aq)\,|\,Ag(s)$

Anode: $Mg(s) \longrightarrow Mg^{2+}(aq) + 2\,e^-$

Cathode: $[Ag^{+}(aq) + e^- \longrightarrow Ag(s)] \times 2$

$Mg(s) + 2Ag^{+}(aq) \longrightarrow Mg^{2+}(aq) + 2Ag(s)$

b.

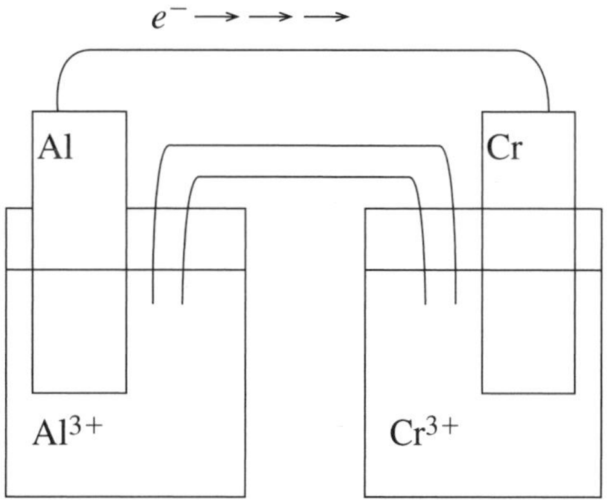

$Al(s)\,|\,Al^{3+}(aq)\,\|\,Cr^{3+}(aq)\,|\,Cr(s)$

Anode: $Al(s) \longrightarrow Al^{3+}(aq) + 3\,e^-$

Cathode: $Cr^{3+}(aq) + 3\,e^- \longrightarrow Cr(s)$

$Cr^{3+}(aq) + Al(s) \longrightarrow Cr(s) + Al^{3+}(aq)$

c.

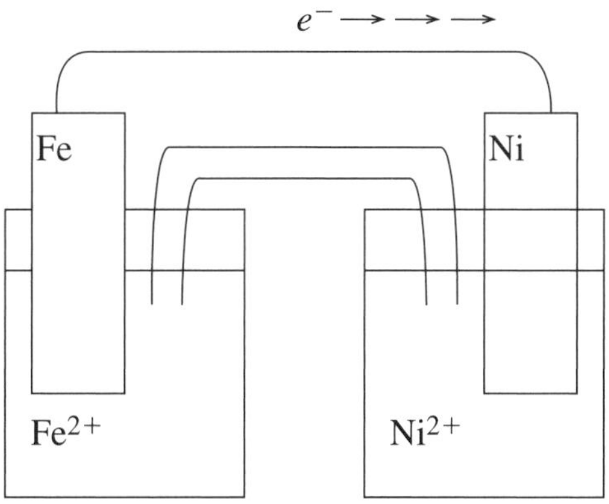

$Fe(s)\,|\,Fe^{2+}(aq)\,\|\,Ni^{2+}(aq)\,|\,Ni(s)$

Anode: $Fe(s) \longrightarrow Fe^{2+}(aq) + 2\,e^-$

Cathode: $Ni^{2+}(aq) + 2\,e^- \longrightarrow Ni(s)$

$Fe(s) + Ni^{2+}(aq) \longrightarrow Fe^{2+}(aq) + Ni(s)$

◆ Learning Exercise 15.5B

Write the half-reactions for the anode and cathode and the overall cell reaction from the following cell notations:

a. $Sr(s)\,|\,Sr^{2+}(aq)\,\|\,Sn^{4+}(aq), Sn^{2+}(aq)\,|\,Pt(s)$

b. $Cr(s)\,|\,Cr^{3+}(aq)\,\|\,Ni^{2+}(aq)\,|\,Ni(s)$

c. $Ag(s)\,|\,Ag^{+}(aq)\,\|\,Au^{3+}(aq)\,|\,Au(s)$

Answers

a. Anode: $Sr(s) \longrightarrow Sr^{2+}(aq) + 2\,e^-$

Cathode: $Sn^{4+}(aq) + 2\,e^- \longrightarrow Sn^{2+}(aq)$

$Sr(s) + Sn^{4+}(aq) \longrightarrow Sr^{2+}(aq) + Sn^{2+}(aq)$

b. Anode: $[Cr(s) \longrightarrow Cr^{3+}(aq) + 3\,e^-] \times 2$

Cathode: $[Ni^{2+}(aq) + 2\,e^- \longrightarrow Ni(s)] \times 3$

$2Cr(s) + 3Ni^{2+}(aq) \longrightarrow 2Cr^{2+}(aq) + 3Ni(s)$

c. Anode: $[Ag(s) \longrightarrow Ag^{+}(aq) + e^-] \times 3$

Cathode: $[Au^{3+}(aq) + 3\,e^- \longrightarrow Au(s)] \times 1$

$3Ag(s) + Au^{3+}(aq) \longrightarrow 3Ag^{+}(aq) + Au(s)$

15.6 Oxidation–Reduction Reactions That Require Electrical Energy

- The activity series arranges elements in order of their ability to oxidize spontaneously when combined with the ions of any metal lower on the list.
- An oxidation–reduction reaction is not spontaneous when a less active metal is combined with the ions of a more active metal.
- In an electrolytic cell, a process called electrolysis uses electrical energy is used to drive a nonspontaneous reaction.
- Electrolytic cells are used in the electrolysis of molten salts and electroplating.

◆ Learning Exercise 15.6A

Use the activity series to identify the more active metal in each of the following pairs:

a. Fe(s) or Pb(s) **b.** Ag(s) or Na(s) **c.** Zn(s) or Ca(s)

d. Al(s) or Pb(s) **e.** Fe(s) or Ni(s)

Answers **a.** Fe(s) **b.** Na(s) **c.** Ca(s) **d.** Al(s) **e.** Fe(s)

◆ Learning Exercise 15.6B

Using the activity series, predict whether each of the following reactions occurs spontaneously or not, and explain:

a. $Cu(s) + Sn^{2+}(aq) \longrightarrow Cu^{2+}(aq) + Sn(s)$

b. $Pb^{2+}(aq) + Mg(s) \longrightarrow Mg^{2+}(aq) + Pb(s)$

c. $2Ag(s) + Ni^{2+}(aq) \longrightarrow Ni(s) + 2Ag^{+}(aq)$

d. $Zn(s) + 2H^{+}(aq) \longrightarrow Zn^{2+}(aq) + H_2(g)$

e. $2Cr(s) + 3Ca^{2+}(aq) \longrightarrow 2Cr^{3+}(aq) + 3Ca(s)$

Answers

a. Not spontaneous: $Cu(s)$ is less active than $Sn(s)$.
b. Spontaneous: $Mg(s)$ is more active than $Pb(s)$.
c. Not spontaneous: $Ag(s)$ is less active than $Ni(s)$.
d. Spontaneous: $Zn(s)$ is more active than $H_2(g)$.
e. Not spontaneous: $Cr(s)$ is less active than $Ca(s)$.

◆ Learning Exercise 15.8C

An electrolytic cell is used to electroplate nickel on the surface an iron object. If the electrolytic cell uses a nickel bar as the anode, and the iron object as the cathode, indicate the following:

a. What happens at the anode when electrical energy flows through the cell?

b. What is the equation for the half-reaction that occurs at the anode?

c. What happens at the cathode when electrical energy flows through the cell?

d. What is the equation for the half-reaction that occurs at the cathode?

Answers

a. At the anode, the nickel metal in the bar is oxidized to nickel(II) ions to make a Ni^{2+} solution.
b. $Ni(s) \longrightarrow Ni^{2+}(aq) + 2\,e^{-}$
c. At the cathode, the nickel(II) ions in the solution are reduced to nickel metal that plates on the surface of the iron object.
d. $Ni^{2+}(aq) + 2\,e^{-} \longrightarrow Ni(s)$

Checklist for Chapter 15

You are ready to take the Practice Test for Chapter 15. Be sure you have accomplished the following learning goals for this chapter. If you are not sure, review the section listed at the end of the goal. Then apply your new skills and understanding to the Practice Test.

After studying Chapter 15, I can successfully:

_____ Identify what is oxidized and reduced in an oxidation–reduction reaction (15.1).

_____ Assign oxidation numbers to the atoms in an equation (15.2).

_____ Use oxidation numbers to determine the oxidized element, reduced element, oxidizing agent, and reducing agent (15.2).

_____ Balance an oxidation–reduction equation using oxidation numbers (15.3).

_____ Balance an oxidation–reduction equation using half-reactions (15.3).

_____ Classify an alcohol as primary, secondary, or tertiary (15.4).

_____ Draw the condensed structural formulas of the products formed by the oxidation of an alcohol (15.4).

_____ Write the half-reactions for the anode and the cathode of a voltaic cell (15.5).

_____ Write the shorthand cell notation for the half-reactions for the anode and the cathode of a voltaic cell (15.5).

_____ Draw the cell diagram for a voltaic cell (15.5).

_____ Write the half-cell reactions for an electrolytic cell (15.6).

_____ Use the activity series for metals to determine whether an oxidation–reduction reaction is spontaneous (15.6).

_____ Describe the process taking place in electroplating (15.6).

Practice Test for Chapter 15

Identify whether the element in each of the following is oxidized (O) or reduced (R):

1. _________ gains electrons

2. _________ an oxidizing agent

3. _________ loses electrons

4. _________ a reducing agent

5. _________ oxidation number increases

For questions 6 through 10, indicate the oxidation number of each of the elements:

A. +1 **B.** +2 **C.** +3 **D.** +4 **E.** +5

6. _________ N in HNO_2

7. _________ P in P_2O_5

8. _________ C in HCO_3^-

9. _________ Mn in MnO_2

10. _________ Ca in $CaSO_4$

For questions 11 through 15, indicate the oxidation number of the indicated element as

A. 0 **B.** −1 **C.** −2 **D.** −3 **E.** −4

11. ________ P in Na_3P

12. ________ N in N_2H_4

13. ________ C in CH_4

14. ________ Br in Br_2

15. ________ F in OF_2

For questions 16 through 18, use the following unbalanced equation:

$NO(g) + Br_2(g) \longrightarrow NOBr(g)$

16. The oxidation numbers of N in the reactants and products are
A. +1, +2 **B.** +2, +3 **C.** +3, +2 **D.** −2, −3 **E.** +2, −3

17. The oxidation numbers of Br in the reactants and products are
A. 0, 0 **B.** −1, 0 **C.** 0, +1 **D.** −1, −1 **E.** 0, −1

18. The oxidation numbers of O in the reactants and products are
A. 0, 0 **B.** −2, 0 **C.** +3, +2 **D.** −2, −2 **E.** 0, −2

19. The change in the oxidation number of N is
A. a decrease of 1 **B.** a decrease of 2 **C.** an increase of 1
D. an increase of 2 **E.** no change

20. The balanced equation is
A. $NO(g) + Br_2(g) \longrightarrow NOBr(g)$
B. $NO(g) + Br_2(g) \longrightarrow 2NOBr(g)$
C. $2NO(g) + Br_2(g) \longrightarrow NOBr(g)$
D. $2NO(g) + Br_2(g) \longrightarrow 2NOBr(g)$
E. $NO(g) + 2Br_2(g) \longrightarrow 2NOBr(g)$

For questions 21 through 23, consider the following half-reaction in acidic solution:

$ClO_2^- \longrightarrow Cl^-$

21. What is the number of H_2O in the balanced half-reaction?
A. 1 **B.** 2 **C.** 3 **D.** 4 **E.** 5

22. What is the number of H^+ in the balanced half-reaction?
A. 1 **B.** 2 **C.** 3 **D.** 4 **E.** 5

23. How many electrons are in the balanced half-reaction?
A. 1 **B.** 2 **C.** 3 **D.** 4 **E.** 5

For questions 24 through 26, consider the following half-reaction in acidic solution:

$Cr_2O_7^{2-} \longrightarrow Cr^{3+}$

24. What is the number of H_2O in the balanced half-reaction?
A. 2 **B.** 4 **C.** 5 **D.** 7 **E.** 14

25. What is the number of H^+ in the balanced half-reaction?
A. 2 **B.** 6 **C.** 7 **D.** 12 **E.** 14

26. How many electrons are in the balanced half-reaction?
A. 2 **B.** 4 **C.** 6 **D.** 10 **E.** 14

27. Which equation has the following cell notation?
$Al(s) \mid Al^{3+}(aq) \parallel Fe^{3+}(aq) \mid Fe(s)$?
A. $Al(s) + Fe^{3+}(aq) \longrightarrow Al^{3+}(aq) + Fe(s)$
B. $Al^{3+}(aq) + Fe(s) \longrightarrow Al(s) + Fe^{3+}(aq)$
C. $Al^{3+}(aq) + Fe(s) \longrightarrow Al(s) + Fe^{3+}(aq)$
D. $Al^{3+}(aq) + Fe^{3+}(aq) \longrightarrow Fe(s) + Al(s)$
E. $Al(s) + Fe(s) \longrightarrow Fe^{3+}(aq) + Al^{3+}(aq)$

For questions 28 through 30, consider the following metals and ions in order from most to least active:
$Mg(s) \longrightarrow Mg^{2+}(aq) + 2\,e^-$
$Cr(s) \longrightarrow Cr^{3+}(aq) + 3\,e^-$
$Pb(s) \longrightarrow Pb^{2+}(aq) + 2\,e^-$
$Ag(s) \longrightarrow Ag^{+}(aq) + e^-$

28. Which of the following reactions is spontaneous?
A. $3Pb(s) + 2Cr^{3+}(aq) \longrightarrow 3Pb^{2+}(aq) + 2Cr(s)$
B. $Mg(s) + 2Ag^{+}(aq) \longrightarrow Mg^{2+}(aq) + 2Ag(s)$
C. $Fe^{2+}(aq) + Cu(s) \longrightarrow Fe(s) + Cu^{2+}(aq)$
D. $2Ag(s) + Pb^{2+}(aq) \longrightarrow Pb(s) + 2Ag^{+}(aq)$
E. $3Ag(s) + Cr^{3+}(aq) \longrightarrow 3Ag^{+}(aq) + Cr(s)$

29. Which of the following reactions requires an external source of energy?
A. $3Mg(s) + 2Cr^{3+}(aq) \longrightarrow 2Cr(s) + 3Mg^{2+}(aq)$
B. $Pb(s) + 2Ag^{+}(aq) \longrightarrow Pb^{2+}(aq) + 2Ag(s)$
C. $Mg(s) + Pb^{2+}(aq) \longrightarrow Mg^{2+}(aq) + Pb(s)$
D. $Mg(s) + 2Ag^{+}(aq) \longrightarrow Mg^{2+}(aq) + 2Ag(s)$
E. $3Pb(s) + 2Cr^{3+}(aq) \longrightarrow 3Pb^{2+}(aq) + 2Cr(s)$

30. In the electroplating of a bowl, an aluminum anode is placed in an Al^{3+} solution, $Al(NO_3)_3$. The bowl to be aluminum-plated is the cathode. The anode and cathode are wired to a battery. The reaction at the anode is
A. $Al^{3+}(aq) + 3\,e^- \longrightarrow Al(s)$
B. $Al^{3+}(aq) \longrightarrow Al(s) + 3\,e^-$
C. $Al(s) + 3\,e^- \longrightarrow Al^{3+}(aq)$
D. $Al(s) \longrightarrow Al^{3+}(aq) + 3\,e^-$
E. $Al(s) + Al^{3+}(aq) \longrightarrow 2Al(s)$

Answers to the Practice Test

1. R	**2.** R	**3.** O	**4.** O	**5.** O
6. C	**7.** E	**8.** D	**9.** D	**10.** B
11. D	**12.** C	**13.** E	**14.** A	**15.** B
16. B	**17.** E	**18.** D	**19.** C	**20.** D
21. B	**22.** D	**23.** D	**24.** D	**25.** E
26. C	**27.** A	**28.** B	**29.** E	**30.** D

16

Nuclear Radiation

Study Goals

- Describe alpha, beta, positron, and gamma radiation.
- Describe the methods required for proper shielding for each type of radiation.
- Write an equation showing the mass numbers and atomic numbers for an atom that undergoes radioactive decay.
- Describe the detection and measurement of radiation.
- Describe the use of radioisotopes in nuclear medicine.
- Calculate the amount of radioisotope that remains after a given number of half-lives.
- Describe nuclear fission and fusion.

Think About It

1. What is nuclear radiation?

2. Why do you receive more radiation if you live in the mountains or travel on an airplane?

3. In nuclear medicine, iodine-125 is used for detecting a tumor in the thyroid. What does the number 125 indicate?

4. How does nuclear fission differ from nuclear fusion?

5. Why is there a concern about radon in our homes?

Key Terms

Match each of the following key terms with the correct definition:

a. radiation **b.** half-life **c.** curie **d.** nuclear fission
e. alpha particle **f.** positron **g.** rem

1. _____ a particle identical to a helium nucleus produced in a radioactive nucleus

2. _____ the time required for one-half of a radioactive sample to undergo radioactive decay

3. _____ a unit of radiation measurement equal to 3.7×10^{10} disintegrations per second

4. _____ a process in which large nuclei split into smaller nuclei with the release of energy

5. _____ energy or particles released by radioactive atoms

6. _____ a measure of biological damage caused by radiation

7. _____ produced when a proton is transformed into a neutron

Answers **1.** e **2.** b **3.** c **4.** d **5.** a **6.** g **7.** f

16.1 Natural Radioactivity

- Radioactive isotopes have unstable nuclei that break down (decay), spontaneously emitting alpha (α), beta (β), positron (β^+), or gamma (γ) radiation.
- An alpha particle is the same as a helium nucleus; it contains 2 protons and 2 neutrons.
- A beta particle is a high-energy electron and a positron is a high-energy positive particle. A gamma ray is high-energy radiation.
- Because radiation can damage cells in the body, proper protection must be used: shielding, time limitation, and distance.

MasteringChemistry

Self Study Activity: Nuclear Chemistry
Self Study Activity: Radiation and Its Biological Effects
Tutorial: Types of Radiation

Study Note

It is important to learn the symbols for the radiation particles in order to describe the different types of radiation:

$^{1}_{1}H$ or p	$^{1}_{0}n$ or n	$^{0}_{-1}e$ or β	$^{4}_{2}He$ or α	$^{0}_{+1}e$ or β^+	$^{0}_{0}\gamma$ or γ
proton	neutron	beta particle	alpha particle	positron	gamma ray

◆ Learning Check 16.1A

Match the description in column B with the terms in column A:

	A		B
1.	____ $^{18}_{8}O$	**a.**	symbol for a beta particle
2.	____ γ	**b.**	symbol for an alpha particle
3.	____ β^{+}	**c.**	an atom that emits radiation
4.	____ radioisotope	**d.**	symbol for a positron
5.	____ $^{4}_{2}He$	**e.**	symbol for an atom of oxygen
6.	____ β	**f.**	symbol for gamma radiation

Answers **1.** e **2.** f **3.** d **4.** c **5.** b **6.** a

◆ Learning Check 16.1B

Discuss some things you can do to minimize the amount of radiation received if you work with a radioactive substance. Describe how each method helps to limit the amount of radiation you would receive.

Answer

Three ways to minimize exposure to radiation are: (1) use shielding, (2) keep time short in the radioactive area, and (3) keep as much distance as possible from the radioactive materials. Shielding such as clothing and gloves stops alpha and beta particles from reaching your skin, whereas lead or concrete will absorb gamma rays. Limiting the time spent near radioactive samples reduces exposure time. Increasing the distance from a radioactive source reduces the intensity of radiation. Wearing a film badge will monitor the amount of radiation you receive.

◆ Learning Check 16.1C

What type(s) of radiation (alpha, beta, and/or gamma) would each of the following shielding materials protect you from?

a. heavy clothing ____________ **b.** skin ____________

c. paper ____________ **d.** concrete ____________

e. lead wall ____________

Answers **a.** alpha, beta **b.** alpha **c.** alpha **d.** alpha, beta, gamma **e.** alpha, beta, gamma

16.2 Nuclear Equations

- A balanced nuclear equation represents the changes in the nuclei of radioisotopes.
- The new isotopes and the type of radiation emitted can be determined from the symbols that show the mass numbers and atomic numbers of the isotopes in the nuclear reaction.

$$\text{Radioisotope} \longrightarrow \text{new nucleus} + \text{radiation}$$

Total of the mass numbers is equal

$$^{11}_{6}C \longrightarrow {}^{7}_{4}Be + {}^{4}_{2}He$$

Total of the atomic numbers is equal

- A new radioactive isotope is produced when a nonradioactive isotope is bombarded by a small particle such as a proton, an alpha particle, or beta particle.

$^{10}_{5}B$	+	$^{4}_{2}He$	$\longrightarrow$	$^{13}_{7}N$	+	$^{1}_{0}n$
Stable nucleus		*Bombarding particle (α)*		*New nucleus*		*Neutron emitted*

Guide to Completing a Nuclear Equation

STEP 1 Write the incomplete nuclear equation.
STEP 2 Determine the missing mass number.
STEP 3 Determine the missing atomic number.
STEP 4 Determine the symbol of the new nucleus.
STEP 5 Complete the nuclear equation.

MasteringChemistry
Tutorial: Types of Radioactivity: Writing Decay Equations
Tutorial: Writing Nuclear Equations
Tutorial: Alpha, Beta, and Gamma Emitters

Study Note

When balancing nuclear equations for radioactive decay, be sure that

1. The mass number of the reactant is equal to the sum of the mass numbers of the products.
2. The atomic number of the reactant is equal to the sum of the atomic numbers of the products.

Changes in Mass Number and Atomic Number Due to Radiation

Decay Process	Radiation Symbol	Change in Mass Number	Change in Atomic Number	Change in Neutron Number
Alpha emission	$^{4}_{2}He$	−4	−2	−2
Beta emission	$^{0}_{-1}e$	0	+1	−1
Positron emission	$^{0}_{+1}e$	0	−1	+1
Gamma emission	$^{0}_{0}\gamma$	0	0	0

◆ Learning Check 16.2A

Write a nuclear symbol that completes each of the following nuclear equations:

a. $^{66}_{29}Cu \longrightarrow {}^{66}_{30}Zn + ?$ **a.** ____________

b. $^{127}_{53}I \longrightarrow {}^{1}_{0}n + ?$ **b.** ____________

c. $^{238}_{92}U \longrightarrow {}^{4}_{2}He + ?$ **c.** ____________

d. $^{24}_{11}Na \longrightarrow {}^{0}_{-1}e + ?$ **d.** ____________

e. $? \longrightarrow {}^{30}_{14}Si + {}^{0}_{-1}e$ **e.** ____________

Answers **a.** $^{0}_{-1}e$ **b.** $^{126}_{53}I$ **c.** $^{234}_{90}Th$ **d.** $^{24}_{12}Mg$ **e.** $^{30}_{13}Al$

Study Note

In balancing nuclear transmutation equations (bombardment by small particles),

1. The sum of the mass numbers of the reactants must equal the sum of the mass numbers of the products.
2. The sum of the atomic numbers of the reactants must equal the sum of the atomic numbers of the products.

◆ Learning Check 16.2B

Complete each of the following equations for bombardment reactions:

a. ${}^{40}_{20}\text{Ca} + ? \longrightarrow {}^{40}_{19}\text{K} + {}^{1}_{1}\text{H}$ **a.** ________

b. ${}^{27}_{13}\text{Al} + {}^{1}_{0}n \longrightarrow {}^{24}_{11}\text{Na} + ?$ **b.** ________

c. ${}^{10}_{5}\text{B} + {}^{1}_{0}n \longrightarrow {}^{4}_{2}\text{He} + ?$ **c.** ________

d. ${}^{23}_{11}\text{Na} + ? \longrightarrow {}^{23}_{12}\text{Mg} + {}^{1}_{0}n$ **d.** ________

e. ${}^{197}_{79}\text{Au} + {}^{1}_{1}\text{H} \longrightarrow ? + {}^{1}_{0}n$ **e.** ________

Answers **a.** ${}^{1}_{0}n$ **b.** ${}^{4}_{2}\text{He}$ **c.** ${}^{7}_{3}\text{Li}$ **d.** ${}^{1}_{1}\text{H}$ **e.** ${}^{197}_{80}\text{Hg}$

16.3 Radiation Measurement

- A Geiger counter is used to detect radiation. When radiation passes through the gas in the counter tube, some atoms of gas are ionized, producing an electrical current.
- The activity of a radioactive sample measures the number of nuclear transformations per second. The curie (Ci) is equal to 3.7×10^{10} disintegrations per second. The SI unit is the becquerel (Bq), which is equal to 1 disintegration per second.
- The radiation dose absorbed by a gram of body tissue is measured in units of rads or the SI unit, grays (Gy), which is equal to 100 rad.
- The biological damage of different types of radiation on the body is measured in radiation units of rems or the SI unit, sieverts (Sv), which is equal to 100 rem.

MasteringChemistry

Case Study: Food Irradiation

◆ Learning Check 16.3A

Match each type of measurement unit with the radiation process measured.

a. curie **b.** becquerel **c.** rad **d.** gray **e.** rem

1. _____ an activity of one disintegration per second

2. _____ the amount of radiation absorbed by 1 g of material

3. _____ an activity of 3.7×10^{10} disintegrations per second

4. _____ the biological damage caused by different kinds of radiation

5. _____ a unit of absorbed dose equal to 100 rad

Answers **1.** b **2.** c **3.** a **4.** e **5.** d

◆ Learning Check 16.3B

a. A sample of Ir-192 has an activity of 35 μCi. What is its activity in becquerels?

b. If an absorbed dose is 15 mrem, what is the absorbed dose in sieverts?

Answers **a.** 1.3×10^6 Bq **b.** 1.5×10^{-4} Sv

16.4 Half-Life of a Radioisotope

- The half-life of a radioactive sample is the time required for one-half of the sample to decay (emit radiation).
- Most radioisotopes used in medicine, such as Tc-99m and I-131, have short half-lives.
- Many naturally occurring radioisotopes, such as C-14, Ra-226, and U-238, have long half-lives.

Study Note

In one half-life, one-half the initial quantity of a radioisotope emits radiation. One-half of the sample remains active. In two half-lives, one-fourth of the initial sample remains active.

Example: Chromium-51 has a half-life of 28 d. How much of a 16-μg sample of chromium-51 remains after 84 d?

Solution:

STEP 1 State the given and needed amounts of radioisotope.

Given 16-μg sample of chromium-51; 84 d ; 28 d/1 half-life
Need μg of Cr-51 remaining

STEP 2 Write a plan to calculate amount of active radioisotope.

Plan $42\text{ d} \xrightarrow{\textit{half-life}} \text{Number of half-lives}$

$$16\text{-}\mu\text{g of Cr-51} \xrightarrow{\textit{Number of half-lives}} \mu\text{g of Cr-51 remaining}$$

STEP 3 Write the half-life equality and conversion factors.

$$1 \text{ half-life Cr-51} = 28 \text{ d}$$

$$\frac{28\text{ d}}{1\text{ half-life}} \quad \text{and} \quad \frac{1\text{ half-life}}{28\text{ d}}$$

STEP 4 Set up problem to calculate amount of active radioisotope.

$$\text{Number of half-lives} = 84\,\cancel{\text{d}} \times \frac{1\text{ half-life}}{28\,\cancel{\text{d}}} = 3\text{ half-lives}$$

Now, we can calculate the quantity of Cr-51 that remains active.

$$16\ \mu\text{g} \xrightarrow{1\textit{ half-life}} 8.0\ \mu\text{g} \xrightarrow{1\textit{ half-life}} 4.0\ \mu\text{g} \xrightarrow{1\textit{ half-life}} 2.0\ \mu\text{g}$$

MasteringChemistry

Tutorial: Radioactive Half-Lives
Tutorial: Radiocarbon Dating

◆ Learning Check 16.4

a. Suppose you have an 80. mg sample of iodine-125. If iodine-125 has a half-life of 60 days, how many mg are radioactive

1. after one half-life?

2. after two half-lives?

3. after 240 days?

b. $^{99m}_{43}Tc$ has a half-life of 6 h. If a technician picked up a 16-mg sample at 8 A.M., how much of the radioactive sample remained at 8 P.M. that same day?

c. Phosphorus-32 has a half-life of 14 days. How much of a 240-μg sample will be radioactive after 56 days?

d. Iodine-131 has a half-life of 8.0 days. How many days will it take for 80. mg of I-131 to decay to 5 mg?

e. Suppose a group of archaeologists digs up some pieces of a wooden boat at an ancient site. When a sample of the wood is analyzed for C-14, scientists determine that 12.5%, or 1/8, of the original amount of C-14 remains. If the half-life of carbon-14 is 5730 years, how long ago was the boat made?

Answers **a.** **1.** 40. mg **2.** 20. mg **3.** 5.0 mg **b.** 4.0 mg
c. 15 μg **d.** 32 days **e.** 17 200 y ago

16.5 Medical Applications Using Radioactivity

- Nuclear medicine uses radioactive isotopes that go to specific sites in the body.
- For diagnostic work, radioisotopes are used that emit gamma rays and produce nonradioactive products.
- By detecting the radiation emitted by medical radioisotopes, evaluations can be made about the location and extent of an injury, disease, or tumor, blood flow, or level of function of a particular organ.

◆ Learning Check 16.5

Write the nuclear symbol for each of the following radioactive isotopes:

a. _____ iodine-131 used to study thyroid gland activity

b. _____ phosphorus-32 used to locate brain tumors

c. _____ sodium-24 used to determine blood flow and to locate a blood clot or embolism

d. _____ nitrogen-13 used in positron emission tomography

Answers **a.** $^{131}_{53}I$ **b.** $^{32}_{15}P$ **c.** $^{24}_{11}Na$ **d.** $^{13}_{7}N$

16.6 Nuclear Fission and Fusion

- In fission, a large nucleus breaks apart into smaller pieces, releasing one or more types of radiation and a great amount of energy.
- A chain reaction is a fission reaction that will continue once initiated.
- In fusion, small nuclei combine to form a larger nucleus, which releases great amounts of energy.
- Nuclear fission is currently used to produce electrical power while nuclear fusion applications are still in the experimental stage.

MasteringChemistry

Tutorial: Fission and Fusion
Tutorial: Nuclear Fission and Fusion Reactions

◆ Learning Check 16.6A

Discuss the nuclear processes of fission and fusion for the production of energy.

Answer

Nuclear fission is a splitting of the atom into two or more nuclei accompanied by the release of large amounts of energy and radiation. In the process of *nuclear fusion*, two or more nuclei combine to form a heavier nucleus and release a large amount of energy. However, fusion requires a considerable amount of energy to initiate the process.

◆ Learning Check 16.6B

Balance each of the following nuclear equations and identify each as a fission or fusion reaction:

a. $^{235}_{92}\text{U} + ^{1}_{0}n \longrightarrow ^{143}_{54}\text{Xe} + 3^{1}_{0}n + ?$

b. $^{2}_{1}\text{H} + ^{1}_{1}\text{H} \longrightarrow ?$

c. $^{3}_{1}\text{H} + ? \longrightarrow ^{4}_{2}\text{He} + ^{1}_{0}n$

d. $^{235}_{92}\text{U} + ^{1}_{0}n \longrightarrow ^{91}_{36}\text{Kr} + 3^{1}_{0}n + ?$

Answers **a.** fission, $^{92}_{38}\text{Sr}$ **b.** fusion, $^{3}_{1}\text{H}$ **c.** fusion, $^{2}_{1}\text{H}$ **d.** fission, $^{142}_{56}\text{Ba}$

Checklist for Chapter 16

You are ready to take the Practice Test for Chapter 16. Be sure you have accomplished the following learning goals for this chapter. If you are not sure, review the section listed at the end of the goal. Then apply your new skills and understanding to the Practice Test.

After studying Chapter 16, I can successfully:

_____ Describe alpha, beta, and gamma radiation (16.1).

_____ Write a nuclear equation showing mass numbers and atomic numbers for radioactive decay (16.2).

_____ Write a nuclear equation for the formation of a radioactive isotope (16.2).

_____ Describe the detection and measurement of radiation (16.3).

_____ Calculate the amount of radioisotope remaining after one or more half-lives (16.4).

_____ Describe the use of radioisotopes in medicine (16.5).

_____ Describe the processes of nuclear fission and fusion (16.6).

Practice Test for Chapter 16

1. The correctly written symbol for an atom of sulfur is
A. $^{30}_{16}\text{Su}$ **B.** $^{14}_{30}\text{Si}$ **C.** $^{30}_{16}\text{S}$ **D.** $^{30}_{16}\text{Si}$ **E.** $^{16}_{30}\text{S}$

2. Alpha particles are composed of
A. protons **B.** neutrons **C.** electrons
D. protons and electrons **E.** protons and neutrons

3. Gamma radiation is a type of radiation that
A. originates in the electron shells
B. is most dangerous
C. is least dangerous
D. is the heaviest
E. travels the shortest distance

4. The charge on an alpha particle is
A. −1 **B.** +1 **C.** −2 **D.** +2 **E.** +4

5. Beta particles formed in a radioactive nucleus are
A. protons **B.** neutrons **C.** electrons
D. protons and electrons **E.** protons and neutrons

For questions 6 through 10, select from the following:
A. ${}^{0}_{-1}X$ **B.** ${}^{4}_{2}X$ **C.** ${}^{1}_{1}X$ **D.** ${}^{1}_{0}X$ **E.** ${}^{0}_{0}X$

6. an alpha particle

7. a beta particle

8. a gamma ray

9. a proton

10. a neutron

11. Shielding from gamma rays is provided by
A. skin. **B.** paper. **C.** clothing. **D.** lead. **E.** air.

12. The skin will provide shielding from
A. alpha particles **B.** beta particles **C.** gamma rays
D. ultraviolet rays **E.** X rays

13. The radioisotope iodine-131 is used as a radioactive tracer for studying thyroid gland activity. The symbol for iodine-131 is
A. I **B.** ${}_{131}I$ **C.** ${}^{131}_{53}I$ **D.** ${}^{53}_{131}I$ **E.** ${}^{78}_{53}I$

14. When an atom emits an alpha particle, its mass number will
A. increase by 1 **B.** increases by 2 **C.** increase by 4
D. decrease by 4 **E.** not change

15. When a nucleus emits a beta particle, the atomic number of the new nucleus
A. increases by 1 **B.** increases by 2 **C.** decreases by 1
D. decreases by 2 **E.** does not change

16. When a nucleus emits a gamma ray, the atomic number of the new nucleus
A. increases by 1 **B.** increases by 2 **C.** decreases by 1
D. decreases by 2 **E.** does not change

For questions 17 through 20, select the particle that completes each of the equations.
A. neutron **B.** alpha particle **C.** beta particle **D.** gamma ray

17. ${}^{126}_{50}Sn \longrightarrow {}^{126}_{51}Sb + ?$

18. ${}^{69}_{30}Zn \longrightarrow {}^{69}_{31}Ga + ?$

19. ${}^{99m}_{43}Tc \longrightarrow {}^{99}_{43}Tc + ?$

20. ${}^{149}_{62}Sm \longrightarrow {}^{145}_{60}Nd + ?$

21. What symbol completes the following reaction?
${}^{14}_{7}N + {}^{1}_{0}n \longrightarrow ? + {}^{1}_{1}H$
A. ${}^{15}_{8}O$ **B.** ${}^{15}_{6}C$ **C.** ${}^{14}_{8}O$ **D.** ${}^{14}_{6}C$ **E.** ${}^{15}_{7}N$

22. To complete this nuclear equation, you need to write
${}^{54}_{26}Fe + ? \longrightarrow {}^{57}_{28}Ni + {}^{1}_{0}n$
A. an alpha particle **B.** a beta particle **C.** gamma
D. neutron **E.** proton

23. The name of the unit used to measure the number of disintegrations per second is
A. curie **B.** rad **C.** rem **D.** gray **E.** sievert

24. The rem and the sievert are units used to measure
- **A.** activity of a radioactive sample
- **B.** biological damage of different types of radiation
- **C.** radiation absorbed
- **D.** background radiation
- **E.** half-life of a radioactive sample

25. Radiation can cause
A. nausea **B.** a lower white cell count **C.** fatigue
D. hair loss **E.** all of these

26. Radioisotopes used in medical diagnosis
A. have short half-lives **B.** emit only gamma rays
C. locate in specific organs **D.** produce nonradioactive nuclei
E. all of these

27. The time required for a radioisotope to decay is measured by its
A. half-life **B.** protons **C.** activity **D.** fusion **E.** radioisotope

28. Oxygen-15 used in imaging has a half-life of 2 min. How many half-lives have occurred in the 10 minutes it takes to prepare the sample?
A. 2 **B.** 3 **C.** 4 **D.** 5 **E.** 6

29. Iodine-131 has a half-life of 8 days. How long will it take for a 160-mg sample to decay to 10 mg?
A. 8 days **B.** 16 days **C.** 32 days **D.** 40 days **E.** 48 days

30. Phosphorus-32 has a half-life of 14 days. After 28 days, how many milligrams of a 100-mg sample will still be radioactive?
A. 75 mg **B.** 50 mg **C.** 40 mg **D.** 25 mg **E.** 12.5 mg

31. The "splitting" of a large nucleus to form smaller particles accompanied by a release of energy is called
A. radioisotope **B.** fission **C.** fusion **D.** rem **E.** half-life

32. The process of combining small nuclei to form larger nuclei is
A. radioisotope **B.** fission **C.** fusion **D.** rem **E.** half-life

33. The fusion reaction
- **A.** occurs in the Sun
- **B.** forms larger nuclei from smaller nuclei
- **C.** requires extremely high temperatures
- **D.** releases a large amount of energy
- **E.** all of the these

Answers to the Practice Test

1. C	**2.** E	**3.** B	**4.** D	**5.** C
6. B	**7.** A	**8.** E	**9.** C	**10.** D
11. D	**12.** A	**13.** C	**14.** D	**15.** A
16. E	**17.** C	**18.** C	**19.** D	**20.** B
21. D	**22.** A	**23.** A	**24.** B	**25.** E
26. E	**27.** A	**28.** D	**29.** C	**30.** D
31. B	**32.** C	**33.** E		

17
Organic Chemistry

Study Goals

- Write the IUPAC names and draw condensed structural formulas of alkanes with substituents.
- Write the IUPAC names and draw condensed structural formulas of alkenes and alkynes.
- Describe the formation of a polymer from alkene monomers.
- Describe the bonding in benzene, draw condensed structural formulas of aromatic compounds, and give their names.
- Write the IUPAC and common names and draw the condensed structural formulas for alcohols, phenols, and ethers.
- Write the IUPAC and common names and draw the condensed structural formulas for aldehydes, ketones, carboxylic acids, esters, amines, and amides.
- Write the IUPAC names and draw the condensed structural formulas for the products from the esterification of carboxylic acids and alcohols.
- Write the IUPAC names and draw the condensed structural formulas for amines and amides.
- Draw the condensed structural formulas of the products from the amidation of amines.

Think About It

1. In a salad dressing, why is the layer of vegetable oil separate from the water layer?

2. What type of compound gives flowers and fruits their pleasant aromas?

3. What are polymers?

4. Fish smell "fishy", but lemon juice removes the "fishy" odor. Why?

Key Terms

Match the following terms with the descriptions shown:

a. branched alkane
b. monomer
c. substituent
d. isomers
e. esterification
f. benzene

1. _____ organic compounds in which identical molecular formulas have different arrangements of atoms

2. _____ a ring of six carbon atoms each of which is attached to one hydrogen atom, C_6H_6

3. _____ the formation of an ester from a carboxylic acid and an alcohol with the elimination of a molecule of water in the presence of an acid catalyst

4. _____ an alkane containing a hydrocarbon substituent bonded to the main chain

5. _____ the small organic molecule that is repeated many times in a polymer

6. _____ groups of atoms such as an alkyl group or a halogen bonded to the main chain or ring of carbon atoms

 Answers **1.** d **2.** f **3.** e **4.** a **5.** b **6.** c

17.1 Alkanes and Naming Substituents

- The IUPAC (International Union of Pure and Applied Chemistry) system for naming organic compounds in used.

- The longest continuous chain is named as the parent alkane, and the substituents attached to the chain are numbered and listed alphabetically in front of the name of the parent chain.

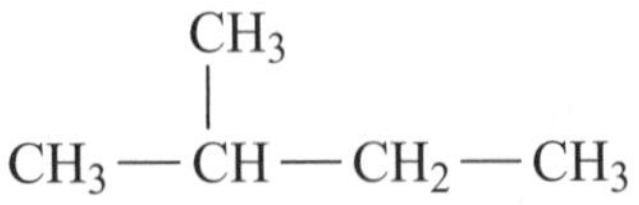

$$CH_3$$
$$|$$
$$CH_3—CH—CH_2—CH_3 \quad \text{2-methylbutane}$$

Guide to Naming Alkanes

STEP 1 Write the alkane name of the longest continuous chain of carbon atoms.
STEP 2 Number the carbon atoms starting from the end nearer a substituent.
STEP 3 Give the location and name of each substituent (alphabetical order) as a prefix to the name of the main chain.

MasteringChemistry
Tutorial: Naming Alkanes with Substituents

◆ Learning Exercise 17.1A

Write the IUPAC name for each of the following compounds:

1. CH_3—$\overset{\overset{\displaystyle CH_3}{|}}{CH}$—$CH_3$

2. CH_3—CH_2—$\overset{\overset{\displaystyle CH_3}{|}}{CH}$—$CH_2$—$\overset{\overset{\displaystyle CH_3}{|}}{CH}$—$CH_3$

3. CH_3—$\overset{\overset{\displaystyle CH_3}{|}}{CH}$—$CH_2$—$CH_2$—$\overset{\overset{\displaystyle CH_3}{|}}{CH}$—$CH_2$—$CH_3$

4. CH_3—$\overset{\overset{\displaystyle CH_3}{|}}{\underset{\underset{\displaystyle CH_3}{|}}{C}}$—$CH_2$—$CH_3$

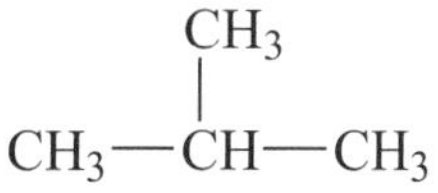

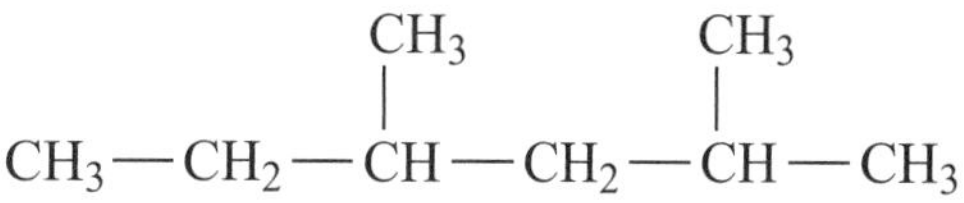

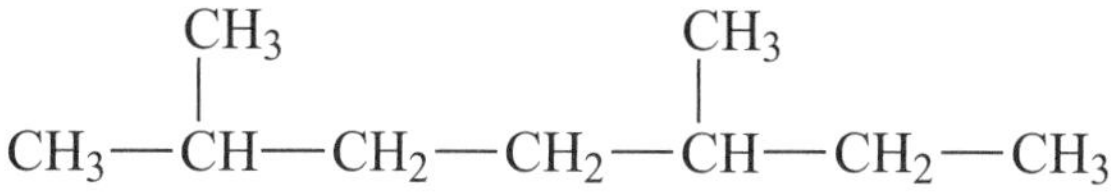

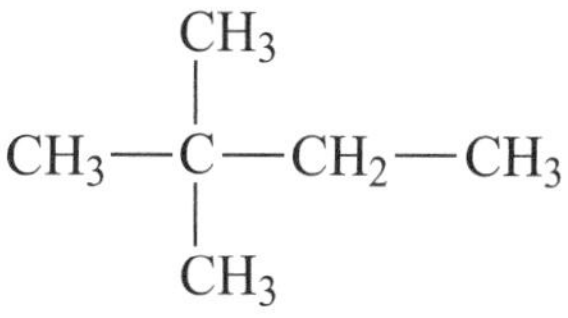

5. Cl—CH_2—CH_2—CH_2—Cl

Answers	1. 2-methylpropane	2. 2,4-dimethylhexane
	3. 2,5-dimethylheptane	4. 2,2-dimethylbutane
	5. 1,3-dichloropropane	

Guide to Drawing Alkane Formulas

STEP 1 Draw the main chain of carbon atoms.
STEP 2 Number the chain and place the substituents on the carbons indicated by the numbers.
STEP 3 Add the correct number of hydrogen atoms to give four bonds to each C atom.

◆ Learning Exercise 17.1B

Draw the condensed structural formula for each of the following compounds:

1. 2,3-dimethylbutane

2. 2,3,4-trimethylpentane

3. 1-bromobutane

Answers

1. CH_3—$\overset{\overset{\displaystyle CH_3}{|}}{CH}$—$\overset{\overset{\displaystyle CH_3}{|}}{CH}$—$CH_3$

2. CH_3—$\overset{\overset{\displaystyle CH_3}{|}}{CH}$—$\overset{\overset{\displaystyle CH_3}{|}}{\underset{\underset{\displaystyle CH_3}{|}}{CH}}$—$\overset{\overset{\displaystyle CH_3}{|}}{CH}$—$CH_3$

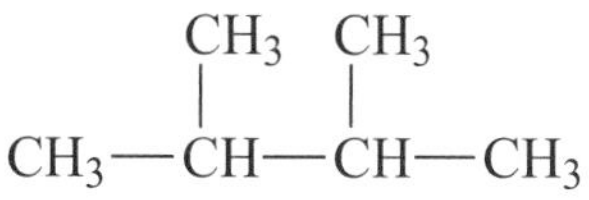

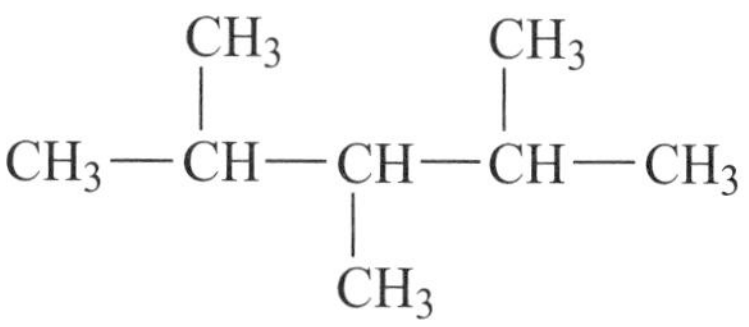

3. CH_3—CH_2—CH_2—CH_2—Br

17.2 Alkenes, Alkynes, and Polymers

- The IUPAC names of alkenes are derived by changing the *ane* ending of the parent alkane to *ene*.
- In alkenes, the longest carbon chain containing the double bond is numbered from the end nearer the double bond.

$CH_3-CH=CH_2$ propene (propylene)

$CH_2=CH-CH_2-CH_3$ 1-butene

$CH_3-CH=\overset{\overset{CH_3}{|}}{C}-CH_3$ 2-methyl-2-butene

- The alkynes are a family of unsaturated hydrocarbons that contain a triple bond. They use naming rules similar to the alkenes, but the parent chain ends with *yne*.

$HC\equiv CH$ Ethyne

$CH_3-C\equiv CH$ Propyne

- *Polymers* are large molecules prepared from the bonding of many small units called *monomers*.
- Many synthetic polymers are made from small alkene monomers.

Guide to Naming Alkenes and Alkynes

STEP 1 Name the longest carbon chain that contains a double or triple bond.
STEP 2 Number the carbon chain starting from the end nearer the double or triple bond.
STEP 3 Give the location and name of each substituent (alphabetical order) as a prefix to the alkene or alkyne name.

MasteringChemistry
Tutorial: Drawing Alkenes and Alkynes
Tutorial: Naming Alkenes and Alkynes

◆ Learning Exercise 17.2A

Write the IUPAC name for each of the following alkenes:

1. $CH_3-CH=CH_2$

2. $CH_3-CH=CH-CH_3$

3. $CH_3-CH=\overset{\overset{CH_3}{|}}{C}-CH_2-CH_3$

4. $CH_2=CH-\overset{\overset{Cl}{|}}{CH}-CH_2-\overset{\overset{CH_3}{|}}{CH}-CH_3$

Answers **1.** propene **2.** 2-butene
3. 3-methyl-2-pentene **4.** 3-chloro-5-methyl-l-hexene

◆ Learning Exercise 17.2B

Write the IUPAC and common name (if any) of each of the following alkynes:

1. $HC\equiv CH$

2. $CH_3-C\equiv CH$

3. $CH_3-CH_2-C\equiv CH$

4. $CH_3-\overset{\overset{CH_3}{|}}{CH}-C\equiv C-CH_3$

Answers **1.** ethyne (acetylene) **2.** propyne
3. 1-butyne **4.** 4-methyl-2-pentyne

◆ **Learning Exercise 17.2C**

Draw the condensed structural formula for each of the following:

1. 2-pentyne

2. 2-chloro-2-butene

3. 2-bromo-3-methyl-2-pentene

4. 2-methyl-3-hexyne

Answers

1. $CH_3-C\equiv C-CH_2-CH_3$

2.
$$CH_3-CH=\overset{\overset{\displaystyle Cl}{|}}{C}-CH_3$$

3.
$$CH_3-\overset{\overset{\displaystyle Br}{|}}{C}=\underset{\underset{\displaystyle CH_3}{|}}{C}-CH_2-CH_3$$

4.
$$CH_3-CH_2-C\equiv C-\overset{\overset{\displaystyle CH_3}{|}}{CH}-CH_3$$

◆ **Learning Check 17.2D**

Write the formula of the alkene monomer that would be used for each of the following polymers:

1.
$$-\overset{\overset{\displaystyle H}{|}}{\underset{\underset{\displaystyle H}{|}}{C}}-\overset{\overset{\displaystyle H}{|}}{\underset{\underset{\displaystyle H}{|}}{C}}-\overset{\overset{\displaystyle H}{|}}{\underset{\underset{\displaystyle H}{|}}{C}}-\overset{\overset{\displaystyle H}{|}}{\underset{\underset{\displaystyle H}{|}}{C}}-\overset{\overset{\displaystyle H}{|}}{\underset{\underset{\displaystyle H}{|}}{C}}-\overset{\overset{\displaystyle H}{|}}{\underset{\underset{\displaystyle H}{|}}{C}}-$$

2.
$$-\overset{\overset{\displaystyle H}{|}}{\underset{\underset{\displaystyle H}{|}}{C}}-\overset{\overset{\displaystyle CH_3}{|}}{\underset{\underset{\displaystyle H}{|}}{C}}-\overset{\overset{\displaystyle H}{|}}{\underset{\underset{\displaystyle H}{|}}{C}}-\overset{\overset{\displaystyle CH_3}{|}}{\underset{\underset{\displaystyle H}{|}}{C}}-\overset{\overset{\displaystyle H}{|}}{\underset{\underset{\displaystyle H}{|}}{C}}-\overset{\overset{\displaystyle CH_3}{|}}{\underset{\underset{\displaystyle H}{|}}{C}}-$$

3.
$$-\overset{\overset{\displaystyle F}{|}}{\underset{\underset{\displaystyle F}{|}}{C}}-\overset{\overset{\displaystyle F}{|}}{\underset{\underset{\displaystyle F}{|}}{C}}-\overset{\overset{\displaystyle F}{|}}{\underset{\underset{\displaystyle F}{|}}{C}}-\overset{\overset{\displaystyle F}{|}}{\underset{\underset{\displaystyle F}{|}}{C}}-\overset{\overset{\displaystyle F}{|}}{\underset{\underset{\displaystyle F}{|}}{C}}-\overset{\overset{\displaystyle F}{|}}{\underset{\underset{\displaystyle F}{|}}{C}}-$$

Answers 1. $H_2C=CH_2$ 2. $H_2C=\overset{\overset{\displaystyle CH_3}{|}}{CH}$ 3. $F_2C=CF_2$

◆ **Learning Check 17.2E**

Write three sections of the polymer that would result when 1,1-difluoroethene is the monomer unit.

Answer
$$-\overset{\overset{\displaystyle F}{|}}{\underset{\underset{\displaystyle F}{|}}{C}}-\overset{\overset{\displaystyle H}{|}}{\underset{\underset{\displaystyle H}{|}}{C}}-\overset{\overset{\displaystyle F}{|}}{\underset{\underset{\displaystyle F}{|}}{C}}-\overset{\overset{\displaystyle H}{|}}{\underset{\underset{\displaystyle H}{|}}{C}}-\overset{\overset{\displaystyle F}{|}}{\underset{\underset{\displaystyle F}{|}}{C}}-\overset{\overset{\displaystyle H}{|}}{\underset{\underset{\displaystyle H}{|}}{C}}-$$

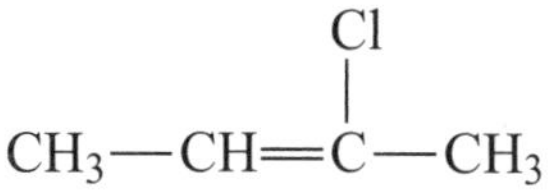

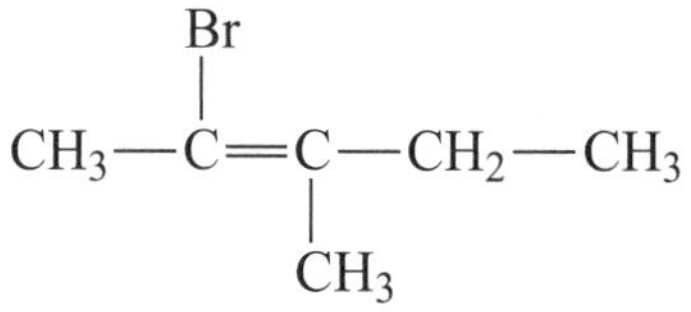

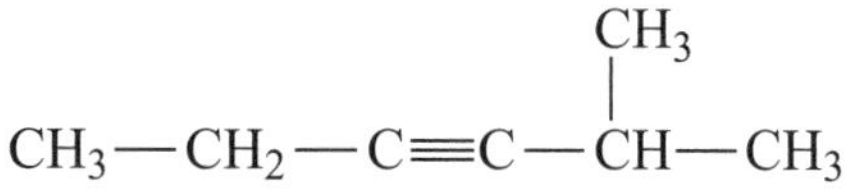

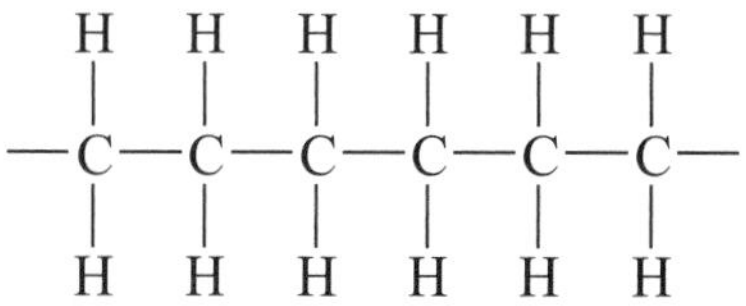

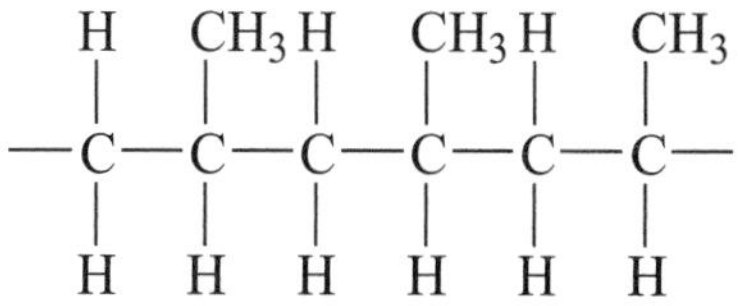

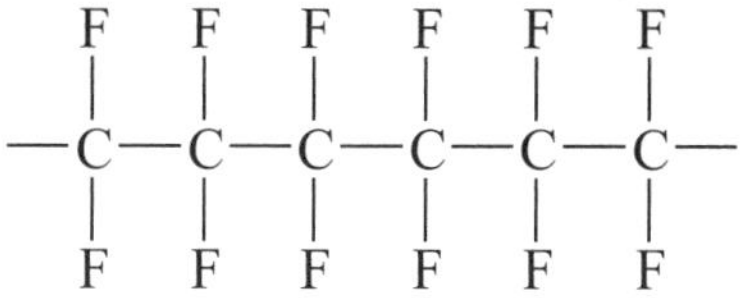

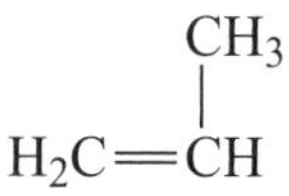

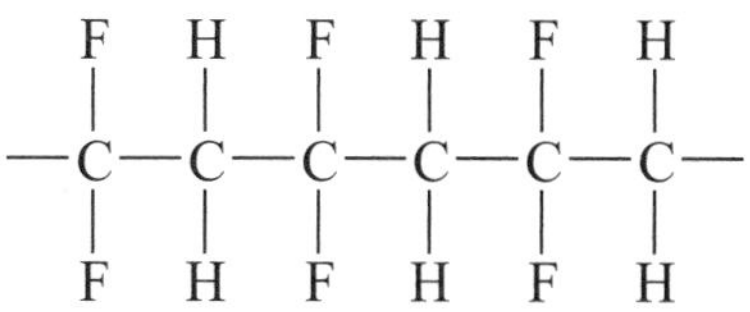

17.3 Aromatic Compounds

- Most aromatic compounds contain benzene, a cyclic structure containing six CH units. The structure of benzene is represented as a hexagon with a circle in the center.
- The names of many aromatic compounds use the parent name benzene, although many common names were retained as IUPAC names, such as toluene.

MasteringChemistry
Tutorial: Naming Aromatic Compounds

◆ Learning Exercise 17.3

Write the IUPAC (or common name) for each of the following:

a. 

b. 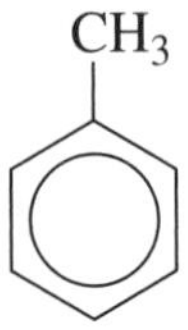

c. 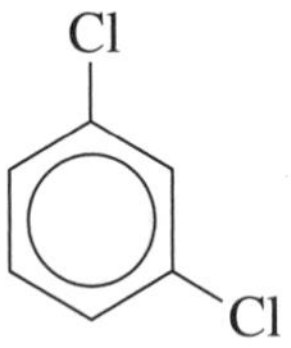

d. 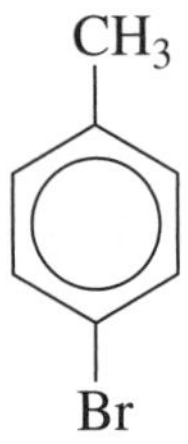

Answers **1.** benzene **2.** methylbenzene; toluene
3. 1,3-dichlorobenzene **4.** 4-bromotoluene

17.4 Alcohols, Phenols, and Ethers

- An alcohol contains the hydroxyl group —OH attached to a carbon chain.
- In the IUPAC system, alcohols are named by replacing the *ane* of the alkane name with *ol*. The location of the —OH group is given by numbering the carbon chain. Simple alcohols are generally named by their common names, with the alkyl name preceding the term *alcohol*. For example, CH_3—OH is methyl alcohol, and CH_3—CH_2—OH is ethyl alcohol.
- When a hydroxyl group is attached to a benzene ring, the compound is a phenol.
- An ether contains an oxygen atom attached by single bonds to two carbon atoms, —O—.
- Simple ethers are named using the names of the alkyl groups attached to the oxygen atom.

Guide to Naming Alcohols

STEP 1 Name the longest carbon chain with the —OH group. Name an aromatic alcohol as a *phenol*.
STEP 2 Number the chain starting at the end closer to the —OH.
STEP 3 Give the location and name of each substituent relative to the —OH group.

MasteringChemistry

Tutorial: Drawing Ethers
Tutorial: Naming Ethers

◆ **Learning Exercise 17.4A**

Give the correct IUPAC and common name (if any) for each of the following compounds:

1. CH_3-CH_2-OH

2. $CH_3-CH_2-CH_2-OH$

3. $CH_3-\overset{\displaystyle OH}{\overset{|}{CH}}-CH_2-CH_2-CH_3$

4. $CH_3-CH_2-\overset{\displaystyle CH_3}{\overset{|}{CH}}-\overset{\displaystyle OH}{\overset{|}{CH}}-CH_3$

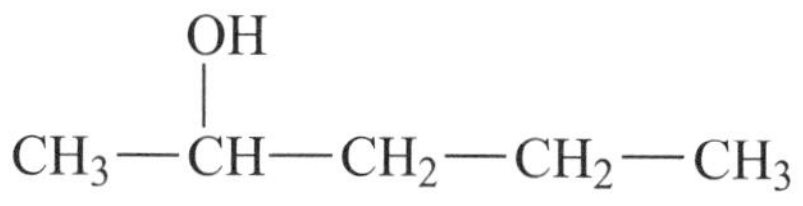

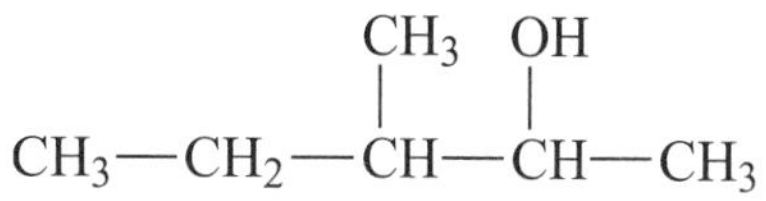

5.

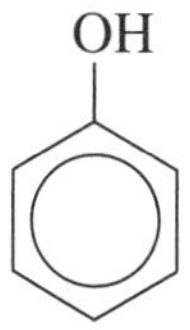

6. $CH_3-CH_2-CH_2-O-CH_3$

Answers	**1.** ethanol (ethyl alcohol)	**2.** 1-propanol (propyl alcohol)
	3. 2-pentanol	**4.** 3-methyl-2-pentanol
	5. phenol	**6.** methyl propyl ether (common)

Learning Exercise 17.4B

Draw the condensed structural formula for each of the following compounds:

1. 2-butanol

2. 2-chloro-1-propanol

3. 2,4-dimethyl-3-pentanol

4. 3-methylphenol

5. ethyl methyl ether

Answers

1. $CH_3-\overset{\displaystyle OH}{\overset{|}{CH}}-CH_2-CH_3$

2. $CH_3-\overset{\displaystyle Cl}{\overset{|}{CH}}-CH_2-OH$

3. $CH_3-\overset{\displaystyle CH_3}{\overset{|}{CH}}-\overset{\displaystyle OH}{\overset{|}{CH}}-\overset{\displaystyle CH_3}{\overset{|}{CH}}-CH_3$

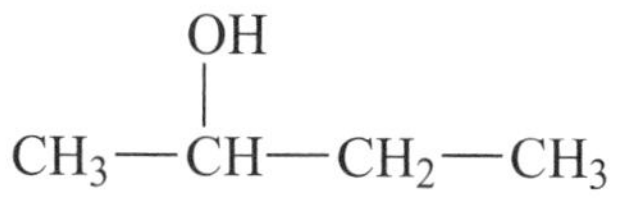

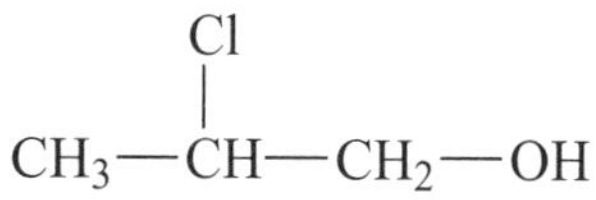

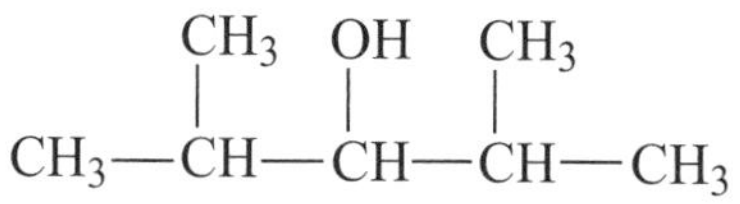

4.

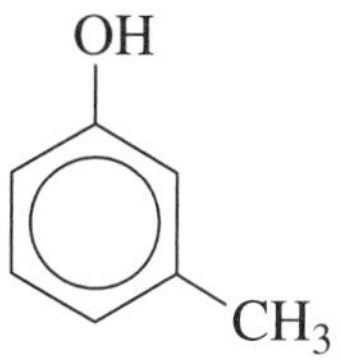

5. $CH_3-CH_2-O-CH_3$

17.5 Aldehydes and Ketones

- In an aldehyde, the carbonyl group (C=O) appears at the end of a carbon chain attached to at least one hydrogen atom.
- In a ketone, the carbonyl group (C=O) occurs between carbon groups and has no hydrogens attached to it.
- In the IUPAC system, aldehydes and ketones are named by replacing the *e* in the longest chain containing the carbonyl group with *al* for aldehydes and *one* for ketones. The location of the carbonyl group in a ketone is given if there are more than four carbon atoms in the chain.
- In the IUPAC system, the location of the aldehyde carbon atom is carbon 1 but is not specified in the name.

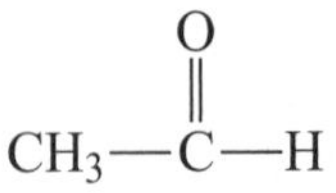

Ethanal
(acetaldehyde)

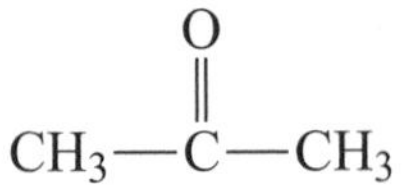

Propanone
(dimethyl ketone, acetone)

Guide to Naming Aldehydes

STEP 1 Name the longest carbon chain containing the carbonyl group by replacing the *e* in the alkane name with *al.*
STEP 2 Name and number substituents by counting the carbonyl group as carbon 1.

Guide to Naming Ketones

STEP 1 Name the longest carbon chain containing the carbonyl group by replacing the *e* in the alkane name with *one.*
STEP 2 Number the carbon chain starting from the end nearer the carbonyl group and indicate its location.
STEP 3 Name and number any substituents on other carbons in the chain.

MasteringChemistry
Tutorial: Naming Aldehydes and Ketones
Self Study Activity: Aldehydes and Ketones

◆ Learning Exercise 17.5A

Write the correct IUPAC (or common name) for the following aldehydes and ketones:

1. 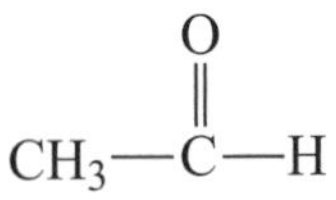

2. 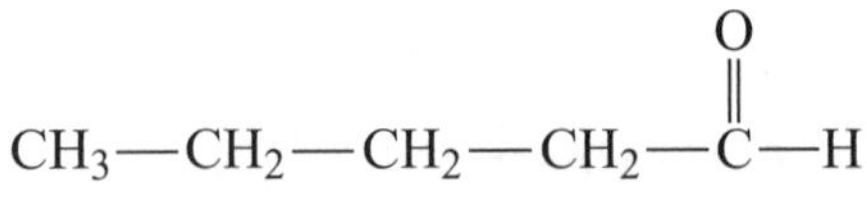

3. 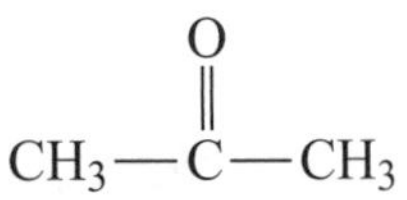

4. 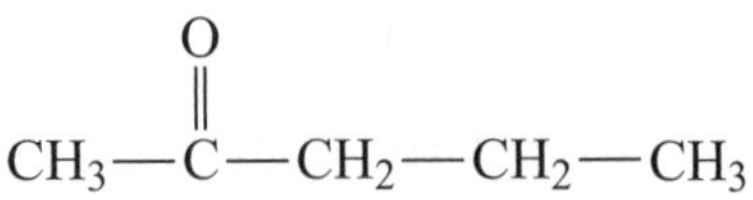

5. 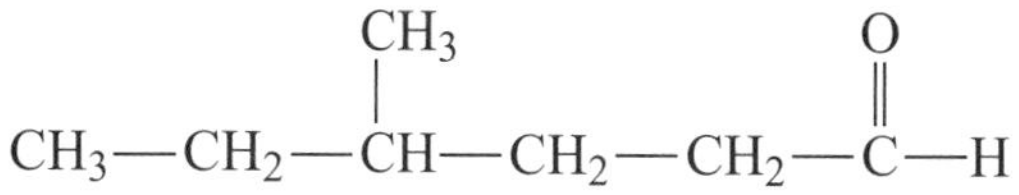 **6.** 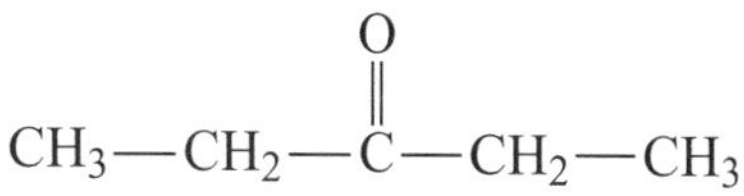

Answers

1. ethanal; acetaldehyde
2. pentanal
3. propanone; dimethyl ketone, acetone
4. 2-pentanone; methyl propyl ketone
5. 4-methylhexanal
6. 3-pentanone; diethyl ketone

◆ Learning Exercise 17.5B

Draw the condensed structural formula for each of the following aldehydes and ketones:

1. ethanal

2. 2-methylbutanal

3. 2-chloropropanal

4. ethyl methyl ketone

5. 3-hexanone

6. benzaldehyde

Answers

1. 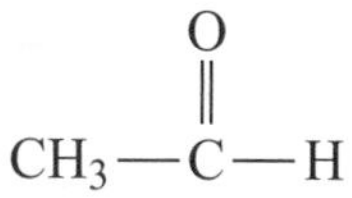

2. 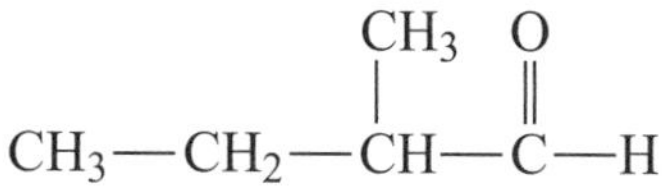

3. 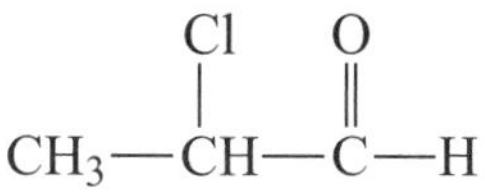

4. 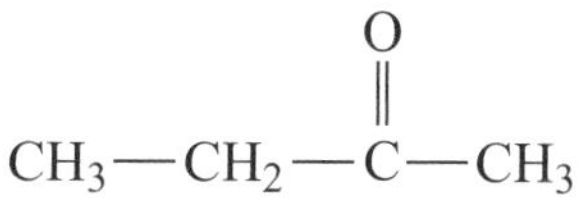

5. 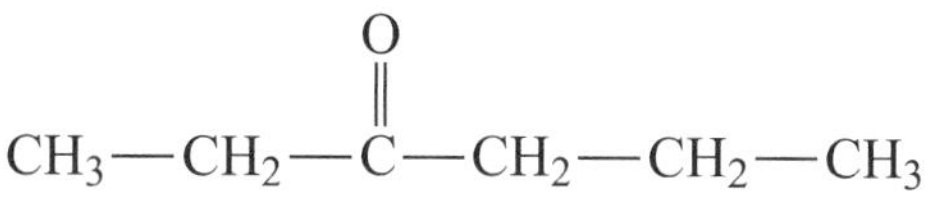

6. 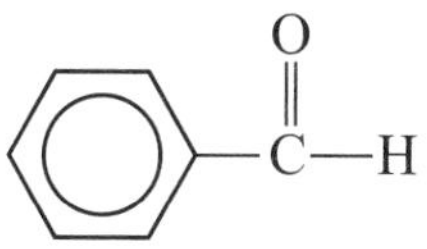

17.6 Carboxylic Acids and Esters

- In the IUPAC system, a carboxylic acid is named by replacing the *e* ending of the alkane name with *oic acid*. The carboxylic acid carbon atom is carbon 1 but it is not specified in the name. Simple acids usually are named by the common naming system using the prefixes *form-* (1C), *acet-* (2C), *propion-* (4C), *butyr-* (4C), followed by *ic acid*.

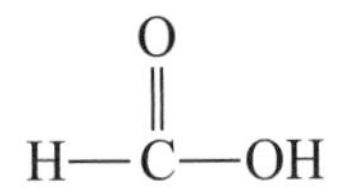

methanoic acid
(formic acid)

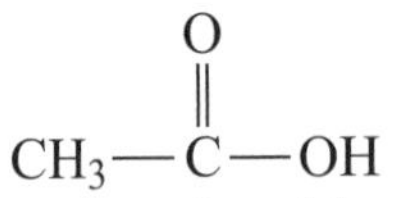

ethanoic acid
(acetic acid)

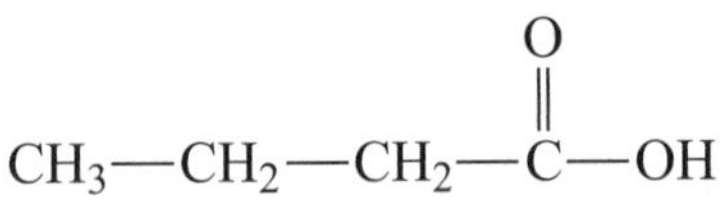

butanoic acid
(butyric acid)

- In the presence of a strong acid, carboxylic acids react with alcohols to produce esters and water (esterification reaction).

Guide to Naming Carboxylic Acids	
STEP 1	Identify the carbon chain containing the carboxyl group and replace the *e* in the alkane name with *oic acid*.
STEP 2	Give the location and name of each substituent on the main chain by counting the carboxyl group as carbon 1.

Guide to Naming Esters	
STEP 1	Write the name of the carbon chain from the alcohol as an *alkyl* group.
STEP 2	Change the *ic acid* of the acid name to *ate* of the alkane name.

MasteringChemistry
Tutorial: Formation of Esters from Carboxylic Acids
Self Study Activity: Carboxylic Acids
Tutorial: Naming and Drawing Carboxylic Acids
Tutorial: Naming Esters

◆ Learning Exercise 17.6A

Give the IUPAC and common names, if any, for each of the following carboxylic acids:

1. 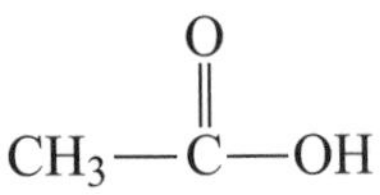

2. 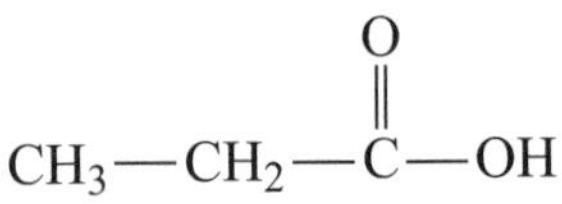

3. 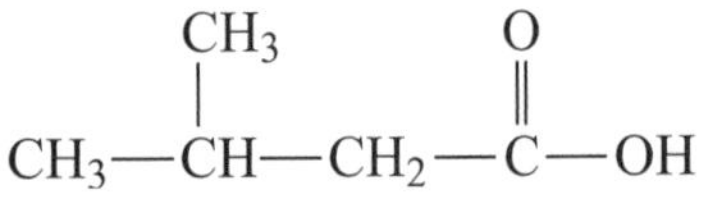

4. 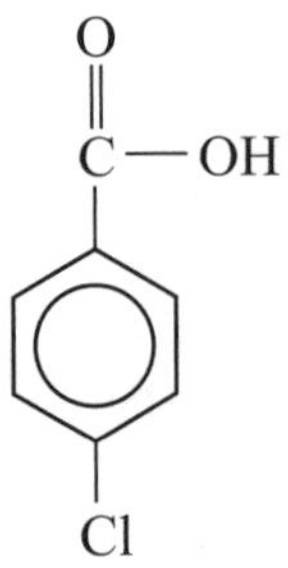

Answers
1. ethanoic acid (acetic acid)
2. propanoic acid (propionic acid)
3. 3-methylbutanoic acid
4. 4-chlorobenzoic acid

◆ Learning Exercise 17.6B

A. Draw the condensed structural formula for each of the following carboxylic acids:

1. acetic acid

2. 3-methylbutanoic acid

3. benzoic acid

4. 3-chloropropanoic acid

5. formic acid

6. 3-chloropentanoic acid

Answers

1. $CH_3-\overset{\overset{\displaystyle O}{\|}}{C}-OH$

2. $CH_3-\overset{\overset{\displaystyle CH_3}{|}}{CH}-CH_2-\overset{\overset{\displaystyle O}{\|}}{C}-OH$

3. 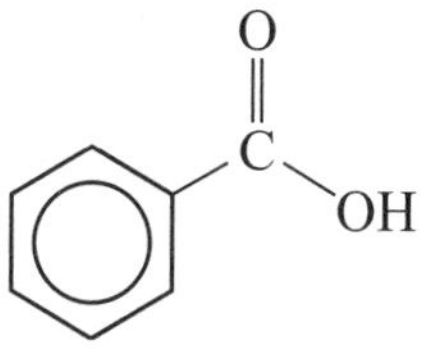

4. $Cl-CH_2-CH_2-\overset{\overset{\displaystyle O}{\|}}{C}-OH$

5. $H-\overset{\overset{\displaystyle O}{\|}}{C}-OH$

6. $CH_3-CH_2-\overset{\overset{\displaystyle Cl}{|}}{CH}-CH_2-\overset{\overset{\displaystyle O}{\|}}{C}-OH$

◆ Learning Exercise 17.6C

Draw the condensed structural formula for the products of each of the following reactions:

1. $CH_3-\overset{\overset{\displaystyle O}{\|}}{C}-OH + CH_3-OH \xrightarrow{H^+}$

2. $H-\overset{\overset{\displaystyle O}{\|}}{C}-OH + CH_3-CH_2-OH \xrightarrow{H^+}$

3. $$CH_3{-}CH_2{-}\overset{\overset{O}{\|}}{C}{-}OH + CH_3{-}CH_2{-}OH \xrightarrow{H^+}$$

Answers

1. $$CH_3{-}\overset{\overset{O}{\|}}{C}{-}O{-}CH_3 + H_2O$$

2. $$H{-}\overset{\overset{O}{\|}}{C}{-}O{-}CH_2{-}CH_3 + H_2O$$

3. $$CH_3{-}CH_2{-}\overset{\overset{O}{\|}}{C}{-}O{-}CH_2{-}CH_3 + H_2O$$

17.7 Amines and Amides

- Aromatic amines are named as derivatives of aniline (aminobenzene).
- Amines are derivatives of ammonia (NH_3), in which alkyl or aromatic groups replace one or more hydrogen atoms.
- Simple amines are named with common names in which the names of the alkyl groups are listed alphabetically preceding the suffix *amine*.

$CH_3{-}NH_2$	$CH_3{-}NH{-}CH_3$	$CH_3{-}\overset{\overset{CH_3}{\vert}}{N}{-}CH_3$
Methylamine	Dimethylamine	trimethylamine

- Amides are derivatives of carboxylic acids in which an amine group replaces the —OH group in the acid.
- Amides are named by replacing the *ic acid* or *oic acid* ending by *amide*.

$$CH_3{-}\overset{\overset{O}{\|}}{C}{-}NH_2$$

ethanamide
(acetamide)

- An amide is produced in an amidation reaction in which a carboxylic acid reacts with ammonia or an amine.

MasteringChemistry

Tutorial: Name that Amine
Tutorial: Drawing Amines
Tutorial: Amidation Reactions

◆ Learning Exercise 17.7A

Name each of the following amines:

1. $CH_3-\overset{\overset{\displaystyle H}{|}}{N}-CH_2-CH_3$

2. $CH_3-CH_2-\overset{\overset{\displaystyle CH_3}{|}}{N}-CH_3$

3. $CH_3-CH_2-CH_2-CH_2-\overset{\overset{\displaystyle H}{|}}{N}-CH_2-CH_3$

4. 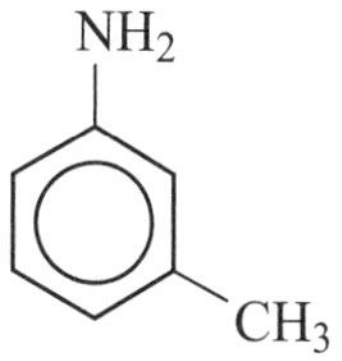

Answers		
	1. ethylmethylamine	**2.** ethyldimethylamine
	3. butylethylamine	**4.** 3-methylaniline

◆ Learning Exercise 17.7B

Draw the condensed structural formula for each of the following amines:

1. propylamine

2. ethylmethylamine

3. dimethylamine

4. aniline

Answers

1. $CH_3-CH_2-CH_2-NH_2$

2. $CH_3-NH-CH_2-CH_3$

3. $CH_3-NH-CH_3$

4. 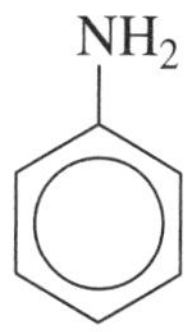

◆ Learning Exercise 17.7C

Write the IUPAC name and common name, if any, for each of the following amides:

1. $CH_3-CH_2-\overset{\displaystyle O}{\overset{\|}{C}}-NH_2$

2. 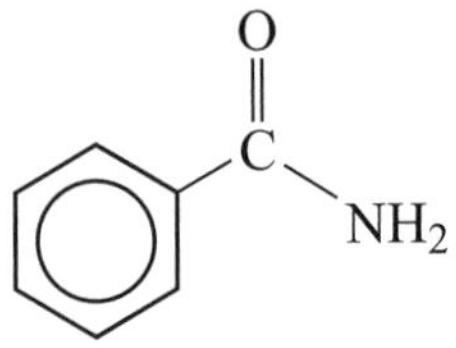

3. $CH_3-CH_2-CH_2-CH_2-\overset{\displaystyle O}{\overset{\|}{C}}-NH_2$

4. $CH_3-\overset{\displaystyle O}{\overset{\|}{C}}-NH_2$

Answers

1. propanamide (propionamide)
2. benzamide
3. pentanamide
4. ethanamide (acetamide)

◆ Learning Exercise 17.7D

Draw the condensed structural formula for each of the following amides:

1. propanamide

2. 2-methylbutanamide

3. 3-chloropentanamide

4. benzamide

Answers

1. $CH_3-CH_2-\overset{\displaystyle O}{\overset{\|}{C}}-NH_2$

2. $CH_3-CH_2-\overset{\displaystyle CH_3}{\overset{|}{CH}}-\overset{\displaystyle O}{\overset{\|}{C}}-NH_2$

3. $CH_3-CH_2-\overset{\displaystyle Cl}{\overset{|}{CH}}-CH_2-\overset{\displaystyle O}{\overset{\|}{C}}-NH_2$

4. 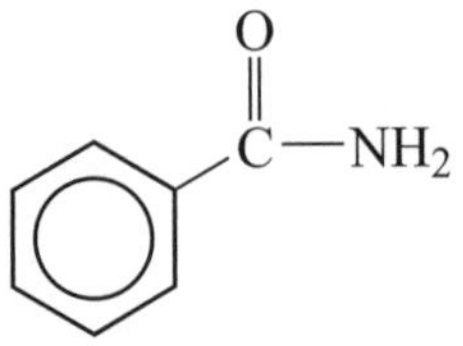

◆ Learning Exercise 17.7E

Draw the condensed structural formula of the amide formed in each of the following reactions:

1. $CH_3-CH_2-\overset{\overset{\large O}{\|}}{C}-OH + NH_3 \xrightarrow{\text{heat}}$

2. $CH_3-\overset{\overset{\large O}{\|}}{C}-OH + NH_3 \xrightarrow{\text{heat}}$

Answers **1.** $CH_3-CH_2-\overset{\overset{\large O}{\|}}{C}-NH_2$ **2.** $CH_3-\overset{\overset{\large O}{\|}}{C}-NH_2$

Checklist for Chapter 17

You are ready to take the Practice Test for Chapter 17. Be sure you have accomplished the following learning goals for this chapter. If you are not sure, review the section listed at the end of the goal. Then apply your new skills and understanding to the Practice Test.

After studying Chapter 17, I can successfully:

_______ Name and draw condensed structural formulas of alkanes with substituents (17.1).

_______ Identify the structural features of alkenes and alkynes (17.2).

_______ Name alkenes and alkynes using IUPAC rules and draw their condensed structural formulas (17.2).

_______ Describe the process of forming polymers from alkene monomers (17.2).

_______ Write the names and draw the condensed structural formulas for compounds that contain a benzene ring (17.3).

_______ Write the IUPAC or common name of an alcohol and draw the condensed structural formula from the name (17.4).

_______ Write the common name of a simple ether and draw the condensed structural formula from the name (17.4).

_______ Write the IUPAC and common names and draw the condensed structural formulas of aldehydes and ketones (17.5).

_______ Write the IUPAC and common names and draw condensed structural formulas of carboxylic acids (17.6).

_______ Write the IUPAC or common names and draw the condensed structural formulas of esters (17.6).

_______ Draw the condensed structural formula of the product of the esterification reaction between an alcohol and a carboxylic acid (17.6).

_______ Write the IUPAC or common names and draw the condensed structural formulas of esters (17.6).

_______ Write the common names of amines and amides and draw their condensed structural formulas (17.7).

_______ Draw the condensed structural formula of the product of the amidation reaction between a carboxylic acid and ammonia or an amine (17.7).

Practice Test for Chapter 17

For questions 1 and 2, match the name of the hydrocarbon with each structure.

A. hexane　**B.** 3-methylbutane　**C.** 2-methylbutane
D. butane　**E.** 2,4-dimethylhexane

1. 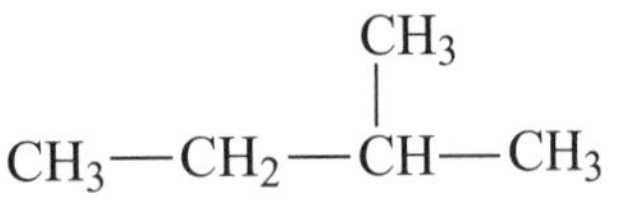

2. 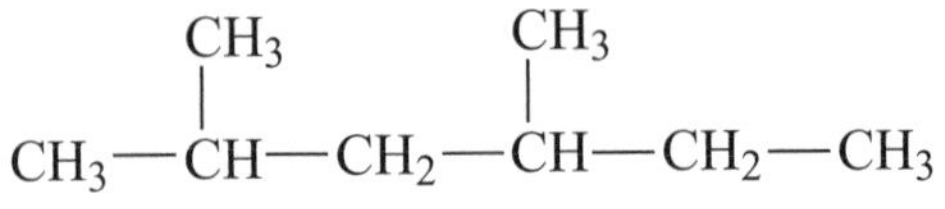

In questions 3 through 5, match the name of the alkene or alkyne with the structural formula.

A. 1-butene　**B.** 2-butene　**C.** 3-butene
D. 2-butyne　**E.** 1-butyne

3. _______ $CH_3-CH_2-C\equiv CH$

4. _______ $CH_3-CH_2-CH=CH_2$

5. _______ $CH_3-CH=CH-CH_3$

In questions 6 through 9, match the name of each aromatic compound with its structural formula:

A. 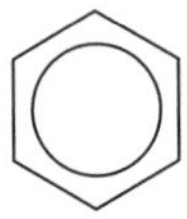

B. 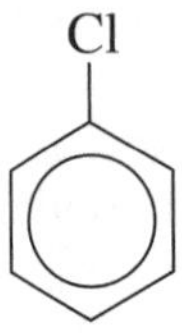

C. 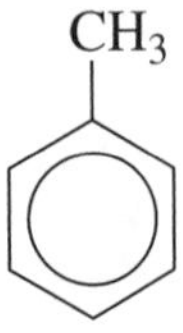

D. 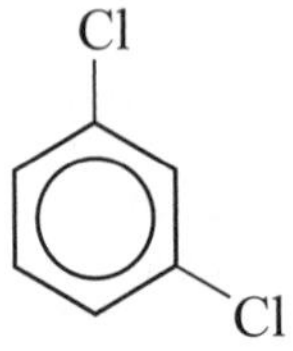

6. _______ chlorobenzene　**7.** _______ benzene

8. _______ toluene　**9.** _______ 1,3-chlorobenzene

For questions 10 through 13, match the names with one of the following condensed structural formulas:

A. 1-propanol　**B.** 3-propanol　**C.** 2-propanol
D. ethyl methyl ether　**E.** diethyl ether

10. 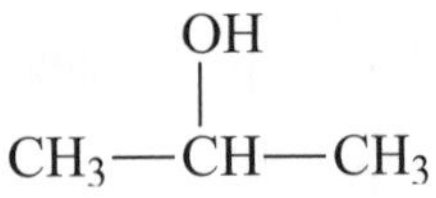

11. $CH_3-CH_2-CH_2-OH$

12. $CH_3-O-CH_2-CH_3$

13. $CH_3-CH_2-O-CH_2-CH_3$

For questions 14 through 17, match each of the condensed structural formulas with its name:

A. CH_3—$\overset{\overset{\displaystyle OH}{|}}{CH}$—$CH_3$

B. CH_3—$\overset{\overset{\displaystyle O}{||}}{C}$—$O$—$CH_2$—$CH_3$

C. CH_3—$\overset{\overset{\displaystyle O}{||}}{C}$—$OH$

D. CH_3—CH_2—CH_2—$\overset{\overset{\displaystyle O}{||}}{C}$—$OH$

E. CH_3—CH_2—CH_2—$\overset{\overset{\displaystyle O}{||}}{C}$—$O$—$CH_3$

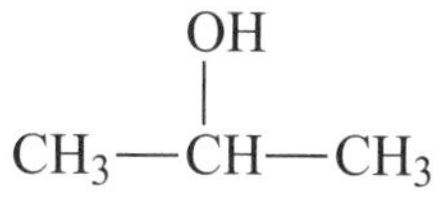

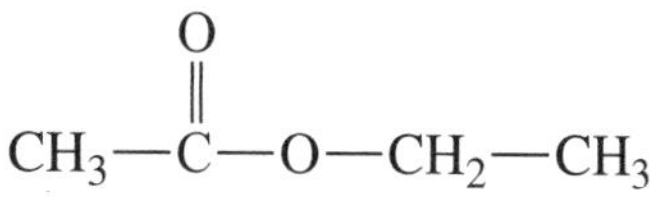

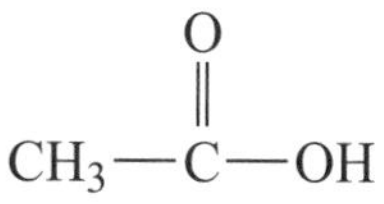

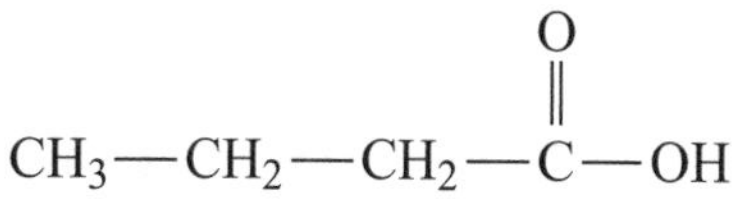

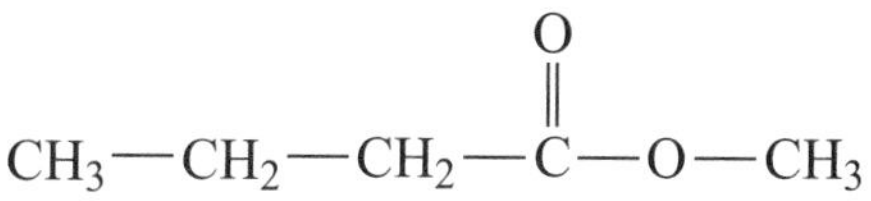

14. _____ butyric acid

15. _____ methyl butanoate

16. _____ ethyl acetate

17. _____ acetic acid

18. Identify the carboxylic acid and alcohol needed to produce

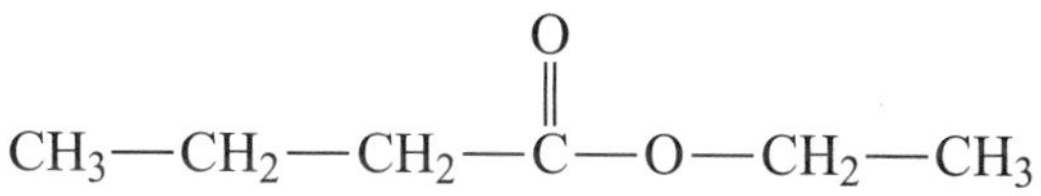

CH_3—CH_2—CH_2—$\overset{\overset{\displaystyle O}{||}}{C}$—$O$—$CH_2$—$CH_3$

 A. propanoic acid and ethanol
 B. acetic acid and 1-pentanol
 C. acetic acid and 1-butanol
 D. butanoic acid and ethanol
 E. hexanoic acid and methanol

For questions 19 through 22, match the amines and amides with the following names:

 A. ethyldimethylamine
 B. ethanamide
 C. propanamide
 D. diethylamine
 E. butanamide

19. _____ CH_3—CH_2—$\overset{\overset{\displaystyle CH_3}{|}}{N}$—$CH_3$

20. _____ CH_3—CH_2—CH_2—$\overset{\overset{\displaystyle O}{||}}{C}$—$NH_2$

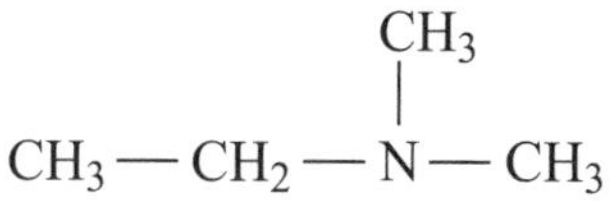

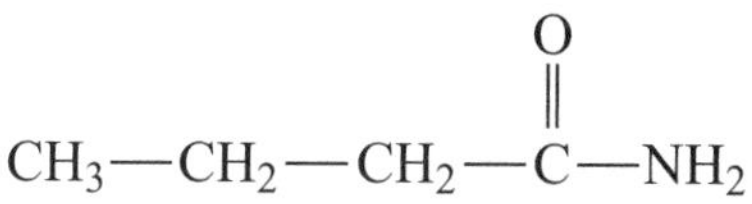

21. _____ CH_3—CH_2—NH—CH_2—CH_3

22. _____ CH_3—CH_2—$\overset{\overset{\displaystyle O}{||}}{C}$—$NH_2$

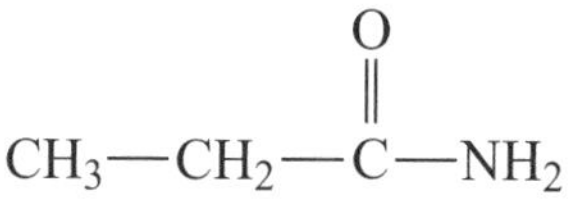

23. Indicate the carboxylic acid and ammonia or amine needed to produce the following:

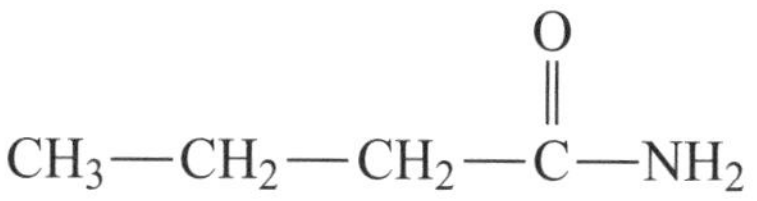

CH_3—CH_2—CH_2—$\overset{\overset{\displaystyle O}{||}}{C}$—$NH_2$

 A. propanoic acid and ammonia
 B. propanoic acid and methylamine
 C. acetic acid and ammonia
 D. butanoic acid and methylamine
 E. butanoic acid and ammonia

24. Indicate the monomer needed to produce polyethylene.
 A. propylene
 B. chloroethene
 C. tetrafluoroethene
 D. ethylene
 E. styrene

25. Indicate the IUPAC name of the following:

$$CH_3{-}CH_2{-}CH_2{-}CH_2{-}\overset{\overset{\displaystyle O}{\|}}{C}{-}H$$

A. methylbutanal　**B.** pentanal　**C.** pentaldehyde
D. hexanal　**E.** pentanone

26. The ester produced from the reaction of 1-butanol and propanoic acid is
A. butyl propanoate　**B.** butyl propanone　**C.** propyl butyrate
D. propyl butanone　**E.** heptanoate

Answers to the Practice Test

1. C	**2.** E	**3.** E	**4.** A	**5.** B
6. B	**7.** A	**8.** C	**9.** D	**10.** C
11. A	**12.** D	**13.** E	**14.** D	**15.** E
16. B	**17.** C	**18.** D	**19.** A	**20.** E
21. D	**22.** C	**23.** E	**24.** D	**25.** B
26. A				

18
Biochemistry

Study Goals

- Distinguish between monosaccharides, disaccharides, and polysaccharides.
- Draw open-chain and cyclic structures for monosaccharides.
- Classify carbohydrates as aldoses or ketoses and by the number of carbon atoms.
- Describe the structural units and bonds in disaccharides and polysaccharides.
- Distinguish between saturated and unsaturated fatty acids.
- Draw the structure of a triacylglycerol obtained from glycerol and fatty acids.
- Draw the structures of the products from hydrogenation and saponification of triacylglycerols.
- Describe steroids and their role in hormones.
- Classify proteins by their functions in the cells.
- Draw the structures of amino acids.
- Write the structural formulas of dipeptides.
- Identify the structural levels of proteins as primary, secondary, tertiary, and quaternary.
- Describe the role of an enzyme in an enzyme-catalyzed reaction.
- Describe the lock-and-key and induced-fit models of enzyme action.
- Identify the bases, sugars, and nucleotides in DNA and RNA.
- Describe the double helix of DNA.
- Explain the process of DNA replication.
- Describe the transcription process during the synthesis of mRNA.
- Use the codons in the genetic code to describe protein synthesis.

Think About It

1. What carbohydrates are present in table sugar, milk, and wood?

2. How are fats used to make soaps?

3. What are some uses of protein in the body?

4. What are some functions of enzymes in the cells of the body?

18.1 Carbohydrates

- Carbohydrates are classified as monosaccharides (simple sugars), disaccharides (two monosaccharide units), and polysaccharides (many monosaccharide units).
- Monosaccharides are aldehydes (aldoses) or ketones (ketoses) with hydroxyl groups on all other carbon atoms.
- Monosaccharides can be classified based on the number of carbons atoms, commonly: triose, tetrose, pentose, or hexose.
- Important monosaccharides are the aldohexoses glucose and galactose and the ketohexose fructose.
- The predominant form of monosaccharides is the cyclic form of five or six atoms. The cyclic structure forms by a reaction between an —OH on carbon 5 in hexoses with the carbonyl group of the same molecule.
- The formation of a new hydroxyl group on carbon 1 (or 2 in fructose) gives α and β forms of the cyclic structure of a monosaccharide.

MasteringChemistry

Tutorial: Carbohydrates
Tutorial: Carbonyls in Carbohydrates
Tutorial: Drawing Cyclic Sugars
Case Study: Calories from Hidden Sugar
Case Study: Diabetes and Blood Glucose

Key Terms

Match the following key terms with the descriptions shown:

A. carbohydrate **B.** glucose **C.** disaccharide **D.** cellulose **E.** polysaccharide

1. _____ a simple or complex sugar composed of a carbon chain with an aldehyde or ketone group and several hydroxyl groups

2. _____ a carbohydrate that contains many monosaccharides linked by glycosidic bonds

3. _____ an unbranched polysaccharide that cannot be digested by humans

4. _____ an aldohexose that is the most prevalent monosaccharide in the diet

5. _____ a carbohydrate that contains two monosaccharides linked by a glycosidic bond

Answers **1.** A **2.** E **3.** D **4.** B **5.** C

◆ **Learning Exercise 18.1A**

Indicate the number of monosaccharide units (one, two, or many) in each of the following carbohydrates:

1. sucrose, a disaccharide _____
2. cellulose, a polysaccharide _____
3. glucose, a monosaccharide _____
4. amylose, a polysaccharide _____
5. maltose, a disaccharide _____

 Answers **1.** two **2.** many **3.** one **4.** many **5.** two

◆ **Learning Exercise 18.1B**

Identify the following monosaccharides as aldo- or ketotrioses, tetroses, pentoses, or hexoses:

1.

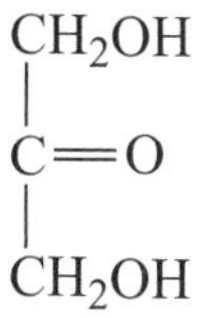

$$
\begin{array}{c}
CH_2OH \\
| \\
C=O \\
| \\
CH_2OH
\end{array}
$$

2.

$$
\begin{array}{c}
H\ \diagdown \\
C=O \\
| \\
H-C-OH \\
| \\
H-C-OH \\
| \\
HO-C-H \\
| \\
CH_2OH
\end{array}
$$

3.

$$
\begin{array}{c}
CH_2OH \\
| \\
C=O \\
| \\
HO-C-H \\
| \\
HO-C-H \\
| \\
H-C-OH \\
| \\
CH_2OH
\end{array}
$$

4.

$$
\begin{array}{c}
H\ \diagdown \\
C=O \\
| \\
H-C-OH \\
| \\
HO-C-H \\
| \\
H-C-OH \\
| \\
H-C-OH \\
| \\
CH_2OH
\end{array}
$$

5.

$$
\begin{array}{c}
H\ \diagdown \\
C=O \\
| \\
H-C-OH \\
| \\
H-C-OH \\
| \\
CH_2OH
\end{array}
$$

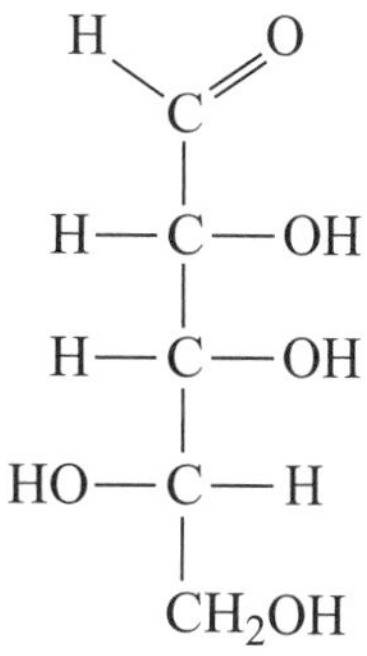

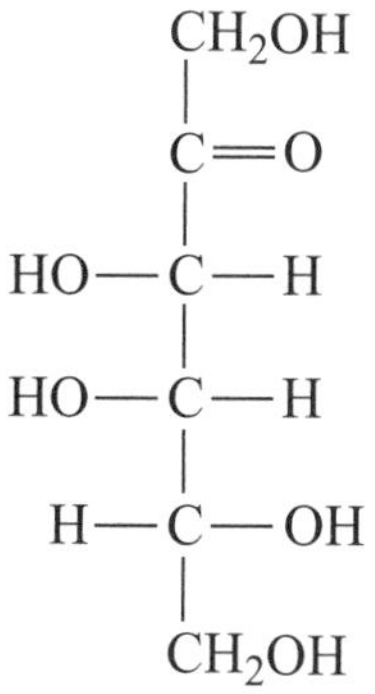

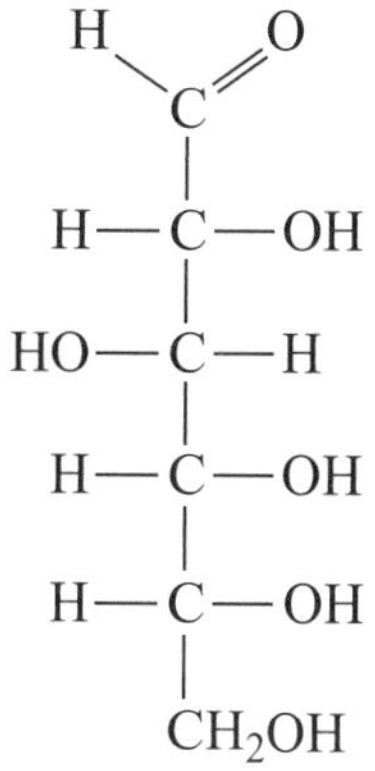

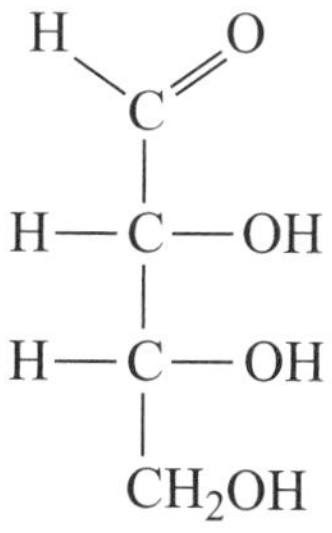

1. _____ **2.** _____ **3.** _____

4. _____ **5.** _____

 Answers **1.** ketotriose **2.** aldopentose **3.** ketohexose
 4. aldohexose **5.** aldotetrose

◆ Learning Exercise 18.1C

Draw the condensed structural formula for each of the following monosaccharides:

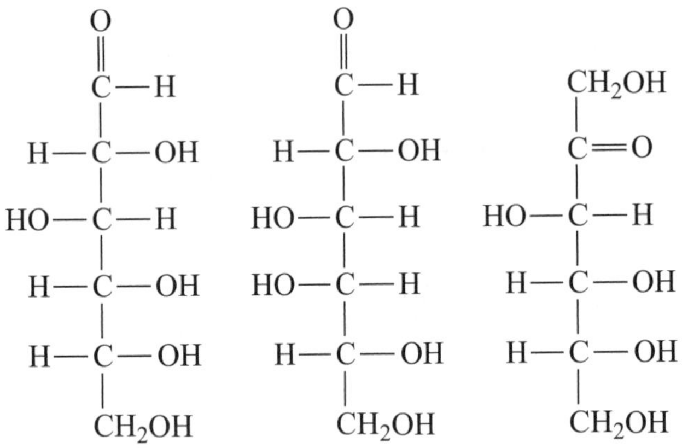

glucose galactose fructose

Answers

$$
\begin{array}{ccc}
\overset{\displaystyle O}{\underset{\displaystyle \|}{}} & \overset{\displaystyle O}{\underset{\displaystyle \|}{}} & \\
\text{C—H} & \text{C—H} & \text{CH}_2\text{OH} \\
\text{H—C—OH} & \text{H—C—OH} & \text{C}=\text{O} \\
\text{HO—C—H} & \text{HO—C—H} & \text{HO—C—H} \\
\text{H—C—OH} & \text{HO—C—H} & \text{H—C—OH} \\
\text{H—C—OH} & \text{H—C—OH} & \text{H—C—OH} \\
\text{CH}_2\text{OH} & \text{CH}_2\text{OH} & \text{CH}_2\text{OH} \\
\text{Glucose} & \text{Galactose} & \text{Fructose}
\end{array}
$$

Guide to Drawing Cyclic Structures

STEP 1 Turn the open-chain structure clockwise 90°.
STEP 2 Fold the chain into a hexagon and bond the O on carbon 5 to the carbonyl group.
STEP 3 Draw the new —OH group on carbon 1 down to give the α form or up to give the β form.

◆ Learning Exercise 18.1D

Draw the cyclic structures (α forms) for the following:

1. glucose **2.** galactose **3.** fructose

Answers

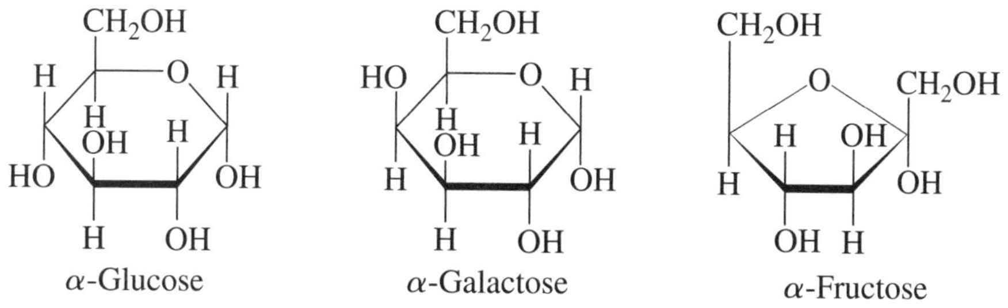

18.2 Disaccharides and Polysaccharides

- Disaccharides are glycosides, two monosaccharide units joined together by a glycosidic bond: monosaccharide (1) +monosaccharide (2) →disaccharide + H_2O
- In maltose, two glucose units are linked by an α-1,4-glycosidic bond. The α-1,4 indicates that the —OH of the alpha form at carbon 1 is bonded to the —OH on carbon 4 of the other glucose molecule.
- A beta (β) linkage involves the —OH of the beta form at carbon 1. A 1,6-glycosidic bond involves bonding to the —OH on carbon 6.
- When a disaccharide is hydrolyzed by water, the products are a glucose unit and one other monosaccharide:

$$\text{Maltose} + H_2O \longrightarrow \text{glucose} + \text{glucose}$$
$$\text{Lactose} + H_2O \longrightarrow \text{glucose} + \text{galactose}$$
$$\text{Sucrose} + H_2O \longrightarrow \text{glucose} + \text{fructose}$$

- Polysaccharides are polymers of monosaccharide units.
- Starches consist of amylose, an unbranched chain of glucose, and amylopectin which is a branched polymer of glucose. Glycogen, the storage form of glucose in animals, is similar to amylopectin, but has more branching.
- Cellulose is also a polymer of glucose, but in cellulose the glycosidic bonds are β bonds rather than α bonds as in the starches. Humans can digest starches to obtain energy, but not cellulose. However, cellulose is important as a source of fiber in our diets.

◆ Learning Exercise 18.2A

For the following disaccharides, state (a) the monosaccharide units, (b) the type of glycosidic bond, and (c) the name of the disaccharide.

1. 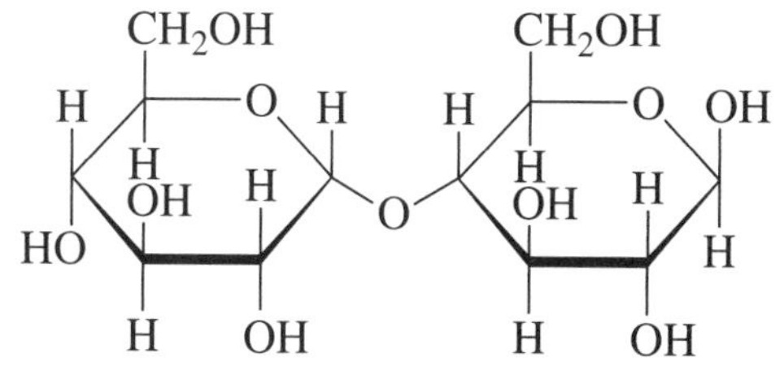

2. 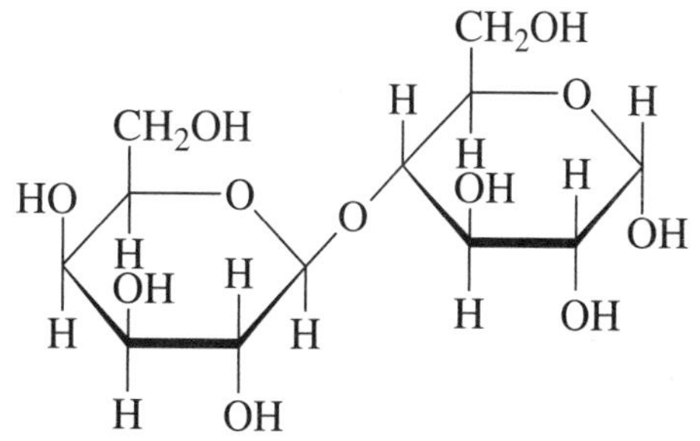

3. 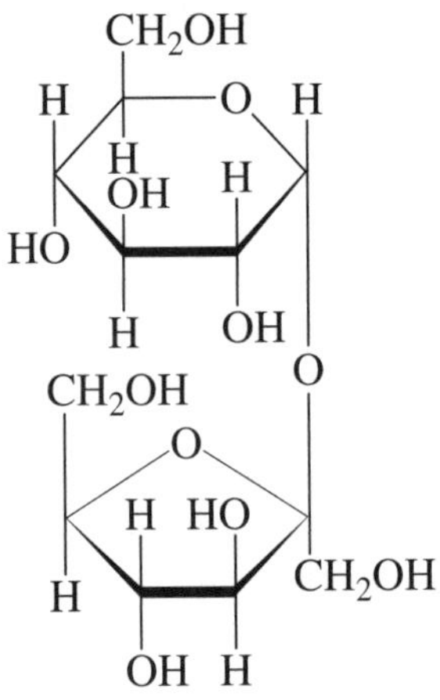

4. 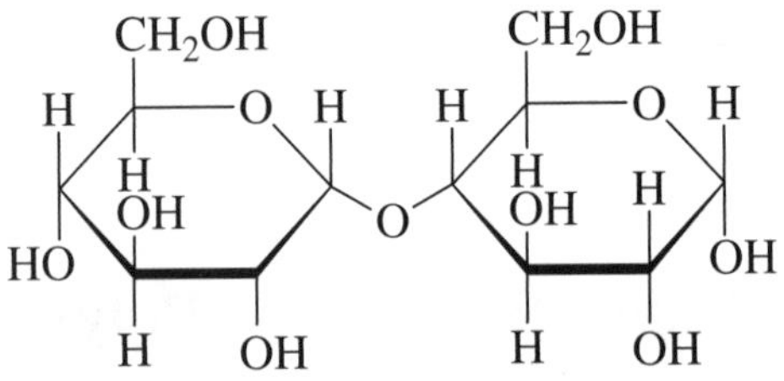

	a. Monosaccharide(s)	b. Type of Glycosidic Bond	c. Name of Disaccharide
1.			
2.			
3.			
4.			

Answers

1. **(a)** two glucose units **(b)** α-1,4-glycosidic bond **(c)** β-maltose
2. **(a)** galactose + glucose **(b)** β-1,4-glycosidic bond **(c)** α-lactose
3. **(a)** fructose + glucose **(b)** α, β-1,2 glycosidic bond **(c)** sucrose
4. **(a)** two glucose units **(b)** α-1,4-glycosidic bond **(c)** α-maltose

◆ Learning Exercise 18.2B

List the monosaccharides and describe the glycosidic bonds in each of the following carbohydrates:

	Monosaccharides	**Type(s) of Glycosidic Bonds**
1. amylose	________________	________________
2. amylopectin	________________	________________
3. glycogen	________________	________________
4. cellulose	________________	________________

Answers
1. glucose; α-1,4-glycosidic bonds
2. glucose; α-1,4- and α-1,6-glycosidic bonds
3. glucose; α-1,4- and α-1,6-glycosidic bonds
4. glucose; β-1,4-glycosidic bonds

18.3 Lipids

- Lipids are nonpolar compounds that are not soluble in water.
- Fatty acids are long-chain carboxylic acids that may be saturated, monounsaturated, or polyunsaturated.
- Unsaturated fatty acids can have cis and trans isomers.
- The hydrogenation of unsaturated fatty acids converts double bonds to single bonds.
- In saponification, a fat heated with a strong base produces glycerol and the salts of the fatty acids or soaps.
- The triacylglycerols in fats and oils are esters of glycerol and three fatty acids.
- Fats from animal sources contain more saturated fatty acids and have higher melting points than fats found in most vegetable oils.
- Steroids are lipids containing the steroid nucleus, which is a fused structure of four rings. Cholesterol is one of the most important and abundant steroids.
- Steroid hormones act as chemical messengers.

MasteringChemistry
Tutorial: Structures and Properties of Fatty Acids
Tutorial: Lipids
Tutorial: Cholesterol

Key Terms

Match each of the following terms with the correct statement:

a. lipid **b.** fatty acid **c.** triacylglycerol **d.** saponification **e.** steroid

1. _______ a type of compound that is not soluble in water, but is in nonpolar solvents
2. _______ the reaction of a triacylglycerol with a strong base producing salts called soaps and glycerol
3. _______ a lipid consisting of glycerol bonded to three fatty acids
4. _______ a lipid composed of a multicyclic ring system
5. _______ long-chain carboxylic acid found in triacylglycerols

Answers **1.** a **2.** d **3.** c **4.** e **5.** b

◆ Learning Exercise 18.3A

Draw the condensed structural formulas of linoleic acid, stearic acid, and oleic acid.

1. linoleic acid

2. stearic acid

3. oleic acid

4. Which of the three fatty acids (1, 2, and/or 3)

a. is the most saturated? ________ **b.** is the most unsaturated? ________

c. has the lowest melting point? ________ **d.** has the highest melting point? ________

e. is found in vegetables? ________ **f.** is from animal sources? ________

Answers

1. linoleic acid

$$CH_3{-}(CH_2)_4{-}CH{=}CH{-}CH_2{-}CH{=}CH{-}(CH_2)_7{-}\overset{\overset{\large O}{\|}}{C}{-}OH$$

2. stearic acid

$$CH_3{-}(CH_2)_{16}{-}\overset{\overset{\large O}{\|}}{C}{-}OH$$

3. oleic acid

$$CH_3{-}(CH_2)_7{-}CH{=}CH{-}(CH_2)_7{-}\overset{\overset{\large O}{\|}}{C}{-}OH$$

4. **a.** 2 **b.** 1 **c.** 1 **d.** 2 **e.** 1 and 3 **f.** 2

◆ Learning Exercise 18.3B

Consider the following fatty acid called oleic acid:

$$CH_3{-}(CH_2)_7{-}CH{=}CH{-}(CH_2)_7{-}\overset{\overset{\large O}{\|}}{C}{-}OH$$

1. Why is oleic acid considered an acid?

2. Is oleic acid a saturated or unsaturated compound? Why?

3. Is oleic acid likely to be a solid or a liquid at room temperature?

4. Why is oleic acid not soluble in water?

Answers	1. contains a carboxylic acid group	2. unsaturated; has a double bond	
	3. liquid	4. It has a long hydrocarbon chain.	

◆ Learning Exercise 18.3C

Draw the condensed structural formula of the triacylglycerol formed from the following:

1. glycerol and three palmitic acids, $CH_3-(CH_2)_{12}-\overset{\overset{\displaystyle O}{\|}}{C}-OH$

2. glycerol and three myristic acids, $CH_3-(CH_2)_{12}-\overset{\overset{\displaystyle O}{\|}}{C}-OH$

Answers

1.

$$CH_2-O-\overset{\overset{\displaystyle O}{\|}}{C}-(CH_2)_{14}-CH_3$$
$$HC-O-\overset{\overset{\displaystyle O}{\|}}{C}-(CH_2)_{14}-CH_3$$
$$CH_2-O-\overset{\overset{\displaystyle O}{\|}}{C}-(CH_2)_{14}-CH_3$$

2.

$$CH_2-O-\overset{\overset{\displaystyle O}{\|}}{C}-(CH_2)_{12}-CH_3$$
$$HC-O-\overset{\overset{\displaystyle O}{\|}}{C}-(CH_2)_{12}-CH_3$$
$$CH_2-O-\overset{\overset{\displaystyle O}{\|}}{C}-(CH_2)_{12}-CH_3$$

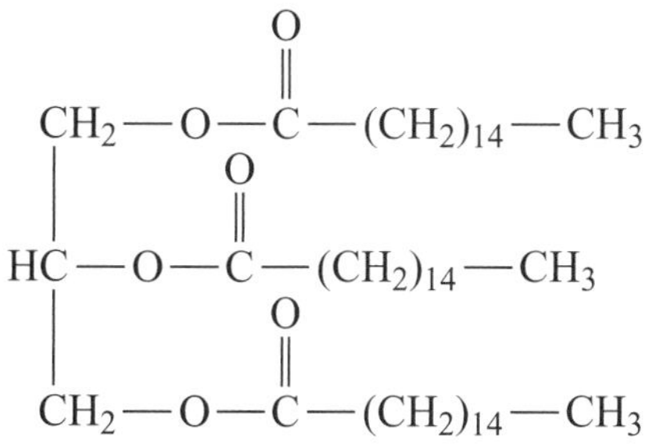

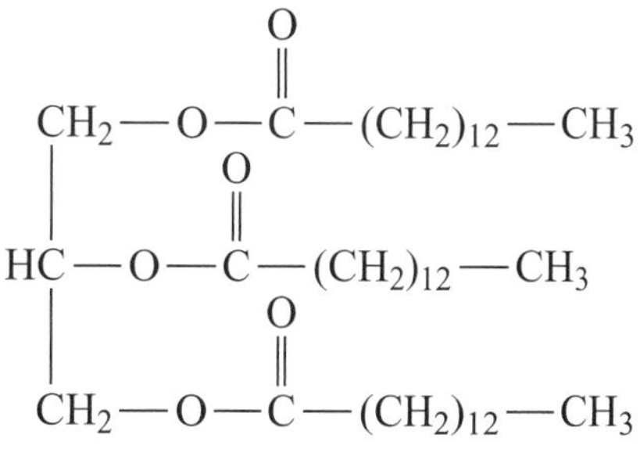

◆ Learning Exercise 18.3D

Draw the condensed structural formula for each of the following triacylglycerols:

1. glyceryl tristearate (tristearin)

2. glyceryl trioleate (triolein)

Answers

1.

$$CH_2-O-\overset{\displaystyle O}{\overset{\|}{C}}-(CH_2)_{16}-CH_3$$
$$HC-O-\overset{\displaystyle O}{\overset{\|}{C}}-(CH_2)_{16}-CH_3$$
$$CH_2-O-\overset{\displaystyle O}{\overset{\|}{C}}-(CH_2)_{16}-CH_3$$

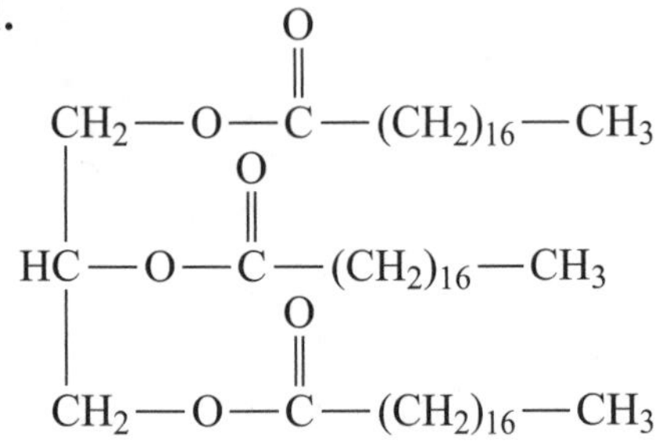

Glyceryl tristearate (tristearin)

2.

$$CH_2-O-\overset{\displaystyle O}{\overset{\|}{C}}-(CH_2)_7-CH=CH-(CH_2)_7-CH_3$$
$$HC-O-\overset{\displaystyle O}{\overset{\|}{C}}-(CH_2)_7-CH=CH-(CH_2)_7-CH_3$$
$$CH_2-O-\overset{\displaystyle O}{\overset{\|}{C}}-(CH_2)_7-CH=CH-(CH_2)_7-CH_3$$

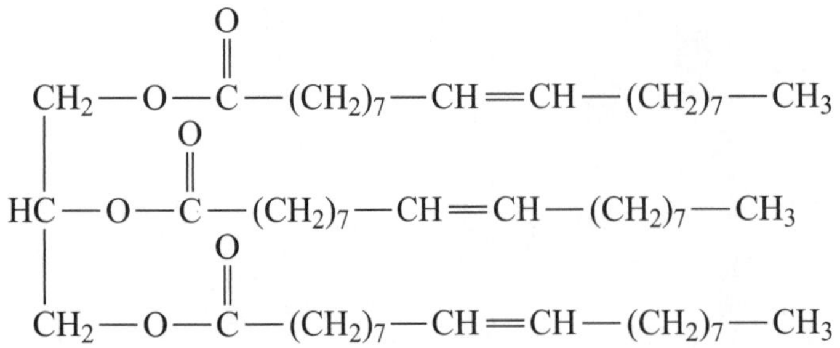

Glyceryl trioleate (triolein)

◆ Learning Exercise 18.3E

Write the equations for the following reactions of glyceryl trioleate (triolein):

1. hydrogenation with a nickel catalyst

2. saponification with NaOH

Answers

1.

$$\begin{array}{l}
CH_2-O-\overset{\displaystyle O}{\overset{\|}{C}}-(CH_2)_7-CH{=}CH-(CH_2)_7-CH_3 \\
HC-O-\overset{\displaystyle O}{\overset{\|}{C}}-(CH_2)_7-CH{=}CH-(CH_2)_7-CH_3 + 3H_2 \xrightarrow{\ Ni\ } \\
CH_2-O-\overset{\displaystyle O}{\overset{\|}{C}}-(CH_2)_7-CH{=}CH-(CH_2)_7-CH_3
\end{array}$$

$$\begin{array}{l}
CH_2-O-\overset{\displaystyle O}{\overset{\|}{C}}-(CH_2)_{16}-CH_3 \\
HC-O-\overset{\displaystyle O}{\overset{\|}{C}}-(CH_2)_{16}-CH_3 \\
CH_2-O-\overset{\displaystyle O}{\overset{\|}{C}}-(CH_2)_{16}-CH_3
\end{array}$$

2.

$$\begin{array}{l}
CH_2-O-\overset{\displaystyle O}{\overset{\|}{C}}-(CH_2)_7-CH{=}CH-(CH_2)_7-CH_3 \\
HC-O-\overset{\displaystyle O}{\overset{\|}{C}}-(CH_2)_7-CH{=}CH-(CH_2)_7-CH_3 + 3NaOH \longrightarrow \\
CH_2-O-\overset{\displaystyle O}{\overset{\|}{C}}-(CH_2)_7-CH{=}CH-(CH_2)_7-CH_3
\end{array}$$

$$\begin{array}{l}
CH_2-OH \\
HC-OH \\
CH_2-OH
\end{array}$$

$$+ 3Na^+ \ ^-O-\overset{\displaystyle O}{\overset{\|}{C}}-(CH_2)_7-CH{=}CH-(CH_2)_7-CH_3$$

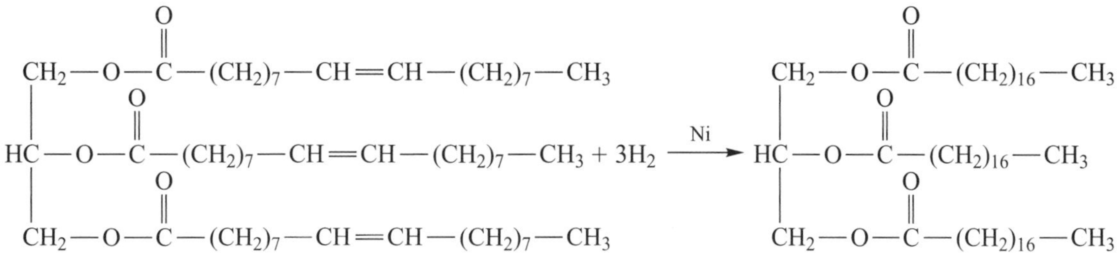

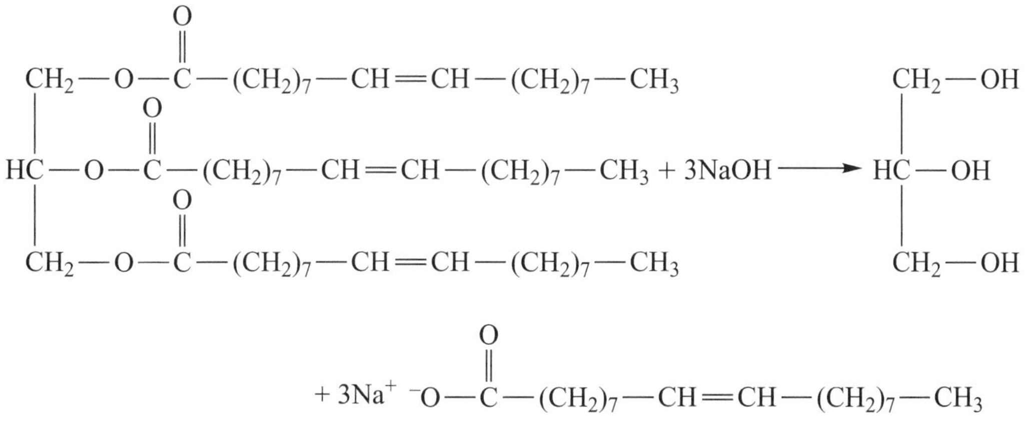

◆ Learning Exercise 18.3F

1. Draw the structure of the steroid nucleus.

2. Draw the structure of cholesterol.

Answers

1.

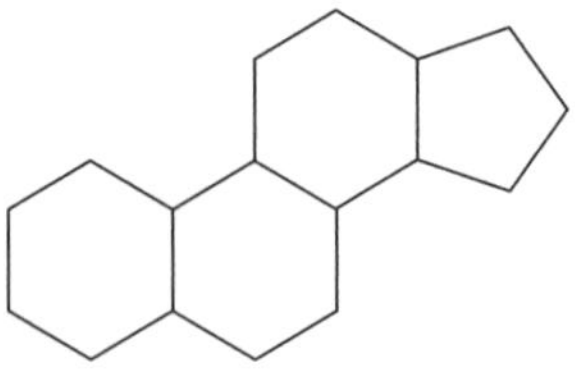

2.

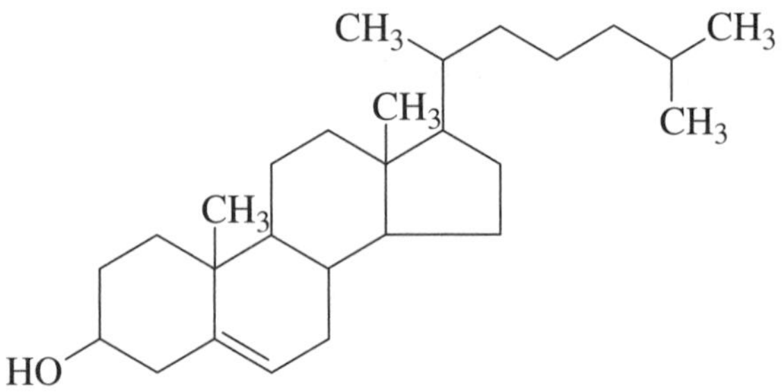

18.4 Proteins

- Some proteins are enzymes or hormones, whereas others are important in structure, transport, protection, storage, and contraction of muscles.
- The molecular building blocks of proteins are a group of 20 amino acids.
- In an amino acid, a central (alpha) carbon is attached to an amino group, a carboxyl group, and a side chain (R group), which is a characteristic group for each amino acid.
- The particular side chain makes each amino acid polar, nonpolar, acidic, or basic. Nonpolar amino acids contain hydrocarbon side chains, whereas polar amino acids contain electronegative atoms such as oxygen (—OH) or sulfur (—SH). Acidic side chains contain a carboxylate group and basic side chains contain an ammonium group.
- A peptide bond is an amide bond between the carboxyl group of one amino acid and the amino group of a second amino acid.

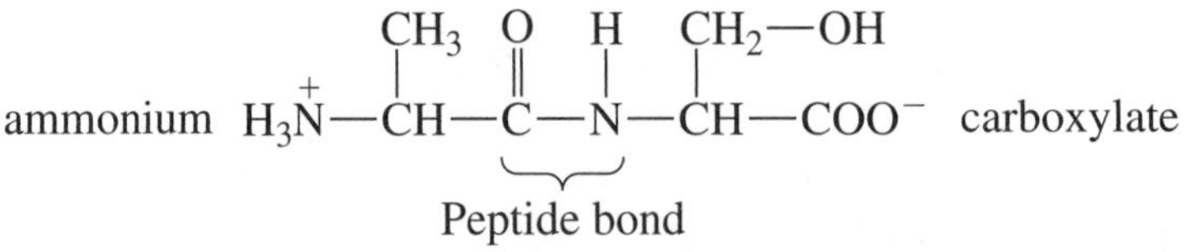

- Short chains of amino acids are called peptides. Long chains of amino acids are called proteins.
- In the name of a peptide, each amino acid beginning from the N terminal end has the *ine* (or *ic acid*) replaced by *yl*. The last amino acid at the C terminal end of the peptide uses its full name.
- Structures of peptides can be shown using the three letter designations for the component amino acids written from the N terminal to the C terminal (left to right).

MasteringChemistry
Self Study Activity: Functions of Proteins
Self Study Activity: Structure of Proteins

Key Terms

Match one of the following functions of a protein with the examples:

a. structural **b.** contractile **c.** storage **d.** transport
e. hormonal **f.** enzyme **g.** protection

1. ______ hemoglobin carries oxygen in blood
2. ______ amylase hydrolyzes starch
3. ______ egg albumin, a protein in egg white
4. ______ hormone that controls growth
5. ______ collagen in connective tissue
6. ______ immunoglobulin
7. ______ keratin, a major protein of hair
8. ______ lipoproteins carry lipids in blood

Answers **1.** d **2.** f **3.** c **4.** e
5. a **6.** g **7.** a **8.** d

◆ Learning Exercise 18.4A

Using the appropriate side chain, complete the structural formula of each of the following amino acids. Indicate whether the amino acid would be polar, nonpolar, acidic, or basic.

Glycine

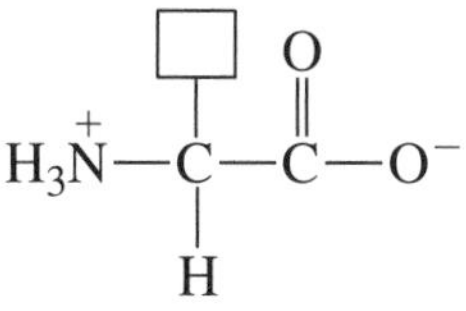

Alanine

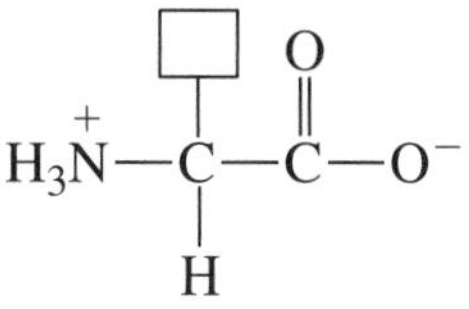

Serine

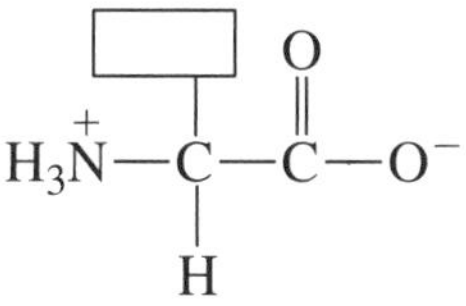

Aspartic acid

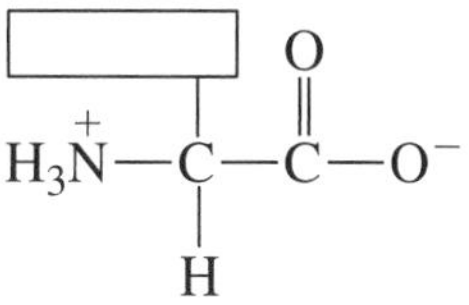

Answers

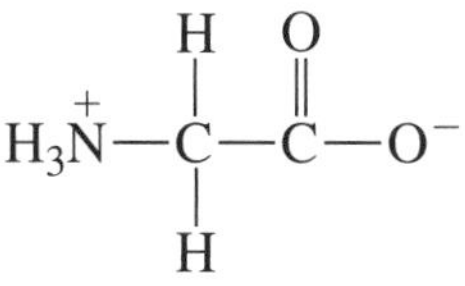

nonpolar

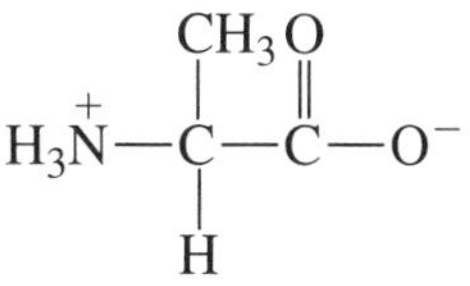

nonpolar

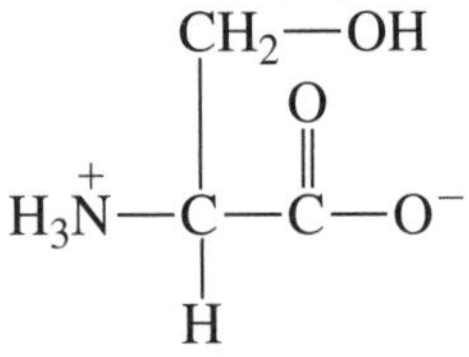

$$\text{H}_3\overset{+}{\text{N}}-\overset{\overset{\displaystyle \text{CH}_2-\text{OH}}{\underset{\underset{\displaystyle \text{H}}{|}}{|}}{\text{C}}-\overset{\overset{\displaystyle \text{O}}{\|}}{\text{C}}-\text{O}^-$$

polar

$$\text{H}_3\overset{+}{\text{N}}-\overset{\overset{\displaystyle \text{CH}_2-\overset{\overset{\displaystyle \text{O}}{\|}}{\text{C}}-\text{O}^-}{\underset{\underset{\displaystyle \text{H}}{|}}{|}}{\text{C}}-\overset{\overset{\displaystyle \text{O}}{\|}}{\text{C}}-\text{O}^-$$

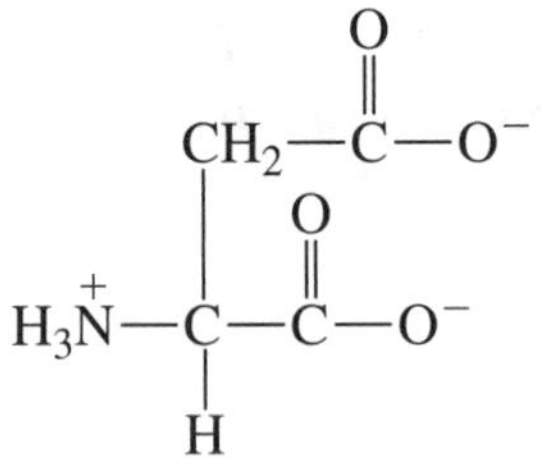

acidic

◆ **Learning Exercise 18.4B**

Draw the structural formulas of the following di- and tripeptides.

1. Ser–Gly

2. Cys–Val

3. Gly–Ser–Cys

Answers

1. $\text{H}_3\overset{+}{\text{N}}-\overset{\overset{\displaystyle \text{OH}}{\overset{\displaystyle |}{\text{CH}_2}}}{\text{CH}}-\overset{\overset{\displaystyle \text{O}}{\|}}{\text{C}}-\overset{\overset{\displaystyle \text{H}}{|}}{\text{N}}-\text{CH}_2-\text{COO}^-$

2. $\text{H}_3\overset{+}{\text{N}}-\overset{\overset{\displaystyle \text{SH}}{\overset{\displaystyle |}{\text{CH}_2}}}{\text{CH}}-\overset{\overset{\displaystyle \text{O}}{\|}}{\text{C}}-\overset{\overset{\displaystyle \text{H}}{|}}{\text{N}}-\overset{\overset{\displaystyle \text{CH}_3}{\overset{\displaystyle |}{\text{CH}-\text{CH}_3}}}{\text{CH}}-\text{COO}^-$

3. $\text{H}_3\overset{+}{\text{N}}-\text{CH}_2-\overset{\overset{\displaystyle \text{O}}{\|}}{\text{C}}-\overset{\underset{\underset{\displaystyle \text{H}}{|}}{|}}{\text{N}}-\overset{\overset{\displaystyle \text{OH}}{\overset{\displaystyle |}{\text{CH}_2}}}{\text{CH}}-\overset{\overset{\displaystyle \text{O}}{\|}}{\text{C}}-\overset{\underset{\underset{\displaystyle \text{H}}{|}}{|}}{\text{N}}-\overset{\overset{\displaystyle \text{SH}}{\overset{\displaystyle |}{\text{CH}_2}}}{\text{CH}}-\text{COO}^-$

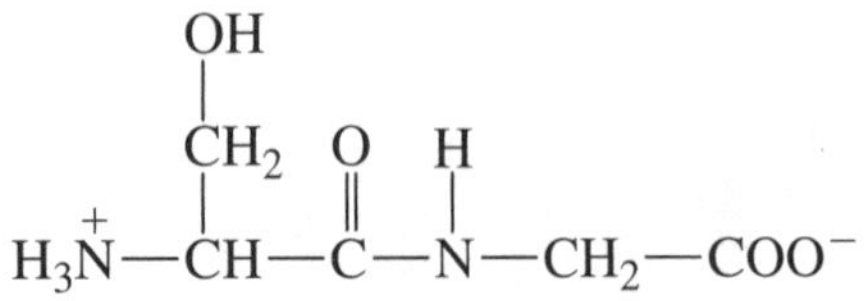

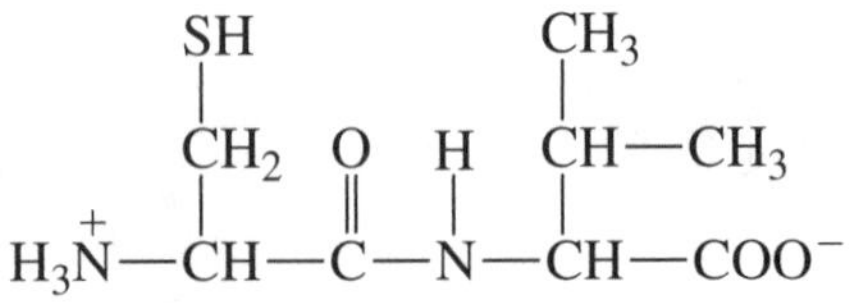

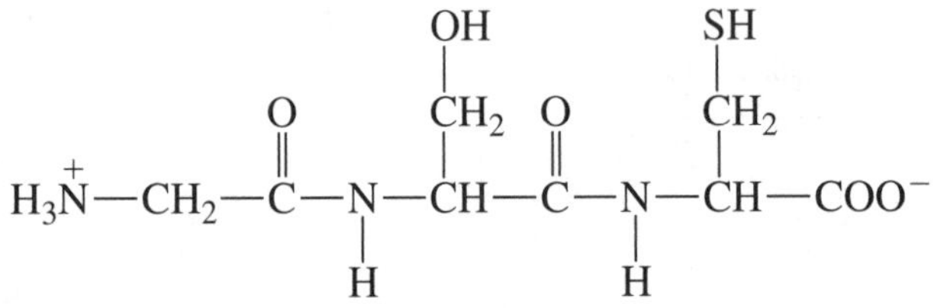

18.5 Protein Structure

- The primary structure of a protein is the sequence of amino acids.

- In the secondary structure, hydrogen bonds between different sections of the polypeptide or between different polypeptides produce a characteristic shape such as an α helix, β-pleated sheet, or a triple helix.

- In the tertiary structure, the polypeptide chain folds upon itself to form a three-dimensional shape with hydrophobic side groups on the inside and hydrophilic side groups on the outside surface. The tertiary structure is stabilized by interactions between side groups.

- The interactions can be hydrophobic, hydrophilic, salt bridges, hydrogen bonds, or disulfide bonds.
- In a quaternary structure, two or more subunits combine for biological activity.

MasteringChemistry
Tutorial: Levels of Structure in Proteins
Self Study Activity: Primary and Secondary Structure
Self Study Activity: Tertiary and Quaternary Structure

◆ Learning Exercise 18.5A

Identify the following as descriptions of primary or secondary protein structure:

1. ______ Hydrogen bonding forms an alpha (α) helix.

2. ______ Hydrogen bonding occurs between C=O and N—H within a peptide chain.

3. ______ The sequence of amino acids is linked by peptide bonds.

4. ______ Hydrogen bonds between protein chains form a pleated-sheet structure.

Answers **1.** secondary **2.** secondary **3.** primary **4.** secondary

◆ Learning Exercise 18.5B

Identify the following descriptions of protein structure as tertiary or quaternary:

1. ______ a disulfide bond joining distant parts of a peptide

2. ______ the combination of four protein subunits

3. ______ hydrophilic side groups attracted to water

4. ______ a salt bridge forms between two oppositely charged side chains

Answers **1.** tertiary **2.** quaternary **3.** tertiary **4.** tertiary

18.6 Proteins as Enzymes

- Enzymes are proteins that accelerate the rate of biological reactions by lowering the activation energy of a reaction.
- Within the structure of the enzyme, a small pocket called the active site has a specific shape that fits a specific substrate.
- In the lock-and-key model or the induced-fit model, an enzyme and substrate form an enzyme–substrate complex, so the reaction of the substrate can be catalyzed at the active site.
- In the lock-and-key model, the active site is rigid and accommodates only a single substrate. In the induced-fit model, the active site is flexible so that both the configurations of the active site and substrate are adjustable for the best fit.

MasteringChemistry
Self Study Activity: How Proteins Work

Key Terms

Match the following terms with the following descriptions:

a. active site **b.** substrate **c.** enzyme–substrate complex

d. lock and key **e.** induced fit

1. _______ the combination of an enzyme with a substrate

2. _______ a model of enzyme action in which the rigid shape of the active site exactly fits the shape of the substrate

3. _______ has a tertiary structure that fits the structure of the active site

4. _______ a model of enzyme action in which the shape of the active site adjusts to fit the shape of a substrate

5. _______ the portion of an enzyme that binds to the substrate and catalyzes the reaction

Answers **1.** c **2.** d **3.** b **4.** e **5.** a

◆ Learning Exercise 18.6

Indicate whether each of the following characteristics of an enzyme is true (T) or false (F).
An enzyme

1. _______ is a biological catalyst.

2. _______ is smaller than a substrate.

3. _______ does not change the equilibrium of a reaction.

4. _______ must be obtained from the diet.

5. _______ increases the rate of a cellular reaction.

6. _______ is needed for every reaction that takes place in the cell.

7. _______ lowers the activation energy of a biological reaction.

Answers **1.** T **2.** F **3.** T **4.** F **5.** T **6.** T **7.** T

18.7 Nucleic Acids

- Nucleic acids are composed of four bases, five-carbon sugars, and a phosphate group.
- Deoxyribonucleic acid (DNA) and ribonucleic acid (RNA) are polymers of nucleotides. Each nucleotide consists of a base, a sugar, and a phosphate group.
- In DNA, the bases are adenine, thymine, guanine, or cytosine. In RNA, uracil replaces thymine. In DNA, the sugar is deoxyribose; in RNA, the sugar is ribose.
- The two strands of DNA are held together in a double helix by hydrogen bonds between complementary base pairs: A with T, and G with C.
- During DNA replication, new DNA strands form along each original DNA strand.
- Complementary base pairing ensures the correct pairing of bases to give identical copies of the original DNA.

MasteringChemistry
Tutorial: Nucleic Acid Building Blocks
Tutorial: DNA Replication
Self Study Activity: DNA Replication
Tutorial: The Double Helix
Self Study Activity: DNA and RNA Structure

Key Terms

Match each of the following items with the statements below:

a. DNA **b.** RNA **c.** double helix
d. adenine **e.** hydrogen bonds **f.** complementary base pair
g. daughter DNA

1. _____ the shape of DNA with a sugar-phosphate backbone and base pairs linked in the center
2. _____ the genetic material containing nucleotides with adenine, cytosine, guanine, and thymine
3. _____ the nucleic acid that is a single strand of nucleotides of adenine, cytosine, guanine, and uracil
4. _____ the attractions between base pairs that connect the two DNA strands
5. _____ the bases guanine and cytosine in the double helix
6. _____ the new DNA strand that forms during DNA replication
7. _____ the base that bonds to thymine

Answers **1.** c **2.** a **3.** b **4.** e
5. f **6.** g **7.** d

◆ Learning Exercise 18.7A

1. Write the names and abbreviations for the bases in each of the following:

 DNA ____

 RNA ____

2. Write the name of the sugar in each of the following nucleotides:

 DNA ____

 RNA ____

Answers
1. DNA: adenine (A), thymine (T), guanine (G), cytosine (C)
RNA: adenine (A), uracil (U), guanine (G), cytosine (C)
2. DNA: deoxyribose
RNA: ribose

◆ Learning Exercise 18.7B

Name and write the abbreviation for each of the following bases:

a.

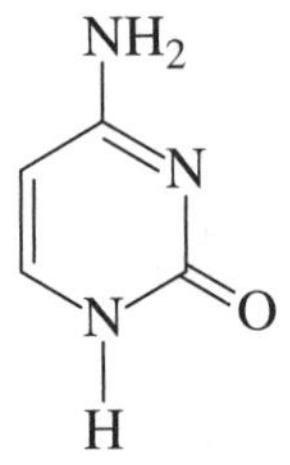

b.

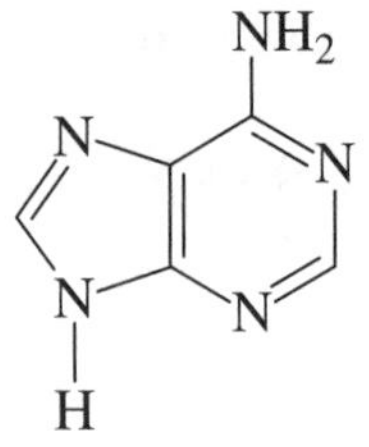

c.

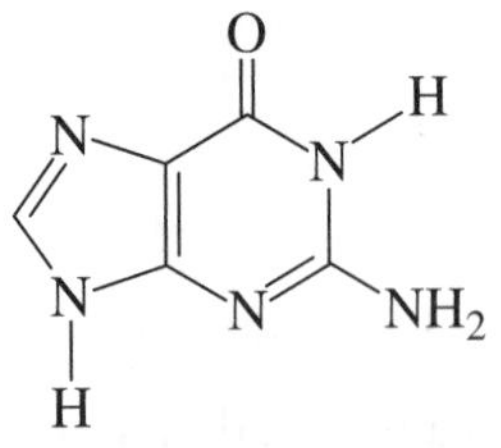

d.

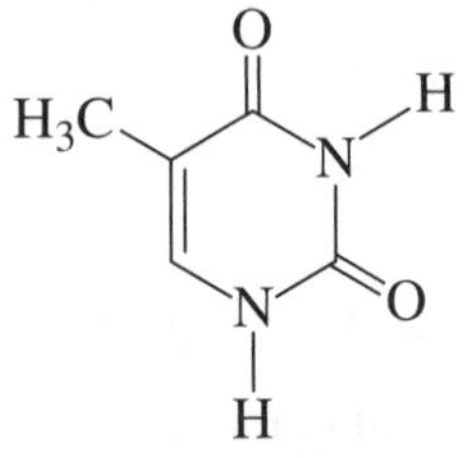

Answers **1.** cytosine (C) **2.** adenine (A)
3. guanine (G) **4.** thymine (T)

◆ Learning Exercise 18.7C

Draw the structural formula for deoxycytidine-5′-monophosphate (dCMP).

Answer

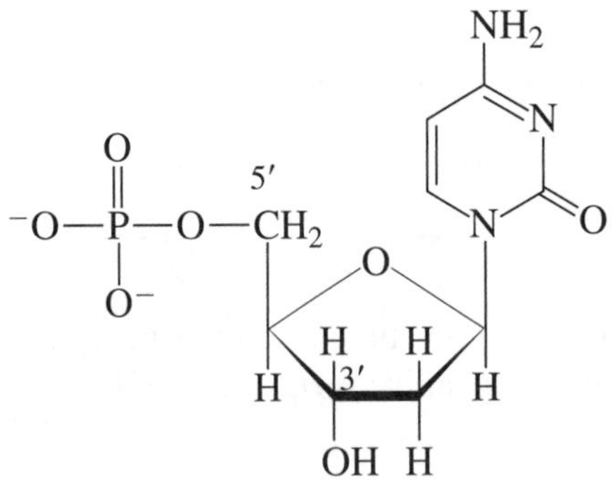

◆ Learning Exercise 18.7D

Identify each of the nucleotides in the following:

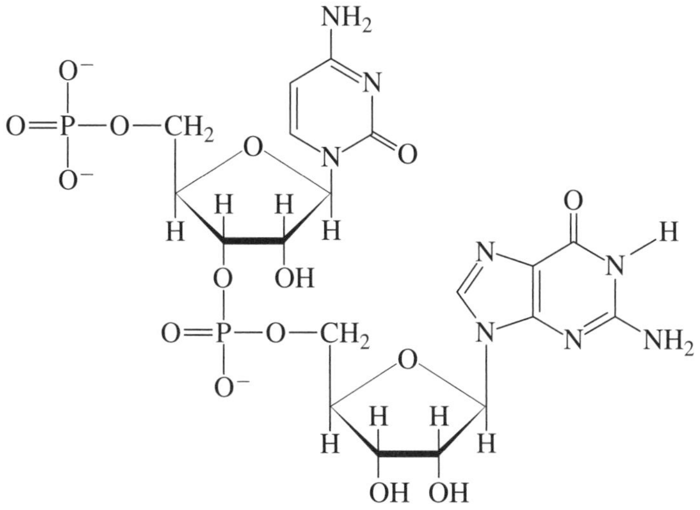

Answer Cytidine-5′-monophosphate and guanosine-5′-monophosphate

◆ Learning Exercise 18.7E

Complete each DNA section by writing the complementary strand:

1. —A—T—G—C—T—T—G—G—C—T—C—C—

2. —A—A—A—T—T—T—C—C—C—G—G—G—

3. —G—C—G—C—T—C—A—A—A—T—G—C—

Answers

1. —T—A—C—G—A—A—C—C—G—A—G—G—
2. —T—T—T—A—A—A—G—G—G—C—C—C—
3. —C—G—C—G—A—G—T—T—T—A—C—G—

◆ Learning Exercise 18.7F

How does the replication of DNA produce identical copies of the DNA?

Answer In the replication process, the bases on each strand of the separated parent DNA are paired with their complementary bases. Because each complementary base is specific for a base in DNA, the new DNA daughter strands exactly duplicate the original strands of DNA.

18.8 Protein Synthesis

- The three types of RNA differ by function in the cell: ribosomal RNA makes up most of the structure of the ribosomes, messenger RNA carries genetic information from the DNA to the ribosomes, and transfer RNA places the correct amino acids in the protein.
- Transcription is the process by which RNA polymerase produces mRNA from one strand of DNA.
- The bases in the mRNA are complementary to the DNA, except U is paired with A in DNA.
- In mRNA, the genetic code consists of a sequence of three bases (triplet or codon) that specifies the order for the amino acids in a protein.
- The codon AUG signals the start of transcription, and codons UAG, UGA, and UAA signal the stop.
- Proteins are synthesized in a translation process in which tRNA molecules bring the appropriate amino acids to the ribosome to which mRNA is bound, where each amino acid is bonded by a peptide bond to a growing peptide chain.
- When the polypeptide is released, it takes on its secondary and tertiary structures to become a functional protein in the cell.

MasteringChemistry

Tutorial: Types of RNA
Self Study Activity: Transcription

◆ Learning Exercise 18.8A

Indicate the type of RNA (mRNA, tRNA, or rRNA) described by each of the following:

1. ________ most abundant type in a cell

2. ________ has the shortest chain of nucleotides

3. ________ carries information from DNA to the ribosomes for protein synthesis

4. ________ major component of ribosomes

5. ________ carries specific amino acids to the ribosome for protein synthesis

Answers **1.** rRNA **2.** tRNA **3.** mRNA **4.** rRNA **5.** tRNA

◆ Learning Exercise 18.8B

Write the corresponding section of an mRNA produced from each of the following sections of DNA:

1. —C—A—T—T—C—G—G—T—A—

2. —G—T—A—C—C—T—A—A—C—G—T—C—C—G—

Answers

1. —G—U—A—A—G—C—C—A—U—
2. —C—A—U—G—G—A—U—U—G—C—A—G—G—C—

◆ Learning Exercise 18.8C

Write the mRNA that would form for the following sections of DNA:

1. DNA strand: —CCC—TCA—GGG—CGC—

   mRNA: _____—_____—_____—_____

2. DNA strand: —ATA—GCC—TTT—GGC—AAC—

   mRNA: _____—_____—_____—_____—_____

Answers

1. mRNA: —GGG—AGU—CCC—GCG—
2. mRNA: —UAU—CGG—AAA—CCG—UUG—

> ***MasteringChemistry***
>
> **Tutorial: Genetic Code**

◆ Learning Exercise 18.8D

Give the abbreviation for each of the amino acids coded for by the following mRNA codons:

1. UUU ______________
2. GCG ______________
3. AGC ______________
4. CCA ______________
5. GGA ______________
6. ACA ______________
7. AUG ______________
8. CUC ______________
9. CAU ______________
10. GUU ______________

Answers

1. Phe	2. Ala	3. Ser	4. Pro	5. Gly
6. Thr	7. Start/Met	8. Leu	9. His	10. Val

> ***MasteringChemistry***
>
> **Self Study Activity: Overview of Protein Synthesis**
> **Self Study Activity: Translation**
> **Tutorial: Following the Instructions in DNA**

◆ Learning Exercise 18.8E

Write the mRNA that would form for the following section of DNA. For each codon in the mRNA, write the amino acid that would be placed in the protein by a tRNA.

1. DNA strand: —CCC—TCA—GGG—CGC—

   mRNA: _____—_____—_____—_____

   Amino acids: _____—_____—_____—_____

2. DNA: —ATA—GCC—TTT—GGC—AAC—

mRNA: ____—____—____—____—____

Amino acids: ____—____—____—____—____

Answers

1. mRNA: —GGG—AGU—CCC—GCG—
—Gly—Ser—Pro—Ala—

2. mRNA: —UAU—CGG—AAA—CCG—UUG—
—Tyr—Arg—Lys—Pro—Leu—

Checklist for Chapter 18

You are ready to take the Practice Test for Chapter 18. Be sure you have accomplished the following learning goals for this chapter. If you are not sure, review the section listed at the end of the goal. Then apply your new skills and understanding to the Practice Test.

After studying Chapter 18, I can successfully:

____ Classify carbohydrates as monosaccharides, disaccharides, and polysaccharides (18.1).

____ Classify a monosaccharide (aldose or ketose), and indicate the number of carbon atoms (18.1).

____ Draw the open-chain structures for glucose, galactose, and fructose (18.1).

____ Draw the cyclic structures of monosaccharides (18.1).

____ Describe the monosaccharide units and linkages in disaccharides (18.2).

____ Describe the structural features of amylose, amylopectin, glycogen, and cellulose (18.2).

____ Identify a fatty acid as saturated or unsaturated (18.3).

____ Draw the triacylglycerol produced by the reaction of glycerol and fatty acids (18.3).

____ Draw the product from the hydrogenation or saponification of a triacylglycerol (18.3).

____ Describe the structure of a steroid and cholesterol (18.3).

____ Classify proteins by their functions in the cells (18.4).

____ Draw the structure for an amino acid (18.4).

____ Classify amino acids by the characteristics of the side chain (R group) (18.4).

____ Describe a peptide bond and draw the structure for a peptide using amino acid abbreviations (18.4).

____ Distinguish between the primary and secondary structures of a protein (18.5).

____ Distinguish between the tertiary and quaternary structures of a protein (18.5).

____ Describe the lock-and-key and induced-fit models of enzyme action (18.6).

____ Identify the components of nucleic acids RNA and DNA (18.7).

____ Describe the nucleotides contained in DNA and RNA (18.7).

____ Describe the primary structure of nucleic acids (18.7).

____ Describe the structures of RNA and DNA (18.7).

____ Explain the process of DNA replication and write a complementary strand for a DNA template (18.7).

____ Describe the structures and characteristics of the three types of RNA (18.8).

_____ Describe the synthesis of mRNA (transcription) and write the mRNA section for a DNA template (18.8).

_____ Describe the role of translation in protein synthesis and write the amino acid sequence for a mRNA section (18.8).

Practice Test for Chapter 18

1. The name *carbohydrate* came from the fact that
- **A.** carbohydrates are hydrates of water.
- **B.** carbohydrates contain hydrogen and oxygen in a 2:1 ratio.
- **C.** carbohydrates contain a great quantity of water.
- **D.** all plants produce carbohydrates.
- **E.** carbon and hydrogen atoms are abundant in carbohydrates.

2. What functional group(s) is (are) in the open chains of monosaccharides?
- **A.** hydroxyl groups
- **B.** aldehyde groups
- **C.** ketone groups
- **D.** hydroxyl and carbonyl groups
- **E.** carbonyl group

3. What is the classification of the following sugar?

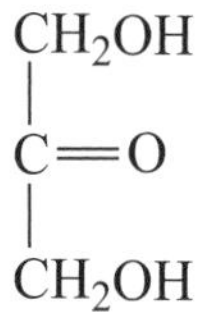

A. aldotriose **B.** ketotriose **C.** aldotetrose
D. ketotetrose **E.** ketopentose

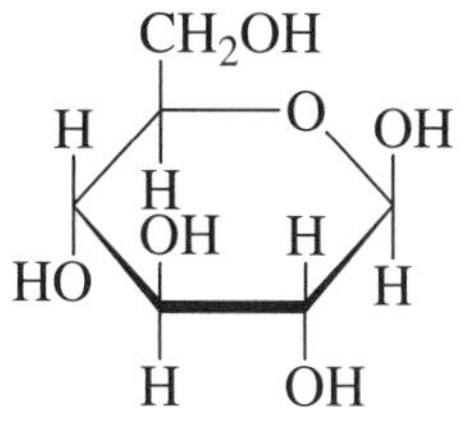

4. The structure is the cyclic structure of a
A. fructose **B.** glucose **C.** ribose
D. glyceraldehydes **E.** galactose

For questions 5 through 10, identify the carbohydrate that each of the following statements describe:

A. amylose **B.** cellulose **C.** glycogen **D.** lactose **E.** sucrose

5. _________ composed of many glucose units linked by α-1,4-glycosidic bonds

6. _________ contains glucose and galactose

7. _________ composed of glucose units joined by both α-1,4- and α-1,6-glycosidic bonds

8. _________ composed of glucose units joined by β-1,4-glycosidic bonds

9. _________ produced as a storage form of energy in plants

10. _________ used for structural purposes by plants

11. A triacylglycerol is a
A. carbohydrate **B.** lipid **C.** protein
D. oxyacid **E.** soap

12. A fatty acid that is unsaturated is usually
A. from animal sources and liquid at room temperature
B. from animal sources and solid at room temperature
C. from vegetable sources and liquid at room temperature
D. from vegetable sources and solid at room temperature
E. from both vegetable and animal sources and solid at room temperature

For questions 13 through 15, consider the following compound:

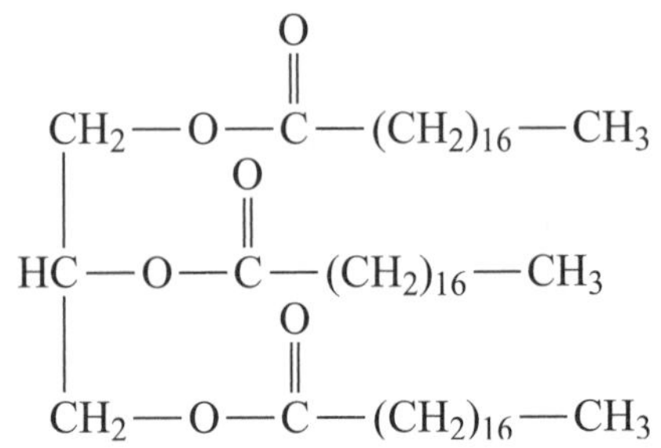

13. The molecule shown above was formed by
A. esterification **B.** combustion **C.** saponification
D. emulsification **E.** oxidation

14. If this molecule reacts with a strong base such as NaOH, the products are
A. glycerol and fatty acids
B. glycerol and water
C. glycerol and soap
D. an ester and salts of fatty acids
E. an ester and fatty acids

15. The compound would be
A. saturated, and a solid at room temperature.
B. saturated, and a liquid at room temperature.
C. unsaturated, and a solid at room temperature.
D. unsaturated, and a liquid at room temperature.
E. supersaturated, and a liquid at room temperature.

16. Which amino acid is nonpolar?
A. serine **B.** aspartic acid **C.** valine **D.** cysteine **E.** glutamine

17. Which amino acid will form disulfide cross-links in a tertiary structure?
A. serine **B.** aspartic acid **C.** valine **D.** cysteine **E.** glutamine

18. Which amino acid has a basic side chain?
A. serine **B.** aspartic acid **C.** valine **D.** cysteine **E.** lysine

19. The sequence Tyr–Ala–Gly
A. is a tripeptide
B. has two peptide bonds
C. has tyrosine with a free $-NH_3^+$ end
D. has glycine with the free $-COO^-$ end
E. all of these

20. What type of interaction is expected between lysine and aspartic acid?
A. salt bridge (ionic bond) **B.** hydrogen bond **C.** disulfide bond
D. hydrophobic interaction **E.** hydrophilic attraction

21. What type of bond is used to form the α helix structure of a protein?
A. peptide bond **B.** hydrogen bond **C.** ionic bond
D. disulfide bond **E.** hydrophobic interaction

22. What type of bonding places portions of the protein chain in the center of a tertiary structure?
A. peptide bonds **B.** ionic bonds **C.** disulfide bonds
D. hydrophobic interaction **E.** hydrophilic attraction

For questions 23 through 27, identify the protein structural levels that each of the following statements describe:
A. primary **B.** secondary **C.** tertiary
D. quaternary **E.** pentenary

23. ________ peptide bonds

24. ________ a pleated sheet

25. ________ two or more protein subunits

26. ________ an α helix

27. ________ disulfide bonds

28. Enzymes are
A. biological catalysts. **B.** polysaccharides. **C.** insoluble in water.
D. lipids. **E.** named with an *ose* ending.

29. The first step in the lock-and-key model of enzyme action is
E = enzyme; S = substrate; P = product
A. $S \longrightarrow P$ **B.** $EPS \longrightarrow E + P$ **C.** $E + S \longrightarrow ES$
D. $ES \longrightarrow E + P$ **E.** $EP \longrightarrow ES$

30. The final step in the lock-and-key model of enzyme action is
E = enzyme; S = substrate; P = product
A. $S \longrightarrow P$ **B.** $EP \longrightarrow E + P$ **C.** $E + S \longrightarrow ES$
D. $ES \longrightarrow E + P$ **E.** $EP \longrightarrow ES$

31. A nucleotide contains
A. a base **B.** a base and a sugar
C. a phosphate and a sugar **D.** a base and a deoxyribose
E. a base, a sugar, and a phosphate

32. The double helix in DNA is held together by
A. hydrogen bonds **B.** ester linkages **C.** peptide bonds
D. salt bridges **E.** disulfide bonds

33. The process of producing DNA in the nucleus is called
A. complementation **B.** replication **C.** translation
D. transcription **E.** mutation

34. Which occurs in RNA but **not** in DNA?
A. thymine **B.** cytosine **C.** adenine
D. phosphate **E.** uracil

For questions 35 through 42, select answers from the following nucleic acids:
A. DNA **B.** mRNA **C.** tRNA **D.** rRNA

35. ________ A double helix consisting of two chains of nucleotides held together by hydrogen bonds between bases.

36. ________ A nucleic acid that uses deoxyribose as the sugar.

37. ________ A nucleic acid produced in the nucleus, which migrates to the ribosomes to direct the formation of a protein.

38. ________ A nucleic acid that brings the proper amino acid to the ribosome to build the peptide chain.

39. ________ A nucleic acid that contains adenine, cytosine, guanine, and thymine.

40. ________ Along with protein, it is a major component of the ribosomes.

41. ________ It contains a triplet called an anticodon.

42. ________ This nucleic acid is replicated during cellular division.

For questions 43 to 46, select answers from the following:

A.
```
A—G—C—C—T—A
|  |  |  |  |  |
T—C—G—G—A—T
```

B. —A—U—U—G—C—U—C—

C.
```
A—G—T—U—G—U—
|  |  |  |  |  |
T—C—A—A—C—A—
```

D. —U—U—U—

E. —A—T—G—T—A—T—

43. ________ a section of mRNA

44. ________ an impossible section of DNA

45. ________ a codon for phenylalanine

46. ________ a section from a DNA molecule

For questions 47 through 50, arrange the following statements:

A. DNA forms mRNA.
B. Protein is formed and breaks away.
C. tRNA picks up specific amino acids.
D. mRNA goes to the ribosomes.

Of the previous statements, select the order in which they occur during protein synthesis.

47. ________ first step **48.** ________ second step **49.** ________ third step

50. ________ fourth step

Answers to the Practice Test

1. B	**2.** D	**3.** B	**4.** B	**5.** A
6. D	**7.** C	**8.** B	**9.** A	**10.** B
11. B	**12.** C	**13.** A	**14.** C	**15.** A
16. C	**17.** D	**18.** E	**19.** E	**20.** A
21. B	**22.** D	**23.** A	**24.** B	**25.** D
26. B	**27.** C	**28.** A	**29.** C	**30.** D
31. E	**32.** A	**33.** B	**34.** E	**35.** A
36. A	**37.** B	**38.** C	**39.** A	**40.** D
41. C	**42.** A	**43.** B	**44.** C	**45.** D
46. A	**47.** A	**48.** D	**49.** C	**50.** B

1

Chemistry in Our Lives

Study Goals

- Define the term chemistry and identify substances as chemicals.
- Describe the activities that are part of the scientific method.
- Develop a study plan for learning chemistry.

Chapter Outline

Answers and Solutions to Text Problems

1.1 **a.** Chemistry is the science of the composition, structure, properties, and reactions of matter.
b. A chemical is a substance that has the same composition and properties wherever it is found.

1.3 Many chemicals are listed on a vitamin bottle such as vitamin A, vitamin B_3, vitamin B_{12}, vitamin C, folic acid, etc.

1.5 Typical items found in a medicine cabinet and some of the chemicals they contain:
Antacid tablets: calcium carbonate, cellulose, starch, stearic acid, silicon dioxide
Mouthwash: water, alcohol, thymol, glycerol, sodium benzoate, benzoic acid
Cough suppressant: menthol, beta-carotene, sucrose, glucose

1.7 No. All of these ingredients are chemicals.

1.9 One advantage of a pesticide is that it gets rid of insects that bite or damage crops. A disadvantage is that a pesticide can destroy beneficial insects or be retained in a crop that is eventually eaten by animals or humans.

1.11 **a.** A hypothesis proposes a possible explanation for a natural phenomenon. A hypothesis must be stated in a way that it can be tested by experiments.
b. An experiment is a procedure that tests the validity of a hypothesis.
c. A theory is a hypothesis that has been validated many times by many scientists.
d. An observation is a description or measurement of a natural phenomenon.

1.13 **a.** An observation (O) is a description or measurement of a natural phenomenon.
b. A hypothesis (H) proposes a possible explanation for a natural phenomenon.
c. An experiment (E) is a procedure that tests the validity of a hypothesis.

d. An observation (O) is a description or measurement of a natural phenomenon.
e. An observation (O) is a description or measurement of a natural phenomenon.
f. A theory (T) is developed from a hypothesis of a possible explanation for a natural phenomenon.

1.15 There are several things a student can do to be successful in chemistry, including forming a study group, going to lecture, working *Sample Problems* and *Study Checks*, working *Questions and Problems* and checking answers, reading the assignment ahead of class, going to the instructor's office hours, keeping a problem notebook, etc.

1.17 Ways you can enhance your learning of chemistry:
a. Form a study group.
c. Visit the professor during office hours.
e. Become an active learner.
f. Work the exercises in the *Study Guide*.

1.19 Yes. Sherlock's investigation includes observations (gathering data), formulating a hypothesis, testing the hypothesis, and modifying it until the hypothesis is validated.

1.21 **a.** Determination of a melting point with a thermometer is an observation (O).
b. Describing a reason for the extinction of dinosaurs is a hypothesis (H).
c. Measuring the completion time of a race is an observation (O).

1.23 A hypothesis is an important part of the scientific process because it is the initial generation of a proposed explanation for a natural phenomenon and can be tested by experiments.

1.25 If experimental results do not support your hypothesis, you should **b.** write another hypothesis, and **c.** do more experiments.

1.27 A successful study plan would include:
b. Working the *Sample Problems* as you go through a chapter.
c. Going to your professor's office hours.

1.29 **a.** An observation (O) is a description or measurement of a natural phenomenon.
b. A hypothesis (H) proposes a possible explanation for a natural phenomenon.
c. An experiment (E) is a procedure that tests the validity of a hypothesis. Also, an observation (O) was made during the experiment.
d. A hypothesis (H) proposes a possible explanation for a natural phenomenon.

2
Measurements

Study Goals

- Write the names and abbreviations for the metric or SI units used in measurements of length, volume, mass, temperature, and time.
- Write a number in scientific notation.
- Identify a number as measured or exact.
- Determine the number of significant figures in a measured number.
- Adjust calculated answers to give the correct number of significant figures.
- Use the numerical values of prefixes to write a metric equality.
- Write a conversion factor for two units that describe the same quantity.
- Use conversion factors to change from one unit to another.
- Calculate the density of a substance; use the density to calculate the mass or volume of a substance.

Chapter Outline

Answers and Solutions to Text Problems

2.1 **a.** The unit is a meter, which is a unit of length.
b. The unit is a gram, which is a unit of mass.
c. The unit is a liter, which is a unit of volume.
d. The unit is a second, which is a unit of time.
e. The unit is a degree Celsius, which is a unit of temperature.

2.3 **a.** The unit is a meter, which is both an SI unit and a metric unit.
b. The unit is a kilogram, which is both an SI unit and a metric unit.
c. The unit is an inch, which is neither an SI unit nor a metric unit.
d. The unit is a second, which is both an SI unit and a metric unit.
e. The unit is a degree Celsius, which is a metric unit.

2.5 **a.** The unit is a gram, which is a metric unit.
b. The unit is a liter, which is a metric unit.
c. The unit is a degree Fahrenheit, which is neither an SI unit nor a metric unit.
d. The unit is a pound, which is neither an SI unit nor a metric unit.
e. The unit is a second, which is both an SI unit and a metric unit.

2.7 **a.** Move the decimal point four places to the left to give 5.5×10^4 m.
b. Move the decimal point two places to the left to give 4.8×10^2 g.
c. Move the decimal point six places to the right to give 5×10^{-6} cm.
d. Move the decimal point four places to the right to give 1.4×10^{-4} s.
e. Move the decimal point three places to the right to give 7.85×10^{-3} L.
f. Move the decimal point five decimal places to the left to give 6.7×10^5 kg.

2.9 **a.** The value 7.2×10^3, which is also 72×10^2, is larger than 8.2×10^2.
b. The value 3.2×10^{-2}, which is also 320×10^{-4}, is larger than 4.5×10^{-4}.
c. The value 1×10^4 or 10 000 is larger than 1×10^{-4} or 0.0001.
d. The value 6.8×10^{-2} or 0.068 is larger than 0.000 52.

2.11 **a.** The standard number is 1.2 times the power of 10^4 or 10 000, which gives 12 000 s.
b. The standard number is 8.25 times the power of 10^{-2} or 0.01, which gives 0.0825 kg.
c. The standard number is 4 times the power of 10^6 or 1 000 000, which gives 4 000 000 g.
d. The standard number is 5.8 times the power of 10^{-3} or 0.001, which gives 0.0058 m^3.

2.13 **a.** The *estimated digit* is the last digit reported in a measurement. In 8.6 m, the 6 in the first decimal (tenths) place was estimated and has some uncertainty.
b. The *estimated digit* is the 5 in the second decimal (hundredths) place.
c. The *estimated digit* is the 0 in the first decimal (tenths) place.

2.15 Measured numbers are obtained using some kind of measuring tool. Exact numbers are numbers obtained by counting items or using a definition that compares two units in the same measuring system.
a. measured **b.** exact **c.** exact **d.** measured

2.17 Measured numbers are obtained using some kind of measuring tool. Exact numbers are numbers obtained by counting items or using a definition that compares two units in the same measuring system.
a. 3 hamburgers is a counted/exact number, whereas the value 6 oz of hamburger meat is obtained by measurement.
b. Neither are measured numbers; both 1 table and 4 chairs are counted/exact numbers.
c. Both 0.75 lb of grapes and 350 g of butter are obtained by measurement.
d. Neither are measured numbers; the values in a definition are exact numbers.

2.19 **a.** Zeros at the beginning of a decimal number are *not significant.*
b. Zeros between nonzero digits are *significant.*
c. Zeros at the end of a decimal number are *significant.*
d. Zeros in the coefficient of a number written in scientific notation are *significant.*
e. Zeros used as placeholders in a large number without a decimal point are *not significant.*

2.21 **a.** All five numbers are significant figures.
b. Only the two nonzero numbers are significant; the preceding zeros are placeholders.
c. Only the two nonzero numbers are significant; the zeros that follow are placeholders.
d. All three numbers in the coefficient of a number written in scientific notation are significant.

e. All four numbers to the right of the decimal point, including the last zero, in a decimal number are significant.
f. All three numbers including the zeros at the end of a decimal number are significant.

2.23 Both measurements in part **c** have two significant figures, and both measurements in part **d** have four significant figures.

2.25 **a.** 5000 L is the same as 5×1000 L, which is written in scientific notation as 5.0×10^3 L with two significant figures.
b. 30 000 g is the same as $3 \times 10\,000$ g, which is written in scientific notation as 3.0×10^4 g with two significant figures.
c. 100 000 m is the same as $1 \times 100\,000$ m, which is written in scientific notation as 1.0×10^5 m with two significant figures.
d. 0.000 25 cm is the same as $2.5 \times \frac{1}{10\,000}$ cm, which is written in scientific notation as 2.5×10^{-4} cm.

2.27 The number of figures expressed in the answer to a calculation is limited by the precision of the measured numbers used in the calculation.

2.29 **a.** 1.85 kg; the last digit is dropped since it is 4 or less.
b. 88.0 L; since the fourth digit is 4 or less, the last three digits are dropped.
c. 0.004 74 cm; since the fourth significant digit (the first digit to be dropped) is 5 or greater, the last retained digit is increased by 1 when the last four digits are dropped.
d. 8810 m; since the fourth significant digit (the first digit to be dropped) is 5 or greater, the last retained digit is increased by 1 when the last digit is dropped (a nonsignificant zero is added at the end as a placeholder).
e. 1.83×10^3 s; since the fourth digit is 4 or less, the last two digits are dropped. The $\times 10^3$ is retained so that the magnitude of the answer is not changed.

2.31 **a.** To round off 56.855 m to three significant figures, drop the final digits 55 and increase the last retained digit by 1 to give 56.9 m.
b. To round off 0.002 282 5 g to three significant figures, drop the final digits 25 to give 0.002 28 g.
c. To round off 11 527 s to three significant figures, drop the final digits 27 and add two zeros as placeholders to give 11 500 s (1.15×10^4 s).
d. To express 8.1 L to three significant figures, add a significant zero to give 8.10 L.

2.33 **a.** $45.7 \times 0.034 = 1.6$ Two significant figures are allowed since 0.034 has 2 SFs.
b. $0.002\,78 \times 5 = 0.01$ One significant figure is allowed since 5 has 1 SF.
c. $\frac{34.56}{1.25} = 27.6$ Three significant figures are allowed since 1.25 has 3 SFs.
d. $\frac{(0.2465)(25)}{1.78} = 3.5$ Two significant figures are allowed since 25 has 2 SFs.
e. $(2.8 \times 10^4)(5.05 \times 10^{-6}) = 0.14$ or 1.4×10^{-1} Two significant figures are allowed since 2.8×10^4 has 2 SFs.
f. $\frac{(3.45 \times 10^{-2})(1.8 \times 10^5)}{(8 \times 10^3)} = 0.8$ or 8×10^{-1} One significant figure is allowed since 8×10^3 has 1 SF.

2.35 **a.** 45.48 cm + 8.057 cm = 53.54 cm Two decimal places are allowed since 45.48 cm has two decimal places.
b. 23.45 g + 104.1 g + 0.025 g = 127.6 g One decimal place is allowed since 104.1 g has one decimal place.
c. 145.675 mL − 24.2 mL = 121.5 mL One decimal place is allowed since 24.2 mL has one decimal place.
d. 1.08 L − 0.585 L = 0.50 L Two decimal places are allowed since 1.08 L has two decimal places.

2.37 The km/h markings indicate how many kilometers (how much distance) will be traversed in 1 hour's time if the speed is held constant. The mph (mi/h) markings indicate the same distance traversed *but measured in miles* during the 1 hour of travel.

2.39 Because the prefix *kilo* means to multiply by 1000, 1 kg is the same mass as 1000 g.

2.41 **a.** mg **b.** dL **c.** km **d.** fg
e. μL **f.** ns

2.43 **a.** 0.01 **b.** 1 000 000 000 000 (or 1×10^{12})
c. 0.001 (or 1×10^{-3}) **d.** 0.1
e. 1 000 000 (or 1×10^{6}) **f.** 0.000 000 001 (or 1×10^{-9})

2.45 **a.** 1 m = 100 cm **b.** 1 nm = 1×10^{-9} m
c. 1 mm = 0.001 m **d.** 1 L = 1000 mL

2.47 **a.** kilogram, since 10^3 g is greater than 10^{-3} g.
b. milliliter, since 10^{-3} L is greater than 10^{-6} L.
c. cm, since 10^{-2} m is greater than 10^{-12} m.
d. kL, since 10^3 L is greater than 10^{-1} L.
e. nanometer, since 10^{-9} m is greater than 10^{-12} m.

2.49 A conversion factor can be inverted to give a second conversion factor: $\frac{1\text{ m}}{100\text{ cm}}$ and $\frac{100\text{ cm}}{1\text{ m}}$

2.51 The numerator and denominator are from the equality 1 kg = 1000 g.

2.53 **a.** 1 yd = 3 ft; $\frac{1\text{ yd}}{3\text{ ft}}$ and $\frac{3\text{ ft}}{1\text{ yd}}$

b. 1 L = 1000 mL; $\frac{1\text{ L}}{1000\text{ mL}}$ and $\frac{1000\text{ mL}}{1\text{ L}}$

c. 1 min = 60 s; $\frac{1\text{ min}}{60\text{ s}}$ and $\frac{60\text{ s}}{1\text{ min}}$

d. 1 gal = 27 mi; $\frac{1\text{ gal}}{27\text{ mi}}$ and $\frac{27\text{ mi}}{1\text{ gal}}$

2.55 **a.** 1 m = 100 cm; $\frac{1\text{ m}}{100\text{ cm}}$ and $\frac{100\text{ cm}}{1\text{ m}}$

b. 1 g = 1×10^9 ng; $\frac{1\text{ g}}{1 \times 10^9\text{ ng}}$ and $\frac{1 \times 10^9\text{ ng}}{1\text{ g}}$

c. 1 kL = 1000 L; $\frac{1\text{ kL}}{1000\text{ L}}$ and $\frac{1000\text{ L}}{1\text{ kL}}$

d. 1 kg = 1×10^6 mg; $\frac{1\text{ kg}}{1 \times 10^6\text{ mg}}$ and $\frac{1 \times 10^6\text{ mg}}{1\text{ kg}}$

e. $(1\text{ m})^3 = (100\text{ cm})^3$; $\frac{(1\text{ m})^3}{(100\text{ cm})^3}$ and $\frac{(100\text{ cm})^3}{(1\text{ m})^3}$

2.57 **a.** 1 s = 3.5 m; $\frac{1\text{ s}}{3.5\text{ m}}$ and $\frac{3.5\text{ m}}{1\text{ s}}$

b. 1 mL = 0.74 g; $\frac{1\text{ mL}}{0.74\text{ g}}$ and $\frac{0.74\text{ g}}{1\text{ mL}}$

c. 1.0 gal of gasoline = 46.0 km; $\frac{1.0\text{ gal gasoline}}{46.0\text{ km}}$ and $\frac{46.0\text{ km}}{1.0\text{ gal gasoline}}$

d. 100 g of sterling = 93 g of silver; $\frac{100\text{ g sterling}}{93\text{ g silver}}$ and $\frac{93\text{ g silver}}{100\text{ g sterling}}$

e. 1 kg of plums = 29 μg of pesticide; $\frac{1\text{ kg plums}}{29\ \mu\text{g pesticide}}$ and $\frac{29\ \mu\text{g pesticide}}{1\text{ kg plums}}$

2.59 When using a conversion factor, you are trying to cancel existing units and arrive at a new (needed) unit. The conversion factor must be properly oriented so that the unit in the denominator cancels the preceding unit in the numerator.

2.61 **a. Given** 175 cm **Need** m

Plan cm → m $\frac{1\text{ m}}{100\text{ cm}}$

Set Up $175\ \cancel{\text{cm}} \times \frac{1\text{ m}}{100\ \cancel{\text{cm}}} = 1.75\text{ m (3 SFs)}$

b. Given 5500 mL **Need** L

Plan mL → L $\frac{1\text{ L}}{1000\text{ mL}}$

Set Up $5500\ \cancel{\text{mL}} \times \frac{1\text{ L}}{1000\ \cancel{\text{mL}}} = 5.5\text{ L (2 SFs)}$

c. Given 0.0055 kg **Need** g

Plan kg → g $\frac{1000\text{ g}}{1\text{ kg}}$

Set Up $0.0055\ \cancel{\text{kg}} \times \frac{1000\text{ g}}{1\ \cancel{\text{kg}}} = 5.5\text{ g (2 SFs)}$

d. Given 350 cm^3 **Need** m^3

Plan $\text{cm}^3 \rightarrow \text{m}^3$ $\frac{(1\text{ m})^3}{(100\text{ cm})^3}$

Set Up $350\ \cancel{\text{cm}^3} \times \frac{(1\text{ m})^3}{(100\ \cancel{\text{cm}})^3} = 3.5 \times 10^{-4}\text{ m}^3\text{ (2 SFs)}$

2.63 **a. Given** 0.750 qt **Need** mL

Plan qt → mL $\frac{946.3\text{ mL}}{1\text{ qt}}$

Set Up $0.750\ \cancel{\text{qt}} \times \frac{946.3\text{ mL}}{1\ \cancel{\text{qt}}} = 710\text{ mL (3 SFs)}$

b. Given 11.8 stones **Need** kg

Plan stones → lb → kg $\frac{14.0\text{ lb}}{1\text{ stone}}$ $\frac{1\text{ kg}}{2.205\text{ lb}}$

Set Up $11.8\ \cancel{\text{stones}} \times \frac{14.0\ \cancel{\text{lb}}}{1\ \cancel{\text{stone}}} \times \frac{1\text{ kg}}{2.205\ \cancel{\text{lb}}} = 74.9\text{ kg (3 SFs)}$

c. Given 19.5 in. **Need** mm

Plan in. → cm → mm $\frac{2.54\text{ cm}}{1\text{ in.}}$ $\frac{10\text{ mm}}{1\text{ cm}}$

Set Up $19.5\ \cancel{\text{in.}} \times \frac{2.54\ \cancel{\text{cm}}}{1\ \cancel{\text{in.}}} \times \frac{10\text{ mm}}{1\ \cancel{\text{cm}}} = 495\text{ mm (3 SFs)}$

d. Given 0.50 μm **Need** in.

Plan μm → m → cm → in. $\frac{1\text{ m}}{10^6\ \mu\text{m}}$ $\frac{100\text{ cm}}{1\text{ m}}$ $\frac{1\text{ in.}}{2.54\text{ cm}}$

Set Up $0.50\ \cancel{\mu\text{m}} \times \frac{1\ \cancel{\text{m}}}{10^6\ \cancel{\mu\text{m}}} \times \frac{100\ \cancel{\text{cm}}}{1\ \cancel{\text{m}}} \times \frac{1\text{ in.}}{2.54\ \cancel{\text{cm}}} = 2.0 \times 10^{-5}\text{ in. (2 SFs)}$

2.65 **a. Given** 78.0 ft (length) **Need** m

Plan ft → in. → cm → m $\frac{12\text{ in.}}{1\text{ ft}}$ $\frac{2.54\text{ cm}}{1\text{ in.}}$ $\frac{1\text{ m}}{100\text{ cm}}$

Set Up $78.0\text{ ft} \times \frac{12\text{ in.}}{1\text{ ft}} \times \frac{2.54\text{ cm}}{1\text{ in.}} \times \frac{1\text{ m}}{100\text{ cm}} = 23.8\text{ m (length) (3 SFs)}$

b. Given 27.0 ft (width), 23.8 m (length) **Need** m^2

Plan ft → in. → cm → m then length (m), width (m) → m^2

$\frac{12\text{ in.}}{1\text{ ft}}$ $\frac{2.54\text{ cm}}{1\text{ in.}}$ $\frac{1\text{ m}}{100\text{ cm}}$ *Area = length × width*

Set Up $27.0\text{ ft} \times \frac{12\text{ in.}}{1\text{ ft}} \times \frac{2.54\text{ cm}}{1\text{ in.}} \times \frac{1\text{ m}}{100\text{ cm}} = 8.23\text{ m (width) (3 SFs)}$

$\therefore$ Area = 23.8 m × 8.23 m = 196 m^2 (3 SFs)

c. Given 23.8 m (length), 185 km/h (speed) **Need** s

Plan m → km → h → min → s $\frac{1\text{ km}}{1000\text{ m}}$ $\frac{1\text{ h}}{185\text{ km}}$ $\frac{60\text{ min}}{1\text{ h}}$ $\frac{60\text{ s}}{1\text{ min}}$

Set Up $23.8\text{ m} \times \frac{1\text{ km}}{1000\text{ m}} \times \frac{1\text{ h}}{185\text{ km}} \times \frac{60\text{ min}}{1\text{ h}} \times \frac{60\text{ s}}{1\text{ min}} = 0.463\text{ s (3 SFs)}$

d. Given 78.0 ft (length), 27.0 ft (width) **Need** L of paint

Plan use length (ft), width (ft) → ft^2 then ft^2 → gal of paint → qt of paint → L of paint

$\frac{1\text{ gal}}{150\text{ ft}^2}$ $\frac{4\text{ qt}}{1\text{ gal}}$ $\frac{1\text{ L}}{1.057\text{ qt}}$

Set Up Area = 78.0 ft × 27.0 ft = 2100 ft^2 (3 SFs)

$2100\text{ ft}^2 \times \frac{1\text{ gal}}{150\text{ ft}^2} \times \frac{4\text{ qt}}{1\text{ gal}} \times \frac{1\text{ L}}{1.057\text{ qt}} = 53\text{ L of paint (2 SFs)}$

2.67 Each of the following requires a percent factor from the problem information.

a. Given 325 g of crust **Need** g of oxygen

Plan g of crust → g of oxygen

(percent equality: 100 g of crust = 46.7 g of oxygen) $\frac{46.7\text{ g oxygen}}{100\text{ g crust}}$

Set Up $325\text{ g crust} \times \frac{46.7\text{ g oxygen}}{100\text{ g crust}} = 152\text{ g of oxygen (3 SFs)}$

b. Given 1.25 g of crust **Need** g of magnesium

Plan g of crust → g of magnesium

(percent equality: 100 g of crust = 2.1 g of magnesium) $\frac{2.1\text{ g magnesium}}{100\text{ g crust}}$

Set Up $1.25\text{ g crust} \times \frac{2.1\text{ g magnesium}}{100\text{ g crust}} = 0.026\text{ g of magnesium (2 SFs)}$

c. Given 10.0 oz of fertilizer **Need** g of nitrogen

Plan oz of fertilizer → lb of fertilizer → g of fertilizer → g of nitrogen

(percent equality: 100 g of fertilizer = 15 g of nitrogen)

$\frac{1\text{ lb}}{16\text{ oz}}$ $\frac{453.6\text{ g}}{1\text{ lb}}$ $\frac{15\text{ g nitrogen}}{100\text{ g fertilizer}}$

Set Up $10.0 \cancel{\text{oz fertilizer}} \times \frac{1 \cancel{\text{lb}}}{16 \cancel{\text{oz}}} \times \frac{453.6 \cancel{\text{g}}}{1 \cancel{\text{lb}}} \times \frac{15 \text{ g nitrogen}}{100 \cancel{\text{g fertilizer}}} = 43 \text{ g of nitrogen (2 SFs)}$

d. Given 5.0 kg of pecans **Need** lb of chocolate bars

Plan kg of pecans → kg of bars → lb of bars

(percent equality: 100 kg of bars = 22.0 kg of pecans) $\frac{100 \text{ kg choc. bars}}{22.0 \text{ kg pecans}}$ $\frac{2.205 \text{ lb}}{1 \text{ kg}}$

Set Up $5.0 \cancel{\text{kg pecans}} \times \frac{100 \cancel{\text{kg}} \text{ choc. bars}}{22.0 \cancel{\text{kg pecans}}} \times \frac{2.205 \text{ lb}}{1 \cancel{\text{kg}}} = 50. \text{ lb of chocolate bars (2 SFs)}$

2.69 Because the density of aluminum is 2.70 g/cm^3, silver is 10.5 g/cm^3, and lead is 11.3 g/cm^3, we can identify the unknown metal by calculating its density as follows:

$$\text{Density} = \frac{\text{mass of metal}}{\text{volume of metal}} = \frac{217 \text{ g}}{19.2 \text{ cm}^3} = 11.3 \text{ g/cm}^3 \text{ (3 SFs)}$$

∴ the metal is lead.

2.71 Density is the mass of a substance divided by its volume. Density $= \frac{\text{mass (grams)}}{\text{volume (mL)}}$

The densities of solids and liquids are usually stated in g/mL or g/cm^3, so in some problems the units will need to be converted.

a. Density $= \frac{\text{mass (grams)}}{\text{volume (mL)}} = \frac{24.0 \text{ g}}{20.0 \text{ mL}} = 1.20 \text{ g/mL (3 SFs)}$

b. Given 0.250 lb of butter, 130.3 mL **Need** density (g/mL)

Plan lb → g then calculate density $\frac{453.6 \text{ g}}{1 \text{ lb}}$

Set Up $0.250 \cancel{\text{lb}} \times \frac{453.6 \text{ g}}{1 \cancel{\text{lb}}} = 113 \text{ g (3 SFs)}$

$\therefore \text{Density} = \frac{\text{mass}}{\text{volume}} = \frac{113 \text{ g}}{130.3 \text{ mL}} = 0.867 \text{ g/mL (3 SFs)}$

c. Given 2.00 mL initial volume, 3.45 mL final volume, 4.50 g **Need** density (g/mL)

Plan calculate volume by difference then calculate density

Set Up volume of gem: 3.45 mL total − 2.00 mL water = 1.45 mL

$\therefore \text{Density} = \frac{\text{mass}}{\text{volume}} = \frac{4.50 \text{ g}}{1.45 \text{ mL}} = 3.10 \text{ g/mL (3 SFs)}$

d. Given 485.6 g, 114 cm^3 **Need** density (g/mL)

Plan convert volume cm^3 → mL then calculate the density

Set Up $114 \cancel{\text{cm}^3} \times \frac{1 \text{ mL}}{1 \cancel{\text{cm}^3}} = 114 \text{ mL}$

$\therefore \text{Density} = \frac{\text{mass}}{\text{volume}} = \frac{485.6 \text{ g}}{114 \text{ mL}} = 4.26 \text{ g/mL (3 SFs)}$

e. Given 0.100 pt, 115.25 g initial, 182.48 g final **Need** density (g/mL)

Plan pt → qt → L → mL then calculate mass by difference

$\frac{1 \text{ qt}}{2 \text{ pt}}$ $\frac{1 \text{ L}}{1.057 \text{ qt}}$ $\frac{1000 \text{ mL}}{1 \text{ L}}$ then calculate the density

Set Up $0.100\ \cancel{\text{pt}} \times \frac{1\ \cancel{\text{qt}}}{2\ \cancel{\text{pt}}} \times \frac{1\ \cancel{\text{L}}}{1.057\ \cancel{\text{qt}}} \times \frac{1000\ \text{mL}}{1\ \cancel{\text{L}}} = 47.3\ \text{mL (3 SFs)}$

mass of syrup = 182.48 g − 115.25 g = 67.23 g

$\therefore\ \text{Density} = \frac{\text{mass}}{\text{volume}} = \frac{67.23\ \text{g}}{47.3\ \text{mL}} = 1.42\ \text{g/mL (3 SFs)}$

2.73 In these problems, the density is used as a conversion factor.

a. **Given** 1.50 kg of ethyl alcohol **Need** L of ethyl alcohol

Plan kg → g → mL → L $\frac{1000\ \text{g}}{1\ \text{kg}}$ $\frac{1\ \text{mL}}{0.785\ \text{g}}$ $\frac{1\ \text{L}}{1000\ \text{mL}}$

Set Up $1.50\ \cancel{\text{kg alcohol}} \times \frac{\cancel{1000\ \text{g}}}{1\ \cancel{\text{kg alcohol}}} \times \frac{1\ \cancel{\text{mL}}}{0.785\ \cancel{\text{g}}} \times \frac{1\ \text{L}}{\cancel{1000\ \text{mL}}}$
= 1.91 L of ethyl alcohol (3 SFs)

b. **Given** 6.5 mL of mercury **Need** g of mercury

Plan mL → g $\frac{13.6\ \text{g}}{1\ \text{mL}}$

Set Up $6.5\ \cancel{\text{mL}} \times \frac{13.6\ \text{g}}{1\ \cancel{\text{mL}}} = 88\ \text{g of mercury (2 SFs)}$

c. **Given** 225 mL of bronze **Need** oz of bronze

Plan mL → g → lb → oz $\frac{7.8\ \text{g}}{1\ \text{mL}}$ $\frac{1\ \text{lb}}{453.6\ \text{g}}$ $\frac{16\ \text{oz}}{1\ \text{lb}}$

Set Up $225\ \cancel{\text{mL}} \times \frac{7.8\ \cancel{\text{g}}}{1\ \cancel{\text{mL}}} \times \frac{1\ \cancel{\text{lb}}}{453.6\ \cancel{\text{g}}} \times \frac{16\ \text{oz}}{1\ \cancel{\text{lb}}} = 62\ \text{oz of bronze (2 SFs)}$

d. **Given** 74.1 cm^3 of copper **Need** g of copper

Plan $\text{cm}^3 \rightarrow \text{g}$ $\frac{8.92\ \text{g}}{1\ \text{cm}^3}$

Set Up $74.1\ \cancel{\text{cm}^3} \times \frac{8.92\ \text{g}}{1\ \cancel{\text{cm}^3}} = 661\ \text{g of copper (3 SFs)}$

e. **Given** 12.0 gal of gasoline **Need** kg of gasoline

Plan gal → qt → mL → g → kg $\frac{4\ \text{qt}}{1\ \text{gal}}$ $\frac{946.3\ \text{mL}}{1\ \text{qt}}$ $\frac{0.74\ \text{g}}{1\ \text{mL}}$ $\frac{1\ \text{kg}}{1000\ \text{g}}$

Set Up $12.0\ \cancel{\text{gal}} \times \frac{4\ \cancel{\text{qt}}}{1\ \cancel{\text{gal}}} \times \frac{946.3\ \cancel{\text{mL}}}{1\ \cancel{\text{qt}}} \times \frac{0.74\ \cancel{\text{g}}}{1\ \cancel{\text{mL}}} \times \frac{1\ \text{kg}}{1000\ \cancel{\text{g}}} = 34\ \text{kg of gasoline (2 SFs)}$

2.75 **a.** The number of legs is a counted number; it is exact.
b. The height is measured with a ruler or tape measure; it is a measured number.
c. The number of chairs is a counted number; it is exact.
d. The area is measured with a ruler or tape measure; it is a measured number.

2.77 **a.** length = 6.96 cm; width = 4.75 cm (Each answer may vary in the estimated digit)
b. length = 69.6 mm; width = 47.5 mm
c. There are three significant figures in the length measurement.
d. There are three significant figures in the width measurement.
e. Area = length × width = 6.96 cm × 4.75 cm = 33.1 cm^2
f. Since there are three significant figures in the width and length measurements, there are three significant figures in the area.

2.79 **Given** 18.5 mL initial volume, 23.1 mL final volume, 8.24 g **Need** density (g/mL)

Plan calculate volume by difference then calculate density

Set Up The volume of the object is 23.1 mL − 18.5 mL = 4.6 mL

$$\therefore \text{ Density} = \frac{\text{mass}}{\text{volume}} = \frac{8.24 \text{ g}}{4.6 \text{ mL}} = 1.8 \text{ g/mL (2 SFs)}$$

2.81 The liquid with the highest density will be at the bottom of the cylinder; the liquid with the lowest density will be at the top of the cylinder:

A is vegetable oil (D = 0.92 g/mL), B is water (D = 1.00 g/mL), C is mercury (D = 13.6 g/mL)

2.83 **a.** To round off 0.000 012 58 L to three significant figures, drop the final digit 8 and increase the last retained digit by 1 to give 0.000 012 6 L or 1.26×10^{-5} L.

b. To round off 3.528×10^2 kg to three significant figures, drop the final digit 8 and increase the last retained digit by 1 to give 3.53×10^2 kg.

c. To round off 125 111 m^3 to three significant figures, drop the final digits 111 and add three zeros as placeholders to give 125 000 m^3 (or 1.25×10^5 m^3).

d. To express 58.703 m to three significant figures, drop the final digits 03 to give 58.7 m.

e. To express 3×10^{-3} s to three significant figures, add two significant zeros to give 3.00×10^{-3} s.

f. To round off 0.010 826 g to three significant figures, drop the final digits 26 to give 0.0108 g or 1.08×10^{-2} g.

2.85 This problem requires several conversion factors. Let's take a look first at a possible unit plan. When you write out the unit plan, be sure you know a conversion factor you can use for each step.

Given 7500 ft **Need** min

Plan ft → in. → cm → m → min $\quad \frac{12 \text{ in.}}{1 \text{ ft}} \quad \frac{2.54 \text{ cm}}{1 \text{ in.}} \quad \frac{1 \text{ m}}{100 \text{ cm}} \quad \frac{1 \text{ min}}{55.0 \text{ m}}$

Set Up $7500 \cancel{\text{ft}} \times \frac{12 \cancel{\text{in.}}}{1 \cancel{\text{ft}}} \times \frac{2.54 \cancel{\text{cm}}}{1 \cancel{\text{in.}}} \times \frac{1 \cancel{\text{m}}}{100 \cancel{\text{cm}}} \times \frac{1 \text{ min}}{55.0 \cancel{\text{m}}} = 42 \text{ min (2 SFs)}$

2.87 **Given** 4.0 lb of onions **Need** number of onions

Plan lb → g → number of onions $\quad \frac{453.6 \text{ g}}{1 \text{ lb}} \quad \frac{1 \text{ onion}}{115 \text{ g}}$

Set Up $4.0 \cancel{\text{lb onions}} \times \frac{453.6 \cancel{\text{g}}}{1 \cancel{\text{lb}}} \times \frac{1 \text{ onion}}{115 \cancel{\text{g}}} = 16 \text{ onions (2 SFs)}$

2.89 **a.** **Given** 8.0 oz **Need** number of crackers

Plan oz → number of crackers $\quad \frac{6 \text{ crackers}}{0.50 \text{ oz}}$

Set Up $8.0 \cancel{\text{oz}} \times \frac{6 \text{ crackers}}{0.50 \cancel{\text{oz}}} = 96 \text{ crackers (2 SFs)}$

b. **Given** 10 crackers **Need** oz of fat

Plan number of crackers → servings → g of fat → lb → oz of fat

$\frac{1 \text{ serving}}{6 \text{ crackers}} \quad \frac{4 \text{ g fat}}{1 \text{ serving}} \quad \frac{1 \text{ lb}}{453.6 \text{ g}} \quad \frac{16 \text{ oz}}{1 \text{ lb}}$

Set Up $10 \cancel{\text{crackers}} \times \frac{1 \cancel{\text{serving}}}{6 \cancel{\text{crackers}}} \times \frac{4 \cancel{\text{g}} \text{ fat}}{1 \cancel{\text{serving}}} \times \frac{1 \cancel{\text{lb}}}{453.6 \cancel{\text{g}}} \times \frac{16 \text{ oz}}{1 \cancel{\text{lb}}} = 0.2 \text{ oz of fat (1 SF)}$

c. **Given** 50 boxes **Need** g of sodium

Plan boxes → oz → servings → mg of sodium → g of sodium

$\frac{8.0 \text{ oz}}{1 \text{ box}} \quad \frac{1 \text{ serving}}{0.50 \text{ oz}} \quad \frac{140 \text{ mg sodium}}{1 \text{ serving}} \quad \frac{1 \text{ g}}{1000 \text{ mg}}$

Set Up $50\ \cancel{\text{boxes}} \times \dfrac{8.0\ \cancel{\text{oz}}}{1\ \cancel{\text{box}}} \times \dfrac{1\ \cancel{\text{serving}}}{0.50\ \cancel{\text{oz}}} \times \dfrac{140\ \cancel{\text{mg}}\ \text{sodium}}{1\ \cancel{\text{serving}}} \times \dfrac{1\ \text{g}}{1000\ \cancel{\text{mg}}}$
$= 110$ g of sodium (2 SFs)

2.91 Given 0.45 lb **Need** cents

Plan lb → kg → pesos → dollars → cents $\dfrac{1\ \text{kg}}{2.205\ \text{lb}}$ $\dfrac{48\ \text{pesos}}{1\ \text{kg}}$ $\dfrac{1\ \text{dollar}}{14.4\ \text{pesos}}$ $\dfrac{100\ \text{cents}}{1\ \text{dollar}}$

Set Up $0.45\ \cancel{\text{lb}} \times \dfrac{1\ \cancel{\text{kg}}}{2.205\ \cancel{\text{lb}}} \times \dfrac{48\ \cancel{\text{pesos}}}{1\ \cancel{\text{kg}}} \times \dfrac{1\ \cancel{\text{dollar}}}{14.4\ \cancel{\text{pesos}}} \times \dfrac{100\ \text{cents}}{1\ \cancel{\text{dollar}}} = 68$ cents (2 SFs)

2.93 Given 325 tubes **Need** kg of benzyl salicylate

Plan tubes → oz of sunscreen → lb of sunscreen → kg of sunscreen → kg of benzyl salicylate

$\dfrac{4\ \text{oz sunscreen}}{1\ \text{tube}}$ $\dfrac{1\ \text{lb}}{16\ \text{oz}}$ $\dfrac{1\ \text{kg}}{2.205\ \text{lb}}$ $\dfrac{2.50\ \text{kg benzyl salicylate}}{100\ \text{kg sunscreen}}$

Set Up $325\ \cancel{\text{tubes}} \times \dfrac{4.0\ \cancel{\text{oz sunscreen}}}{1\ \cancel{\text{tube}}} \times \dfrac{1\ \cancel{\text{lb}}}{16\ \cancel{\text{oz}}} \times \dfrac{1\ \cancel{\text{kg}}}{2.205\ \cancel{\text{lb}}} \times \dfrac{2.50\ \text{kg benzyl salicylate}}{100\ \cancel{\text{kg sunscreen}}}$
$= 0.92$ kg of benzyl salicylate (2 SFs)

2.95 Given 1.85 g/L **Need** mg/dL

Plan g/L → mg/L → mg/dL $\dfrac{1000\ \text{mg}}{1\ \text{g}}$ $\dfrac{1\ \text{L}}{10\ \text{dL}}$

Set Up This problem involved two unit conversions. Convert g to mg in the numerator, and convert L in the denominator to dL. $\dfrac{1.85\ \cancel{\text{g}}}{1\ \cancel{\text{L}}} \times \dfrac{1000\ \text{mg}}{1\ \cancel{\text{g}}} \times \dfrac{1\ \cancel{\text{L}}}{10\ \text{dL}} = 185$ mg/dL (3 SFs)

2.97 Given 215 mL initial, 285 mL final **Need** g of lead

Plan calculate the volume by difference and mL → g $\dfrac{11.3\ \text{g}}{1\ \text{mL}}$

Set Up The difference between the initial volume of the water and its volume with the lead object will give us the volume of the lead object: 285 mL total − 215 mL water = 70. mL lead then

$70.\ \cancel{\text{mL lead}} \times \dfrac{11.3\ \text{g lead}}{1\ \cancel{\text{mL lead}}} = 790$ g of lead (2 SFs)

2.99 Given 1.00 L of gasoline **Need** cm^3 of olive oil

Plan L of gasoline → mL of gasoline → g of gasoline → g of olive oil → cm^3 of olive oil

$\dfrac{1000\ \text{mL gasoline}}{1\ \text{L gasoline}}$ $\dfrac{0.74\ \text{g gasoline}}{1\ \text{mL gasoline}}$ $\dfrac{1\ \text{g olive oil}}{1\ \text{g gasoline}}$ $\dfrac{1\ \text{cm}^3\ \text{olive oil}}{0.92\ \text{g olive oil}}$

Set Up $1.00\ \cancel{\text{L gasoline}} \times \dfrac{1000\ \cancel{\text{mL}}}{1\ \cancel{\text{L}}} \times \dfrac{0.74\ \cancel{\text{g gasoline}}}{1\ \cancel{\text{mL gasoline}}} \times \dfrac{1\ \cancel{\text{g olive oil}}}{1\ \cancel{\text{g gasoline}}} \times \dfrac{1\ \text{cm}^3\ \text{olive oil}}{0.92\ \cancel{\text{g olive oil}}}$
$= 800\ \text{cm}^3\ (8.0 \times 10^2\ \text{cm}^3)$ of olive oil (2 SFs)

2.101 a. Given 65 kg of body mass **Need** lb of fat

Plan kg of body mass → kg of fat → lb of fat

(percent equality: 100 kg of body mass = 3.0 kg of fat) $\dfrac{3.0\ \text{kg fat}}{100\ \text{kg body mass}}$ $\dfrac{2.205\ \text{lb fat}}{1\ \text{kg fat}}$

Set Up $65\ \cancel{\text{kg body mass}} \times \dfrac{3.0\ \cancel{\text{kg fat}}}{100\ \cancel{\text{kg body mass}}} \times \dfrac{2.205\ \text{lb fat}}{1\ \cancel{\text{kg fat}}} = 4.3$ lb of fat (2 SFs)

b. Given 3.0 L of fat **Need** lb of fat

Plan L → mL → g → lb $\quad \dfrac{1000\text{ mL}}{1\text{ L}} \quad \dfrac{0.94\text{ g}}{1\text{ mL}} \quad \dfrac{1\text{ lb}}{453.6\text{ g}}$

Set Up $3.0\,\cancel{\text{L}} \times \dfrac{1000\,\cancel{\text{mL}}}{1\,\cancel{\text{L}}} \times \dfrac{0.94\,\cancel{\text{g}}}{1\,\cancel{\text{mL}}} \times \dfrac{1\text{ lb}}{453.6\,\cancel{\text{g}}} = 6.2$ lb of fat (2 SFs)

2.103 Given 27.0 cm^3 of sterling silver **Need** oz of pure silver

Plan cm^3 of sterling → g of sterling → g of silver → lb → oz of silver

(percent equality: 100 g of sterling = 92.5 g of silver)

$\dfrac{10.3\text{ g sterling}}{1\text{ cm}^3\text{ sterling}} \quad \dfrac{92.5\text{ g silver}}{100\text{ g sterling}} \quad \dfrac{1\text{ lb}}{453.6\text{ g}} \quad \dfrac{16\text{ oz}}{1\text{ lb}}$

Set Up $27.0\,\cancel{\text{cm}^3\text{ sterling}} \times \dfrac{10.3\,\cancel{\text{g sterling}}}{1\,\cancel{\text{cm}^3\text{ sterling}}} \times \dfrac{92.5\,\cancel{\text{g silver}}}{100\,\cancel{\text{g sterling}}} \times \dfrac{1\,\cancel{\text{lb}}}{453.6\,\cancel{\text{g}}} \times \dfrac{16\text{ oz}}{1\,\cancel{\text{lb}}}$

= 9.07 oz of pure silver (3 SFs)

2.105 Because the balance can measure mass to 0.001 g, the mass should be reported to 0.001 g. You should record the mass of the object as 32.075 g.

2.107 Given 3.0 h **Need** gal of gasoline

Plan h → mi → km → L → qt → gal $\quad \dfrac{55\text{ mi}}{1\text{ h}} \quad \dfrac{1\text{ km}}{0.6214\text{ mi}} \quad \dfrac{1\text{ L}}{11\text{ km}} \quad \dfrac{1.057\text{ qt}}{1\text{ L}} \quad \dfrac{1\text{ gal}}{4\text{ qt}}$

Set Up $3.0\,\cancel{\text{h}} \times \dfrac{55\,\cancel{\text{mi}}}{1\,\cancel{\text{h}}} \times \dfrac{1\,\cancel{\text{km}}}{0.6214\,\cancel{\text{mi}}} \times \dfrac{1\,\cancel{\text{L}}}{11\,\cancel{\text{km}}} \times \dfrac{1.057\,\cancel{\text{qt}}}{1\,\cancel{\text{L}}} \times \dfrac{1\text{ gal}}{4\,\cancel{\text{qt}}}$

= 6.4 gal of gasoline (2 SFs)

2.109 Given 66.7 yd long, 12 in. wide, 0.000 30 in. thick **Need** g of aluminum

Plan Length: yd → ft → in. → cm $\quad \dfrac{3\text{ ft}}{1\text{ yd}} \quad \dfrac{12\text{ in.}}{1\text{ ft}} \quad \dfrac{2.54\text{ cm}}{1\text{ in.}}$

and Width and Thickness: in. → cm $\quad \dfrac{2.54\text{ cm}}{1\text{ in.}}$

then calculate Volume: cm^3 → g $\quad \dfrac{2.70\text{ g}}{1\text{ cm}^3}$

$V = l \times w \times h$

Set Up Length: $66.7\,\cancel{\text{yd}} \times \dfrac{3\,\cancel{\text{ft}}}{1\,\cancel{\text{yd}}} \times \dfrac{12\,\cancel{\text{in.}}}{1\,\cancel{\text{ft}}} \times \dfrac{2.54\text{ cm}}{1\,\cancel{\text{in.}}} = 6100$ cm (2 SFs)

Width: $12\,\cancel{\text{in.}} \times \dfrac{2.54\text{ cm}}{1\,\cancel{\text{in.}}} = 30$ cm (2 SFs)

Thickness: $0.000\,30\,\cancel{\text{in.}} \times \dfrac{2.54\text{ cm}}{1\,\cancel{\text{in.}}} = 0.000\,76$ cm (2 SFs)

Volume = 6100 cm × 30 cm × 0.000 76 cm = 140 cm^3

then $140\,\cancel{\text{cm}^3} \times \dfrac{2.70\text{ g}}{1\,\cancel{\text{cm}^3}} = 3.8 \times 10^2$ g of aluminum (2 SFs)

2.111 a. Given 0.24 oz of silver **Need** g of necklace

Plan oz of silver → lb of silver → g of silver → g of necklace

$\dfrac{1\text{ lb silver}}{16\text{ oz silver}} \quad \dfrac{453.6\text{ g silver}}{1\text{ lb silver}} \quad \dfrac{100\text{ g necklace}}{16\text{ g silver}}$

Set Up

$0.24\,\cancel{\text{oz silver}} \times \dfrac{1\,\cancel{\text{lb silver}}}{16\,\cancel{\text{oz silver}}} \times \dfrac{453.6\,\cancel{\text{g silver}}}{1\,\cancel{\text{lb silver}}} \times \dfrac{100\text{ g necklace}}{16\,\cancel{\text{g silver}}} = 43$ g of necklace (2 SFs)

b. Given g of necklace from part **a** **Need** g of copper

Plan g of necklace → g of copper $\frac{9.0\text{ g copper}}{100\text{ g necklace}}$

Set Up $43\ \cancel{\text{g necklace}} \times \frac{9.0\text{ g copper}}{100\ \cancel{\text{g necklace}}} = 3.9\text{ g of copper (2 SFs)}$

c. Given g of necklace from part **a** **Need** cm^3

Plan g of necklace → cm^3 $\frac{1\text{ cm}^3}{15.5\text{ g necklace}}$

Set Up $43\ \cancel{\text{g necklace}} \times \frac{1\text{ cm}^3}{15.5\ \cancel{\text{g necklace}}} = 2.8\text{ cm}^3\text{ (2 SFs)}$

2.113 Given 75.5 mL initially, 50.0 g of silver, 50.0 g of gold **Need** final volume (mL)

Plan for both silver and gold g → mL then calculate the final volume using addition

$\frac{1\text{ mL}}{10.5\text{ g silver}}$ $\frac{1\text{ mL}}{19.3\text{ g gold}}$

Set Up $50.0\ \cancel{\text{g silver}} \times \frac{1\text{ mL silver}}{10.5\ \cancel{\text{g silver}}} = 4.76\text{ mL of silver (3 SFs)}$

$50.0\ \cancel{\text{g gold}} \times \frac{1\text{ mL gold}}{19.3\ \cancel{\text{g gold}}} = 2.59\text{ mL of gold (3 SFs)}$

∴ final volume = 75.5 mL water + 4.76 mL silver + 2.59 mL gold

= 82.9 mL (one place to right of decimal)

3
Matter and Energy

Study Goals

- Classify examples of matter as pure substances or mixtures.
- Identify the states and the physical and chemical properties of matter.
- Given a temperature, calculate a corresponding value on another temperature scale.
- Identify energy as potential or kinetic; convert between units of energy.
- Use specific heat to calculate heat loss or gain, temperature change, or mass of a sample.
- Use the energy values to calculate the kilocalories (kcal) or kilojoules (kJ) in a food.

Chapter Outline

Answers and Solutions to Text Problems

3.1 A *pure substance* is matter that has a fixed or definite composition: either elements or compounds. In a *mixture*, two or more substances are physically mixed but not chemically combined.
a. Baking soda is composed of one type of matter ($NaHCO_3$), which makes it a pure substance.
b. A blueberry muffin is composed of several substances mixed together, which makes it a mixture.
c. Ice is composed of one type of matter (H_2O molecules), which makes it a pure substance.
d. Zinc is composed of one type of matter (Zn atoms), which makes it a pure substance.
e. Trimix is a physical mixture of oxygen, nitrogen, and helium gases, which makes it a mixture.

3.3 *Elements* are the simplest type of pure substance, containing only one type of atom. *Compounds* contain two or more elements chemically combined in a specific proportion.
a. A silicon (Si) chip is an element since it contains only one type of atom (Si).
b. Hydrogen peroxide (H_2O_2) is a compound since it contains two elements (H, O) chemically combined.

c. Oxygen (O_2) is an element since it contains only one type of atom (O).
d. Rust (Fe_2O_3) is a compound since it contains two elements (Fe, O) chemically combined.
e. Methane (CH_4) in natural gas is a compound since it contains two elements (C, H) chemically combined.

3.5 A *homogeneous mixture* has a uniform composition; a *heterogeneous mixture* does not have a uniform composition throughout the mixture.
a. Vegetable soup is a heterogeneous mixture since it has chunks of vegetables.
b. Seawater is a homogeneous mixture since it has a uniform composition.
c. Tea is a homogeneous mixture since it has a uniform composition.
d. Tea with ice and lemon slices is a heterogeneous mixture since it has chunks of ice and lemon.
e. Fruit salad is a heterogeneous mixture since it has chunks of fruit.

3.7 **a.** A gas has no definite volume or shape.
b. In a gas, the particles do not interact with each other.
c. In a solid, the particles are held in a rigid structure.

3.9 A *physical property* is a characteristic of the substance such as color, shape, odor, luster, size, melting point, and density. A *chemical property* is a characteristic that indicates the ability of a substance to change into a new substance.
a. Color and physical state are physical properties.
b. The ability to react with oxygen is a chemical property.
c. The melting point of a substance is a physical property.
d. Milk souring describes chemical reactions and so is a chemical property.
e. Burning butane gas in oxygen forms new substances, which makes it a chemical property.

3.11 A *physical change* describes a change in a physical property that retains the identity of the substance. A *chemical change* occurs when the atoms of the initial substances rearrange to form new substances.
a. Water vapor condensing is a physical change, since the physical form of the water changes, but not the substance.
b. Cesium metal reacting is a chemical change, since new substances form.
c. Gold melting is a physical change, since the physical state changes, but not the substance.
d. Cutting a puzzle is a physical change, since the size and shape change, but not the substance.
e. Dissolving sugar is a physical change, since the size of the sugar particles changes, but no new substances are formed.

3.13 **a.** The high reactivity of fluorine is a chemical property, since new substances will be formed.
b. The physical state of fluorine is a physical property.
c. The color of fluorine is a physical property.
d. The reactivity of fluorine with hydrogen is a chemical property, since new substances will be formed.
e. The melting point of fluorine is a physical property.

3.15 The Fahrenheit temperature scale is still used in the United States. A normal body temperature is 98.6 °F on this scale. To convert her 99.8 °F temperature to the equivalent reading on the Celsius scale, the following calculation must be performed:

$$T_C = \frac{(99.8 - 32)}{1.8} = \frac{67.8}{1.8} = 37.7\ °C \text{ (3 SFs) (1.8 and 32 are exact)}$$

Because a normal body temperature is 37.0 on the Celsius scale, her temperature of 37.7 °C would be a mild fever.

3.17 To convert Celsius to Fahrenheit: $T_F = 1.8(T_C) + 32$

To convert Fahrenheit to Celsius: $T_C = \dfrac{(T_F - 32)}{1.8}$ (1.8 and 32 are exact numbers)

To convert Celsius to Kelvin: $T_K = T_C + 273$

To convert Kelvin to Celsius: $T_C = T_K - 273$

a. $T_F = 1.8(T_C) + 32 = 1.8(37.0) + 32 = 66.6 + 32 = 98.6\ °F$

b. $T_C = \dfrac{(T_F - 32)}{1.8} = \dfrac{(65.3 - 32)}{1.8} = \dfrac{33.3}{1.8} = 18.5\ °C$

c. $T_K = T_C + 273 = -27 + 273 = 246\ K$

d. $T_K = T_C + 273 = 62 + 273 = 335\ K$

e. $T_C = \dfrac{(T_F - 32)}{1.8} = \dfrac{(114 - 32)}{1.8} = \dfrac{82}{1.8} = 46\ °C$

f. $T_C = \dfrac{(T_F - 32)}{1.8} = \dfrac{(72 - 32)}{1.8} = \dfrac{40.}{1.8} = 22\ °C$; $T_K = T_C + 273 = 22 + 273 = 295\ K$

3.19 **a.** $T_C = \dfrac{(T_F - 32)}{1.8} = \dfrac{(106 - 32)}{1.8} = \dfrac{74}{1.8} = 41\ °C$

b. $T_C = \dfrac{(T_F - 32)}{1.8} = \dfrac{(103 - 32)}{1.8} = \dfrac{71}{1.8} = 39\ °C$

No, there is no need to phone the doctor. The child's temperature is less than 40.0 °C.

3.21 When the car is at the top of the ramp, it has its maximum potential energy. As it descends, potential energy is converted into kinetic energy. At the bottom, all of its energy is kinetic.

3.23 **a.** Potential energy is stored in the water at the top of the waterfall.
b. Kinetic energy is displayed as the kicked ball moves.
c. Potential energy is stored in the chemical bonds in the coal.
d. Potential energy is stored when the skier is at the top of the hill.

3.25 **a. Given** 20 matches, 1.1×10^3 J/match **Need** kJ

Plan J → kJ $\dfrac{1\ \text{kJ}}{1000\ \text{J}}$

Set Up $20\ \cancel{\text{matches}} \times \dfrac{1.1 \times 10^3\ \cancel{\text{J}}}{1\ \cancel{\text{match}}} \times \dfrac{1\ \text{kJ}}{1000\ \cancel{\text{J}}} = 22\ \text{kJ}$ (2 SFs)

b. Given 20 matches, 1.1×10^3 J/match **Need** cal

Plan J → cal $\dfrac{1\ \text{cal}}{4.184\ \text{J}}$

Set Up $20\ \cancel{\text{matches}} \times \dfrac{1.1 \times 10^3\ \cancel{\text{J}}}{1\ \cancel{\text{match}}} \times \dfrac{1\ \text{cal}}{4.184\ \cancel{\text{J}}} = 5300\ \text{cal}$ (2 SFs)

c. Given 20 matches, 1.1×10^3 J/match **Need** kcal

Plan J → cal → kcal $\dfrac{1\ \text{cal}}{4.184\ \text{J}}$ $\dfrac{1\ \text{kcal}}{1000\ \text{cal}}$

Set Up $20\ \cancel{\text{matches}} \times \dfrac{1.1 \times 10^3\ \cancel{\text{J}}}{1\ \cancel{\text{match}}} \times \dfrac{1\ \cancel{\text{cal}}}{4.184\ \cancel{\text{J}}} \times \dfrac{1\ \text{kcal}}{1000\ \cancel{\text{cal}}} = 5.3\ \text{kcal}$ (2 SFs)

3.27 **a. Given** 3500 cal **Need** kcal

Plan cal → kcal $\dfrac{1\ \text{kcal}}{1000\ \text{cal}}$

Set Up $3500\ \cancel{\text{cal}} \times \dfrac{1\ \text{kcal}}{1000\ \cancel{\text{cal}}} = 3.5\ \text{kcal}$ (2 SFs)

b. Given 415 J **Need** cal

Plan J → cal $\dfrac{1\ \text{cal}}{4.184\ \text{J}}$

Set Up $415\ \cancel{J} \times \dfrac{1\ \text{cal}}{4.184\ \cancel{J}} = 99.2\ \text{cal (3 SFs)}$

c. Given 28 cal **Need** J

Plan cal → J $\dfrac{4.184\ \text{J}}{1\ \text{cal}}$

Set Up $28\ \cancel{\text{cal}} \times \dfrac{4.184\ \text{J}}{1\ \cancel{\text{cal}}} = 120\ \text{J (2 SFs)}$

d. Given 4.5 kJ **Need** cal

Plan kJ → J → cal $\dfrac{1000\ \text{J}}{1\ \text{kJ}}$ $\dfrac{1\ \text{cal}}{4.184\ \text{J}}$

Set Up $4.5\ \cancel{\text{kJ}} \times \dfrac{1000\ \cancel{J}}{1\ \cancel{\text{kJ}}} \times \dfrac{1\ \text{cal}}{4.184\ \cancel{J}} = 1100\ \text{cal (2 SFs)}$

3.29 Copper, which has the lowest specific heat, would reach the highest temperature.

3.31 **a. Given** heat = 312 J; mass = 13.5 g; ΔT = 83.6 °C − 24.2 °C = 59.4 °C
Need specific heat (J/g °C)

Plan $SH = \dfrac{\text{heat}}{\text{mass} \times \Delta T}$

Set Up Specific Heat (SH) $= \dfrac{312\ \text{J}}{13.5\ \text{g} \times 59.4\ °\text{C}} = 0.389\ \dfrac{\text{J}}{\text{g °C}}$ (3 SFs)

b. Given heat = 345 J; mass = 48.2 g; ΔT = 57.9 °C − 35.0 °C = 22.9 °C
Need specific heat (J/g °C)

Plan $SH = \dfrac{\text{heat}}{\text{mass} \times \Delta T}$

Set Up Specific Heat (SH) $= \dfrac{345\ \text{J}}{48.2\ \text{g} \times 22.9\ °\text{C}} = 0.313\ \dfrac{\text{J}}{\text{g °C}}$ (3 SFs)

3.33 **a. Given** SH_{water} = 4.184 J/g °C; mass = 25.0 g; ΔT = 25.7 °C − 12.5 °C = 13.2 °C

Need heat (q) in J and cal

Plan Heat (q) = mass × ΔT × SH then J → cal $\dfrac{1\ \text{cal}}{4.184\ \text{J}}$

Set Up Heat $= 25.0\ \cancel{g} \times 13.2\ \cancel{°C} \times \dfrac{4.184\ \text{J}}{\cancel{g}\ \cancel{°C}} = 1380\ \text{J (3 SFs)}$

or Heat $= 25.0\ \cancel{g} \times 13.2\ \cancel{°C} \times \dfrac{1\ \text{cal}}{\cancel{g}\ \cancel{°C}} = 330.\ \text{cal (3 SFs)}$

b. Given SH_{copper} = 0.385 J/g °C; mass = 38.0 g; ΔT = 246 °C − 122 °C = 124 °C

Need heat (q) in J and cal

Plan Heat (q) = mass × ΔT × SH then J → cal $\dfrac{1\ \text{cal}}{4.184\ \text{J}}$

Set Up Heat $= 38.0\ \cancel{g} \times 124\ \cancel{°C} \times \dfrac{0.385\ \text{J}}{\cancel{g}\ \cancel{°C}} = 1810\ \text{J (3 SFs)}$

or Heat $= 38.0\ \cancel{g} \times 124\ \cancel{°C} \times \dfrac{0.385\ \cancel{J}}{\cancel{g}\ \cancel{°C}} \times \dfrac{1\ \text{cal}}{4.184\ \cancel{J}} = 434\ \text{cal (3 SFs)}$

c. Given SH_{ethanol} = 2.46 J/g °C; mass = 15.0 g; ΔT = −42.0 °C − 60.5 °C = −102.5 °C

Need heat (q) in J and cal

Plan Heat (q) = mass × ΔT × SH then J → cal $\dfrac{1\ \text{cal}}{4.184\ \text{J}}$

Set Up $\text{Heat} = 15.0\,\cancel{g} \times (-102.5\,\cancel{°C}) \times \dfrac{2.46\text{ J}}{\cancel{g}\,\cancel{°C}} = -3780\text{ J (3 SFs)}$

$\text{or Heat} = 15.0\,\cancel{g} \times (-102.5\,\cancel{°C}) \times \dfrac{2.46\,\cancel{J}}{\cancel{g}\,\cancel{°C}} \times \dfrac{1\text{ cal}}{4.184\,\cancel{J}} = -904\text{ cal (3 SFs)}$

d. Given $SH_{\text{iron}} = 0.452$ J/g °C; mass = 125 g; $\Delta T = 55\text{ °C} - 118\text{ °C} = -63\text{ °C}$

Need heat (q) in J and cal

Plan $\text{Heat } (q) = \text{mass} \times \Delta T \times SH$ then $\text{J} \rightarrow \text{cal}$ $\quad \dfrac{1\text{ cal}}{4.184\text{ J}}$

Set Up $\text{Heat} = 125\,\cancel{g} \times (-63\,\cancel{°C}) \times \dfrac{0.452\text{ J}}{\cancel{g}\,\cancel{°C}} = -3600\text{ J (3 SFs)}$

$\text{or Heat} = 125\,\cancel{g} \times (-63\,\cancel{°C}) \times \dfrac{0.452\,\cancel{J}}{\cancel{g}\,\cancel{°C}} \times \dfrac{1\text{ cal}}{4.184\,\cancel{J}} = -850\text{ cal (2 SFs)}$

3.35 a. Given $SH_{\text{gold}} = 0.129$ J/g °C; heat = 225 J; $\Delta T = 47.0\text{ °C} - 15.0\text{ °C} = 32.0\text{ °C}$

Need mass (g)

Plan $\text{Heat } (q) = \text{mass} \times \Delta T \times SH \quad \therefore \text{mass} = \dfrac{\text{heat}}{\Delta T \times SH}$

Set Up $\text{Mass} = \dfrac{\text{heat}}{\Delta T \times SH} = \dfrac{225\,\cancel{J}}{32.0\,\cancel{°C} \times \dfrac{0.129\,\cancel{J}}{\text{g}\,\cancel{°C}}} = 54.5\text{ g of gold (3 SFs)}$

b. Given $SH_{\text{iron}} = 0.452$ J/g °C; heat = −8.40 kJ; $\Delta T = 82.0\text{ °C} - 168.0\text{ °C} = -86.0\text{ °C}$

Need mass (g)

Plan $\text{Heat } (q) = \text{mass} \times \Delta T \times SH \quad \therefore \text{mass} = \dfrac{\text{heat}}{\Delta T \times SH} \quad \dfrac{1000\text{ J}}{1\text{ kJ}}$

Set Up $\text{Mass} = \dfrac{\text{heat}}{\Delta T \times SH} = \dfrac{(-8.40\,\cancel{kJ}) \times \dfrac{1000\,\cancel{J}}{1\,\cancel{kJ}}}{(-86.0\,\cancel{°C}) \times \dfrac{0.452\,\cancel{J}}{\text{g}\,\cancel{°C}}} = 216\text{ g of iron (3 SFs)}$

c. Given $SH_{\text{aluminum}} = 0.897$ J/g °C; heat = 8.80 kJ; $\Delta T = 26.8\text{ °C} - 12.5\text{ °C} = 14.3\text{ °C}$

Need mass (g)

Plan $\text{Heat } (q) = \text{mass} \times \Delta T \times SH \quad \therefore \text{mass} = \dfrac{\text{heat}}{\Delta T \times SH} \quad \dfrac{1000\text{ J}}{1\text{ kJ}}$

Set Up $\text{Mass} = \dfrac{\text{heat}}{\Delta T \times SH} = \dfrac{8.80\,\cancel{kJ} \times \dfrac{1000\,\cancel{J}}{1\,\cancel{kJ}}}{14.3\,\cancel{°C} \times \dfrac{0.897\,\cancel{J}}{\text{g}\,\cancel{°C}}} = 686\text{ g of aluminum (3 SFs)}$

d. Given $SH_{\text{titanium}} = 0.523$ J/g °C; heat = −14 200 J; $\Delta T = 42\text{ °C} - 185\text{ °C} = -143\text{ °C}$

Need mass (g)

Plan $\text{Heat } (q) = \text{mass} \times \Delta T \times SH \quad \therefore \text{mass} = \dfrac{\text{heat}}{\Delta T \times SH}$

Set Up $\text{Mass} = \dfrac{\text{heat}}{\Delta T \times SH} = \dfrac{(-14\,200\,\cancel{J})}{(-143\,\cancel{°C}) \times \dfrac{0.523\,\cancel{J}}{\text{g}\,\cancel{°C}}} = 190.\text{ g of titanium (3 SFs)}$

3.37 **a. Given** $SH_{\text{iron}} = 0.452 \text{ J/g °C}$; heat = 1580 J; mass = 20.0 g **Need** ΔT

Plan Heat (q) = mass × ΔT × SH $\therefore \Delta T = \dfrac{\text{heat}}{\text{mass} \times SH}$

Set Up $$\Delta T = \frac{\text{heat}}{\text{mass} \times SH} = \frac{1580 \text{ J}}{20.0 \text{ g} \times \dfrac{0.452 \text{ J}}{\text{g °C}}} = 175 \text{ °C (3 SFs)}$$

b. Given $SH_{\text{water}} = 4.184 \text{ J/g °C}$; heat = 7.10 kJ; mass = 150.0 g **Need** ΔT

Plan Heat (q) = mass × ΔT × SH $\therefore \Delta T = \dfrac{\text{heat}}{\text{mass} \times SH}$ $\dfrac{1000 \text{ J}}{1 \text{ kJ}}$

Set Up $$\Delta T = \frac{\text{heat}}{\text{mass} \times SH} = \frac{7.10 \text{ kJ} \times \dfrac{1000 \text{ J}}{1 \text{ kJ}}}{150.0 \text{ g} \times \dfrac{4.184 \text{ J}}{\text{g °C}}} = 11.3 \text{ °C (3 SFs)}$$

c. Given $SH_{\text{gold}} = 0.129 \text{ J/g °C}$; heat = 7680 J; mass = 85.0 g **Need** ΔT

Plan Heat (q) = mass × ΔT × SH $\therefore \Delta T = \dfrac{\text{heat}}{\text{mass} \times SH}$

Set Up $$\Delta T = \frac{\text{heat}}{\text{mass} \times SH} = \frac{7680 \text{ J}}{85.0 \text{ g} \times \dfrac{0.129 \text{ J}}{\text{g °C}}} = 700. \text{ °C (3 SFs)}$$

d. Given $SH_{\text{copper}} = 0.385 \text{ J/g °C}$; heat = 6.75 kJ; mass = 50.0 g **Need** ΔT

Plan Heat (q) = mass × ΔT × SH $\therefore \Delta T = \dfrac{\text{heat}}{\text{mass} \times SH}$ $\dfrac{1000 \text{ J}}{1 \text{ kJ}}$

Set Up $$\Delta T = \frac{\text{heat}}{\text{mass} \times SH} = \frac{6.75 \text{ kJ} \times \dfrac{1000 \text{ J}}{1 \text{ kJ}}}{50.0 \text{ g} \times \dfrac{0.385 \text{ J}}{\text{g °C}}} = 351 \text{ °C (3 SFs)}$$

3.39 **a. Given** mass = 505 g of water; ΔT = 35.7 °C − 25.2 °C = 10.5 °C;
$SH_{\text{water}} = 4.184 \text{ J/g °C} = 1.00 \text{ cal/g °C}$

Need energy (q) in kJ and kcal

Plan Heat (q) = mass × ΔT × SH

Set Up $$505 \text{ g} \times 10.5 \text{ °C} \times \frac{4.184 \text{ J}}{\text{g °C}} \times \frac{1 \text{ kJ}}{1000 \text{ J}} = 22.2 \text{ kJ (3 SFs)}$$

$$505 \text{ g} \times 10.5 \text{ °C} \times \frac{1 \text{ cal}}{\text{g °C}} \times \frac{1 \text{ kcal}}{1000 \text{ cal}} = 5.30 \text{ kcal (3 SFs)}$$

b. Given mass = 4980 g of water; ΔT = 62.4 °C − 20.6 °C = 41.8 °C;
$SH_{\text{water}} = 4.184 \text{ J/g °C} = 1.00 \text{ cal/g °C}$

Need energy (q) in kJ and kcal

Plan Heat (q) = mass × ΔT × SH

Set Up $$4980 \text{ g} \times 41.8 \text{ °C} \times \frac{4.184 \text{ J}}{\text{g °C}} \times \frac{1 \text{ kJ}}{1000 \text{ J}} = 871 \text{ kJ (3 SFs)}$$

$$4980 \text{ g} \times 41.8 \text{ °C} \times \frac{1 \text{ cal}}{\text{g °C}} \times \frac{1 \text{ kcal}}{1000 \text{ cal}} = 208 \text{ kcal (3 SFs)}$$

3.41 **a.** **Given** 1 cup of orange juice contains 26 g of carbohydrate, 2 g of protein, and no fat
Need total energy (q) in kJ and kcal

Food Type	Mass	Energy Values	Energy
Carbohydrate	$26\ \cancel{g} \times$	$\frac{17\text{ kJ (or 4 kcal)}}{1\ \cancel{g}}$ =	440 kJ (or 100 kcal)
Protein	$2\ \cancel{g} \times$	$\frac{17\text{ kJ (or 4 kcal)}}{1\ \cancel{g}}$ =	30 kJ (or 10 kcal)
		Total energy content =	470 kJ (or 110 kcal)

b. **Given** one apple provides 72 kcal of energy and contains no fat or protein
Need grams of carbohydrate

Plan kcal $\rightarrow$ g of carbohydrate $\quad \frac{1\text{ g of carbohydrate}}{4\text{ kcal}}$

Set Up $72\ \cancel{\text{kcal}} \times \frac{1\text{ g carbohydrate}}{4\ \cancel{\text{kcal}}} = 18\text{ g of carbohydrate (2 SFs)}$

c. **Given** 1 tablespoon of vegetable oil contains 14 g of fat, and no carbohydrate or protein
Need total energy (q) in kJ and kcal

Food Type	Mass	Energy Values	Energy
Fat	$14\ \cancel{g} \times$	$\frac{38\text{ kJ (or 9 kcal)}}{1\ \cancel{g}}$ =	530 kJ (or 130 kcal)

d. **Given** a diet that consists of 68 g of carbohydrate, 150 g of protein, and 9.0 g of fat
Need total energy (q) in kJ and kcal

Food Type	Mass	Energy Values	Energy
Carbohydrate	$68\ \cancel{g} \times$	$\frac{17\text{ kJ (or 4 kcal)}}{1\ \cancel{g}}$ =	1160 kJ (or 270 kcal)
Protein	$150\ \cancel{g} \times$	$\frac{17\text{ kJ (or 4 kcal)}}{1\ \cancel{g}}$ =	2550 kJ (or 600 kcal)
Fat	$9.0\ \cancel{g} \times$	$\frac{38\text{ kJ (or 9 kcal)}}{1\ \cancel{g}}$ =	340 kJ (or 80 kcal)
		Total energy content =	4050 kJ (or 950 kcal)

3.43 **a.** The diagram shows two different types of atoms chemically combined in a specific proportion; it is a compound.
b. The diagram shows two different types of atoms physically mixed, not chemically combined; it is a mixture.
c. The diagram contains only one type of atom; it represents an element.

3.45 A *homogeneous mixture* has a uniform composition; a *heterogeneous mixture* does not have a uniform composition throughout the mixture.
a. Lemon-flavored water is a homogeneous mixture since it has a uniform composition (as long as there are no lemon pieces).
b. Stuffed mushrooms are a heterogeneous mixture since there are mushrooms and chunks of filling.
c. Chicken noodle soup is a heterogeneous mixture since it has noodles and chunks of chicken and vegetables.

3.47 $T_C = \frac{(T_F - 32)}{1.8} = \frac{(155 - 32)}{1.8} = \frac{123}{1.8} = 68.3\ °C$ (3 SFs) (1.8 and 32 are exact numbers)

3.49 **Given** 10.0 cm^3 cubes of gold (SH_{gold} = 0.129 J/g °C), aluminum ($SH_{aluminum}$ = 0.897 J/g °C), and silver (SH_{silver} = 0.235 J/g °C); ΔT = 25 °C − 15 °C = 10. °C

Need energy (q) in J and cal

Plan $cm^3 \rightarrow g$ Heat (q) = mass × ΔT × SH then J → cal

$$\frac{19.3 \text{ g gold}}{1 \text{ cm}^3} \quad \frac{2.70 \text{ g aluminum}}{1 \text{ cm}^3} \quad \frac{10.5 \text{ g silver}}{1 \text{ cm}^3} \quad \frac{1 \text{ cal}}{4.184 \text{ J}}$$

Set Up for gold: $10.0 \cancel{\text{cm}^3} \times \frac{19.3 \cancel{\text{g}}}{1 \cancel{\text{cm}^3}} \times 10. \cancel{°\text{C}} \times \frac{0.129 \text{ J}}{\cancel{\text{g}} \cancel{°\text{C}}} = 250 \text{ J (2 SFs)}$

or $10.0 \cancel{\text{cm}^3} \times \frac{19.3 \cancel{\text{g}}}{1 \cancel{\text{cm}^3}} \times 10. \cancel{°\text{C}} \times \frac{0.129 \cancel{\text{J}}}{\cancel{\text{g}} \cancel{°\text{C}}} \times \frac{1 \text{ cal}}{4.184 \cancel{\text{J}}} = 60. \text{ cal (2 SFs)}$

for aluminum: $10.0 \cancel{\text{cm}^3} \times \frac{2.70 \cancel{\text{g}}}{1 \cancel{\text{cm}^3}} \times 10. \cancel{°\text{C}} \times \frac{0.897 \text{ J}}{\cancel{\text{g}} \cancel{°\text{C}}} = 240 \text{ J (2 SFs)}$

or $10.0 \cancel{\text{cm}^3} \times \frac{2.70 \cancel{\text{g}}}{1 \cancel{\text{cm}^3}} \times 10. \cancel{°\text{C}} \times \frac{0.897 \cancel{\text{J}}}{\cancel{\text{g}} \cancel{°\text{C}}} \times \frac{1 \text{ cal}}{4.184 \cancel{\text{J}}} = 58 \text{ cal (2 SFs)}$

for silver: $10.0 \cancel{\text{cm}^3} \times \frac{10.5 \cancel{\text{g}}}{1 \cancel{\text{cm}^3}} \times 10. \cancel{°\text{C}} \times \frac{0.235 \text{ J}}{\cancel{\text{g}} \cancel{°\text{C}}} = 250 \text{ J (2 SFs)}$

or $10.0 \cancel{\text{cm}^3} \times \frac{10.5 \cancel{\text{g}}}{1 \cancel{\text{cm}^3}} \times 10. \cancel{°\text{C}} \times \frac{0.235 \cancel{\text{J}}}{\cancel{\text{g}} \cancel{°\text{C}}} \times \frac{1 \text{ cal}}{4.184 \cancel{\text{J}}} = 59 \text{ cal (2 SFs)}$

Thus, the heat needed for each of the three samples of metals is almost the same.

3.51 *Elements* are the simplest type of pure substance, containing only one type of atom. *Compounds* contain two or more elements chemically combined in a specific proportion. In a *mixture*, two or more substances are physically mixed but not chemically combined.

a. Carbon in pencils is an element since it contains only one type of atom (C).
b. Carbon dioxide is a compound since it contains two elements (C, O) chemically combined.
c. Orange juice is composed of several substances mixed together (e.g. water, sugar, citric acid), which makes it a mixture.

3.53 A *homogeneous mixture* has a uniform composition; a *heterogeneous mixture* does not have a uniform composition throughout the mixture.

a. A hot fudge sundae is a heterogeneous mixture since it has ice cream, fudge sauce, and perhaps a cherry.
b. Herbal tea is a homogeneous mixture since it has a uniform composition.
c. Vegetable oil is a homogeneous mixture since it has a uniform composition.

3.55 **a.** A vitamin tablet is a solid. **b.** Helium in a balloon is a gas.
c. Milk is a liquid. **d.** Air is a mixture of gases.
e. Charcoal is a solid.

3.57 A *physical property* is a characteristic of the substance such as color, shape, odor, luster, size, melting point, and density. A *chemical property* is a characteristic that indicates the ability of a substance to change into a new substance.

a. The luster of gold is a *physical* property.
b. The melting point of gold is a *physical* property.
c. The ability of gold to conduct electricity is a *physical* property.
d. The ability of gold to form a new substance with sulfur is a *chemical* property.

3.59 A *physical change* describes a change in a physical property that retains the identity of the substance. A *chemical change* occurs when the atoms of the initial substances rearrange to form new substances.

a. Plant growth produces new substances, so it is a *chemical* change.
b. A change of state from solid to liquid is a *physical* change.

c. Chopping wood into smaller pieces is a *physical* change.
d. Burning wood, which forms new substances, is a *chemical* change.

3.61 **a.** $T_C = \frac{(T_F - 32)}{1.8} = \frac{(134 - 32)}{1.8} = \frac{102}{1.8} = 56.7\ °C$

$T_K = T_C + 273 = 56.7 + 273 = 330.\ K$

b. $T_C = \frac{(T_F - 32)}{1.8} = \frac{(-69.7 - 32)}{1.8} = \frac{-101.7}{1.8} = -56.5\ °C$

$T_K = T_C + 273 = -56.5 + 273 = 217\ K$

3.63 $T_C = \frac{(T_F - 32)}{1.8} = \frac{(-15 - 32)}{1.8} = \frac{-47}{1.8} = -26\ °C$

$T_K = T_C + 273 = -26 + 273 = 247\ K$

3.65 **Given** 1 lb of body fat; 15% (m/m) water in body fat **Need** kcal to "burn off"

Plan Because each gram of body fat contains 15% water, a person actually loses 85 grams of fat per hundred grams of body fat. (We considered 1 lb of fat as exactly 1 lb.)

lb of body fat → g of body fat → g of fat → kcal

$\frac{453.6 \text{ g body fat}}{1 \text{ lb body fat}} \quad \frac{85 \text{ g fat}}{100 \text{ g body fat}} \quad \frac{9 \text{ kcal}}{1 \text{ g fat}}$

Set Up $1 \cancel{\text{lb body fat}} \times \frac{453.6 \cancel{\text{g body fat}}}{1 \cancel{\text{lb body fat}}} \times \frac{85 \cancel{\text{g fat}}}{100 \cancel{\text{g body fat}}} \times \frac{9 \text{ kcal}}{1 \cancel{\text{g fat}}} = 3500 \text{ kcal (2 SFs)}$

3.67 Water has a higher specific heat than sand, which means that a large amount of energy is required to cause a significant temperature change. Even a small amount of energy will cause a significant temperature change in the sand.

3.69 **Given** mass = 725 g of water; $SH_{water} = 4.184$ J/g °C; $\Delta T = 37\ °C - 65\ °C = -28\ °C$

Need heat (q) in kJ

Plan Heat (q) = mass × ΔT × SH then J → kJ $\frac{1 \text{ kJ}}{1000 \text{ J}}$

Set Up Heat $= 725 \cancel{\text{g}} \times (-28 \cancel{°C}) \times \frac{4.184 \cancel{\text{J}}}{\cancel{\text{g}}\ \cancel{°C}} \times \frac{1 \text{ kJ}}{1000 \cancel{\text{J}}} = -85 \text{ kJ (2 SFs)}$

∴ 85 kJ are lost from the water bottle and are transferred to the muscles

3.71 **Given**

Metal	Water
Mass = 25.0 g	mass = 50.0 g
Initial temperature = 98.0 °C	initial temperature = 18.0 °C
Final temperature = 27.4 °C	final temperature = 27.4 °C
SH = ? J/g °C	SH = 4.184 J/g °C

Need specific heat (SH) of metal

Plan Heat lost by metal = heat gained by water

For both, heat (q) = mass × ΔT × SH

Set Up

For water: $q = m \times \Delta T \times SH = 50.0 \cancel{\text{g}} \times (27.4 \cancel{°C} - 18.0 \cancel{°C}) \times \frac{4.184 \text{ J}}{\cancel{\text{g}}\ \cancel{°C}} = 1970 \text{ J}$

For metal: $SH = \frac{-q}{\Delta T \times \text{mass}} = \frac{(-1970 \text{ J})}{(27.4\ °C - 98.0\ °C) \times 25.0 \text{ g}} = 1.1 \frac{\text{J}}{\text{g °C}}$ (2 SFs)

3.73 **Given** 0.66 g of olive oil; 370 g of water; $SH_{water} = 4.184$ J/g °C;

$\Delta T = 38.8\ °C - 22.7\ °C = 16.1\ °C$

Need energy value (in kJ/g and kcal/g)

Plan Heat (q) = mass × ΔT × SH then J → kJ $\dfrac{1\ \text{kJ}}{1000\ \text{J}}$

Set Up $\text{Heat} = 370\ \cancel{\text{g}} \times 16.1\ \cancel{°\text{C}} \times \dfrac{4.184\ \cancel{\text{J}}}{\cancel{\text{g}}\ \cancel{°\text{C}}} \times \dfrac{1\ \text{kJ}}{1000\ \cancel{\text{J}}} = 25\ \text{kJ}$

$\therefore \text{energy value} = \dfrac{\text{heat produced}}{\text{mass of oil}} = \dfrac{25\ \text{kJ}}{0.66\ \text{g oil}} = 38\ \text{kJ/g of olive oil (2 SFs)}$

$\text{or Heat} = 370\ \cancel{\text{g}} \times 16.1\ \cancel{°\text{C}} \times \dfrac{1\ \cancel{\text{cal}}}{\cancel{\text{g}}\ \cancel{°\text{C}}} \times \dfrac{1\ \text{kcal}}{1000\ \cancel{\text{cal}}} = 6.0\ \text{kcal}$

$\therefore \text{energy value} = \dfrac{\text{heat produced}}{\text{mass of oil}} = \dfrac{6.0\ \text{kcal}}{0.66\ \text{g oil}} = 9.1\ \text{kcal/g of olive oil (2 SFs)}$

3.75 **Given**

Copper	Water
Mass = ? g	mass = 50.0 g
Initial temperature = 86.0 °C	initial temperature = 16.0 °C
Final temperature = 24.0 °C	final temperature = 24.0 °C
SH = 0.385 J/g °C	SH = 4.184 J/g °C

Need mass of copper (g)

Plan Heat lost by copper = heat gained by water

For both, heat (q) = mass × ΔT × SH

Set Up

For water: $q = m \times \Delta T \times SH = 50.0\ \cancel{\text{g}} \times (24.0\ \cancel{°\text{C}} - 16.0\ \cancel{°\text{C}}) \times \dfrac{4.184\ \text{J}}{\cancel{\text{g}}\ \cancel{°\text{C}}} = 1670\ \text{J}$

For copper: $\text{mass} = \dfrac{-q}{\Delta T \times SH} = \dfrac{(-1670\ \cancel{\text{J}})}{(24.0\ \cancel{°\text{C}} - 86.0\ \cancel{°\text{C}}) \times \left(\dfrac{0.385\ \cancel{\text{J}}}{\text{g}\ \cancel{°\text{C}}}\right)} = 70.\ \text{g of copper (2 SFs)}$

3.77 **a.** **Given** 1200 kcal diet, 15% of calories from protein, 45% of calories from carbohydrates

Need g of protein, g of carbohydrate, and g of fat

Plan kcal in diet → kcal of food type → g of food type

$\dfrac{15\ \text{kcal protein}}{100\ \text{kcal diet}}\quad \dfrac{45\ \text{kcal carbohydrate}}{100\ \text{kcal diet}}\quad \dfrac{40\ \text{kcal fat}}{100\ \text{kcal diet}}\quad \dfrac{1\ \text{g protein}}{4\ \text{kcal}}\quad \dfrac{1\ \text{g carbohydrate}}{4\ \text{kcal}}\quad \dfrac{1\ \text{g fat}}{9\ \text{kcal}}$

Set Up $1200\ \cancel{\text{kcal diet}} \times \dfrac{15\ \cancel{\text{kcal protein}}}{100\ \cancel{\text{kcal diet}}} \times \dfrac{1\ \text{g protein}}{4\ \cancel{\text{kcal}}} = 45\ \text{g of protein (2 SFs)}$

$1200\ \cancel{\text{kcal diet}} \times \dfrac{45\ \cancel{\text{kcal carbohydrate}}}{100\ \cancel{\text{kcal diet}}} \times \dfrac{1\ \text{g carbohydrate}}{4\ \cancel{\text{kcal}}} = 140\ \text{g of carbohydrate (2 SFs)}$

$1200\ \cancel{\text{kcal diet}} \times \dfrac{40\ \cancel{\text{kcal fat}}}{100\ \cancel{\text{kcal diet}}} \times \dfrac{1\ \text{g fat}}{9\ \cancel{\text{kcal}}} = 53\ \text{g of fat (2 SFs)}$

b. Given 1900 kcal diet, 15% of calories from protein, 45% of calories from carbohydrates
Need g of protein, g of carbohydrate, and g of fat
Plan kcal in diet → kcal of food type → g of food type

$$\frac{15\text{ kcal protein}}{100\text{ kcal diet}}\ \frac{45\text{ kcal carbohydrate}}{100\text{ kcal diet}}\ \frac{40\text{ kcal fat}}{100\text{ kcal diet}}\ \frac{1\text{ g protein}}{4\text{ kcal}}\ \frac{1\text{ g carbohydrate}}{4\text{ kcal}}\ \frac{1\text{ g fat}}{9\text{ kcal}}$$

Set Up $1900\ \cancel{\text{kcal diet}} \times \frac{15\ \cancel{\text{kcal protein}}}{100\ \cancel{\text{kcal diet}}} \times \frac{1\text{ g protein}}{4\ \cancel{\text{kcal}}} = 71\text{ g of protein (2 SFs)}$

$$1900\ \cancel{\text{kcal diet}} \times \frac{45\ \cancel{\text{kcal carbohydrate}}}{100\ \cancel{\text{kcal diet}}} \times \frac{1\text{ g carbohydrate}}{4\ \cancel{\text{kcal}}} = 210\text{ g of carbohydrate (2 SFs)}$$

$$1900\ \cancel{\text{kcal diet}} \times \frac{40\ \cancel{\text{kcal fat}}}{100\ \cancel{\text{kcal diet}}} \times \frac{1\text{ g fat}}{9\ \cancel{\text{kcal}}} = 84\text{ g of fat (2 SFs)}$$

c. Given 2600 kcal diet, 15% of calories from protein, 45% of calories from carbohydrates
Need g of protein, g of carbohydrate, and g of fat
Plan kcal in diet → kcal of food type → g of food type

$$\frac{15\text{ kcal protein}}{100\text{ kcal diet}}\ \frac{45\text{ kcal carbohydrate}}{100\text{ kcal diet}}\ \frac{40\text{ kcal fat}}{100\text{ kcal diet}}\ \frac{1\text{ g protein}}{4\text{ kcal}}\ \frac{1\text{ g carbohydrate}}{4\text{ kcal}}\ \frac{1\text{ g fat}}{9\text{ kcal}}$$

Set Up $2600\ \cancel{\text{kcal diet}} \times \frac{15\ \cancel{\text{kcal protein}}}{100\ \cancel{\text{kcal diet}}} \times \frac{1\text{ g protein}}{4\ \cancel{\text{kcal}}} = 98\text{ g of protein (2 SFs)}$

$$2600\ \cancel{\text{kcal diet}} \times \frac{45\ \cancel{\text{kcal carbohydrate}}}{100\ \cancel{\text{kcal diet}}} \times \frac{1\text{ g carbohydrate}}{4\ \cancel{\text{kcal}}} = 290\text{ g of carbohydrate (2 SFs)}$$

$$2600\ \cancel{\text{kcal diet}} \times \frac{40\ \cancel{\text{kcal fat}}}{100\ \cancel{\text{kcal diet}}} \times \frac{1\text{ g fat}}{9\ \cancel{\text{kcal}}} = 120\text{ g of fat (2 SFs)}$$

3.79 a. Given $SH_{\text{water}} = 4.184\text{ J/g °C}$; heat $= 8250\text{ J}$; $\Delta T = 92.6\text{ °C} - 18.3\text{ °C} = 74.3\text{ °C}$
Need mass (g)

Plan Heat $(q) = \text{mass} \times \Delta T \times SH \quad \therefore \text{mass} = \frac{\text{heat}}{\Delta T \times SH}$

Set Up $\text{Mass} = \frac{\text{heat}}{\Delta T \times SH} = \frac{8250\ \cancel{\text{J}}}{74.3\ \cancel{\text{°C}} \times \frac{4.184\ \cancel{\text{J}}}{\text{g}\ \cancel{\text{°C}}}} = 26.5\text{ g of water (3 SFs)}$

b. Given $SH_{\text{gold}} = 0.129\text{ J/g °C}$; heat $= 225\text{ J}$; $\Delta T = 47.0\text{ °C} - 15.0\text{ °C} = 32.0\text{ °C}$
Need mass (g)

Plan Heat $(q) = \text{mass} \times \Delta T \times SH \quad \therefore \text{mass} = \frac{\text{heat}}{\Delta T \times SH}$

Set Up $\text{Mass} = \frac{\text{heat}}{\Delta T \times SH} = \frac{225\ \cancel{\text{J}}}{32.0\ \cancel{\text{°C}} \times \frac{0.129\ \cancel{\text{J}}}{\text{g}\ \cancel{\text{°C}}}} = 54.5\text{ g of gold (3 SFs)}$

c. Given $SH_{\text{iron}} = 0.452\text{ J/g °C}$; heat $= 1580\text{ J}$; mass $= 20.0\text{ g}$ **Need** ΔT

Plan Heat $(q) = \text{mass} \times \Delta T \times SH \quad \therefore \Delta T = \frac{\text{heat}}{\text{mass} \times SH}$

Set Up $\Delta T = \frac{\text{heat}}{\text{mass} \times SH} = \frac{1580\ \cancel{\text{J}}}{20.0\ \cancel{\text{g}} \times \frac{0.452\ \cancel{\text{J}}}{\cancel{\text{g}}\text{ °C}}} = 175\text{ °C (3 SFs)}$

d. Given heat $= 28\text{ cal}$; mass $= 8.50\text{ g}$; $\Delta T = 24\text{ °C} - 12\text{ °C} = 12\text{ °C}$
Need specific heat (cal/g °C)

Plan $SH = \frac{\text{heat}}{\text{mass} \times \Delta T}$

Set Up $\text{Specific Heat }(SH) = \frac{28\text{ cal}}{8.50\text{ g} \times 12\text{ °C}} = 0.27\ \frac{\text{cal}}{\text{g °C}}\text{ (2 SFs)}$

Answers to Combining Ideas from Chapters 1 to 3

CI.1 **a.** There are 5 significant figures in the weight of 294.10 troy ounces.

b. $294.10\ \cancel{\text{troy ounces}} \times \dfrac{31.1035\text{ g}}{1\ \cancel{\text{troy ounce}}} = 9147.5\text{ g (5 SFs)}$

$294.10\ \cancel{\text{troy ounces}} \times \dfrac{31.1035\ \cancel{\text{g}}}{1\ \cancel{\text{troy ounce}}} \times \dfrac{1\text{ kg}}{1000\ \cancel{\text{g}}} = 9.1475\text{ kg (5 SFs)}$

c. $9147.5\ \cancel{\text{g}} \times \dfrac{1\text{ cm}^3}{19.3\ \cancel{\text{g}}} = 474\text{ cm}^3\text{ (3 SFs)}$

d. Volume = Area × height (i.e. thickness),

$\therefore\ \text{Area of foil} = \dfrac{\text{Volume of gold}}{\text{thickness of foil}} = \dfrac{474\ \cancel{\text{cm}^3}}{0.0035\ \cancel{\text{in.}}} \times \dfrac{1\ \cancel{\text{in.}}}{2.54\ \cancel{\text{cm}}} \times \left(\dfrac{1\text{ m}}{100\ \cancel{\text{cm}}}\right)^2$

$= 5.3\text{ m}^2\text{ (2 SFs)}$

e. Melting point of gold = 1064 °C

in degrees Fahrenheit:

$T_F = 1.8(T_C) + 32 = 1.8(1064) + 32 = 1947\ °\text{F}$ (answer to the ones place)

in kelvins:

$T_K = T_C + 273 = 1064 + 273 = 1337\text{ K}$ (answer to the ones place)

f. **Given** $SH_{gold} = 0.129$ J/g °C; mass = 9147.5 g of gold; initial temperature = 63 °F; final temperature = 85 °F

Need heat (q) in kilojoules (kJ)

Plan Heat (q) = mass × ΔT × SH

Set Up Convert temperatures to Celsius:

Initial temperature: $T_C = \dfrac{T_F - 32}{1.8} = \dfrac{(63 - 32)}{1.8} = 17\ °\text{C}$

Final temperature: $T_C = \dfrac{T_F - 32}{1.8} = \dfrac{(85 - 32)}{1.8} = 29\ °\text{C}$

$\Delta T = T_f - T_i = 29\ °\text{C} - 17\ °\text{C} = 12\ °\text{C}$

$\text{Heat} = 9147.5\ \cancel{\text{g}} \times 12\ \cancel{°\text{C}} \times \dfrac{0.129\ \cancel{\text{J}}}{\cancel{\text{g}}\ \cancel{°\text{C}}} \times \dfrac{1\text{ kJ}}{1000\ \cancel{\text{J}}} = 14\text{ kJ (2 SFs)}$

CI.3 **a.** Since it is in the solid state, sample B has its own shape.

b. When transferred to a new container, sample A takes the shape of the new container, but its volume does not change. Sample B retains its own shape and its own volume.

c. Sample A is represented as particles in diagram 2; the particles are close together but in a random arrangement. Sample B is represented as particles in diagram 1; the particles are very close together and in a rigid fixed arrangement.

d. diagram 1, solid; diagram 2, liquid; diagram 3, gas

e. The change from a liquid to a solid state is a physical change.

CI.5 **a.** $0.250\ \cancel{\text{lb}} \times \dfrac{453.6\ \cancel{\text{g}}}{1\ \cancel{\text{lb}}} \times \dfrac{1\text{ cm}^3}{7.86\ \cancel{\text{g}}} = 14.4\text{ cm}^3\text{ (3 SFs)}$

b. $30\ \cancel{\text{nails}} \times \dfrac{14.4\text{ cm}^3}{75\ \cancel{\text{nails}}} = 5.76\text{ cm}^3 = 5.76\text{ mL (3 SFs)}$

New water level = 17.6 mL water + 5.76 mL = 23.4 mL (3 SFs)

c. $\Delta T = T_f - T_i = 125\ °C - 16\ °C = 109\ °C$

Mass of iron nails $= 0.250\ \cancel{lb} \times \dfrac{453.6\ g}{1\ \cancel{lb}} = 113\ g$ (3 SFs)

Heat = mass × ΔT × SH

$$= 0.250\ \cancel{lb} \times \frac{453.6\ \cancel{g}}{1\ \cancel{lb}} \times 109\ \cancel{°C} \times \frac{0.452\ J}{\cancel{g}\ \cancel{°C}} = 5590\ J \text{ (3 SFs)}$$

d. Given

Iron nails	Water
Mass = 113 g	mass = 325 g
Initial temperature = 55.0 °C	initial temperature = 4.0 °C
Final temperature = ?	final temperature = ?
SH = 0.452 J/g °C	SH = 4.184 J/g °C

Need Final temperature

Plan Heat ($-q$) lost by nails at 55.0 °C = heat (q) gained by water at 4.0 °C

$-q_{iron} = q_{water}$

Set Up $-(m \times \Delta T \times SH)_{iron} = (m \times \Delta T \times SH)_{water}$

$$-\left(\left(0.250\ \cancel{lb} \times \frac{453.6\ \cancel{g}}{1\ \cancel{lb}}\right) \times (T_f - 55.0) \times \frac{0.452\ J}{\cancel{g}\ \cancel{°C}}\right) = \left(325\ \cancel{g} \times (T_f - 4.0)\ \cancel{°C} \times \frac{4.184\ J}{\cancel{g}\ \cancel{°C}}\right)$$

$$2820 - 51.3T_f = 1360T_f - 5440$$

$$1411T_f = 8260$$

$$T_f = 5.8\ °C$$

4

Atoms and Elements

Study Goals

- Given the name of an element, write its correct symbol; from the symbol, write the correct name.
- Use the periodic table to identify the group and the period of an element and decide whether it is a metal, a metalloid, or a nonmetal.
- Describe the electrical charge and location in an atom for a proton, a neutron, and an electron.
- Given the atomic number and the mass number of an atom, state the number of protons, neutrons, and electrons.
- Give the number of protons, electrons, and neutrons in an isotope of an element.
- Calculate the atomic mass of an element.

Chapter Outline

Answers and Solutions to Text Problems

4.1
a. Cu is the symbol for copper.
b. Pt is the symbol for platinum.
c. Ca is the symbol for calcium.
d. Mn is the symbol for manganese.
e. Fe is the symbol for iron.
f. Ba is the symbol for barium.
g. Pb is the symbol for lead.
h. Sr is the symbol for strontium.

4.3
a. Carbon is the element with the symbol C.
b. Chlorine is the element with the symbol Cl.
c. Iodine is the element with the symbol I.
d. Mercury is the element with the symbol Hg.
e. Silver is the element with the symbol Ag.
f. Argon is the element with the symbol Ar.
g. Boron is the element with the symbol B.
h. Nickel is the element with the symbol Ni.

4.5
a. Sodium (Na) and chlorine (Cl) are in NaCl.
b. Calcium (Ca), sulfur (S), and oxygen (O) are in $CaSO_4$.
c. Carbon (C), hydrogen (H), chlorine (Cl), nitrogen (N), and oxygen (O) are in $C_{15}H_{22}ClNO_2$.
d. Calcium (Ca), carbon (C), and oxygen (O) are in $CaCO_3$.

4.7 **a.** C, N, and O are in Period 2.
b. He is the element at the top of Group 8A (18).
c. The alkali metals are the elements in Group 1A (1).
d. Period 2 is the horizontal row of elements that ends with neon (Ne).

4.9 **a.** Ca is an alkaline earth metal.
b. Fe is a transition element.
c. Xe is a noble gas.
d. K is an alkali metal.
e. Cl is a halogen.

4.11 **a.** C is Group 4A (14), Period 2.
b. He is the noble gas in Period 1.
c. Na is the alkali metal in Period 3.
d. Ca is Group 2A (2), Period 4.
e. Al is Group 3A (13), Period 3.

4.13 On the periodic table, *metals* are located to the left of the heavy zigzag line, *nonmetals* are elements to the right, and *metalloids* (B, Si, Ge, As, Sb, Te, Po, and At) are located along the line.
a. Ca is a metal.
b. S is a nonmetal.
c. Metals are shiny.
d. An element that is a gas at room temperature is a nonmetal.
e. Group 8A (18) elements are nonmetals.
f. Br is a nonmetal.
g. B is a metalloid.
h. Ag is a metal.

4.15 **a.** The electron has the smallest mass.
b. The proton has a 1+ charge.
c. The electron is found outside the nucleus.
d. The neutron is electrically neutral.

4.17 Rutherford determined that positively charged particles called protons are located in a very small, dense region of the atom called the nucleus.

4.19 All of the statements are true except for selection **d.** Since a neutron has no charge, it is not attracted to a proton (a proton is attracted to an electron).

4.21 Your hair is more likely to fly away on a dry day; because the hair strands repel one another, there must be like electrical charges on each strand.

4.23 **a.** The atomic number is the same as the number of protons in an atom.
b. Both are needed since the number of neutrons is the (mass number) − (atomic number).
c. The mass number is the number of particles (protons + neutrons) in the nucleus.
d. The atomic number is the same as the number of electrons in a neutral atom.

4.25 The atomic number defines the element and is found above the symbol of the element in the periodic table.
a. Lithium, Li, has an atomic number of 3.
b. Fluorine, F, has an atomic number of 9.
c. Calcium, Ca, has an atomic number of 20.
d. Zinc, Zn, has an atomic number of 30.
e. Neon, Ne, has an atomic number of 10.
f. Silicon, Si, has an atomic number of 14.
g. Iodine, I, has an atomic number of 53.
h. Oxygen, O, has an atomic number of 8.

4.27 The atomic number gives the number of protons in the nucleus of an atom. Since atoms are neutral, the atomic number also gives the number of electrons in the neutral atom.
a. There are 18 protons and 18 electrons in a neutral argon atom.
b. There are 30 protons and 30 electrons in a neutral zinc atom.
c. There are 53 protons and 53 electrons in a neutral iodine atom.
d. There are 19 protons and 19 electrons in a neutral potassium atom.

4.29 The atomic number is the same as the number of protons in an atom and the number of electrons in a neutral atom; the atomic number defines the element. The number of neutrons is the (mass number) − (atomic number).

Name of Element	Symbol	Atomic Number	Mass Number	Number of Protons	Number of Neutrons	Number of Electrons
Aluminum	Al	13	27	13	27 − 13 = 14	13
Magnesium	Mg	12	12 + 12 = 24	12	12	12
Potassium	K	19	19 + 20 = 39	19	20	19
Sulfur	S	16	16 + 15 = 31	16	15	16
Iron	Fe	26	56	26	56 − 26 = 30	26

4.31 **a.** Since the atomic number of aluminum is 13, every Al atom has 13 protons. An atom of aluminum (mass number 27) has 14 neutrons (27 − 13 = 14 n). Neutral atoms have the same number of protons and electrons. Therefore, 13 protons, 14 neutrons, 13 electrons.

b. Since the atomic number of chromium is 24, every Cr atom has 24 protons. An atom of chromium (mass number 52) has 28 neutrons (52 − 24 = 28 n). Neutral atoms have the same number of protons and electrons. Therefore, 24 protons, 28 neutrons, 24 electrons.

c. Since the atomic number of sulfur is 16, every S atom has 16 protons. An atom of sulfur (mass number 34) has 18 neutrons (34 − 16 = 18 n). Neutral atoms have the same number of protons and electrons. Therefore, 16 protons, 18 neutrons, 16 electrons.

d. Since the atomic number of bromine is 35, every Br atom has 35 protons. An atom of bromine (mass number 81) has 46 neutrons (81 − 35 = 46 n). Neutral atoms have the same number of protons and electrons. Therefore, 35 protons, 46 neutrons, 35 electrons.

4.33 **a.** Since the number of protons is 15, the atomic number is 15 and the element symbol is P. The mass number is the sum of the number of protons and the number of neutrons, 15 + 16 = 31. The atomic symbol for this isotope is $^{31}_{15}P$.

b. Since the number of protons is 35, the atomic number is 35 and the element symbol is Br. The mass number is the sum of the number of protons and the number of neutrons, 35 + 45 = 80. The atomic symbol for this isotope is $^{80}_{35}Br$.

c. Since the number of electrons is 50, there must be 50 protons in a neutral atom. Since the number of protons is 50, the atomic number is 50, and the element symbol is Sn. The mass number is the sum of the number of protons and the number of neutrons, 50 + 72 = 122. The atomic symbol for this isotope is $^{122}_{50}Sn$.

d. Since the element is chlorine, the element symbol is Cl, the atomic number is 17, and the number of protons is 17. The mass number is the sum of the number of protons and the number of neutrons, 17 + 18 = 35. The atomic symbol for this isotope is $^{35}_{17}Cl$.

e. Since the element is mercury, the element symbol is Hg, the atomic number is 80, and the number of protons is 80. The mass number is the sum of the number of protons and the number of neutrons, 80 + 122 = 202. The atomic symbol for this isotope is $^{202}_{80}Hg$.

4.35 **a.** Since the element is argon, the element symbol is Ar, the atomic number is 18, and the number of protons is 18. The atomic symbols for the isotopes with mass numbers of 36, 38, and 40, are $^{36}_{18}Ar$, $^{38}_{18}Ar$, and $^{40}_{18}Ar$, respectively.

b. They all have the same atomic number (the same number of protons and electrons).

c. They have different numbers of neutrons, which is reflected in their mass numbers.

d. The atomic mass of argon listed on the periodic table is the weighted average atomic mass of all the naturally occurring isotopes of argon.

e. The isotope Ar-40 ($^{40}_{18}Ar$) is the most abundant in a sample of argon because its mass is closest to the average atomic mass of argon listed on the periodic table (39.95 amu).

4.37 The mass of an isotope is the mass of an individual atom. The atomic mass is the weighted average of all the naturally occurring isotopes of that element.

4.39 Since the atomic mass of copper (63.55 amu) is closer to 63 amu, there are more atoms of $^{63}_{29}Cu$ in a sample of copper.

4.41 Since the atomic mass of neon (20.18 amu) is closest to 20 amu, the most abundant isotope of neon is $^{20}_{10}Ne$.

4.43

$$^{69}_{31}Ga \quad 68.93 \times \frac{60.11}{100} = 41.43 \text{ amu}$$

$$^{71}_{31}Ga \quad 70.92 \times \frac{39.89}{100} = 28.29 \text{ amu}$$

$$\text{Atomic mass of Ga} = \overline{69.72 \text{ amu}} \text{ (4 SFs)}$$

4.45 Statements **b** and **c** are true. According to Dalton's atomic theory, atoms of an element are different than atoms of other elements, and atoms do not appear and disappear in a chemical reaction.

4.47 **a.** The atomic mass is the weighted average of the masses of all of the naturally occurring isotopes of the element. Those isotope masses are based on the number of protons (1) and the number of neutrons (2) in the isotopes.
b. The number of protons (1) is the atomic number.
c. The protons (1) are positively charged.
d. The electrons (3) are negatively charged.
e. The number of neutrons (2) is the (mass number) − (atomic number).

4.49 **a.** $^{16}_{8}X$, $^{17}_{8}X$, and $^{18}_{8}X$ all have an atomic number of 8, so all have eight protons.
b. $^{16}_{8}X$, $^{17}_{8}X$, and $^{18}_{8}X$ all have an atomic number of 8, so all are isotopes of oxygen.
c. $^{16}_{8}X$ and $^{16}_{9}X$ have mass numbers of 16, whereas $^{18}_{8}X$ and $^{18}_{10}X$ have mass numbers of 18.
d. $^{16}_{8}X$ ($16 - 8 = 8\ n$) and $^{18}_{10}X$ ($18 - 10 = 8\ n$) both have eight neutrons.

4.51 **a.** $^{37}_{17}Cl$ and $^{38}_{18}Ar$ both have 20 neutrons ($37 - 17 = 20\ n$; $38 - 18 = 20\ n$).
b. Since both $^{36}_{16}S$ and $^{34}_{16}S$ have an atomic number of 16, they will both have 16 protons in the nucleus. Since the number of protons is 16, there must be 16 electrons in the neutral atom for both. They have different numbers of neutrons ($36 - 16 = 20\ n$; $34 - 16 = 18\ n$).
c. $^{79}_{34}Se$ (34 protons, $79 - 34 = 45$ neutrons, 34 electrons) and $^{81}_{35}Br$ (35 protons, $81 - 35 = 46$ neutrons, 35 electrons) do not share common numbers of protons, neutrons, or electrons.
d. $^{40}_{18}Ar$ and $^{39}_{17}Cl$ both have 22 neutrons ($40 - 18 = 22\ n$; $39 - 17 = 22\ n$).

4.53 **a.** The diagram shows 4 protons and 5 neutrons in the nucleus of the element. Since the number of protons is 4, the atomic number is 4, and the element symbol is Be. The mass number is the sum of the number of protons and the number of neutrons, $4 + 5 = 9$. The atomic symbol is $^{9}_{4}Be$.
b. The diagram shows 5 protons and 6 neutrons in the nucleus of the element. Since the number of protons is 5, the atomic number is 5, and the element symbol is B. The mass number is the sum of the number of protons and the number of neutrons, $5 + 6 = 11$. The atomic symbol is $^{11}_{5}B$.
c. The diagram shows 6 protons and 7 neutrons in the nucleus of the element. Since the number of protons is 6, the atomic number is 6, and the element symbol is C. The mass number is the sum of the number of protons and the number of neutrons, $6 + 7 = 13$. The atomic symbol is $^{13}_{6}C$.
d. The diagram shows 5 protons and 5 neutrons in the nucleus of the element. Since the number of protons is 5, the atomic number is 5, and the element symbol is B. The mass number is the sum of the number of protons and the number of neutrons, $5 + 5 = 10$. The atomic symbol is $^{10}_{5}B$.

e. The diagram shows 6 protons and 6 neutrons in the nucleus of the element. Since the number of protons is 6, the atomic number is 6, and the element symbol is C. The mass number is the sum of the number of protons and the number of neutrons, 6 + 6 = 12. The atomic symbol is ${}^{12}_{6}C$.
Both **b** (${}^{11}_{5}B$) and **d** (${}^{10}_{5}B$) are isotopes of boron; **c** (${}^{13}_{6}C$) and **e** (${}^{12}_{6}C$) are isotopes of carbon.

4.55 **a.** The lightest alkali metal (Group 1A (1)) is lithium, which has an atomic number 3 and the symbol Li.
b. The heaviest noble gas (Group 8A (18)) is radon, which has an atomic number 86 and the symbol Rn.
c. The alkaline earth metal (Group 2A (2)) in Period 3 is magnesium, which has an atomic mass of 24.31 amu and the symbol Mg.
d. The halogen (Group 7A (17)) with the fewest electrons is fluorine, which has an atomic mass of 19.00 amu and the symbol F.

4.57 The first letter of a symbol is a capital, but a second letter is lowercase. The symbol for cobalt is Co, but the symbols in CO are for the elements carbon and oxygen.

4.59 **a.** Mg, magnesium is in Group 2A (2), Period 3.
b. Br, bromine is in Group 7A (17), Period 4.
c. Al, aluminum is in Group 3A (13), Period 3.
d. O, oxygen is in Group 6A (16), Period 2.

4.61 **a.** The halogens are the elements in Group 7A (17): fluorine, chlorine, bromine, iodine, astatine
b. The noble gases are the elements in Group 8A (18): helium, neon, argon, krypton, xenon, radon
c. The alkali metals are the elements in Group 1A (1), excluding hydrogen: lithium, sodium, potassium, rubidium, cesium, francium
d. The alkaline earth metals are the elements in Group 2A (2): beryllium, magnesium, calcium, strontium, barium, radium

4.63 **a.** False. A proton is a positively charged particle.
b. False. The neutron has about the same mass as a proton.
c. True
d. False. The nucleus is the tiny, dense central core of an atom.
e. True

4.65 **a.** The atomic number gives the number of protons in the nucleus.
b. In an atom, the number of electrons is equal to the number of protons.
c. Sodium and potassium are examples of elements called alkali metals.

4.67 The atomic number defines the element and is found above the symbol of the element in the periodic table.
a. Nickel, Ni, has an atomic number of 28.
b. Barium, Ba, has an atomic number of 56.
c. Radium, Ra, has an atomic number of 88.
d. Arsenic, As, has an atomic number of 33.
e. Tin, Sn, has an atomic number of 50.
f. Cesium, Cs, has an atomic number of 55.
g. Gold, Au, has an atomic number of 79.
h. Mercury, Hg, has an atomic number of 80.

4.69 The atomic number gives the number of protons in the nucleus of an atom. Since atoms are neutral, the atomic number also gives the number of electrons in the neutral atom.
a. There are 25 protons and 25 electrons in a neutral manganese (Mn) atom.
b. There are 15 protons and 15 electrons in a neutral phosphorus atom.
c. There are 38 protons and 38 electrons in a neutral strontium (Sr) atom.
d. There are 27 protons and 27 electrons in a neutral cobalt (Co) atom.
e. There are 92 protons and 92 electrons in a neutral uranium atom.

4.71 **a.** Since the atomic number of silver is 47, every Ag atom has 47 protons. An atom of silver (mass number 107) has 60 neutrons ($107 - 47 = 60\ n$). Neutral atoms have the same number of protons and electrons. Therefore, 47 protons, 60 neutrons, 47 electrons.

b. Since the atomic number of technetium is 43, every Tc atom has 43 protons. An atom of technetium (mass number 98) has 55 neutrons ($98 - 43 = 55\ n$). Neutral atoms have the same number of protons and electrons. Therefore, 43 protons, 55 neutrons, 43 electrons.

c. Since the atomic number of lead is 82, every Pb atom has 82 protons. An atom of lead (mass number 208) has 126 neutrons ($208 - 82 = 126\ n$). Neutral atoms have the same number of protons and electrons. Therefore, 82 protons, 126 neutrons, 82 electrons.

d. Since the atomic number of radon is 86, every Rn atom has 86 protons. An atom of radon (mass number 222) has 136 neutrons ($222 - 86 = 136\ n$). Neutral atoms have the same number of protons and electrons. Therefore, 86 protons, 136 neutrons, 86 electrons.

e. Since the atomic number of xenon is 54, every Xe atom has 54 protons. An atom of xenon (mass number 136) has 82 neutrons ($136 - 54 = 82\ n$). Neutral atoms have the same number of protons and electrons. Therefore, 54 protons, 82 neutrons, 54 electrons.

4.73 **a.** Since the number of protons is 4, the atomic number is 4, and the element symbol is Be. The mass number is the sum of the number of protons and the number of neutrons, $4 + 5 = 9$. The atomic symbol for this isotope is $^{9}_{4}\text{Be}$.

b. Since the number of protons is 12, the atomic number is 12, and the element symbol is Mg. The mass number is the sum of the number of protons and the number of neutrons, $12 + 14 = 26$. The atomic symbol for this isotope is $^{26}_{12}\text{Mg}$.

c. Since the element is calcium, the element symbol is Ca, and the atomic number is 20. The mass number is given as 46. The atomic symbol for this isotope is $^{46}_{20}\text{Ca}$.

d. Since the number of electrons is 30, there must be 30 protons in a neutral atom. Since the number of protons is 30, the atomic number is 30, and the element symbol is Zn. The mass number is the sum of the number of protons and the number of neutrons, $30 + 40 = 70$. The atomic symbol for this isotope is $^{70}_{30}\text{Zn}$.

4.75

Name	Atomic Symbol	Number of Protons	Number of Neutrons	Number of Electrons
Sulfur	$^{34}_{16}\text{S}$	16	$34 - 16 = 18$	16
Nickel	$^{28+34}_{28}\text{Ni}$ or $^{62}_{28}\text{Ni}$	28	34	28
Magnesium	$^{12+14}_{12}\text{Mg}$ or $^{26}_{12}\text{Mg}$	12	14	12
Radon	$^{220}_{86}\text{Rn}$	86	$220 - 86 = 134$	86

4.77 **a.** Since the element is gold (Au), the atomic number is 79, and the number of protons is 79. In a neutral atom, the number of electrons is equal to the number of protons, so there will be 79 electrons. The number of neutrons is the (mass number) − (atomic number) $= 197 - 79 = 118\ n$. In Au-197, there are 79 protons, 118 neutrons, 79 electrons.

b. Since the element is gold (Au), the atomic number is 79, and the number of protons is 79. The mass number is the sum of the number of protons and the number of neutrons, $79 + 116 = 195$. The atomic symbol for this isotope is $^{195}_{79}\text{Au}$.

c. Since the atomic number is 78, the number of protons is 78, and the element symbol is Pt. The mass number is the sum of the number of protons and the number of neutrons, $78 + 116 = 194$. The atomic symbol for this isotope is $^{194}_{78}\text{Pt}$.

4.79 **a.** Since the element is lead (Pb), the atomic number is 82, and the number of protons is 82. In a neutral atom, the number of electrons is equal to the number of protons, so there will be 82 electrons. The number of neutrons is the (mass number) − (atomic number) = 208 − 82 = 126 *n*. In $^{208}_{82}Pb$, there are 82 protons, 126 neutrons, 82 electrons.

b. Since the element is lead (Pb), the atomic number is 82, and the number of protons is 82. The mass number is the sum of the number of protons and the number of neutrons, 82 + 132 = 214. The atomic symbol for this isotope is $^{214}_{82}Pb$.

c. Since the mass number is 214 (as in part **b**) and the number of neutrons is 131, the number of protons is the (mass number) − (number of neutrons) = 214 − 131 = 83 *p*. Since there are 83 protons, the atomic number is 83, and the element symbol is Bi (bismuth). The atomic symbol for this isotope is $^{214}_{83}Bi$.

4.81

$$^{204}_{82}Pb \quad 203.97 \times \frac{1.40}{100} = 2.86 \text{ amu}$$

$$^{206}_{82}Pb \quad 205.97 \times \frac{24.10}{100} = 49.64 \text{ amu}$$

$$^{207}_{82}Pb \quad 206.98 \times \frac{22.10}{100} = 45.74 \text{ amu}$$

$$^{208}_{82}Pb \quad 207.98 \times \frac{52.40}{100} = 109.0 \text{ amu}$$

Atomic mass of Pb = $\overline{207.2 \text{ amu}}$ (4 SFs)

4.83

$$^{28}_{14}Si \quad 27.977 \times \frac{92.23}{100} = 25.80 \text{ amu}$$

$$^{29}_{14}Si \quad 28.976 \times \frac{4.68}{100} = 1.36 \text{ amu}$$

$$^{30}_{14}Si \quad 29.974 \times \frac{3.09}{100} = 0.926 \text{ amu}$$

Atomic mass of Si = $\overline{28.09 \text{ amu}}$ (4 SFs)

4.85 **Given** length = 1 in., sodium atoms (3.14×10^{-8} cm diameter) **Need** number of Na atoms

Plan in. → cm → number of Na atoms

$$\frac{2.54 \text{ cm}}{1 \text{ in.}} \qquad \frac{1 \text{ atom Na}}{3.14 \times 10^{-8} \text{ cm}}$$

Set Up $1 \cancel{\text{in.}} \times \frac{2.54 \cancel{\text{cm}}}{1 \cancel{\text{in.}}} \times \frac{1 \text{ atom Na}}{3.14 \times 10^{-8} \cancel{\text{cm}}} = 8.09 \times 10^{7}$ atoms of Na (3 SFs)

5

Electronic Structure and Periodic Trends

Study Goals

- Compare the wavelength of radiation with its energy.
- Explain how atomic spectra correlate with the energy levels in atoms.
- Describe the sublevels and orbitals in atoms.
- Draw the orbital diagram and write the electron configuration for an element.
- Write the electron configuration for an atom using the sublevel blocks on the periodic table.
- Use the electron configurations of elements to explain periodic trends.

Chapter Outline

Answers and Solutions to Text Problems

5.1 The wavelength of UV light is the distance between crests of the wave.

5.3 "White" light has all the colors and wavelengths of the visible spectrum, including red and blue light. Red and blue light are visible light with specific wavelengths.

5.5 **Given** $\nu = 650$ kHz **Need** Hz

Plan kHz → Hz $\frac{1000 \text{ Hz}}{1 \text{ kHz}}$

Set Up $650 \cancel{\text{kHz}} \times \frac{1000 \text{ Hz}}{1 \cancel{\text{kHz}}} = 6.5 \times 10^5$ Hz (2 SFs)

5.7 **Given** $\lambda = 6.3 \times 10^{-5}$ cm **Need** m and nm

Plan cm → m $\frac{1\text{ m}}{100\text{ cm}}$; cm → m → nm $\frac{1\text{ m}}{100\text{ cm}}$ $\frac{1 \times 10^9\text{ nm}}{1\text{ m}}$

Set Up $6.3 \times 10^{-5}\ \cancel{\text{cm}} \times \frac{1\text{ m}}{100\ \cancel{\text{cm}}} = 6.3 \times 10^{-7}$ m (2 SFs)

or $6.3 \times 10^{-5}\ \cancel{\text{cm}} \times \frac{1\ \cancel{\text{m}}}{100\ \cancel{\text{cm}}} \times \frac{1 \times 10^9\text{ nm}}{1\ \cancel{\text{m}}} = 630$ nm (2 SFs)

5.9 Microwaves have longer wavelengths than ultraviolet light or X-rays.

5.11 From shortest to longest wavelengths: X-rays, blue light, infrared, microwaves

5.13 Atomic spectra consist of a series of lines separated by dark sections, indicating that the energy emitted by the elements is not continuous.

5.15 Electrons can jump to higher energy levels when they absorb a photon.

5.17 A photon in the infrared region of the spectrum is likely to be emitted when an excited electron drops to the third energy level.

5.19 **a.** A photon of green light has greater energy than a photon of yellow light.
b. A photon of blue light has greater energy than a photon of red light.

5.21 **a.** A 1*s* orbital is spherical.
b. A 2*p* orbital has two lobes.
c. A 5*s* orbital is spherical.

5.23 **a.** All *s* orbitals are spherical.
b. A 3*s* sublevel and a 3*p* sublevel are in the energy level $n = 3$.
c. All *p* sublevels contain three *p* orbitals.
d. The 3*p* orbitals all have two lobes and are in the energy level $n = 3$.

5.25 **a.** There are five orbitals in the 3*d* sublevel.
b. There is one sublevel in the $n = 1$ energy level.
c. There is one orbital in the 6*s* sublevel.
d. There are nine orbitals in the $n = 3$ energy level: one 3*s* orbital, three 3*p* orbitals, and five 3*d* orbitals.

5.27 **a.** Any orbital can hold a maximum of 2 electrons. Thus, a 2*p* orbital has a maximum of 2 electrons.
b. The 3*p* sublevel contains three *p* orbitals, each of which can hold a maximum of 2 electrons, which gives a maximum of 6 electrons in the 3*p* sublevel.
c. Using $2n^2$, the calculation for the maximum number of electrons in the $n = 4$ energy level is $2(4)^2 = 2(16) = 32$ electrons.
d. The 5*d* sublevel contains five *d* orbitals, each of which can hold a maximum of 2 electrons, which gives a maximum of 10 electrons in the 5*d* sublevel.

5.29 The electron configuration shows the number of electrons in each sublevel of an atom. The abbreviated electron configuration uses the symbol of the preceding noble gas to show completed sublevels.

5.31 **a.**

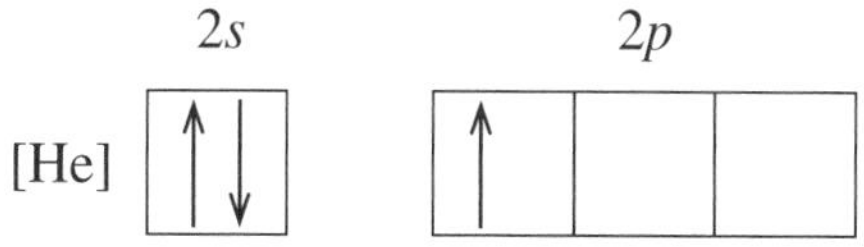

b.

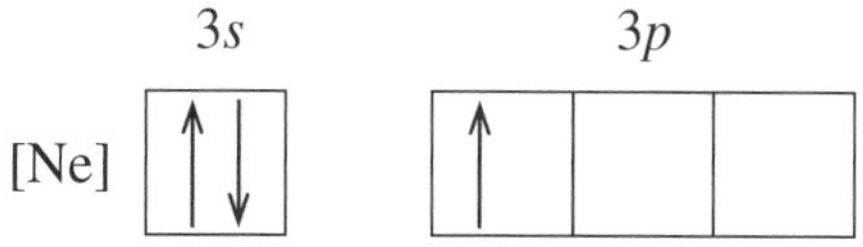

c.

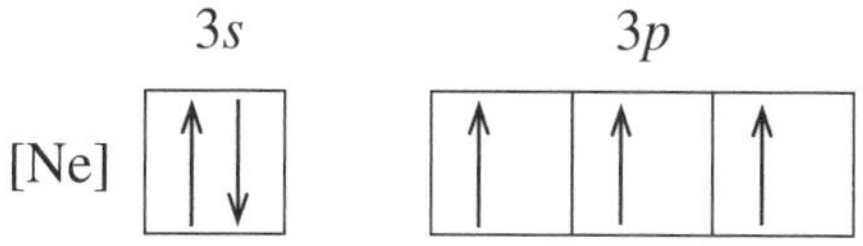

d.

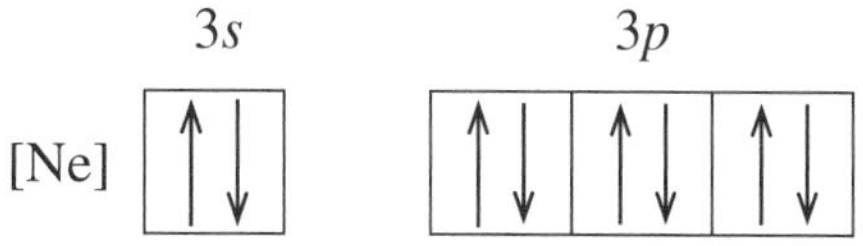

5.33 **a.** N $1s^22s^22p^3$ **b.** Na $1s^22s^22p^63s^1$
c. Br $1s^22s^22p^63s^23p^64s^23d^{10}4p^5$ **d.** Ni $1s^22s^22p^63s^23p^64s^23d^8$

5.35 **a.** Ca $[Ar]4s^2$ **b.** Sr $[Kr]5s^2$
c. Ga $[Ar]4s^23d^{10}4p^1$ **d.** Zn $[Ar]4s^23d^{10}$

5.37 **a.** Li is the element with 1 electron in the 2*s* sublevel.
b. S is the element with 2 electrons in the 3*s* and 4 electrons in the 3*p* sublevels.
c. Si is the element with 2 electrons in the 3*s* and 2 electrons in the 3*p* sublevels.
d. F is the element with 2 electrons in the 2*s* and 5 electrons in the 2*p* sublevels.

5.39 **a.** Al has 3 electrons in the energy level $n = 3$, $3s^23p^1$.
b. C has two 2*p* electrons.
c. Ar completes the 3*p* sublevel, $3p^6$.
d. Be completes the 2*s* sublevel, $2s^2$.

5.41 Using the periodic table, the *s* sublevel block is on the left, the *p* sublevel block is on the right, and the *d* sublevel block is in the center between the *s* and *p* blocks.
a. As $1s^22s^22p^63s^23p^64s^23d^{10}4p^3$
b. Fe $1s^22s^22p^63s^23p^64s^23d^6$
c. Pd $1s^22s^22p^63s^23p^64s^23d^{10}4p^65s^24d^8$
d. I $1s^22s^22p^63s^23p^64s^23d^{10}4p^65s^24d^{10}5p^5$

5.43 Using the periodic table, the *s* sublevel block is on the left, the *p* sublevel block is on the right, and the *d* sublevel block is in the center between the *s* and *p* blocks. The abbreviated electron configuration consists of the symbol of the preceding noble gas, followed by the electron configuration in the next period.
a. Ti $[Ar]4s^23d^2$
b. Sr $[Kr]5s^2$
c. Ba $[Xe]6s^2$
d. Pb $[Xe]6s^24f^{14}5d^{10}6p^2$

5.45 Use the final sublevel notation in the electron configuration to locate the element.
a. P (ends in $3p^3$) **b.** Co (ends in $3d^7$)
c. Zn (ends in $3d^{10}$) **d.** Br (ends in $4p^5$)

5.47 **a.** Ga has three electrons in energy level $n = 4$; 2 are in the 4*s* block, and 1 is in the 4*p* block.
b. N is the third element in the 2*p* block; it has three 2*p* electrons.
c. Xe is the final element (6 electrons) in the 5*p* block.
d. Zr is the second element in the 4*d* block; it has two 4*d* electrons.

5.49 **a.** Zn is the tenth element in the 3*d* block; it has ten 3*d* electrons.
b. Na has 1 electron in the 3*s* block; the 2*p* block in Na is complete with six electrons.
c. As is the third element in the 4*p* block; it has three 4*p* electrons.
d. Rb is the first element in the 5*s* block; Rb has one 5*s* electron.

5.51 The group numbers 1A–8A (or the ones digit in 1, 2, and 13–18) indicate the number of valence (outer) electrons for the elements in each vertical column. The group numbers 3–12 give the electrons in the *s* and *d* sublevels. The Group B indicates that the *d* sublevel is filling.

5.53 **a.** An element with 2 valence electrons is in Group 2A (2).
b. An element with 5 valence electrons is in Group 5A (15).
c. An element with 2 valence electrons and 5 electrons in the *d* block is in Group 7B (7).
d. An element with 6 valence electrons is in Group 6A (16).

5.55 **a.** Alkali metals, which are in Group 1A (1), have a valence electron configuration ns^1.
b. Elements in Group 4A (14) have a valence electron configuration ns^2np^2.
c. Elements in Group 3A (13) have a valence electron configuration ns^2np^1.
d. Elements in Group 5A (15) have a valence electron configuration ns^2np^3.

5.57 **a.** Aluminum in Group 3A (13) has 3 valence electrons.
b. Any element in Group 5A (15) has 5 valence electrons.
c. The 2 valence electrons for nickel are $4s^2$ (the 8 electrons in the 3*d* sublevel are not considered valence electrons).
d. Each halogen in Group 7A (17) has 7 valence electrons.

5.59 **a.** Sulfur is in Group 6A (16); $\cdot\ddot{\underset{\cdot}{\text{S}}}:$
b. Nitrogen is in Group 5A (15); $\cdot\ddot{\underset{\cdot}{\text{N}}}\cdot$
c. Calcium is in Group 2A (2); $\dot{\text{C}}\text{a}\cdot$
d. Sodium is in Group 1A (1); $\text{Na}\cdot$
e. Gallium is in Group 3A (13); $\cdot\dot{\text{G}}\text{a}\cdot$

5.61 **a.** The atomic radius of representative elements decreases from Group 1A to 8A: Mg, Al, Si.
b. The atomic radius of representative elements increases going down a group: I, Br, Cl.
c. The atomic radius of representative elements decreases from Group 1A to 8A: Sr, Sb, I.
d. The atomic radius of representative elements decreases from Group 1A to 8A: Na, Si, P.

5.63 The atomic radius of representative elements decreases going across a period from Group 1A to 8A and increases going down a group.
a. Na is larger than O; Na is in Period 3 and is to the left of O in Period 2.
b. In Group 1A (1), Rb, which is farther down the group, is larger than Na.
c. In Period 3, Na, which is on the left, is larger than Mg.
d. In Period 3, Na, which is on the left, is larger than Cl.

5.65 **a.** Br, Cl, F; ionization energy decreases going down a group.
b. Na, Al, Cl; going across a period from left to right, ionization energy generally increases.
c. Cs, K, Na; ionization energy decreases going down a group.
d. Sn, Sb, As; going across a period from left to right, ionization energy generally increases. Ionization energy decreases going down a group.

5.67 **a.** Br, which is above I in Group 7A (17), has a higher ionization energy than I.
b. Ionization energy decreases from Group 2A (2) to Group 3A (13), which gives Mg a higher ionization energy than Al.
c. Ionization energy decreases from Group 5A (15) to Group 6A (16), which gives P a higher ionization energy than S.
d. The noble gases have the highest ionization energy in each period, which gives Xe a higher ionization energy than I.

5.69 When a potassium ion is formed, a potassium atom loses the only valence electron in its outermost energy level; this makes a potassium ion smaller than a potassium atom.

5.71 **a.** A Na atom with a valence electron in energy level $n = 3$ is larger than Na^+, which has lost that valence electron.
b. A chloride ion (Cl^-) is larger than a Cl atom because the addition of a valence electron increases repulsion between electrons, which increases the size.
c. A sulfide ion (S^{2-}) is larger than a S atom because the addition of two valence electrons increases repulsion between electrons, which increases the size.

5.73 **a.** C has the longest wavelength.
b. A has the shortest wavelength.
c. A has the highest frequency.
d. C has the lowest frequency.

5.75 **a.** An orbital with two lobes is a *p* orbital.
b. A spherical orbital is an *s* orbital.
c. An orbital with two lobes is a *p* orbital.

5.77 **a.** This orbital diagram is possible. The orbitals up to $3s^2$ are filled with electrons having opposite spins. The element represented would be magnesium.
b. Not possible. The 2*p* sublevel would fill completely before the 3*s*, and only 2 electrons are allowed in an orbital (not 3 as shown in the 3*s*).

5.79 Atomic radius increases going down a group. Li is D because it would be smallest. Na is A, K is C, and Rb is B.

5.81 A continuous spectrum from white light contains wavelengths of all energies. Atomic spectra are line spectra in which a series of lines corresponds to energy emitted when electrons drop from a higher energy level to a lower level.

5.83 The Pauli exclusion principle states that two electrons in the same orbital must have opposite spins.

5.85 **a.** A 2*p* and a 3*p* orbital have the same shape with two lobes; each *p* orbital can hold up to two electrons with opposite spins. However, the 3*p* orbital is larger because a 3*p* electron has a higher energy level and is more likely to be found farther from the nucleus.
b. A 3*s* and a 3*p* orbital are found in the same energy level, $n = 3$, and each can hold up to 2 electrons with opposite spins. However, the shapes of a 3*s* orbital and a 3*p* orbital are different.
c. The orbitals in the 4*p* sublevel all have the same energy level and the same two-lobed shape. However, there are three 4*p* orbitals directed along the *x*, *y*, and *z* axes around the nucleus.

5.87 A 4*p* orbital is possible because $n = 4$ has four sublevels, including a *p* sublevel. A 2*d* orbital is not possible because $n = 2$ has only *s* and *p* sublevels. There are no 3*f* orbitals because only *s*, *p*, and *d* sublevels are allowed for $n = 3$. A 5*f* sublevel is possible in $n = 5$ because five sublevels are allowed, including 5*f*.

5.89 **a.** The 3*p* sublevel starts to fill after completion of the 3*s* sublevel.
b. The 5*s* sublevel starts to fill after completion of the 4*p* sublevel.
c. The 4*p* sublevel starts to fill after completion of the 3*d* sublevel.
d. The 4*s* sublevel starts to fill after completion of the 3*p* sublevel.

5.91 **a.** Iron is the sixth element in the 3*d* block; iron has six 3*d* electrons.
b. Barium has a completely filled 5*p* sublevel, which is six 5*p* electrons.
c. Iodine has a completely filled 4*d* sublevel, which is ten 4*d* electrons.
d. Barium has a filled 6*s* sublevel, which is two 6*s* electrons.

5.93 Ca, Sr, and Ba all have 2 valence electrons, ns^2, which places them in Group 2A (2), the alkaline earth metals.

5.95 **a.** X is a metal; Y and Z are nonmetals. **b.** X has the largest atomic radius.
c. Y has the highest ionization energy. **d.** Y has the smallest atomic radius.

5.97 **a.** Phosphorus in Group 5A (15) has an electron configuration that ends with $3s^23p^3$.
b. Lithium is the alkali metal that is highest in Group 1A (1) and has the smallest atomic radius (H in Group 1A (1) is a nonmetal).
c. Cadmium in Period 5 has a complete 4*d* sublevel with 10 electrons.
d. Nitrogen at the top of Group 5A (15) has the highest ionization energy in that group.
e. Sodium, the first element in Period 3, has the largest atomic radius of that period.

5.99 When two electrons are added to the outermost level in an oxygen atom, there is an increase in electron repulsion that pushes the electrons apart and increases the size of the oxide ion compared to the oxygen atom.

5.101 **a.** Sr^{2+} A positive ion is smaller than its corresponding atom.
b. Se An atom is smaller than its corresponding negative ion.
c. Br^- Bromide ion is smaller than iodide ion because bromide ion has one less energy level than the iodide ion.

5.103 Calcium has a greater number of protons than K; this greater nuclear charge makes it more difficult to remove an electron from Ca. The least tightly bound (valence) electron in Ca is farther from the nucleus than in Mg and less energy is needed to remove it.

5.105 In Group 3A (13), ns^2np^1, the *p* electron is farther from the nucleus and easier to remove.

5.107 **a.** Na is on the far left of the heavy zigzag line. Na is a metal.
b. Na at the beginning of Period 3 has the largest atomic radius.
c. F at the top of Group 7A (17) and to the far right in Period 2 has the highest ionization energy.
d. Na has the lowest ionization energy and loses an electron most easily.
e. Cl is found in Period 3 in Group 7A.

5.109 **a.** Si $[Ne]3s^23p^2$; Group 4A (14) **b.** Se $[Ar]4s^23d^{10}4p^4$; Group 6A (16)
c. Mn $[Ar]4s^23d^5$; Group 7B (7) **d.** Sb $[Kr]5s^24d^{10}5p^3$; Group 5A (15)

5.111 **a.** Atomic radius increases going down a group, which gives O the smallest atomic radius in Group 6A.
b. Atomic radius generally decreases going from left to right across a period, which gives Ar the smallest atomic radius in Period 3.
c. Ionization energy decreases going down a group, which gives B the highest ionization energy in Group 13.
d. Ionization energy generally increases going from left to right across a period, which gives Na the lowest ionization energy in Period 3.
e. Ru in Period 5 has 6 electrons in the 4*d* sublevel.

5.113 Ultraviolet light and microwaves both travel at the same speed: 3.00×10^8 m/s. The wavelength of ultraviolet light is shorter than the wavelength of microwaves, and the frequency of ultraviolet light is higher than the frequency of microwaves.

5.115 The series of lines separated by dark sections in atomic spectra indicate that the energy emitted from heated elements is not continuous and that electrons are moving between discrete energy levels.

5.117 An energy level contains all the electrons with similar energy. A sublevel contains electrons with the same energy, while an orbital is the region around the nucleus where electrons of a certain energy are most likely to be found.

5.119 S has a larger atomic radius than Cl; Cl is larger than F: S > Cl > F.
F has a higher ionization energy than Cl; Cl has a higher ionization energy than S: F > Cl > S.

6

Inorganic and Organic Compounds: Names and Formulas

Study Goals

- Using the octet rule, write the symbols of the single ions for the representative elements.
- Using charge balance, write the correct formula for an ionic compound.
- Given the formula of an ionic compound, write the correct name; given the name of an ionic compound, write the correct formula.
- Write the name and formula of a compound containing a polyatomic ion.
- Given the formula of a covalent compound, write its correct name; given the name of a covalent compound, write its formula.
- Identify properties characteristic of organic and inorganic compounds.
- Write the IUPAC name and draw the condensed structural formula for an alkane.

Chapter Outline

Chapter Opener: Dentist

6.1 Octet Rule and Ions
Chemistry and Industry: Some Uses for Noble Gases
Chemistry and Health: Some Important Ions in the Body

6.2 Ionic Compounds

6.3 Naming and Writing Ionic Formulas

6.4 Polyatomic Ions

6.5 Covalent Compounds and Their Names

6.6 Organic Compounds

6.7 Names and Formulas of Hydrocarbons: Alkanes
Chemistry and Industry: Crude Oil
Career Focus: Geologist
Chemistry and Health: Fatty Acids in Lipids, Soaps, and Cell Membranes
Answers

Answers and Solutions to Text Problems

6.1 **a.** When a sodium atom loses its valence electron, its second energy level has a complete octet.
b. Atoms of Group 1A (1) and 2A (2) elements can lose 1 or 2 electrons respectively to attain a noble gas electron configuration. Atoms of Group 8A (18) elements already have an octet of valence electrons (two for helium), so they do not lose or gain electrons and are not normally found in compounds.

6.3 Atoms with one, two, or three valence electrons will lose those electrons to acquire a noble gas electron configuration.
 a. one **b.** two
 c. three **d.** one
 e. two

6.5 Atoms form ions by losing or gaining electrons to achieve the same electron configuration as the nearest noble gas.
 a. Li^+ has an electron configuration $1s^2$, which is the same as helium (He).
 b. Mg^{2+} has an electron configuration $1s^22s^22p^6$, which is the same as neon (Ne).
 c. K^+ has an electron configuration $1s^22s^22p^63s^23p^6$, which is the same as argon (Ar).
 d. O^{2-} has an electron configuration $1s^22s^22p^6$, which is the same as neon (Ne).
 e. Br^- has an electron configuration $1s^22s^22p^63s^23p^64s^23d^{10}4p^6$, which is the same as krypton (Kr).

6.7 Atoms form ions by losing or gaining valence electrons to achieve the same electron configuration as the nearest noble gas. Elements in Groups 1A (1), 2A (2), and 3A (13) lose valence electrons, whereas elements in Groups 5A (15), 6A (16), and 7A (17) gain valence electrons to complete octets.
 a. Sr loses $2\,e^-$. **b.** P gains $3\,e^-$.
 c. Elements in Group 7A (17) gain $1\,e^-$. **d.** Na loses $1\,e^-$.
 e. Ga loses $3\,e^-$.

6.9 **a.** Li^+ **b.** F^-
 c. Mg^{2+} **d.** Fe^{3+}
 e. Zn^{2+}

6.11 **a** (Li and Cl) and **c** (K and O) will form ionic compounds.

6.13 **a.** Potassium loses $1\,e^-$, and fluorine gains $1\,e^-$.

$$K\!\cdot\; +\; \cdot\ddot{\underset{..}{F}}\!: \;\rightarrow\; K^+ \;+\; \left[:\ddot{\underset{..}{F}}\!:\right]^- \;\rightarrow\; KF$$

 b. Barium loses $2\,e^-$, and two chlorine atoms each gain $1\,e^-$.

$$\cdot Ba\!\cdot\; +\; \cdot\ddot{\underset{..}{Cl}}\!: \;+\; \cdot\ddot{\underset{..}{Cl}}\!: \;\rightarrow\; Ba^{2+} + 2\left[:\ddot{\underset{..}{Cl}}\!:\right]^- \;\rightarrow\; BaCl_2$$

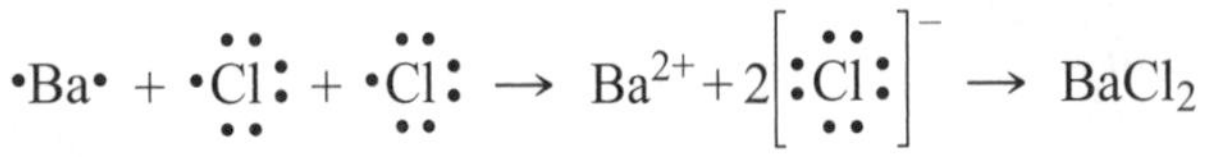

 c. Each of three sodium atoms lose $1\,e^-$, and the nitrogen gains $3\,e^-$.

$$Na\!\cdot\; +\; Na\!\cdot\; +\; Na\!\cdot\; +\; \cdot\ddot{\underset{\cdot}{N}}\!\cdot \;\rightarrow\; 3Na^+ \;+\; \left[:\ddot{\underset{..}{N}}\!:\right]^{3-} \;\rightarrow\; Na_3N$$

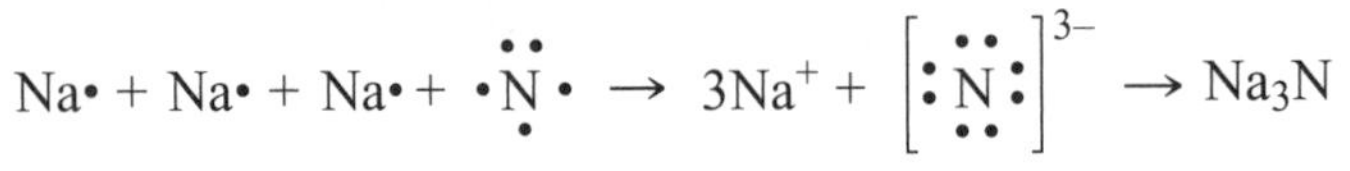

6.15 **a.** Na^+ and $O^{2-} \rightarrow Na_2O$ **b.** Al^{3+} and $Br^- \rightarrow AlBr_3$
 c. Ba^{2+} and $N^{3-} \rightarrow Ba_3N_2$ **d.** Mg^{2+} and $F^- \rightarrow MgF_2$
 e. Al^{3+} and $S^{2-} \rightarrow Al_2S_3$

6.17 **a.** Ions: Na^+ and $S^{2-} \rightarrow Na_2S$ Check: $2(1+) + 1(2-) = 0$
 b. Ions: K^+ and $N^{3-} \rightarrow K_3N$ Check: $3(1+) + 1(3-) = 0$
 c. Ions: Al^{3+} and $I^- \rightarrow AlI_3$ Check: $1(3+) + 3(1-) = 0$
 d. Ions: Ga^{3+} and $O^{2-} \rightarrow Ga_2O_3$ Check: $2(3+) + 3(2-) = 0$

6.19 **a.** Chlorine in Group 7A (17) gains 1 electron to form chloride ion Cl^-.
 b. Potassium in Group 1A (1) loses 1 electron to form potassium ion K^+.
 c. Oxygen in Group 6A (16) gains 2 electrons to form oxide ion O^{2-}.
 d. Aluminum in Group 3A (13) loses 3 electrons to form aluminum ion Al^{3+}.

6.21 **a.** lithium **b.** sulfide
 c. calcium **d.** nitride

6.23 **a.** aluminum oxide **b.** calcium chloride
c. sodium oxide **d.** magnesium phosphide
e. potassium iodide **f.** barium fluoride

6.25 Most of the transition elements form more than one positive ion. The specific ion is indicated in a name by writing a Roman numeral that is the same as the ionic charge. For example, iron forms Fe^{2+} and Fe^{3+} ions, which are named iron(II) and iron(III).

6.27 **a.** iron(II) **b.** copper(II)
c. zinc **d.** lead(IV)
e. chromium(III) **f.** manganese(II)

6.29 **a.** Ions: Sn^{2+} and $Cl^- \rightarrow$ tin(II) chloride
b. Ions: Fe^{2+} and $O^{2-} \rightarrow$ iron(II) oxide
c. Ions: Cu^+ and $S^{2-} \rightarrow$ copper(I) sulfide
d. Ions: Cu^{2+} and $S^{2-} \rightarrow$ copper(II) sulfide
e. Ions: Cd^{2+} and $Br^- \rightarrow$ cadmium bromide
f. Ions: Hg^{2+} and $Cl^- \rightarrow$ mercury(II) chloride

6.31 **a.** Au^{3+} **b.** Fe^{3+}
c. Pb^{4+} **d.** Sn^{2+}

6.33 **a.** Ions: Mg^{2+} and $Cl^- \rightarrow MgCl_2$ **b.** Ions: Na^+ and $S^{2-} \rightarrow Na_2S$
c. Ions: Cu^+ and $O^{2-} \rightarrow Cu_2O$ **d.** Ions: Zn^{2+} and $P^{3-} \rightarrow Zn_3P_2$
e. Ions: Au^{3+} and $N^{3-} \rightarrow AuN$

6.35 **a.** Ions: Co^{3+} and $Cl^- \rightarrow CoCl_3$ **b.** Ions: Pb^{4+} and $O^{2-} \rightarrow PbO_2$
c. Ions: Ag^+ and $I^- \rightarrow AgI$ **d.** Ions: Ca^{2+} and $N^{3-} \rightarrow Ca_3N_2$
e. Ions: Cu^+ and $P^{3-} \rightarrow Cu_3P$ **f.** Ions: Cr^{2+} and $Cl^- \rightarrow CrCl_2$

6.37 **a.** HCO_3^- **b.** NH_4^+
c. PO_4^{3-} **d.** HSO_4^-
e. ClO^-

6.39 **a.** sulfate **b.** carbonate
c. phosphate **d.** nitrate
e. perchlorate

6.41

	NO_2^-	CO_3^{2-}	HSO_4^-	PO_4^{3-}
Li^+	$LiNO_2$ Lithium nitrite	Li_2CO_3 Lithium carbonate	$LiHSO_4$ Lithium hydrogen sulfate	Li_3PO_4 Lithium phosphate
Cu^{2+}	$Cu(NO_2)_2$ Copper(II) nitrite	$CuCO_3$ Copper(II) carbonate	$Cu(HSO_4)_2$ Copper(II) hydrogen sulfate	$Cu_3(PO_4)_2$ Copper(II) phosphate
Ba^{2+}	$Ba(NO_2)_2$ Barium nitrite	$BaCO_3$ Barium carbonate	$Ba(HSO_4)_2$ Barium hydrogen sulfate	$Ba_3(PO_4)_2$ Barium phosphate

6.43 **a.** CO_3^{2-}, sodium carbonate **b.** NH_4^+, ammonium chloride
c. PO_3^{3-}, sodium phosphite **d.** NO_2^-, manganese(II) nitrite
e. SO_3^{2-}, iron(II) sulfite **f.** $C_2H_3O_2^-$, potassium acetate

6.45 **a.** Ions: Ba^{2+} and OH^- → $Ba(OH)_2$ **b.** Ions: Na^+ and SO_4^{2-} → Na_2SO_4
c. Ions: Fe^{2+} and NO_3^- → $Fe(NO_3)_2$ **d.** Ions: Zn^{2+} and PO_4^{3-} → $Zn_3(PO_4)_2$
e. Ions: Fe^{3+} and CO_3^{2-} → $Fe_2(CO_3)_3$

6.47 **a.** Ions: Al^{3+} and SO_4^{2-} → aluminum sulfate
b. Ions: Ca^{2+} and CO_3^{2-} → calcium carbonate
c. Ions: Cr^{3+} and O^{2-} → chromium(III) oxide
d. Ions: Na^+ and PO_4^{3-} → sodium phosphate
e. Ions: NH_4^+ and SO_4^{2-} → ammonium sulfate
f. Ions: Fe^{3+} and O^{2-} → iron(III) oxide

6.49 The nonmetallic elements that are not noble gases are most likely to form covalent bonds.

6.51 **a.** Each H atom has one valence electron, for a total of two valence electrons in the molecule. In H_2, there is one bonding pair and no (0) lone pairs.
b. The Br atom has seven valence electrons and the H atom has one valence electron, for a total of eight valence electrons in the molecule. The Br atom achieves an octet by sharing a valence electron with one H atom to give eight valence electrons; the H atom has two valence electrons. The molecule has one bonding pair and there are three lone pairs on the Br atom.
c. Each Br atom has seven valence electrons, for a total of 14 valence electrons in the molecule. Each Br atom achieves an octet by sharing one valence electron with the other Br atom. There is one bonding pair between the Br atoms, and six lone pairs (three lone pairs on each Br atom).

6.53 When naming covalent compounds, prefixes are used to indicate the number of each atom as shown in the subscripts of the formula. The first nonmetal is named by its elemental name; the second nonmetal is named by using its elemental name with the ending changed to *ide*.
a. 1 P and 3 Br → phosphorus tribromide **b.** 1 C and 4 Br → carbon tetrabromide
c. 1 Si and 2 O → silicon dioxide **d.** 1 H and 1 F → hydrogen fluoride
e. 1 N and 3 I → nitrogen triiodide

6.55 When naming covalent compounds, prefixes are used to indicate the number of each atom as shown in the subscripts of the formula. The first nonmetal is named by its elemental name; the second nonmetal is named by using its elemental name with the ending changed to *ide*.
a. 2 N and 3 O → dinitrogen trioxide **b.** 1 N and 3 Cl → nitrogen trichloride
c. 1 Si and 4 Br → silicon tetrabromide **d.** 1 P and 5 Cl → phosphorus pentachloride
e. 2 N and 3 S → dinitrogen trisulfide

6.57 **a.** 1 C and 4 Cl → CCl_4 **b.** 1 C and 1 O → CO
c. 1 P and 3 Cl → PCl_3 **d.** 2 N and 4 O → N_2O_4

6.59 **a.** 1 O and 2 F → OF_2 **b.** 1 B and 3 F → BF_3
c. 2 N and 3 O → N_2O_3 **d.** 1 S and 6 F → SF_6

6.61 **a.** Ions: Al^{3+} and Cl^- → aluminum chloride **b.** 1 B and 3 O → diboron trioxide
c. 2 N and 4 O → dinitrogen tetroxide **d.** Ions: Sn^{2+} and NO_3^- → tin(II) nitrate
e. Ions: Cu^{2+} and ClO_2^- → copper(II) chlorite

6.63 Organic compounds contain C and H and sometimes O, N, or a halogen atom. Inorganic compounds usually contain elements other than C and H.
a. inorganic **b.** organic
c. organic **d.** inorganic
e. inorganic **f.** organic

6.65 **a.** Inorganic compounds are usually soluble in water.
b. Organic compounds have lower boiling points than most inorganic compounds.
c. Organic compounds often burn in air.
d. Inorganic compounds are more likely to have high melting points.

6.67 **a.** Ethane; organic compounds have low boiling points.
b. Ethane; organic compounds burn vigorously in air.
c. Sodium bromide; inorganic compounds have high melting points.
d. Sodium bromide; inorganic compounds (ionic) typically dissolve in water.

6.69 The four bonds from carbon to hydrogen in CH_4 are as far apart as possible in order to give the lowest amount of repulsion between the H atoms. This means that the hydrogen atoms are directed to the corners of a tetrahedron.

6.71 **a.** Pentane has a carbon chain of five carbon atoms.
b. Ethane has a carbon chain of two carbon atoms.
c. Hexane has a carbon chain of six carbon atoms.

6.73 **a.** CH_4
b. $CH_3—CH_3$
c. $CH_3—CH_2—CH_2—CH_2—CH_3$

6.75 **a.** $CH_3—CH_2—CH_2—CH_2—CH_2—CH_2—CH_3$
b. Alkanes having 5–8 carbon atoms are liquid at room temperature.
c. Since alkanes are nonpolar compounds, they are insoluble in water.
d. Since the density of heptane (0.68 g/mL) is less than the density of water (1.00 g/mL), heptane will float in water.

6.77 **a.** By losing two valence electrons from the fourth energy level, calcium achieves an octet in the third energy level.
b. The calcium ion, Ca^{2+}, has an electron configuration $1s^22s^22p^63s^23p^6$, which is the same as argon (Ar).
c. Atoms of Group 1A (1) and 2A (2) elements can lose 1 or 2 electrons respectively to attain a noble gas electron configuration. Atoms of Group 8A (18) elements already have an octet of valence electrons (two for helium), so they do not lose or gain electrons and are not normally found in compounds.

6.79 **a.** P^{3-} ion **b.** O atom
c. Zn^{2+} ion **d.** Fe^{3+} ion

6.81 **a.** X is in Group 1A (1); Y is in Group 6A (16) **b.** ionic
c. X^+, Y^{2-} **d.** X_2Y
e. X^+ and $Cl^- \rightarrow XCl$ **f.** YCl_2

6.83

Period	Electron-Dot Symbols	Formula of Compound	Name of Compound
2	X· and ·Ÿ· (with one dot below Y)	Li_3N	Lithium nitride
4	Ẋ· and :Ÿ· (with two dots below Y)	$CaBr_2$	Calcium bromide
4	·Ẋ· and :Ÿ· (with one dot below Y)	Ga_2Se_3	Gallium selenide

6.85

Electron Configurations		Symbols of Ions			
Metal	Nonmetal	Cation	Anion	Formula of Compound	Name of Compound
$1s^22s^22p^63s^2$	$1s^22s^22p^3$	Mg^{2+}	N^{3-}	Mg_3N_2	Magnesium nitride
$1s^22s^22p^63s^23p^64s^1$	$1s^22s^22p^4$	K^+	O^{2-}	K_2O	Potassium oxide
$1s^22s^22p^63s^23p^1$	$1s^22s^22p^5$	Al^{3+}	F^-	AlF_3	Aluminum fluoride

6.87 **a.** Butane; organic compounds have low melting points.
b. Butane; organic compounds burn vigorously in air.
c. Potassium chloride; inorganic compounds have high melting points.
d. Potassium chloride; inorganic compounds (ionic) contain ionic bonds.
e. Butane; organic compounds are more likely to be gases at room temperature.

6.89 **a.** Organic, covalent, heptane
b. Inorganic, covalent, 1 N and 2 O $\rightarrow$ nitrogen dioxide
c. Inorganic, ionic, ions are K^+ and PO_4^{3-} $\rightarrow$ potassium phosphate
d. Organic, covalent, ethane

6.91 **a.** N^{3-} has an electron configuration $1s^22s^22p^6$, which is the same as neon (Ne).
b. Sr^{2+} has an electron configuration $1s^22s^22p^63s^23p^64s^23d^{10}4p^6$, which is the same as krypton (Kr).
c. I^- has an electron configuration $1s^22s^22p^63s^23p^64s^23d^{10}4p^65s^24d^{10}5p^6$, which is the same as xenon (Xe).
d. O^{2-} has an electron configuration $1s^22s^22p^6$, which is the same as neon (Ne).
e. Cs^+ has an electron configuration $1s^22s^22p^63s^23p^64s^23d^{10}4p^65s^24d^{10}5p^6$, which is the same as xenon (Xe).

6.93 **a.** An element that forms an ion with a 2+ charge would be in Group 2A (2).
b. The electron-dot symbol for an element in Group 2A (2) is $\dot{X}\cdot$.
c. Be is the Group 2A (2) element in Period 2.
d. Ions: X^{2+} and $N^{3-} \rightarrow X_3N_2$

6.95 **a.** Tin(IV) is Sn^{4+}.
b. The Sn^{4+} ion has 50 protons and $50 - 4 = 46$ electrons.
c. Ions: Sn^{4+} and $O^{2-} \rightarrow SnO_2$
d. Ions: Sn^{4+} and $PO_4^{3-} \rightarrow Sn_3(PO_4)_4$

6.97 **a.** X as a X^{3+} ion would be in Group 3A (13).
b. X as a X^{2-} ion would be in Group 6A (16).
c. X as a X^{2+} ion would be in Group 2A (2).

6.99 **a.** Ions: Fe^{3+} and $Cl^- \rightarrow$ iron(III) chloride
b. Ions: Ca^{2+} and $PO_4^{3-} \rightarrow$ calcium phosphate
c. Ions: Al^{3+} and $CO_3^{2-} \rightarrow$ aluminum carbonate
d. Ions: Pb^{4+} and $Cl^- \rightarrow$ lead(IV) chloride
e. Ions: Mg^{2+} and $CO_3^{2-} \rightarrow$ magnesium carbonate
f. Ions: Sn^{2+} and $SO_4^{2-} \rightarrow$ tin(II) sulfate
g. Ions: Cu^{2+} and $S^{2-} \rightarrow$ copper(II) sulfide

6.101 **a.** Ions: Cu^+ and $N^{3-} \rightarrow Cu_3N$
b. Ions: K^+ and $HSO_3^- \rightarrow KHSO_3$
c. Ions: Pb^{4+} and $S^{2-} \rightarrow PbS_2$
d. Ions: Au^{3+} and $CO_3^{2-} \rightarrow Au_2(CO_3)_3$
e. Ions: Zn^{2+} and $ClO_4^- \rightarrow Zn(ClO_4)_2$

6.103 **a.** Ions: Ni^{3+} and Cl^- → $NiCl_3$
b. Ions: Pb^{4+} and O^{2-} → PbO_2
c. Ions: Ag^+ and Br^- → $AgBr$
d. Ions: Ca^{2+} and N^{3-} → Ca_3N_2
e. Ions: Cu^+ and P^{3-} → Cu_3P
f. Ions: Cr^{2+} and S^{2-} → CrS

6.105 **a.** Ions: Mg^{2+} and O^{2-} → magnesium oxide
b. Ions: Cr^{3+} and HCO_3^- → chromium(III) hydrogen carbonate or chromium(III) bicarbonate
c. Ions: Mn^{3+} and CrO_4^{2-} → manganese(III) chromate

6.107 **a.** 1 N and 3 Cl → nitrogen trichloride
b. 1 S and 2 Cl → sulfur dichloride
c. 2 N and 1 O → dinitrogen oxide
d. 2 F → fluorine (named as the element)
e. 1 P and 5 F → phosphorus pentafluoride
f. 2 P and 5 O → diphosphorus pentoxide

6.109 **a.** 1 I and 7 F → IF_7
b. 1 B and 3 F → BF_3
c. 1 H and 1 I → HI
d. 1 S and 2 Cl → SCl_2

6.111 **a.** Ionic, ions are Co^{3+} and Cl^- → cobalt(III) chloride
b. Ionic, ions are Na^+ and SO_4^{2-} → sodium sulfate
c. Covalent, 2 N and 3 O → dinitrogen trioxide
d. Covalent, diatomic element → iodine
e. Covalent, 1 I and 3 Cl → iodine trichloride
f. Covalent, 1 C and 4 F → carbon tetrafluoride

6.113 **a.** Ions: Sn^{2+} and CO_3^{2-} → $SnCO_3$
b. Ions: Li^+ and P^{3-} → Li_3P
c. Covalent, 1 Si and 4 Cl → $SiCl_4$
d. Ions: Cr^{2+} and S^{2-} → CrS
e. Covalent, 1 C and 2 O → CO_2
f. Ions: Ca^{2+} and Br^- → $CaBr_2$
g. Pentane is an alkane with five carbon atoms; its formula can be written as C_5H_{12} or $CH_3{-}CH_2{-}CH_2{-}CH_2{-}CH_3$.

6.115 The valence electrons are the electrons in the highest energy levels. These are the electrons most easily lost when forming a positive ion. When an atom forms a negative ion, the electrons it gains must occupy these high energy levels because all lower levels are already occupied by the electrons of the neutral atom.

6.117 Elements in Group 2A (2) will lose two electrons to attain an octet; elements in Group 6A (16) will gain two electrons to attain an octet. The ions formed are similar in that they all contain an octet in the outermost electron level.

6.119

Formula of Compound	Type of Compound	Name of Compound
$FeSO_4$	Ionic	Iron(II) sulfate
SiO_2	Covalent	Silicon dioxide
NH_4NO_3	Ionic	Ammonium nitrate
$Al_2(SO_4)_3$	Ionic	Aluminum sulfate
Co_2S_3	Ionic	Cobalt(III) sulfide

6.121 Compounds with a metal and nonmetal are classified as ionic; compounds with two nonmetals are covalent.

a. Inorganic, ionic, ions are Li^+ and O^{2-} → lithium oxide
b. Organic, covalent, propane
c. Inorganic, covalent, 1 C and 4 F → carbon tetrafluoride
d. Organic, covalent, methane
e. Inorganic, ionic, ions are Mg^{2+} and F^- → magnesium fluoride
f. Organic, covalent, butane

6.123 Ions: Na^+ and S^{2-} → Na_2S

$$4.8 \times 10^{22}\ \cancel{Na^+\ ions} \times \frac{1\ S^{2-}\ ion}{2\ \cancel{Na^+\ ions}} = 2.4 \times 10^{22}\ S^{2-}\ ions\ (2\ SFs)$$

7
Chemical Quantities

Study Goals

- Use Avogadro's number to determine the number of particles in a given number of moles.
- Given the chemical formula of a substance, calculate its molar mass.
- Given the number of moles of a substance, calculate the mass in grams; given the mass, calculate the number of moles.
- Given the formula of a compound, calculate the percent composition; from the percent composition, determine the empirical formula of a compound.
- Given a quantity in moles of reactant or product, use a mole–mole factor from the balanced equation to calculate the moles of another substance in the reaction.
- Determine the molecular formula of a substance from the empirical formula and molar mass.

Chapter Outline

Answers and Solutions to Text Problems

7.1 **a.** $0.200 \text{ \sout{mol Ag}} \times \dfrac{6.022 \times 10^{23} \text{ atoms Ag}}{1 \text{ \sout{mol Ag}}} = 1.20 \times 10^{23}$ atoms of Ag (3 SFs)

b. $0.750 \text{ \sout{mol C}}_3\text{\sout{H}}_8\text{\sout{O}} \times \dfrac{6.022 \times 10^{23} \text{ molecules } C_3H_8O}{1 \text{ \sout{mol C}}_3\text{\sout{H}}_8\text{\sout{O}}} = 4.52 \times 10^{23}$ molecules of C_3H_8O (3 SFs)

c. $2.88 \times 10^{23} \text{ \sout{atoms Au}} \times \dfrac{1 \text{ mol Au}}{6.022 \times 10^{23} \text{ \sout{atoms Au}}} = 0.478$ mol of Au (3 SFs)

7.3 **a.** $1.0 \text{ \sout{mol C}}_{20}\text{\sout{H}}_{24}\text{\sout{N}}_2\text{\sout{O}}_2 \times \dfrac{24 \text{ mol H}}{1 \text{ \sout{mol C}}_{20}\text{\sout{H}}_{24}\text{\sout{N}}_2\text{\sout{O}}_2} = 24$ mol of H (2 SFs)

b. $5.0 \text{ \sout{mol C}}_{20}\text{\sout{H}}_{24}\text{\sout{N}}_2\text{\sout{O}}_2 \times \dfrac{20 \text{ mol C}}{1 \text{ \sout{mol C}}_{20}\text{\sout{H}}_{24}\text{\sout{N}}_2\text{\sout{O}}_2} = 1.0 \times 10^2$ mol of C (2 SFs)

c. $0.020\ \cancel{\text{mol } C_{20}H_{24}N_2O_2} \times \dfrac{2 \text{ mol N}}{1\ \cancel{\text{mol } C_{20}H_{24}N_2O_2}} = 0.040 \text{ mol of N (2 SFs)}$

7.5 **a.** $0.500\ \cancel{\text{mol C}} \times \dfrac{6.022 \times 10^{23} \text{ atoms C}}{1\ \cancel{\text{mol C}}} = 3.01 \times 10^{23} \text{ atoms of C (3 SFs)}$

b. $1.28\ \cancel{\text{mol } SO_2} \times \dfrac{6.022 \times 10^{23} \text{ molecules } SO_2}{1\ \cancel{\text{mol } SO_2}} = 7.71 \times 10^{23} \text{ molecules of } SO_2 \text{ (3 SFs)}$

c. $5.22 \times 10^{22}\ \cancel{\text{atoms Fe}} \times \dfrac{1 \text{ mol Fe}}{6.022 \times 10^{23}\ \cancel{\text{atoms Fe}}} = 0.0867 \text{ mol of Fe (3 SFs)}$

d. $8.50 \times 10^{24}\ \cancel{\text{molecules } C_2H_5OH} \times \dfrac{1 \text{ mol } C_2H_5OH}{6.022 \times 10^{23}\ \cancel{\text{molecules } C_2H_5OH}} = 14.1 \text{ mol of } C_2H_5OH$ (3 SFs)

7.7 1 mol of H_3PO_4 molecules contains 3 mol of H atoms, 1 mol of P atoms, and 4 mol of O atoms.

a. $2.00\ \cancel{\text{mol } H_3PO_4} \times \dfrac{3 \text{ mol H}}{1\ \cancel{\text{mol } H_3PO_4}} = 6.00 \text{ mol of H (3 SFs)}$

b. $2.00\ \cancel{\text{mol } H_3PO_4} \times \dfrac{4 \text{ mol O}}{1\ \cancel{\text{mol } H_3PO_4}} = 8.00 \text{ mol of O (3 SFs)}$

c. $2.00\ \cancel{\text{mol } H_3PO_4} \times \dfrac{1\ \cancel{\text{mol P}}}{1\ \cancel{\text{mol } H_3PO_4}} \times \dfrac{6.022 \times 10^{23} \text{ atoms P}}{1\ \cancel{\text{mol P}}} = 1.20 \times 10^{24} \text{ atoms of P (3 SFs)}$

d. $2.00\ \cancel{\text{mol } H_3PO_4} \times \dfrac{4\ \cancel{\text{mol O}}}{1\ \cancel{\text{mol } H_3PO_4}} \times \dfrac{6.022 \times 10^{23} \text{ atoms O}}{1\ \cancel{\text{mol O}}} = 4.82 \times 10^{24} \text{ atoms of O (3 SFs)}$

7.9 **a.** $1\ \cancel{\text{mol Na}} \times \dfrac{22.99 \text{ g Na}}{1\ \cancel{\text{mol Na}}} = 22.99 \text{ g of Na}$

$1\ \cancel{\text{mol Cl}} \times \dfrac{35.45 \text{ g Cl}}{1\ \cancel{\text{mol Cl}}} = 35.45 \text{ g of Cl}$

1 mol of Na = 22.99 g of Na
1 mol of Cl = 35.45 g of Cl
∴ Molar mass of NaCl = 58.44 g

b. $2\ \cancel{\text{mol Fe}} \times \dfrac{55.85 \text{ g Fe}}{1\ \cancel{\text{mol Fe}}} = 111.7 \text{ g of Fe}$

$3\ \cancel{\text{mol O}} \times \dfrac{16.00 \text{ g O}}{1\ \cancel{\text{mol O}}} = 48.00 \text{ g of O}$

2 mol of Fe = 111.7 g of Fe
3 mol of O = 48.00 g of O
∴ Molar mass of Fe_2O_3 = 159.7 g

c. $19\ \cancel{\text{mol C}} \times \dfrac{12.01 \text{ g C}}{1\ \cancel{\text{mol C}}} = 228.2 \text{ g of C}$

$20\ \cancel{\text{mol H}} \times \dfrac{1.008 \text{ g H}}{1\ \cancel{\text{mol H}}} = 20.16 \text{ g of H}$

$1\ \cancel{\text{mol F}} \times \dfrac{19.00 \text{ g F}}{1\ \cancel{\text{mol F}}} = 19.00 \text{ g of F}$

$1\ \cancel{\text{mol N}} \times \dfrac{14.01 \text{ g N}}{1\ \cancel{\text{mol N}}} = 14.01 \text{ g of N}$

$3\ \cancel{\text{mol O}} \times \dfrac{16.00 \text{ g O}}{1\ \cancel{\text{mol O}}} = 48.00 \text{ g of O}$

$$\begin{aligned}
&\text{19 mol of C} &&= 228.2 \text{ g of C}\\
&\text{20 mol of H} &&= 20.16 \text{ g of H}\\
&\text{1 mol of F} &&= 19.00 \text{ g of F}\\
&\text{1 mol of N} &&= 14.01 \text{ g of N}\\
&\text{3 mol of O} &&= \underline{48.00 \text{ g of O}}\\
&\therefore \text{Molar mass of } C_{19}H_{20}FNO_3 &&= 329.4 \text{ g}
\end{aligned}$$

d. $2 \cancel{\text{mol Al}} \times \dfrac{26.98 \text{ g Al}}{1 \cancel{\text{mol Al}}} = 53.96 \text{ g of Al}$

$3 \cancel{\text{mol S}} \times \dfrac{32.07 \text{ g S}}{1 \cancel{\text{mol S}}} = 96.21 \text{ g of S}$

$12 \cancel{\text{mol O}} \times \dfrac{16.00 \text{ g O}}{1 \cancel{\text{mol O}}} = 192.0 \text{ g of O}$

$$\begin{aligned}
&\text{2 mol of Al} &&= 53.96 \text{ g of Al}\\
&\text{3 mol of S} &&= 96.21 \text{ g of S}\\
&\text{12 mol of O} &&= \underline{192.0 \text{ g of O}}\\
&\therefore \text{Molar mass of } Al_2(SO_4)_3 &&= 342.2 \text{ g}
\end{aligned}$$

e. $1 \cancel{\text{mol K}} \times \dfrac{39.10 \text{ g K}}{1 \cancel{\text{mol K}}} = 39.10 \text{ g of K}$

$4 \cancel{\text{mol C}} \times \dfrac{12.01 \text{ g C}}{1 \cancel{\text{mol C}}} = 48.04 \text{ g of C}$

$5 \cancel{\text{mol H}} \times \dfrac{1.008 \text{ g H}}{1 \cancel{\text{mol H}}} = 5.040 \text{ g of H}$

$6 \cancel{\text{mol O}} \times \dfrac{16.00 \text{ g O}}{1 \cancel{\text{mol O}}} = 96.00 \text{ g of O}$

$$\begin{aligned}
&\text{1 mol of K} &&= 39.10 \text{ g of K}\\
&\text{4 mol of C} &&= 48.04 \text{ g of C}\\
&\text{5 mol of H} &&= 5.040 \text{ g of H}\\
&\text{6 mol of O} &&= \underline{96.00 \text{ g of O}}\\
&\therefore \text{Molar mass of } KC_4H_5O_6 &&= 188.18 \text{ g}
\end{aligned}$$

f. $16 \cancel{\text{mol C}} \times \dfrac{12.01 \text{ g C}}{1 \cancel{\text{mol C}}} = 192.2 \text{ g of C}$

$19 \cancel{\text{mol H}} \times \dfrac{1.008 \text{ g H}}{1 \cancel{\text{mol H}}} = 19.15 \text{ g of H}$

$3 \cancel{\text{mol N}} \times \dfrac{14.01 \text{ g N}}{1 \cancel{\text{mol N}}} = 42.03 \text{ g of N}$

$5 \cancel{\text{mol O}} \times \dfrac{16.00 \text{ g O}}{1 \cancel{\text{mol O}}} = 80.00 \text{ g of O}$

$1 \cancel{\text{mol S}} \times \dfrac{32.07 \text{ g S}}{1 \cancel{\text{mol S}}} = 32.07 \text{ g of S}$

$$\begin{aligned}
&\text{16 mol of C} &&= 192.2 \text{ g of C}\\
&\text{19 mol of H} &&= 19.15 \text{ g of H}\\
&\text{3 mol of N} &&= 42.03 \text{ g of N}\\
&\text{5 mol of O} &&= 80.00 \text{ g of O}\\
&\text{1 mol of S} &&= \underline{32.07 \text{ g of S}}\\
&\therefore \text{Molar mass of } C_{16}H_{19}N_3O_5S &&= 365.5 \text{ g}
\end{aligned}$$

7.11 **a.** $2\ \cancel{\text{mol Cl}} \times \dfrac{35.45\text{ g Cl}}{1\ \cancel{\text{mol Cl}}} = 70.90\text{ g of Cl}$

$\therefore$ Molar mass of Cl_2 = 70.90 g

b. $3\ \cancel{\text{mol C}} \times \dfrac{12.01\text{ g C}}{1\ \cancel{\text{mol C}}} = 36.03\text{ g of C}$

$6\ \cancel{\text{mol H}} \times \dfrac{1.008\text{ g H}}{1\ \cancel{\text{mol H}}} = 6.048\text{ g of H}$

$3\ \cancel{\text{mol O}} \times \dfrac{16.00\text{ g O}}{1\ \cancel{\text{mol O}}} = 48.00\text{ g of O}$

3 mol of C = 36.03 g of C
6 mol of H = 6.048 g of H
3 mol of O = 48.00 g of O

$\therefore$ Molar mass of $C_3H_6O_3$ = 90.08 g

c. $3\ \cancel{\text{mol Mg}} \times \dfrac{24.31\text{ g Mg}}{1\ \cancel{\text{mol Mg}}} = 72.93\text{ g of Mg}$

$2\ \cancel{\text{mol P}} \times \dfrac{30.97\text{ g P}}{1\ \cancel{\text{mol P}}} = 61.94\text{ g of P}$

$8\ \cancel{\text{mol O}} \times \dfrac{16.00\text{ g O}}{1\ \cancel{\text{mol O}}} = 128.0\text{ g of O}$

3 mol of Mg = 72.93 g of Mg
2 mol of P = 61.94 g of P
8 mol of O = 128.0 g of O

$\therefore$ Molar mass of $Mg_3(PO_4)_2$ = 262.9 g

d. $1\ \cancel{\text{mol Al}} \times \dfrac{26.98\text{ g Al}}{1\ \cancel{\text{mol Al}}} = 26.98\text{ g of Al}$

$3\ \cancel{\text{mol F}} \times \dfrac{19.00\text{ g F}}{1\ \cancel{\text{mol F}}} = 57.00\text{ g of F}$

1 mol of Al = 26.98 g of Al
3 mol of F = 57.00 g of F

$\therefore$ Molar mass of AlF_3 = 83.98 g

e. $2\ \cancel{\text{mol C}} \times \dfrac{12.01\text{ g C}}{1\ \cancel{\text{mol C}}} = 24.02\text{ g of C}$

$4\ \cancel{\text{mol H}} \times \dfrac{1.008\text{ g H}}{1\ \cancel{\text{mol H}}} = 4.032\text{ g of H}$

$2\ \cancel{\text{mol Cl}} \times \dfrac{35.45\text{ g Cl}}{1\ \cancel{\text{mol Cl}}} = 70.90\text{ g of Cl}$

2 mol of C = 24.02 g of C
4 mol of H = 4.032 g of H
2 mol of Cl = 70.90 g of Cl

$\therefore$ Molar mass of $C_2H_4Cl_2$ = 98.95 g

f. $1\ \cancel{\text{mol Sn}} \times \dfrac{118.7\text{ g Sn}}{1\ \cancel{\text{mol Sn}}} = 118.7\text{ g of Sn}$

$2\ \cancel{\text{mol F}} \times \dfrac{19.00\text{ g F}}{1\ \cancel{\text{mol F}}} = 38.00\text{ g of F}$

1 mol of Sn = 118.7 g of Sn
2 mol of F = 38.00 g of F

$\therefore$ Molar mass of SnF_2 = 156.7 g

7.13 **a.** $1.50\ \cancel{\text{mol Na}} \times \dfrac{22.99\ \text{g Na}}{1\ \cancel{\text{mol Na}}} = 34.5\ \text{g of Na (3 SFs)}$

b. $2.80\ \cancel{\text{mol Ca}} \times \dfrac{40.08\ \text{g Ca}}{1\ \cancel{\text{mol Ca}}} = 112\ \text{g of Ca (3 SFs)}$

c. Molar mass of CO_2

$= 1\ \text{mol of C} + 2\ \text{mol of O} = 12.01\ \text{g} + 2(16.00\ \text{g}) = 44.01\ \text{g}$

$0.125\ \cancel{\text{mol CO}_2} \times \dfrac{44.01\ \text{g CO}_2}{1\ \cancel{\text{mol CO}_2}} = 5.50\ \text{g of CO}_2\ \text{(3 SFs)}$

d. Molar mass of Na_2CO_3

$= 2\ \text{mol of Na} + 1\ \text{mol of C} + 3\ \text{mol of O} = 2(22.99\ \text{g}) + 12.01\ \text{g} + 3(16.00\ \text{g})$

$= 105.99\ \text{g}$

$0.0485\ \cancel{\text{mol Na}_2\text{CO}_3} \times \dfrac{105.99\ \text{g Na}_2\text{CO}_3}{1\ \cancel{\text{mol Na}_2\text{CO}_3}} = 5.14\ \text{g of Na}_2\text{CO}_3\ \text{(3 SFs)}$

e. Molar mass of PCl_3

$= 1\ \text{mol of P} + 3\ \text{mol of Cl} = 30.97\ \text{g} + 3(35.45\ \text{g}) = 137.32\ \text{g}$

$7.14 \times 10^2\ \cancel{\text{mol PCl}_3} \times \dfrac{137.32\ \text{g PCl}_3}{1\ \cancel{\text{mol PCl}_3}} = 9.80 \times 10^4\ \text{g of PCl}_3\ \text{(3 SFs)}$

7.15 **a.** $0.150\ \cancel{\text{mol Ne}} \times \dfrac{20.18\ \text{g Ne}}{1\ \cancel{\text{mol Ne}}} = 3.03\ \text{g of Ne}$

b. Molar mass of I_2

$= 2\ \text{mol of I} = 2(126.9\ \text{g}) = 253.8\ \text{g}$

$0.150\ \cancel{\text{mol I}_2} \times \dfrac{253.8\ \text{g I}_2}{1\ \cancel{\text{mol I}_2}} = 38.1\ \text{g of I}_2\ \text{(3 SFs)}$

c. Molar mass of Na_2O

$= 2\ \text{mol of Na} + 1\ \text{mol of O} = 2(22.99\ \text{g}) + 16.00\ \text{g} = 61.98\ \text{g}$

$0.150\ \cancel{\text{mol Na}_2\text{O}} \times \dfrac{61.98\ \text{g Na}_2\text{O}}{1\ \cancel{\text{mol Na}_2\text{O}}} = 9.30\ \text{g of Na}_2\text{O (3 SFs)}$

d. Molar mass of $Ca(NO_3)_2$

$= 1\ \text{mol of Ca} + 2\ \text{mol of N} + 6\ \text{mol of O} = 40.08\ \text{g} + 2(14.01\ \text{g}) + 6(16.00\ \text{g}) = 164.10\ \text{g}$

$0.150\ \cancel{\text{mol Ca(NO}_3)_2} \times \dfrac{164.10\ \text{g Ca(NO}_3)_2}{1\ \cancel{\text{mol Ca(NO}_3)_2}} = 24.6\ \text{g of Ca(NO}_3)_2\ \text{(3 SFs)}$

e. Molar mass of C_6H_{14}

$= 6\ \text{mol of C} + 14\ \text{mol of H} = 6(12.01\ \text{g}) + 14(1.008\ \text{g}) = 86.17\ \text{g}$

$0.150\ \cancel{\text{mol C}_6\text{H}_{14}} \times \dfrac{86.17\ \text{g C}_6\text{H}_{14}}{1\ \cancel{\text{mol C}_6\text{H}_{14}}} = 12.9\ \text{g of C}_6\text{H}_{14}\ \text{(3 SFs)}$

7.17 **a.** $82.0\ \cancel{\text{g Ag}} \times \dfrac{1\ \text{mol Ag}}{107.9\ \cancel{\text{g Ag}}} = 0.760\ \text{mol of Ag (3 SFs)}$

b. $0.288\ \cancel{\text{g C}} \times \dfrac{1\ \text{mol C}}{12.01\ \cancel{\text{g C}}} = 0.0240\ \text{mol of C (3 SFs)}$

c. $15.0\ \cancel{\text{g NH}_3} \times \dfrac{1\ \text{mol NH}_3}{17.03\ \cancel{\text{g NH}_3}} = 0.881\ \text{mol of NH}_3\ \text{(3 SFs)}$

d. $7.25\ \cancel{\text{g C}_3\text{H}_8} \times \dfrac{1\ \text{mol C}_3\text{H}_8}{44.09\ \cancel{\text{g C}_3\text{H}_8}} = 0.164\ \text{mol of C}_3\text{H}_8\ \text{(3 SFs)}$

e. Molar mass of Fe_2O_3

$= 2\ \text{mol of Fe} + 3\ \text{mol of O} = 2(55.85\ \text{g}) + 3(16.00\ \text{g}) = 159.7\ \text{g}$

$245\ \cancel{\text{g Fe}_2\text{O}_3} \times \dfrac{1\ \text{mol Fe}_2\text{O}_3}{159.7\ \cancel{\text{g Fe}_2\text{O}_3}} = 1.53\ \text{mol of Fe}_2\text{O}_3\ \text{(3 SFs)}$

7.19 **a.** $25.0\ \cancel{\text{g He}} \times \dfrac{1\ \text{mol He}}{4.003\ \cancel{\text{g He}}} = 6.25\ \text{mol of He (3 SFs)}$

b. Molar mass of O_2
$= 2\ \text{mol of O} = 2(16.00\ \text{g}) = 32.00\ \text{g}$
$25.0\ \cancel{\text{g O}_2} \times \dfrac{1\ \text{mol O}_2}{32.00\ \cancel{\text{g O}_2}} = 0.781\ \text{mol of O}_2\ \text{(3 SFs)}$

c. Molar mass of $Al(OH)_3$
$= 1\ \text{mol of Al} + 3\ \text{mol of O} + 3\ \text{mol of H} = 26.98\ \text{g} + 3(16.00\ \text{g}) + 3(1.008\ \text{g}) = 78.00\ \text{g}$
$25.0\ \cancel{\text{g Al(OH)}_3} \times \dfrac{1\ \text{mol Al(OH)}_3}{78.00\ \cancel{\text{g Al(OH)}_3}} = 0.321\ \text{mol of Al(OH)}_3\ \text{(3 SFs)}$

d. Molar mass of Ga_2S_3
$= 2\ \text{mol of Ga} + 3\ \text{mol of S} = 2(69.72\ \text{g}) + 3(32.07\ \text{g}) = 235.7\ \text{g}$
$25.0\ \cancel{\text{g Ga}_2\text{S}_3} \times \dfrac{1\ \text{mol Ga}_2\text{S}_3}{235.7\ \cancel{\text{g Ga}_2\text{S}_3}} = 0.106\ \text{mol of Ga}_2\text{S}_3\ \text{(3 SFs)}$

e. Molar mass of C_4H_{10}
$= 4\ \text{mol of C} + 10\ \text{mol of H} = 4(12.01\ \text{g}) + 10(1.008\ \text{g}) = 58.12\ \text{g}$
$25.0\ \cancel{\text{g C}_4\text{H}_{10}} \times \dfrac{1\ \text{mol C}_4\text{H}_{10}}{58.12\ \cancel{\text{g C}_4\text{H}_{10}}} = 0.430\ \text{mol of C}_4\text{H}_{10}\ \text{(3 SFs)}$

7.21 **a.** $25.0\ \cancel{\text{g C}} \times \dfrac{1\ \cancel{\text{mol C}}}{12.01\ \cancel{\text{g C}}} \times \dfrac{6.022 \times 10^{23}\ \text{atoms C}}{1\ \cancel{\text{mol C}}} = 1.25 \times 10^{24}\ \text{atoms of C}$

b. $0.688\ \cancel{\text{mol CO}_2} \times \dfrac{1\ \cancel{\text{mol C}}}{1\ \cancel{\text{mol CO}_2}} \times \dfrac{6.022 \times 10^{23}\ \text{atoms C}}{1\ \cancel{\text{mol C}}} = 4.14 \times 10^{23}\ \text{atoms of C}$

c. Molar mass of C_3H_8
$= 3\ \text{mol of C} + 8\ \text{mol of H} = 3(12.01\ \text{g}) + 8(1.008\ \text{g}) = 44.09\ \text{g}$
$275\ \cancel{\text{g C}_3\text{H}_8} \times \dfrac{1\ \cancel{\text{mol C}_3\text{H}_8}}{44.09\ \cancel{\text{g C}_3\text{H}_8}} \times \dfrac{3\ \cancel{\text{mol C}}}{1\ \cancel{\text{mol C}_3\text{H}_8}} \times \dfrac{6.022 \times 10^{23}\ \text{atoms C}}{1\ \cancel{\text{mol C}}} = 1.13 \times 10^{25}\ \text{atoms of C}$

d. $1.84\ \cancel{\text{mol C}_2\text{H}_6\text{O}} \times \dfrac{2\ \cancel{\text{mol C}}}{1\ \cancel{\text{mol C}_2\text{H}_6\text{O}}} \times \dfrac{6.022 \times 10^{23}\ \text{atoms C}}{1\ \cancel{\text{mol C}}} = 2.22 \times 10^{24}\ \text{atoms of C}$

e. $7.5 \times 10^{24}\ \cancel{\text{molecules CH}_4} \times \dfrac{1\ \text{atom C}}{1\ \cancel{\text{molecule CH}_4}} = 7.5 \times 10^{24}\ \text{atoms of C (2 SFs)}$

7.23 **a.** Molar mass of C_3H_8
$= 3\ \text{mol of C} + 8\ \text{mol of H} = 3(12.01\ \text{g}) + 8(1.008\ \text{g}) = 44.09\ \text{g}$
$1.50\ \cancel{\text{mol C}_3\text{H}_8} \times \dfrac{44.09\ \text{g C}_3\text{H}_8}{1\ \cancel{\text{mol C}_3\text{H}_8}} = 66.1\ \text{g of C}_3\text{H}_8\ \text{(3 SFs)}$

b. $34.0\ \cancel{\text{g C}_3\text{H}_8} \times \dfrac{1\ \text{mol C}_3\text{H}_8}{44.09\ \cancel{\text{g C}_3\text{H}_8}} = 0.771\ \text{mol of C}_3\text{H}_8\ \text{(3 SFs)}$

c. $34.0\ \cancel{\text{g C}_3\text{H}_8} \times \dfrac{1\ \cancel{\text{mol C}_3\text{H}_8}}{44.09\ \cancel{\text{g C}_3\text{H}_8}} \times \dfrac{3\ \cancel{\text{mol C}}}{1\ \cancel{\text{mol C}_3\text{H}_8}} \times \dfrac{12.01\ \text{g C}}{1\ \cancel{\text{mol C}}} = 27.8\ \text{g of C (3 SFs)}$

d. $0.254\ \cancel{\text{g C}_3\text{H}_8} \times \dfrac{1\ \cancel{\text{mol C}_3\text{H}_8}}{44.09\ \cancel{\text{g C}_3\text{H}_8}} \times \dfrac{8\ \cancel{\text{mol H}}}{1\ \cancel{\text{mol C}_3\text{H}_8}} \times \dfrac{6.022 \times 10^{23}\ \text{atoms H}}{1\ \cancel{\text{mol H}}}$
$= 2.78 \times 10^{22}\ \text{atoms of H (3 SFs)}$

7.25 **a.** $1\ \cancel{\text{mol Mg}} \times \frac{24.31\text{ g Mg}}{1\ \cancel{\text{mol Mg}}} = 24.31\text{ g of Mg}$

$2\ \cancel{\text{mol F}} \times \frac{19.00\text{ g F}}{1\ \cancel{\text{mol F}}} = 38.00\text{ g of F}$

$\therefore$ Molar mass of $MgF_2 = 62.31\text{ g}$

$$\text{Mass \% Mg} = \frac{24.31\text{ g Mg}}{62.31\text{ g MgF}_2} \times 100\% = 39.01\%\text{ Mg (4 SFs)}$$

$$\text{Mass \% F} = \frac{38.00\text{ g F}}{62.31\text{ g MgF}_2} \times 100\% = 60.99\%\text{ F (4 SFs)}$$

b. $1\ \cancel{\text{mol Ca}} \times \frac{40.08\text{ g Ca}}{1\ \cancel{\text{mol Ca}}} = 40.08\text{ g of Ca}$

$2\ \cancel{\text{mol O}} \times \frac{16.00\text{ g O}}{1\ \cancel{\text{mol O}}} = 32.00\text{ g of O}$

$2\ \cancel{\text{mol H}} \times \frac{1.008\text{ g H}}{1\ \cancel{\text{mol H}}} = 2.016\text{ g of H}$

$\therefore$ Molar mass of $Ca(OH)_2 = 74.10\text{ g}$

$$\text{Mass \% Ca} = \frac{40.08\text{ g Ca}}{74.10\text{ g Ca(OH)}_2} \times 100\% = 54.09\%\text{ Ca (4 SFs)}$$

$$\text{Mass \% O} = \frac{32.00\text{ g O}}{74.10\text{ g Ca(OH)}_2} \times 100\% = 43.18\%\text{ O (4 SFs)}$$

$$\text{Mass \% H} = \frac{2.016\text{ g H}}{74.10\text{ g Ca(OH)}_2} \times 100\% = 2.72\%\text{ H (rounded to hundredths)}$$

c. $4\ \cancel{\text{mol C}} \times \frac{12.01\text{ g C}}{1\ \cancel{\text{mol C}}} = 48.04\text{ g of C}$

$8\ \cancel{\text{mol H}} \times \frac{1.008\text{ g H}}{1\ \cancel{\text{mol H}}} = 8.064\text{ g of H}$

$4\ \cancel{\text{mol O}} \times \frac{16.00\text{ g O}}{1\ \cancel{\text{mol O}}} = 64.00\text{ g of O}$

$\therefore$ Molar mass of $C_4H_8O_4 = 120.10\text{ g}$

$$\text{Mass \% C} = \frac{48.04\text{ g C}}{120.10\text{ g C}_4\text{H}_8\text{O}_4} \times 100\% = 40.00\%\text{ C (4 SFs)}$$

$$\text{Mass \% H} = \frac{8.064\text{ g H}}{120.10\text{ g C}_4\text{H}_8\text{O}_4} \times 100\% = 6.71\%\text{ H (rounded to hundredths)}$$

$$\text{Mass \% O} = \frac{64.00\text{ g O}}{120.10\text{ g C}_4\text{H}_8\text{O}_4} \times 100\% = 53.29\%\text{ O (4 SFs)}$$

d. $3\ \cancel{\text{mol N}} \times \frac{14.01\text{ g N}}{1\ \cancel{\text{mol N}}} = 42.03\text{ g of N}$

$12\ \cancel{\text{mol H}} \times \frac{1.008\text{ g H}}{1\ \cancel{\text{mol H}}} = 12.10\text{ g of H}$

$1\ \cancel{\text{mol P}} \times \frac{30.97\text{ g P}}{1\ \cancel{\text{mol P}}} = 30.97\text{ g of P}$

$4\ \cancel{\text{mol O}} \times \frac{16.00\text{ g O}}{1\ \cancel{\text{mol O}}} = 64.00\text{ g of O}$

$\therefore$ Molar mass of $(NH_4)_3PO_4 = 149.10\text{ g}$

$$\text{Mass \% N} = \frac{42.03\text{ g N}}{149.10\text{ g (NH}_4)_3\text{PO}_4} \times 100\% = 28.19\%\text{ N (4 SFs)}$$

$$\text{Mass \% H} = \frac{12.10 \text{ g H}}{149.10 \text{ g } (NH_4)_3PO_4} \times 100\% = 8.12\% \text{ H (rounded to hundredths)}$$

$$\text{Mass \% P} = \frac{30.97 \text{ g P}}{149.10 \text{ g } (NH_4)_3PO_4} \times 100\% = 20.77\% \text{ P (4 SFs)}$$

$$\text{Mass \% O} = \frac{64.00 \text{ g O}}{149.10 \text{ g } (NH_4)_3PO_4} \times 100\% = 42.92\% \text{ O (4 SFs)}$$

e. $$17 \cancel{\text{mol C}} \times \frac{12.01 \text{ g C}}{1 \cancel{\text{mol C}}} = 204.2 \text{ g of C}$$

$$19 \cancel{\text{mol H}} \times \frac{1.008 \text{ g H}}{1 \cancel{\text{mol H}}} = 19.15 \text{ g of H}$$

$$1 \cancel{\text{mol N}} \times \frac{14.01 \text{ g N}}{1 \cancel{\text{mol N}}} = 14.01 \text{ g of N}$$

$$3 \cancel{\text{mol O}} \times \frac{16.00 \text{ g O}}{1 \cancel{\text{mol O}}} = 48.00 \text{ g of O}$$

$$\therefore \text{Molar mass of } C_{17}H_{19}NO_3 = 285.4 \text{ g}$$

$$\text{Mass \% C} = \frac{204.2 \text{ g C}}{285.4 \text{ g } C_{17}H_{19}NO_3} \times 100\% = 71.55\% \text{ C (4 SFs)}$$

$$\text{Mass \% H} = \frac{19.15 \text{ g H}}{285.4 \text{ g } C_{17}H_{19}NO_3} \times 100\% = 6.71\% \text{ H (rounded to hundredths)}$$

$$\text{Mass \% N} = \frac{14.01 \text{ g N}}{285.4 \text{ g } C_{17}H_{19}NO_3} \times 100\% = 4.91\% \text{ N (rounded to hundredths)}$$

$$\text{Mass \% O} = \frac{48.00 \text{ g O}}{285.4 \text{ g } C_{17}H_{19}NO_3} \times 100\% = 16.82\% \text{ O (4 SFs)}$$

7.27 a. $$2 \cancel{\text{mol N}} \times \frac{14.01 \text{ g N}}{1 \cancel{\text{mol N}}} = 28.02 \text{ g of N}$$

$$5 \cancel{\text{mol O}} \times \frac{16.00 \text{ g O}}{1 \cancel{\text{mol O}}} = 80.00 \text{ g of O}$$

$$\therefore \text{Molar mass of } N_2O_5 = 108.02 \text{ g}$$

$$\text{Mass \% N} = \frac{28.02 \text{ g N}}{108.02 \text{ g } N_2O_5} \times 100\% = 25.94\% \text{ N (4 SFs)}$$

b. $$2 \cancel{\text{mol N}} \times \frac{14.01 \text{ g N}}{1 \cancel{\text{mol N}}} = 28.02 \text{ g of N}$$

$$4 \cancel{\text{mol H}} \times \frac{1.008 \text{ g H}}{1 \cancel{\text{mol H}}} = 4.032 \text{ g of H}$$

$$3 \cancel{\text{mol O}} \times \frac{16.00 \text{ g O}}{1 \cancel{\text{mol O}}} = 48.00 \text{ g of O}$$

$$\therefore \text{Molar mass of } NH_4NO_3 = 80.05 \text{ g}$$

$$\text{Mass \% N} = \frac{28.02 \text{ g N}}{80.05 \text{ g } NH_4NO_3} \times 100\% = 35.00\% \text{ N (4 SFs)}$$

c. $$2 \cancel{\text{mol C}} \times \frac{12.01 \text{ g C}}{1 \cancel{\text{mol C}}} = 24.02 \text{ g of C}$$

$$8 \cancel{\text{mol H}} \times \frac{1.008 \text{ g H}}{1 \cancel{\text{mol H}}} = 8.064 \text{ g of H}$$

$$2 \cancel{\text{mol N}} \times \frac{14.01 \text{ g N}}{1 \cancel{\text{mol N}}} = 28.02 \text{ g of N}$$

$$\therefore \text{Molar mass of } C_2H_8N_2 = 60.10 \text{ g}$$

$$\text{Mass \% N} = \frac{28.02 \text{ g N}}{60.10 \text{ g } C_2H_8N_2} \times 100\% = 46.62\% \text{ N (4 SFs)}$$

d. $9 \cancel{\text{mol C}} \times \frac{12.01 \text{ g C}}{1 \cancel{\text{mol C}}} = 108.1 \text{ g of C}$

$15 \cancel{\text{mol H}} \times \frac{1.008 \text{ g H}}{1 \cancel{\text{mol H}}} = 15.12 \text{ g of H}$

$5 \cancel{\text{mol N}} \times \frac{14.01 \text{ g N}}{1 \cancel{\text{mol N}}} = 70.05 \text{ g of N}$

$1 \cancel{\text{mol O}} \times \frac{16.00 \text{ g O}}{1 \cancel{\text{mol O}}} = 16.00 \text{ g of O}$

$\therefore$ Molar mass of $C_9H_{15}N_5O$ = 209.3 g

$$\text{Mass \% N} = \frac{70.05 \text{ g N}}{209.3 \text{ g } C_9H_{15}N_5O} \times 100\% = 33.47\% \text{ N (4 SFs)}$$

e. $14 \cancel{\text{mol C}} \times \frac{12.01 \text{ g C}}{1 \cancel{\text{mol C}}} = 168.1 \text{ g of C}$

$22 \cancel{\text{mol H}} \times \frac{1.008 \text{ g H}}{1 \cancel{\text{mol H}}} = 22.18 \text{ g of H}$

$2 \cancel{\text{mol N}} \times \frac{14.01 \text{ g N}}{1 \cancel{\text{mol N}}} = 28.02 \text{ g of N}$

$1 \cancel{\text{mol O}} \times \frac{16.00 \text{ g O}}{1 \cancel{\text{mol O}}} = 16.00 \text{ g of O}$

$\therefore$ Molar mass of $C_{14}H_{22}N_2O$ = 234.3 g

$$\text{Mass \% N} = \frac{28.02 \text{ g N}}{234.3 \text{ g } C_{14}H_{22}N_2O} \times 100\% = 11.96\% \text{ N (4 SFs)}$$

7.29 **a.** $3.57 \cancel{\text{g N}} \times \frac{1 \text{ mol N}}{14.01 \cancel{\text{g N}}} = 0.255 \text{ mol of N}$

$2.04 \cancel{\text{g O}} \times \frac{1 \text{ mol O}}{16.00 \cancel{\text{g O}}} = 0.128 \text{ mol of O (smaller number of moles)}$

$\frac{0.255 \text{ mol N}}{0.128} = 1.99 \text{ mol N}$ $\quad\quad$ $\frac{0.128 \text{ mol O}}{0.128} = 1.00 \text{ mol of O}$

$\therefore$ Empirical formula = $N_{1.99}O_{1.00} \rightarrow N_2O$

b. $7.00 \cancel{\text{g C}} \times \frac{1 \text{ mol C}}{12.01 \cancel{\text{g C}}} = 0.583 \text{ mol of C (smaller number of moles)}$

$1.75 \cancel{\text{g H}} \times \frac{1 \text{ mol H}}{1.008 \cancel{\text{g H}}} = 1.74 \text{ mol of H}$

$\frac{0.583 \text{ mol C}}{0.583} = 1.00 \text{ mol of C}$ $\quad\quad$ $\frac{1.74 \text{ mol H}}{0.583} = 2.98 \text{ mol of H}$

$\therefore$ Empirical formula = $C_{1.00}H_{2.98} \rightarrow CH_3$

c. $0.175 \cancel{\text{g H}} \times \frac{1 \text{ mol H}}{1.008 \cancel{\text{g H}}} = 0.174 \text{ mol of H (smallest number of moles)}$

$2.44 \cancel{\text{g N}} \times \frac{1 \text{ mol N}}{14.01 \cancel{\text{g N}}} = 0.174 \text{ mol of N}$

$8.38 \cancel{\text{g O}} \times \frac{1 \text{ mol O}}{16.00 \cancel{\text{g O}}} = 0.524 \text{ mol of O}$

$\frac{0.174 \text{ mol H}}{0.174} = 1.00 \text{ mol of H}$ $\quad\quad$ $\frac{0.174 \text{ mol N}}{0.174} = 1.00 \text{ mol of N}$

$$\frac{0.524 \text{ mol O}}{0.174} = 3.01 \text{ mol of O}$$

$\therefore$ Empirical formula $= H_{1.00}N_{1.00}O_{3.01} \rightarrow HNO_3$

d. $$2.06 \text{ g Ca} \times \frac{1 \text{ mol Ca}}{40.08 \text{ g Ca}} = 0.0514 \text{ mol of Ca}$$

$$2.66 \text{ g Cr} \times \frac{1 \text{ mol Cr}}{52.00 \text{ g Cr}} = 0.0512 \text{ mol of Cr (smallest number of moles)}$$

$$3.28 \text{ g O} \times \frac{1 \text{ mol O}}{16.00 \text{ g O}} = 0.205 \text{ mol of O}$$

$$\frac{0.0514 \text{ mol Ca}}{0.0512} = 1.00 \text{ mol of Ca} \qquad \frac{0.0512 \text{ mol Cr}}{0.0512} = 1.00 \text{ mol of Cr}$$

$$\frac{0.205 \text{ mol O}}{0.0512} = 4.00 \text{ mol of O}$$

$\therefore$ Empirical formula $= Ca_{1.00}Cr_{1.00}O_{4.00} \rightarrow CaCrO_4$

7.31 11.44 g compound (S and F) − 2.51 g S = 8.93 g of F

$$2.51 \text{ g S} \times \frac{1 \text{ mol S}}{32.07 \text{ g S}} = 0.0783 \text{ mol of S (smaller number of moles)}$$

$$8.93 \text{ g F} \times \frac{1 \text{ mol F}}{19.00 \text{ g F}} = 0.470 \text{ mol of F}$$

$$\frac{0.0783 \text{ mol S}}{0.0783} = 1.00 \text{ mol of S} \qquad \frac{0.470 \text{ mol F}}{0.0783} = 6.00 \text{ mol of F}$$

$\therefore$ Empirical formula $= S_{1.00}F_{6.00} \rightarrow SF_6$

7.33 **a.** In exactly 100 g of compound, there are 70.9 g of K and 29.1 g of S.

$$70.9 \text{ g K} \times \frac{1 \text{ mol K}}{39.01 \text{ g K}} = 1.81 \text{ mol of K}$$

$$29.1 \text{ g S} \times \frac{1 \text{ mol S}}{32.07 \text{ g S}} = 0.907 \text{ mol of S (smaller number of moles)}$$

$$\frac{1.81 \text{ mol K}}{0.907} = 2.00 \text{ mol of K} \qquad \frac{0.907 \text{ mol S}}{0.907} = 1.00 \text{ mol of S}$$

$\therefore$ Empirical formula $= K_{2.00}S_{1.00} \rightarrow K_2S$

b. In exactly 100 g of compound, there are 55.0 g of Ga and 45.0 g of F.

$$55.0 \text{ g Ga} \times \frac{1 \text{ mol Ga}}{69.72 \text{ g Ga}} = 0.789 \text{ mol of Ga (smaller number of moles)}$$

$$45.0 \text{ g F} \times \frac{1 \text{ mol F}}{19.00 \text{ g F}} = 2.37 \text{ mol of F}$$

$$\frac{0.789 \text{ mol Ga}}{0.789} = 1.00 \text{ mol of Ga} \qquad \frac{2.37 \text{ mol F}}{0.789} = 3.00 \text{ mol of F}$$

$\therefore$ Empirical formula $= Ga_{1.00}F_{3.00} \rightarrow GaF_3$

c. In exactly 100 g of compound, there are 31.0 g of B and 69.0 g of O.

$$31.0 \text{ g B} \times \frac{1 \text{ mol B}}{10.81 \text{ g B}} = 2.87 \text{ mol of B (smaller number of moles)}$$

$$69.0 \text{ g O} \times \frac{1 \text{ mol O}}{16.00 \text{ g O}} = 4.31 \text{ mol of O}$$

$$\frac{2.87 \text{ mol B}}{2.87} = 1.00 \text{ mol of B} \qquad \frac{4.31 \text{ mol O}}{2.87} = 1.50 \text{ mol of O}$$

$\therefore$ Empirical formula $= B_{1.00}O_{1.50} \rightarrow B_{(1.00\times2)}O_{(1.50\times2)} = B_{2.00}O_{3.00} \rightarrow B_2O_3$

d. In exactly 100 g of compound, there are 18.8 g of Li, 16.3 g of C, and 64.9 g of O.

$$18.8\,\cancel{g\,Li} \times \frac{1 \text{ mol Li}}{6.941\,\cancel{g\,Li}} = 2.71 \text{ mol of Li}$$

$$16.3\,\cancel{g\,C} \times \frac{1 \text{ mol C}}{12.01\,\cancel{g\,C}} = 1.36 \text{ mol of C (smallest number of moles)}$$

$$64.9\,\cancel{g\,O} \times \frac{1 \text{ mol O}}{16.00\,\cancel{g\,O}} = 4.06 \text{ mol of O}$$

$$\frac{2.71 \text{ mol Li}}{1.36} = 1.99 \text{ mol of Li} \qquad \frac{1.36 \text{ mol C}}{1.36} = 1.00 \text{ mol of C}$$

$$\frac{4.06 \text{ mol O}}{1.36} = 2.99 \text{ mol of O}$$

$\therefore$ Empirical formula $= Li_{1.99}C_{1.00}O_{2.99} \rightarrow Li_2CO_3$

e. In exactly 100 g of compound, there are 51.7 g of C, 6.95 g of H, and 41.3 g of O.

$$51.7\,\cancel{g\,C} \times \frac{1 \text{ mol C}}{12.01\,\cancel{g\,C}} = 4.30 \text{ mol of C}$$

$$6.95\,\cancel{g\,H} \times \frac{1 \text{ mol H}}{1.008\,\cancel{g\,H}} = 6.89 \text{ mol of H}$$

$$41.3\,\cancel{g\,O} \times \frac{1 \text{ mol O}}{16.00\,\cancel{g\,O}} = 2.58 \text{ mol of O (smallest number of moles)}$$

$$\frac{4.30 \text{ mol C}}{2.58} = 1.67 \text{ mol of C} \qquad \frac{6.89 \text{ mol H}}{2.58} = 2.67 \text{ mol of H}$$

$$\frac{2.58 \text{ mol O}}{2.58} = 1.00 \text{ mol of O}$$

$\therefore$ Empirical formula $= C_{1.67}H_{2.67}O_{1.00} \rightarrow C_{(1.67\times3)}H_{(2.67\times3)}O_{(1.00\times3)}$

$= C_{5.01}H_{8.01}O_{3.00} \rightarrow C_5H_8O_3$

7.35 **a.** $H_2O_2 \rightarrow H_{(2\div2)}O_{(2\div2)} = HO$ (empirical formula)

b. $C_{18}H_{12} \rightarrow C_{(18\div6)}H_{(12\div6)} = C_3H_2$ (empirical formula)

c. $C_{10}H_{16}O_2 \rightarrow C_{(10\div2)}H_{(16\div2)}O_{(2\div2)} = C_5H_8O$ (empirical formula)

d. $C_9H_{18}N_6 \rightarrow C_{(9\div3)}H_{(18\div3)}N_{(6\div3)} = C_3H_6N_2$ (empirical formula)

e. $C_2H_4N_2O_2 \rightarrow C_{(2\div2)}H_{(4\div2)}N_{(2\div2)}O_{(2\div2)} = CH_2NO$ (empirical formula)

7.37 Empirical formula mass of CH_2O = 12.01 + 2(1.008) + 16.00 = 30.03 g

$$\text{Small integer} = \frac{\text{molar mass of fructose}}{\text{empirical formula mass of } CH_2O} = \frac{180\,\cancel{g}}{30.03\,\cancel{g}} = 6$$

$\therefore$ Molecular formula of fructose $= C_{(1\times6)}H_{(2\times6)}O_{(1\times6)} = C_6H_{12}O_6$

7.39 Empirical formula mass of CH = 12.01 + 1.008 = 13.02 g

For benzene: $$\text{Small integer} = \frac{\text{molar mass of benzene}}{\text{empirical formula mass of CH}} = \frac{78\,\cancel{g}}{13.02\,\cancel{g}} = 6$$

$\therefore$ Molecular formula of benzene $= C_{(1\times6)}H_{(1\times6)} = C_6H_6$

For acetylene: $$\text{Small integer} = \frac{\text{molar mass of acetylene}}{\text{empirical formula mass of CH}} = \frac{26\,\cancel{g}}{13.02\,\cancel{g}} = 2$$

$\therefore$ Molecular formula of acetylene $= C_{(1\times2)}H_{(1\times2)} = C_2H_2$

7.41 In exactly 100 g of mevalonic acid, there are 48.64 g of C, 8.16 g of H, and 43.20 g of O.

$$48.64\,\cancel{g\,C} \times \frac{1 \text{ mol C}}{12.01\,\cancel{g\,C}} = 4.050 \text{ mol of C}$$

$$8.16\,\cancel{g\,H} \times \frac{1 \text{ mol H}}{1.008\,\cancel{g\,H}} = 8.10 \text{ mol of H}$$

$$43.20\ \cancel{g\ O} \times \frac{1\ \text{mol O}}{16.00\ \cancel{g\ O}} = 2.70\ \text{mol of O (smallest number of moles)}$$

$$\frac{4.050\ \text{mol C}}{2.70} = 1.50\ \text{mol of C} \qquad \frac{8.10\ \text{mol H}}{2.70} = 3.00\ \text{mol of H}$$

$$\frac{2.70\ \text{mol O}}{2.70} = 1.00\ \text{mol of O}$$

$$\therefore \text{Empirical formula} = C_{1.50}H_{3.00}O_{1.00} \rightarrow C_{(1.50\times2)}H_{(3.00\times2)}O_{(1.00\times2)}$$
$$= C_{3.00}H_{6.00}O_{2.00} \rightarrow C_3H_6O_2$$

Empirical formula mass of $C_3H_6O_2 = 3(12.01) + 6(1.008) + 2(16.00) = 74.08\ g$

$$\text{Small integer} = \frac{\text{molar mass of mevalonic acid}}{\text{empirical formula mass of } C_3H_6O_2} = \frac{148\ \cancel{g}}{74.08\ \cancel{g}} = 2$$

$$\therefore \text{Molecular formula of mevalonic acid} = C_{(3\times2)}H_{(6\times2)}O_{(2\times2)} = C_6H_{12}O_4$$

7.43 In exactly 100 g of vanillic acid, there are 57.14 g of C, 4.80 g of H, and 38.06 g of O.

$$57.14\ \cancel{g\ C} \times \frac{1\ \text{mol C}}{12.01\ \cancel{g\ C}} = 4.758\ \text{mol of C}$$

$$4.80\ \cancel{g\ H} \times \frac{1\ \text{mol H}}{1.008\ \cancel{g\ H}} = 4.76\ \text{mol of H}$$

$$38.06\ \cancel{g\ O} \times \frac{1\ \text{mol O}}{16.00\ \cancel{g\ O}} = 2.379\ \text{mol of O (smallest number of moles)}$$

$$\frac{4.758\ \text{mol C}}{2.379} = 2.000\ \text{mol of C} \qquad \frac{4.76\ \text{mol H}}{2.379} = 2.00\ \text{mol of H}$$

$$\frac{2.379\ \text{mol O}}{2.379} = 1.000\ \text{mol of O}$$

$$\therefore \text{Empirical formula} = C_{2.000}H_{2.00}O_{1.000} \rightarrow C_2H_2O$$

Empirical formula mass of $C_2H_2O = 2(12.01) + 2(1.008) + 16.00 = 42.04\ g$

$$\text{Small integer} = \frac{\text{molar mass of vanillic acid}}{\text{empirical formula mass of } C_2H_2O} = \frac{168\ \cancel{g}}{42.04\ \cancel{g}} = 4$$

$$\therefore \text{Molecular formula of vanillic acid} = C_{(2\times4)}H_{(2\times4)}O_{(1\times4)} = C_8H_8O_4$$

7.45 In exactly 100 g of nicotine, there are 74.0 g of C and 8.7 g of H.

$\therefore$ Mass of N $= 100\ g - (74.0\ g + 8.7\ g) = 17.3\ g$ of N

$$74.0\ \cancel{g\ C} \times \frac{1\ \text{mol C}}{12.01\ \cancel{g\ C}} = 6.16\ \text{mol of C}$$

$$8.7\ \cancel{g\ H} \times \frac{1\ \text{mol H}}{1.008\ \cancel{g\ H}} = 8.6\ \text{mol of H}$$

$$17.3\ \cancel{g\ N} \times \frac{1\ \text{mol N}}{14.01\ \cancel{g\ N}} = 1.23\ \text{mol of N (smallest number of moles)}$$

$$\frac{6.16\ \text{mol C}}{1.23} = 5.00\ \text{mol of C} \qquad \frac{8.6\ \text{mol H}}{1.23} = 7.0\ \text{mol of H}$$

$$\frac{1.23\ \text{mol N}}{1.23} = 1.00\ \text{mol of N}$$

$$\therefore \text{Empirical formula} = C_{5.00}H_{7.0}N_{1.00} \rightarrow C_5H_7N$$

Empirical formula mass of $C_5H_7N = 5(12.01) + 7(1.008) + 14.01 = 81.12\ g$

$$\text{Small integer} = \frac{\text{molar mass of nicotine}}{\text{empirical formula mass of } C_5H_7N} = \frac{162\ \cancel{g}}{81.12\ \cancel{g}} = 2$$

$$\therefore \text{Molecular formula of nicotine} = C_{(5\times2)}H_{(7\times2)}N_{(1\times2)} = C_{10}H_{14}N_2$$

7.47 **a.** $C_{10}H_8N_2O_2S_2 \rightarrow C_{(10\div2)}H_{(8\div2)}N_{(2\div2)}O_{(2\div2)}S_{(2\div2)} = C_5H_4NOS$ (empirical formula)

b. $10\ \cancel{\text{mol C}} \times \dfrac{12.01\text{ g C}}{1\ \cancel{\text{mol C}}} = 120.1\text{ g of C}$

$8\ \cancel{\text{mol H}} \times \dfrac{1.008\text{ g H}}{1\ \cancel{\text{mol H}}} = 8.064\text{ g of H}$

$2\ \cancel{\text{mol N}} \times \dfrac{14.01\text{ g N}}{1\ \cancel{\text{mol N}}} = 28.02\text{ g of N}$

$2\ \cancel{\text{mol O}} \times \dfrac{16.00\text{ g O}}{1\ \cancel{\text{mol O}}} = 32.00\text{ g of O}$

$2\ \cancel{\text{mol S}} \times \dfrac{32.07\text{ g S}}{1\ \cancel{\text{mol S}}} = 64.14\text{ g of S}$

$\therefore$ Molar mass of $C_{10}H_8N_2O_2S_2 = 252.3\text{ g}$

c. $\text{Mass \% C} = \dfrac{120.1\text{ g C}}{252.3\text{ g } C_{10}H_8N_2O_2S_2} \times 100\% = 47.60\%\text{ C (4 SFs)}$

$\text{Mass \% H} = \dfrac{8.064\text{ g H}}{252.3\text{ g } C_{10}H_8N_2O_2S_2} \times 100\% = 3.20\%\text{ H (rounded to hundredths)}$

$\text{Mass \% N} = \dfrac{28.02\text{ g N}}{252.3\text{ g } C_{10}H_8N_2O_2S_2} \times 100\% = 11.11\%\text{ N (4 SFs)}$

$\text{Mass \% O} = \dfrac{32.00\text{ g O}}{252.3\text{ g } C_{10}H_8N_2O_2S_2} \times 100\% = 12.68\%\text{ O (4 SFs)}$

$\text{Mass \% S} = \dfrac{64.14\text{ g S}}{252.3\text{ g } C_{10}H_8N_2O_2S_2} \times 100\% = 25.42\%\text{ S (4 SFs)}$

d. $25.0\ \cancel{\text{g } C_{10}H_8N_2O_2S_2} \times \dfrac{1\ \cancel{\text{mol } C_{10}H_8N_2O_2S_2}}{252.3\ \cancel{\text{g } C_{10}H_8N_2O_2S_2}}$

$\times \dfrac{10\ \cancel{\text{mol C}}}{1\ \cancel{\text{mol } C_{10}H_8N_2O_2S_2}} \times \dfrac{6.022 \times 10^{23}\text{ atoms C}}{1\ \cancel{\text{mol C}}}$

$= 5.97 \times 10^{23}\text{ atoms of C (3 SFs)}$

e. $8.2 \times 10^{24}\ \cancel{\text{atoms of N}} \times \dfrac{1\ \cancel{\text{molecule } C_{10}H_8N_2O_2S_2}}{2\ \cancel{\text{atoms of N}}}$

$\times \dfrac{1\text{ mol } C_{10}H_8N_2O_2S_2}{6.022 \times 10^{23}\ \cancel{\text{molecules } C_{10}H_8N_2O_2S_2}}$

$= 6.8\text{ mol of } C_{10}H_8N_2O_2S_2\text{ (2 SFs)}$

7.49 **(1) a.** molecular formula $= S_2Cl_2$

b. $S_2Cl_2 \rightarrow S_{(2\div2)}Cl_{(2\div2)} = SCl$ (empirical formula)

c. $2\ \cancel{\text{mol S}} \times \dfrac{32.07\text{ g S}}{1\ \cancel{\text{mol S}}} = 64.14\text{ g of S}$

$2\ \cancel{\text{mol Cl}} \times \dfrac{35.45\text{ g Cl}}{1\ \cancel{\text{mol Cl}}} = 70.90\text{ g of Cl}$

$\therefore$ Molar mass of $S_2Cl_2 = 135.04\text{ g}$

d. $\text{Mass \% S} = \dfrac{64.14\text{ g S}}{135.04\text{ g } S_2Cl_2} \times 100\% = 47.50\%\text{ S (4 SFs)}$

$\text{Mass \% Cl} = \dfrac{70.90\text{ g Cl}}{135.04\text{ g } S_2Cl_2} \times 100\% = 52.50\%\text{ Cl (4 SFs)}$

(2) a. molecular formula $= C_6H_6$

b. $C_6H_6 \rightarrow C_{(6\div6)}H_{(6\div6)}$ = CH (empirical formula)

c. $6\ \cancel{\text{mol C}} \times \dfrac{12.01\text{ g C}}{1\ \cancel{\text{mol C}}} = 72.06\text{ g of C}$

$6\ \cancel{\text{mol H}} \times \dfrac{1.008\text{ g H}}{1\ \cancel{\text{mol H}}} = 6.048\text{ g of H}$

$\therefore$ Molar mass of C_6H_6 = 78.11 g

d. $\text{Mass \% C} = \dfrac{72.06\text{ g C}}{78.11\text{ g } C_6H_6} \times 100\% = 92.25\%$ C (4 SFs)

$\text{Mass \% H} = \dfrac{6.048\text{ g H}}{78.11\text{ g } C_6H_6} \times 100\% = 7.74\%$ H (rounded to hundredths)

7.51 **a.** $1\ \cancel{\text{mol Zn}} \times \dfrac{65.41\text{ g Zn}}{1\ \cancel{\text{mol Zn}}} = 65.41\text{ g of Zn}$

$1\ \cancel{\text{mol S}} \times \dfrac{32.07\text{ g S}}{1\ \cancel{\text{mol S}}} = 32.07\text{ g of S}$

$4\ \cancel{\text{mol O}} \times \dfrac{16.00\text{ g O}}{1\ \cancel{\text{mol O}}} = 64.00\text{ g of O}$

1 mol of Zn = 65.41 g of Zn
1 mol of S = 32.07 g of S
4 mol of O = 64.00 g of O
$\therefore$ Molar mass of $ZnSO_4$ = 161.48 g

b. $1\ \cancel{\text{mol Ca}} \times \dfrac{40.08\text{ g Ca}}{1\ \cancel{\text{mol Ca}}} = 40.08\text{ g of Ca}$

$2\ \cancel{\text{mol I}} \times \dfrac{126.9\text{ g I}}{1\ \cancel{\text{mol I}}} = 253.8\text{ g of I}$

$6\ \cancel{\text{mol O}} \times \dfrac{16.00\text{ g O}}{1\ \cancel{\text{mol O}}} = 96.00\text{ g of O}$

1 mol of Ca = 40.08 g of Ca
2 mol of I = 253.8 g of I
6 mol of O = 96.00 g of O
$\therefore$ Molar mass of $Ca(IO_3)_2$ = 389.9 g

c. $5\ \cancel{\text{mol C}} \times \dfrac{12.01\text{ g C}}{1\ \cancel{\text{mol C}}} = 60.05\text{ g of C}$

$8\ \cancel{\text{mol H}} \times \dfrac{1.008\text{ g H}}{1\ \cancel{\text{mol H}}} = 8.064\text{ g of H}$

$1\ \cancel{\text{mol N}} \times \dfrac{14.01\text{ g N}}{1\ \cancel{\text{mol N}}} = 14.01\text{ g of N}$

$1\ \cancel{\text{mol Na}} \times \dfrac{22.99\text{ g Na}}{1\ \cancel{\text{mol Na}}} = 22.99\text{ g of Na}$

$4\ \cancel{\text{mol O}} \times \dfrac{16.00\text{ g O}}{1\ \cancel{\text{mol O}}} = 64.00\text{ g of O}$

5 mol of C = 60.05 g of C
8 mol of H = 8.064 g of H
1 mol of N = 14.01 g of N
1 mol of Na = 22.99 g of Na
4 mol of O = 64.00 g of O
$\therefore$ Molar mass of $C_5H_8NNaO_4$ = 169.11 g

d. $6\ \cancel{\text{mol C}} \times \dfrac{12.01\text{ g C}}{1\ \cancel{\text{mol C}}} = 72.06\text{ g of C}$

$12\ \cancel{\text{mol H}} \times \dfrac{1.008\text{ g H}}{1\ \cancel{\text{mol H}}} = 12.10\text{ g of H}$

$2\ \cancel{\text{mol O}} \times \dfrac{16.00\text{ g O}}{1\ \cancel{\text{mol O}}} = 32.00\text{ g of O}$

$$\begin{aligned} 6\text{ mol of C} &= 72.06\text{ g of C} \\ 12\text{ mol of H} &= 12.10\text{ g of H} \\ 2\text{ mol of O} &= \underline{32.00\text{ g of O}} \\ \therefore \text{Molar mass of } C_6H_{12}O_2 &= 116.16\text{ g} \end{aligned}$$

7.53 **a.** $2\ \cancel{\text{mol K}} \times \dfrac{39.10\text{ g K}}{1\ \cancel{\text{mol K}}} = 78.20\text{ g of K}$

$1\ \cancel{\text{mol Cr}} \times \dfrac{52.00\text{ g Cr}}{1\ \cancel{\text{mol Cr}}} = 52.00\text{ g of Cr}$

$4\ \cancel{\text{mol O}} \times \dfrac{16.00\text{ g O}}{1\ \cancel{\text{mol O}}} = \underline{64.00\text{ g of O}}$

$\therefore$ Molar mass of $K_2CrO_4 = 194.20\text{ g}$

$\text{Mass \% K} = \dfrac{78.20\text{ g K}}{194.20\text{ g } K_2CrO_4} \times 100\% = 40.27\%\text{ K (4 SFs)}$

$\text{Mass \% Cr} = \dfrac{52.00\text{ g Cr}}{194.20\text{ g } K_2CrO_4} \times 100\% = 26.78\%\text{ Cr (4 SFs)}$

$\text{Mass \% O} = \dfrac{64.00\text{ g O}}{194.20\text{ g } K_2CrO_4} \times 100\% = 32.96\%\text{ O (4 SFs)}$

b. $1\ \cancel{\text{mol Al}} \times \dfrac{26.98\text{ g Al}}{1\ \cancel{\text{mol Al}}} = 26.98\text{ g of Al}$

$3\ \cancel{\text{mol H}} \times \dfrac{1.008\text{ g H}}{1\ \cancel{\text{mol H}}} = 3.024\text{ g of H}$

$3\ \cancel{\text{mol C}} \times \dfrac{12.01\text{ g C}}{1\ \cancel{\text{mol C}}} = 36.03\text{ g of C}$

$9\ \cancel{\text{mol O}} \times \dfrac{16.00\text{ g O}}{1\ \cancel{\text{mol O}}} = \underline{144.0\text{ g of O}}$

$\therefore$ Molar mass of $Al(HCO_3)_3 = 210.0\text{ g}$

$\text{Mass \% Al} = \dfrac{26.98\text{ g Al}}{210.0\text{ g } Al(HCO_3)_3} \times 100\% = 12.85\%\text{ Al (4 SFs)}$

$\text{Mass \% H} = \dfrac{3.024\text{ g H}}{210.0\text{ g } Al(HCO_3)_3} \times 100\% = 1.44\%\text{ H (rounded to hundredths)}$

$\text{Mass \% C} = \dfrac{36.03\text{ g C}}{210.0\text{ g } Al(HCO_3)_3} \times 100\% = 17.16\%\text{ C (4 SFs)}$

$\text{Mass \% O} = \dfrac{144.0\text{ g O}}{210.0\text{ g } Al(HCO_3)_3} \times 100\% = 68.57\%\text{ O (4 SFs)}$

c. $6\ \cancel{\text{mol C}} \times \dfrac{12.01\text{ g C}}{1\ \cancel{\text{mol C}}} = 72.06\text{ g of C}$

$12\ \cancel{\text{mol H}} \times \dfrac{1.008\text{ g H}}{1\ \cancel{\text{mol H}}} = 12.10\text{ g of H}$

$6\ \cancel{\text{mol O}} \times \dfrac{16.00\text{ g O}}{1\ \cancel{\text{mol O}}} = \underline{96.00\text{ g of O}}$

$\therefore$ Molar mass of $C_6H_{12}O_6 = 180.16\text{ g}$

$$\text{Mass \% C} = \frac{72.06 \text{ g C}}{180.16 \text{ g } C_6H_{12}O_6} \times 100\% = 40.00\% \text{ C (4 SFs)}$$

$$\text{Mass \% H} = \frac{12.10 \text{ g H}}{180.16 \text{ g } C_6H_{12}O_6} \times 100\% = 6.72\% \text{ H (rounded to hundredths)}$$

$$\text{Mass \% O} = \frac{96.00 \text{ g O}}{180.16 \text{ g } C_6H_{12}O_6} \times 100\% = 53.29\% \text{ O (4 SFs)}$$

7.55 **a.** $3 \cancel{\text{mol C}} \times \frac{12.01 \text{ g C}}{1 \cancel{\text{mol C}}} = 36.03 \text{ g of C}$

$6 \cancel{\text{mol H}} \times \frac{1.008 \text{ g H}}{1 \cancel{\text{mol H}}} = 6.048 \text{ g of H}$

$3 \cancel{\text{mol O}} \times \frac{16.00 \text{ g O}}{1 \cancel{\text{mol O}}} = 48.00 \text{ g of O}$

$\therefore$ Molar mass of $C_3H_6O_3$ = 90.08 g

$$\text{Mass \% O} = \frac{48.00 \text{ g O}}{90.08 \text{ g } C_3H_6O_3} \times 100\% = 53.29\% \text{ O (4 SFs)}$$

b. $125 \cancel{\text{g } C_3H_6O_3} \times \frac{1 \cancel{\text{mol } C_3H_6O_3}}{90.08 \cancel{\text{g } C_3H_6O_3}} \times \frac{3 \cancel{\text{mol C}}}{1 \cancel{\text{mol } C_3H_6O_3}} \times \frac{6.022 \times 10^{23} \text{ atoms C}}{1 \cancel{\text{mol C}}}$

$= 2.51 \times 10^{24}$ atoms of C

c. $3.50 \cancel{\text{g H}} \times \frac{1 \cancel{\text{mol H}}}{1.008 \cancel{\text{g H}}} \times \frac{1 \cancel{\text{mol } C_3H_6O_3}}{6 \cancel{\text{mol H}}} \times \frac{90.08 \text{ g } C_3H_6O_3}{1 \cancel{\text{mol } C_3H_6O_3}} = 52.1 \text{ g of } C_3H_6O_3 \text{ (3 SFs)}$

d. $C_3H_6O_3 \rightarrow C_{(3\div3)}H_{(6\div3)}O_{(3\div3)} = CH_2O$ (empirical formula)

7.57 **a.** $9 \cancel{\text{mol C}} \times \frac{12.01 \text{ g C}}{1 \cancel{\text{mol C}}} = 108.1 \text{ g of C}$

$8 \cancel{\text{mol H}} \times \frac{1.008 \text{ g H}}{1 \cancel{\text{mol H}}} = 8.064 \text{ g of H}$

$4 \cancel{\text{mol O}} \times \frac{16.00 \text{ g O}}{1 \cancel{\text{mol O}}} = 64.00 \text{ g of O}$

$\therefore$ Molar mass of $C_9H_8O_4$ = 180.2 g

$$\text{Mass \% C} = \frac{108.1 \text{ g C}}{180.2 \text{ g } C_9H_8O_4} \times 100\% = 59.99\% \text{ C (4 SFs)}$$

$$\text{Mass \% H} = \frac{8.064 \text{ g H}}{180.2 \text{ g } C_9H_8O_4} \times 100\% = 4.48\% \text{ H (rounded to hundredths)}$$

$$\text{Mass \% O} = \frac{64.00 \text{ g O}}{180.2 \text{ g } C_9H_8O_4} \times 100\% = 35.52\% \text{ O (4 SFs)}$$

b. $5.0 \times 10^{24} \cancel{\text{atoms C}} \times \frac{1 \cancel{\text{mol C}}}{6.022 \times 10^{23} \cancel{\text{atoms C}}} \times \frac{1 \text{ mol } C_9H_8O_4}{9 \cancel{\text{mol C}}}$

$= 0.92$ mol of $C_9H_8O_4$ (aspirin) (2 SFs)

c. $7.50 \cancel{\text{g } C_9H_8O_4} \times \frac{1 \cancel{\text{mol } C_9H_8O_4}}{180.2 \cancel{\text{g } C_9H_8O_4}} \times \frac{4 \cancel{\text{mol O}}}{1 \cancel{\text{mol } C_9H_8O_4}} \times \frac{6.022 \times 10^{23} \text{ atoms O}}{1 \cancel{\text{mol O}}}$

$= 1.00 \times 10^{23}$ atoms of O (3 SFs)

d. $2.50 \cancel{\text{g H}} \times \frac{1 \cancel{\text{mol H}}}{1.008 \cancel{\text{g H}}} \times \frac{1 \cancel{\text{mol } C_9H_8O_4}}{8 \cancel{\text{mol H}}} \times \frac{6.022 \times 10^{23} \text{ molecules } C_9H_8O_4}{1 \cancel{\text{mol } C_9H_8O_4}}$

$= 1.87 \times 10^{23}$ molecules of $C_9H_8O_4$ (3 SFs)

7.59 **a.** $0.250\ \cancel{\text{mol Mn}_2\text{O}_3} \times \frac{3\text{ mol O}}{1\ \cancel{\text{mol Mn}_2\text{O}_3}} = 0.750\text{ mol of O}$

Molar mass of MnO_2

$= 1\text{ mol of Mn} + 2\text{ mol of O} = 54.94\text{ g} + 2(16.00\text{ g}) = 86.94\text{ g}$

$20.0\ \cancel{\text{g MnO}_2} \times \frac{1\ \cancel{\text{mol MnO}_2}}{86.94\ \cancel{\text{g MnO}_2}} \times \frac{2\text{ mol O}}{1\ \cancel{\text{mol MnO}_2}} = 0.460\text{ mol of O}$

$\therefore$ Total moles of O $= 0.750\text{ mol O} + 0.460\text{ mol O} = 1.210\text{ mol of O}$

$1.210\ \cancel{\text{mol O}} \times \frac{6.022 \times 10^{23}\text{ atoms O}}{1\ \cancel{\text{mol O}}} = 7.29 \times 10^{23}\text{ atoms of O (3 SFs)}$

b. $0.250\ \cancel{\text{mol Mn}_2\text{O}_3} \times \frac{2\ \cancel{\text{mol Mn}}}{1\ \cancel{\text{mol Mn}_2\text{O}_3}} \times \frac{54.94\text{ g Mn}}{1\ \cancel{\text{mol Mn}}} = 27.5\text{ g of Mn}$

$20.0\ \cancel{\text{g MnO}_2} \times \frac{1\ \cancel{\text{mol MnO}_2}}{86.94\ \cancel{\text{g MnO}_2}} \times \frac{1\ \cancel{\text{mol Mn}}}{1\ \cancel{\text{mol MnO}_2}} \times \frac{54.94\text{ g Mn}}{1\ \cancel{\text{mol Mn}}} = 12.6\text{ g of Mn}$

$\therefore$ Total grams of Mn $= 27.5\text{ g Mn} + 12.6\text{ g Mn} = 40.1\text{ g of Mn (3 SFs)}$

7.61 **a.** $C_5H_5N_5 \rightarrow C_{(5\div5)}H_{(5\div5)}N_{(5\div5)} = CHN$ (empirical formula)

b. $FeC_2O_4 \rightarrow Fe_{(1\div1)}C_{(2\div1)}O_{(4\div1)} = FeC_2O_4$ (empirical formula)

c. $C_{16}H_{16}N_4 \rightarrow C_{(16\div4)}H_{(16\div4)}N_{(4\div4)} = C_4H_4N$ (empirical formula)

d. $C_6H_{14}N_2O_2 \rightarrow C_{(6\div2)}H_{(14\div2)}N_{(2\div2)}O_{(2\div2)} = C_3H_7NO$ (empirical formula)

7.63 **a.** $2.20\ \cancel{\text{g S}} \times \frac{1\text{ mol S}}{32.07\ \cancel{\text{g S}}} = 0.0686\text{ mol of S}$ (smaller number of moles)

$7.81\ \cancel{\text{g F}} \times \frac{1\text{ mol F}}{19.00\ \cancel{\text{g F}}} = 0.411\text{ mol of F}$

$\frac{0.0686\text{ mol S}}{0.0686} = 1.00\text{ mol of S}$ $\qquad \frac{0.411\text{ mol F}}{0.0686} = 5.99\text{ mol of F}$

$\therefore$ Empirical formula $= S_{1.00}F_{5.99} \rightarrow SF_6$

b. $6.35\ \cancel{\text{g Ag}} \times \frac{1\text{ mol Ag}}{107.9\ \cancel{\text{g Ag}}} = 0.0589\text{ mol of Ag}$ (smallest number of moles)

$0.825\ \cancel{\text{g N}} \times \frac{1\text{ mol N}}{14.01\ \cancel{\text{g N}}} = 0.0589\text{ mol of N}$

$2.83\ \cancel{\text{g O}} \times \frac{1\text{ mol O}}{16.00\ \cancel{\text{g O}}} = 0.177\text{ mol of O}$

$\frac{0.0589\text{ mol Ag}}{0.0589} = 1.00\text{ mol of Ag}$ $\qquad \frac{0.0589\text{ mol N}}{0.0589} = 1.00\text{ mol of N}$

$\frac{0.177\text{ mol O}}{0.0589} = 3.01\text{ mol of O}$

$\therefore$ Empirical formula $= Ag_{1.00}N_{1.00}O_{3.01} \rightarrow AgNO_3$

c. $89.2\ \cancel{\text{g Au}} \times \frac{1\text{ mol Au}}{197.0\ \cancel{\text{g Au}}} = 0.453\text{ mol of Au}$ (smaller number of moles)

$10.9\ \cancel{\text{g O}} \times \frac{1\text{ mol O}}{16.00\ \cancel{\text{g O}}} = 0.681\text{ mol of O}$

$\frac{0.453\text{ mol Au}}{0.453} = 1.00\text{ mol Au}$ $\qquad \frac{0.681\text{ mol O}}{0.453} = 1.50\text{ mol O}$

$\therefore$ Empirical formula $= Au_{1.00}O_{1.50} \rightarrow Au_{(1.00\times2)}O_{(1.50\times2)} = Au_{2.00}O_{3.00} \rightarrow Au_2O_3$

7.65 **a.** In exactly 100 g of oleic acid, there are 76.54 g of C, 12.13 g of H, and 11.33 g of O.

$$76.54\ \cancel{g\,C} \times \frac{1\text{ mol C}}{12.01\ \cancel{g\,C}} = 6.373\text{ mol of C}$$

$$12.13\ \cancel{g\,H} \times \frac{1\text{ mol H}}{1.008\ \cancel{g\,H}} = 12.03\text{ mol of H}$$

$$11.33\ \cancel{g\,O} \times \frac{1\text{ mol O}}{16.00\ \cancel{g\,O}} = 0.7081\text{ mol of O (smallest number of moles)}$$

$$\frac{6.373\text{ mol C}}{0.7081} = 9.000\text{ mol of C} \qquad \frac{12.03\text{ mol H}}{0.7081} = 16.99\text{ mol of H}$$

$$\frac{0.7081\text{ mol O}}{0.7081} = 1.000\text{ mol of O}$$

$\therefore$ Empirical formula $= C_{9.000}H_{16.99}O_{1.000} \rightarrow C_9H_{17}O$

Empirical formula mass of $C_9H_{17}O = 9(12.01) + 17(1.008) + 16.00 = 141.2$ g

$$\text{Small integer} = \frac{\text{molar mass of oleic acid}}{\text{empirical formula mass of } C_9H_{17}O} = \frac{282\ \cancel{g}}{141.2\ \cancel{g}} = 2$$

$\therefore$ Molecular formula of oleic acid $= C_{(9\times2)}H_{(17\times2)}O_{(1\times2)} = C_{18}H_{34}O_2$

b. Molar mass of oleic acid ($C_{18}H_{34}O_2$)

$= 18$ mol of C $+ 34$ mol of H $+ 2$ mol of O

$= 18(12.01\text{ g}) + 34(1.008\text{ g}) + 2(16.00\text{ g}) = 282.5$ g

$$3.00\ \cancel{mL\ C_{18}H_{34}O_2} \times \frac{0.895\ \cancel{g\ C_{18}H_{34}O_2}}{1\ \cancel{mL\ C_{18}H_{34}O_2}}$$

$$\times \frac{1\ \cancel{mol\ C_{18}H_{34}O_2}}{282.5\ \cancel{g\ C_{18}H_{34}O_2}} \times \frac{6.022 \times 10^{23}\text{ molecules } C_{18}H_{34}O_2}{1\ \cancel{mol\ C_{18}H_{34}O_2}}$$

$= 5.72 \times 10^{21}$ molecules of $C_{18}H_{34}O_2$ (3 SFs)

7.67 In exactly 100 g of succinic acid, there are 40.7 g of C, 5.12 g of H, and 54.2 g of O.

$$40.7\ \cancel{g\,C} \times \frac{1\text{ mol C}}{12.01\ \cancel{g\,C}} = 3.39\text{ mol of C}$$

$$5.12\ \cancel{g\,H} \times \frac{1\text{ mol H}}{1.008\ \cancel{g\,H}} = 5.08\text{ mol of H}$$

$$54.2\ \cancel{g\,O} \times \frac{1\text{ mol O}}{16.00\ \cancel{g\,O}} = 3.39\text{ mol of O (smallest number of moles)}$$

$$\frac{3.39\text{ mol C}}{3.39} = 1.00\text{ mol of C} \qquad \frac{5.08\text{ mol H}}{3.39} = 1.50\text{ mol of H}$$

$$\frac{3.39\text{ mol O}}{3.39} = 1.00\text{ mol of O}$$

$\therefore$ Empirical formula $= C_{1.00}H_{1.50}O_{1.00} \rightarrow C_{(1.00\times2)}H_{(1.50\times2)}O_{(1.00\times2)}$

$= C_{2.00}H_{3.00}O_{2.00} \rightarrow C_2H_3O_2$

Empirical formula mass of $C_2H_3O_2 = 2(12.01) + 3(1.008) + 2(16.00) = 59.04$ g

$$\text{Small integer} = \frac{\text{molar mass of succinic acid}}{\text{empirical formula mass of } C_2H_3O_2} = \frac{118\ \cancel{g}}{59.04\ \cancel{g}} = 2$$

$\therefore$ Molecular formula of succinic acid $= C_{(2\times2)}H_{(3\times2)}O_{(2\times2)} = C_4H_6O_4$

7.69 In the sample of compound, there are 1.65×10^{23} atoms of C, 0.552 g of H, and 4.39 g of O.

$$1.65 \times 10^{23} \cancel{\text{atoms C}} \times \frac{1 \text{ mol C}}{6.022 \times 10^{23} \cancel{\text{atoms C}}} = 0.274 \text{ mol of C (smallest number of moles)}$$

$$0.552 \cancel{\text{g H}} \times \frac{1 \text{ mol H}}{1.008 \cancel{\text{g H}}} = 0.548 \text{ mol of H}$$

$$4.39 \cancel{\text{g O}} \times \frac{1 \text{ mol O}}{16.00 \cancel{\text{g O}}} = 0.274 \text{ mol of O}$$

$$\frac{0.274 \text{ mol C}}{0.274} = 1.00 \text{ mol of C} \qquad \frac{0.548 \text{ mol H}}{0.274} = 2.00 \text{ mol of H}$$

$$\frac{0.274 \text{ mol O}}{0.274} = 1.00 \text{ mol of O}$$

$\therefore$ Empirical formula $= C_{1.00}H_{2.00}O_{1.00} \rightarrow CH_2O$

Since 1 mol of compound contains 4 mol of O,

$\therefore$ Molecular formula $= C_{(1\times4)}H_{(2\times4)}O_{(1\times4)} = C_4H_8O_4$

Molar mass of compound ($C_4H_8O_4$)

$= 4$ mol of C $+ 8$ mol of H $+ 4$ mol of O

$= 4(12.01 \text{ g}) + 8(1.008 \text{ g}) + 4(16.00 \text{ g}) = 120.10 \text{ g}$

7.71 **a.** Molar mass of NaF

$= 1$ mol of Na $+ 1$ mol of F $= 22.99 \text{ g} + 19.00 \text{ g} = 41.99 \text{ g}$

$$1 \cancel{\text{tube}} \times \frac{119 \cancel{\text{g toothpaste}}}{1 \cancel{\text{tube}}} \times \frac{0.240 \cancel{\text{g NaF}}}{100 \cancel{\text{g toothpaste}}} \times \frac{1 \text{ mol NaF}}{41.99 \cancel{\text{g NaF}}}$$

$= 0.00680$ mol of NaF (3 SFs)

b. $$1 \cancel{\text{tube}} \times \frac{119 \cancel{\text{g toothpaste}}}{1 \cancel{\text{tube}}} \times \frac{0.240 \cancel{\text{g NaF}}}{100 \cancel{\text{g toothpaste}}} \times \frac{1 \cancel{\text{mol NaF}}}{41.99 \cancel{\text{g NaF}}} \times \frac{1 \cancel{\text{mol F}^- \text{ions}}}{1 \cancel{\text{mol NaF}}}$$

$$\times \frac{6.022 \times 10^{23} \text{ F}^- \text{ ions}}{1 \cancel{\text{mol F}^- \text{ions}}} = 4.10 \times 10^{21} \text{ F}^- \text{ ions (3 SFs)}$$

c. $$1.50 \cancel{\text{g toothpaste}} \times \frac{0.240 \cancel{\text{g NaF}}}{100 \cancel{\text{g toothpaste}}} \times \frac{1 \cancel{\text{mol NaF}}}{41.99 \cancel{\text{g NaF}}} \times \frac{1 \cancel{\text{mol Na}^+}}{1 \cancel{\text{mol NaF}}} \times \frac{22.99 \text{ g Na}^+}{1 \cancel{\text{mol Na}^+}}$$

$= 0.00197$ g of Na^+ ions (3 SFs)

d. $$12 \cancel{\text{mol C}} \times \frac{12.01 \text{ g C}}{1 \cancel{\text{mol C}}} = 144.1 \text{ g of C}$$

$$7 \cancel{\text{mol H}} \times \frac{1.008 \text{ g H}}{1 \cancel{\text{mol H}}} = 7.056 \text{ g of H}$$

$$3 \cancel{\text{mol Cl}} \times \frac{35.45 \text{ g Cl}}{1 \cancel{\text{mol Cl}}} = 106.4 \text{ g of Cl}$$

$$2 \cancel{\text{mol O}} \times \frac{16.00 \text{ g O}}{1 \cancel{\text{mol O}}} = 32.00 \text{ g of O}$$

$\therefore$ Molar mass of $C_{12}H_7Cl_3O_2 = 289.6 \text{ g}$

$$1 \cancel{\text{tube}} \times \frac{119 \cancel{\text{g toothpaste}}}{1 \cancel{\text{tube}}} \times \frac{0.30 \cancel{\text{g C}_{12}\text{H}_7\text{Cl}_3\text{O}_2}}{100 \cancel{\text{g toothpaste}}} \times \frac{1 \cancel{\text{mol C}_{12}\text{H}_7\text{Cl}_3\text{O}_2}}{289.6 \cancel{\text{g C}_{12}\text{H}_7\text{Cl}_3\text{O}_2}}$$

$$\times \frac{6.022 \times 10^{23} \text{ molecules C}_{12}\text{H}_7\text{Cl}_3\text{O}_2}{1 \cancel{\text{mol C}_{12}\text{H}_7\text{Cl}_3\text{O}_2}}$$

$= 7.4 \times 10^{20}$ molecules of $C_{12}H_7Cl_3O_2$ (triclosan) (2 SFs)

e. $\text{Mass \% C} = \dfrac{144.1 \text{ g C}}{289.6 \text{ g } C_{12}H_7Cl_3O_2} \times 100\% = 49.76\% \text{ C (4 SFs)}$

$\text{Mass \% H} = \dfrac{7.056 \text{ g H}}{289.6 \text{ g } C_{12}H_7Cl_3O_2} \times 100\% = 2.44\% \text{ H (rounded to hundredths)}$

$\text{Mass \% Cl} = \dfrac{106.4 \text{ g Cl}}{289.6 \text{ g } C_{12}H_7Cl_3O_2} \times 100\% = 36.74\% \text{ Cl (4 SFs)}$

$\text{Mass \% O} = \dfrac{32.00 \text{ g O}}{289.6 \text{ g } C_{12}H_7Cl_3O_2} \times 100\% = 11.05\% \text{ O (4 SFs)}$

7.73 In exactly 100 g of iron(III) chromate, there are 24.3 g of Fe, 33.9 g of Cr, and 41.8 g of O.

$24.3 \cancel{\text{g Fe}} \times \dfrac{1 \text{ mol Fe}}{55.85 \cancel{\text{g Fe}}} = 0.435 \text{ mol of Fe (smallest number of moles)}$

$33.9 \cancel{\text{g Cr}} \times \dfrac{1 \text{ mol Cr}}{52.00 \cancel{\text{g Cr}}} = 0.652 \text{ mol of Cr}$

$41.8 \cancel{\text{g O}} \times \dfrac{1 \text{ mol O}}{16.00 \cancel{\text{g O}}} = 2.61 \text{ mol of O}$

$\dfrac{0.435 \text{ mol Fe}}{0.435} = 1.00 \text{ mol of Fe}$ $\qquad \dfrac{0.652 \text{ mol Cr}}{0.435} = 1.50 \text{ mol of Cr}$

$\dfrac{2.61 \text{ mol O}}{0.435} = 6.00 \text{ mol of O}$

$\therefore$ Empirical formula $= Fe_{1.00}Cr_{1.50}O_{6.00} \rightarrow Fe_{(1.00\times2)}Cr_{(1.50\times2)}O_{(6.00\times2)}$

$= Fe_{2.00}Cr_{3.00}O_{12.0} \rightarrow Fe_2Cr_3O_{12}$

Empirical formula mass of $Fe_2Cr_3O_{12} = 2(55.85) + 3(52.00) + 12(16.00) = 459.7 \text{ g}$

$\text{Small integer} = \dfrac{\text{molar mass of compound}}{\text{empirical formula mass of } Fe_2Cr_3O_{12}} = \dfrac{460 \cancel{\text{g}}}{459.7 \cancel{\text{g}}} = 1$

$\therefore$ Molecular formula $= Fe_{(2\times1)}Cr_{(3\times1)}O_{(12\times1)} = Fe_2Cr_3O_{12}$

Answers to Combining Ideas from Chapters 4 to 7

CI.7 **a.** X is a metal. When an atom of a metal loses electrons, the size of the ion is smaller than the corresponding atom.

b. Y is a nonmetal. When an atom of a nonmetal gains electrons, the size of the ion is larger than the corresponding atom.

c. X^{2+}, Y^-

d. $X = 1s^22s^22p^63s^2$ $\quad Y = 1s^22s^22p^63s^23p^5$

e. $X^{2+} = 1s^22s^22p^6$ $\quad Y^- = 1s^22s^22p^63s^23p^6$

f. $MgCl_2$, magnesium chloride

CI.9 **a.** In exactly 100 g of oxalic acid, there are 26.7 g of C, 2.24 g of H, and 71.1 g of O.

$$26.7\ \cancel{g\ C} \times \frac{1\ \text{mol C}}{12.01\ \cancel{g\ C}} = 2.22\ \text{mol of C}$$

$$2.24\ \cancel{g\ H} \times \frac{1\ \text{mol H}}{1.008\ \cancel{g\ H}} = 2.22\ \text{mol of H (smallest number of moles)}$$

$$71.1\ \cancel{g\ O} \times \frac{1\ \text{mol O}}{16.00\ \cancel{g\ O}} = 4.44\ \text{mol of O}$$

$$\frac{2.22\ \text{mol C}}{2.22} = 1.00\ \text{mol of C} \qquad \frac{2.22\ \text{mol H}}{2.22} = 1.00\ \text{mol of H}$$

$$\frac{4.44\ \text{mol O}}{2.22} = 2.00\ \text{mol of O}$$

$\therefore$ Empirical formula $= C_{1.00}H_{1.00}O_{2.00} \rightarrow CHO_2$

b. Empirical formula mass of $CHO_2 = 12.01 + 1.008 + 2(16.00) = 45.02$ g

$$\text{Small integer} = \frac{\text{molar mass of oxalic acid}}{\text{empirical formula mass of } CHO_2} = \frac{90\ \cancel{g}}{45.02\ \cancel{g}} = 2$$

$\therefore$ Molecular formula $= C_{(1\times2)}H_{(1\times2)}O_{(2\times2)} = C_2H_2O_4$

c. $$160\ \cancel{lb} \times \frac{1\ \cancel{kg}}{2.205\ \cancel{lb}} \times \frac{375\ \cancel{\text{mg oxalic acid}}}{1\ \cancel{kg}} \times \frac{1\ \text{g oxalic acid}}{1000\ \cancel{\text{mg oxalic acid}}}$$

= 27 g of oxalic acid (2 SFs)

d. $$160\ \cancel{lb} \times \frac{1\ \cancel{kg}}{2.205\ \cancel{lb}} \times \frac{375\ \cancel{\text{mg oxalic acid}}}{1\ \cancel{kg}} \times \frac{1\ \cancel{\text{g oxalic acid}}}{1000\ \cancel{\text{mg oxalic acid}}}$$

$$\times \frac{100\ \cancel{\text{g rhubarb leaves}}}{0.5\ \cancel{\text{g oxalic acid}}} \times \frac{1\ \text{kg}}{1000\ \cancel{g}}$$

= 5 kg of rhubarb leaves (1 SF)

CI.11 **a.** $C_{16}H_{28}N_2O_4 \rightarrow C_{(16\div2)}H_{(28\div2)}N_{(2\div2)}O_{(4\div2)} = C_8H_{14}NO_2$ (empirical formula)

b. $$16\ \cancel{\text{mol C}} \times \frac{12.01\ \text{g C}}{1\ \cancel{\text{mol C}}} = 192.2\ \text{g of C}$$

$$28\ \cancel{\text{mol H}} \times \frac{1.008\ \text{g H}}{1\ \cancel{\text{mol H}}} = 28.22\ \text{g of H}$$

$$2\ \cancel{\text{mol N}} \times \frac{14.01\ \text{g N}}{1\ \cancel{\text{mol N}}} = 28.02\ \text{g of N}$$

$$4\ \cancel{\text{mol O}} \times \frac{16.00\ \text{g O}}{1\ \cancel{\text{mol O}}} = 64.00\ \text{g of O}$$

$\therefore$ Molar mass of $C_{16}H_{28}N_2O_4 = 312.4$ g

$$\text{Mass \% C} = \frac{192.2 \text{ g C}}{312.4 \text{ g } C_{16}H_{28}N_2O_4} \times 100\% = 61.52\% \text{ C (4 SFs)}$$

$$\text{Mass \% H} = \frac{28.22 \text{ g H}}{312.4 \text{ g } C_{16}H_{28}N_2O_4} \times 100\% = 9.03\% \text{ H (rounded to hundredths)}$$

$$\text{Mass \% N} = \frac{28.02 \text{ g N}}{312.4 \text{ g } C_{16}H_{28}N_2O_4} \times 100\% = 8.97\% \text{ N (rounded to hundredths)}$$

$$\text{Mass \% O} = \frac{64.00 \text{ g O}}{312.4 \text{ g } C_{16}H_{28}N_2O_4} \times 100\% = 20.49\% \text{ O (4 SFs)}$$

c. $C_7H_{10}O_5$

d. Molar mass of shikimic acid ($C_7H_{10}O_5$)
= 7 mol of C + 10 mol of H + 5 mol of O
= 7(12.01 g) + 10(1.008 g) + 5(16.00 g) = 174.15 g

$$1.3 \cancel{\text{g } C_7H_{10}O_5} \times \frac{1 \text{ mol } C_7H_{10}O_5}{174.15 \cancel{\text{g } C_7H_{10}O_5}} = 0.0075 \text{ mol of } C_7H_{10}O_5 \text{ (shikimic acid) (2 SFs)}$$

e. $$155 \cancel{\text{g star anise}} \times \frac{0.13 \cancel{\text{g } C_7H_{10}O_5}}{2.6 \cancel{\text{g star anise}}} \times \frac{75 \cancel{\text{mg Tamiflu}}}{0.13 \cancel{\text{g } C_7H_{10}O_5}} \times \frac{1 \text{ capsule}}{75 \cancel{\text{mg Tamiflu}}}$$
= 59 capsules of Tamiflu (2 SFs)

f. $$1 \cancel{\text{capsule}} \times \frac{75 \cancel{\text{mg } C_{16}H_{28}N_2O_4}}{1 \cancel{\text{capsule}}} \times \frac{1 \cancel{\text{g } C_{16}H_{28}N_2O_4}}{1000 \cancel{\text{mg } C_{16}H_{28}N_2O_4}} \times \frac{1 \cancel{\text{mol } C_{16}H_{28}N_2O_4}}{312.4 \cancel{\text{g } C_{16}H_{28}N_2O_4}}$$
$$\times \frac{16 \cancel{\text{mol C}}}{1 \cancel{\text{mol } C_{16}H_{28}N_2O_4}} \times \frac{12.01 \text{ g C}}{1 \cancel{\text{mol C}}} = 0.046 \text{ g of C (2 SFs)}$$

g. $$500\,000 \cancel{\text{people}} \times \frac{2 \cancel{\text{capsules}}}{1 \cancel{\text{day}}\ 1 \cancel{\text{person}}} \times 5 \cancel{\text{days}} \times \frac{75 \cancel{\text{mg Tamiflu}}}{1 \cancel{\text{capsule}}} \times \frac{1 \cancel{\text{g Tamiflu}}}{1000 \cancel{\text{mg Tamiflu}}}$$
$$\times \frac{1 \text{ kg Tamiflu}}{1000 \cancel{\text{g Tamiflu}}} = 4 \times 10^2 \text{ kg of Tamiflu (1 SF)}$$

8
Chemical Reactions of Inorganic and Organic Compounds

Study Goals

- Identify a balanced chemical equation and determine the number of atoms in the reactants and products.
- Write a balanced chemical equation from the formulas of the reactants and products for a chemical reaction.
- Identify a reaction as a combination, decomposition, single replacement, or double replacement.
- Classify organic molecules according to their functional groups.
- Draw the condensed structural formulas of the products of reactions of organic compounds.

Chapter Outline

Answers and Solutions to Text Problems

8.1 **a.** Reactant side: 2 N atoms, 4 O atoms
Product side: 2 N atoms, 4 O atoms
b. Reactant side: 5 C atoms, 2 S atoms, 4 O atoms
Product side: 5 C atoms, 2 S atoms, 4 O atoms
c. Reactant side: 4 C atoms, 4 H atoms, 10 O atoms
Product side: 4 C atoms, 4 H atoms, 10 O atoms
d. Reactant side: 2 N atoms, 8 H atoms, 4 O atoms
Product side: 2 N atoms, 8 H atoms, 4 O atoms

8.3 An equation is balanced when there are an equal number of atoms of each element on the reactant side and on the product side.

a. not balanced **b.** balanced
c. not balanced **d.** balanced

8.5 **a.** Reactant side: 2 Na atoms, 2 Cl atoms
Product side: 2 Na atoms, 2 Cl atoms
b. Reactant side: 1 P atom, 3 Cl atoms, 6 H atoms
Product side: 1 P atom, 3 Cl atoms, 6 H atoms
c. Reactant side: 4 P atoms, 16 O atoms, 12 H atoms
Product side: 4 P atoms, 16 O atoms, 12 H atoms

8.7 Place coefficients in front of formulas until you make the atoms of each element equal on each side of the equation. We typically begin with the formula that has the highest subscript values.

a. $N_2(g) + O_2(g) \rightarrow 2NO(g)$ **b.** $2HgO(s) \rightarrow 2Hg(l) + O_2(g)$
c. $4Fe(s) + 3O_2(g) \rightarrow 2Fe_2O_3(s)$ **d.** $2Na(s) + Cl_2(g) \rightarrow 2NaCl(s)$
e. $2Cu_2O(s) + O_2(g) \rightarrow 4CuO(s)$

8.9 **a.** Since it has the highest subscript values, we begin with $Mg(NO_3)_2$. Balance the nitrate ions (NO_3^-) by placing a 2 in front of $AgNO_3$. A 2 in front of Ag completes the balancing.
$Mg(s) + 2AgNO_3(aq) \rightarrow Mg(NO_3)_2(aq) + 2Ag(s)$
b. $CuCO_3(s) \rightarrow CuO(s) + CO_2(g)$
c. Since it has the highest subscript values, we begin with $Al_2(SO_4)_3$. Balance the sulfate ions (SO_4^{2-}) by placing a 3 in front of $CuSO_4$. A 2 in front of Al and a 3 in front of Cu completes the balancing.
$2Al(s) + 3CuSO_4(aq) \rightarrow 3Cu(s) + Al_2(SO_4)_3(aq)$
d. $Pb(NO_3)_2(aq) + 2NaCl(aq) \rightarrow PbCl_2(s) + 2NaNO_3(aq)$
e. $2Al(s) + 6HCl(aq) \rightarrow 2AlCl_3(aq) + 3H_2(g)$

8.11 **a.** $Fe_2O_3(s) + 3CO(g) \rightarrow 2Fe(s) + 3CO_2(g)$
b. $2Li_3N(s) \rightarrow 6Li(s) + N_2(g)$
c. $2Al(s) + 6HBr(aq) \rightarrow 2AlBr_3(aq) + 3H_2(g)$
d. $3Ba(OH)_2(aq) + 2Na_3PO_4(aq) \rightarrow Ba_3(PO_4)_2(s) + 6NaOH(aq)$
e. $As_4S_6(s) + 9O_2(g) \rightarrow As_4O_6(s) + 6SO_2(g)$

8.13 **a.** $2Li(s) + 2H_2O(l) \rightarrow H_2(g) + 2LiOH(aq)$
b. $2P(s) + 5Cl_2(g) \rightarrow 2PCl_5(s)$
c. $FeO(s) + CO(g) \rightarrow Fe(s) + CO_2(g)$
d. $2C_5H_{10}(l) + 15O_2(g) \xrightarrow{\Delta} 10CO_2(g) + 10H_2O(g)$
e. $3H_2S(g) + 2FeCl_3(s) \rightarrow Fe_2S_3(s) + 6HCl(g)$

8.15 **a.** This is a decomposition reaction because a single reactant splits into two simpler substances.
b. This is a single replacement reaction because one element in the reacting compound (I in BaI_2) is replaced by the other reactant (Br in Br_2).

8.17 **a.** combination **b.** single replacement
c. decomposition **d.** double replacement
e. decomposition **f.** double replacement
g. combination

8.19 **a.** Since this is a combination reaction, the product is a single compound.
$Mg(s) + Cl_2(g) \rightarrow MgCl_2(s)$
b. Since this is a decomposition reaction, the product is two species.
$2HBr(g) \rightarrow H_2(g) + Br_2(g)$

c. Since this is a single replacement reaction, the uncombined metal element takes the place of the metal element in the compound.
$Mg(s) + Zn(NO_3)_2(aq) \rightarrow Zn(s) + Mg(NO_3)_2(aq)$

d. Since this is a double replacement reaction, the positive ions in the reacting compounds switch places.
$K_2S(aq) + Pb(NO_3)_2(aq) \rightarrow PbS(s) + 2KNO_3(aq)$

8.21 **a.** Alcohols contain a hydroxyl group (—OH) attached to a carbon chain.
b. Alkenes have carbon–carbon double bonds.
c. Aldehydes contain a carbonyl group (C═O) attached to at least one H atom.
d. Esters contain a —COO— group attached to two carbon atoms.

8.23 **a.** Ethers contain a —O— group attached to two carbon atoms.
b. Alcohols have a hydroxyl group (—OH).
c. Ketones have a carbonyl group (C═O) between two carbon atoms.
d. Carboxylic acids have a carboxyl group (—COOH).
e. Amines contain a nitrogen atom bonded to at least one carbon atom.

8.25 **a.** $C_3H_8(g) + 5O_2(g) \xrightarrow{\Delta} 3CO_2(g) + 4H_2O(g) + \text{energy}$
b. $2C_8H_{18}(l) + 25O_2(g) \xrightarrow{\Delta} 16CO_2(g) + 18H_2O(g) + \text{energy}$
c. $2C_3H_8O(l) + 9O_2(g) \xrightarrow{\Delta} 6CO_2(g) + 8H_2O(g) + \text{energy}$
d. $C_6H_{12}(l) + 9O_2(g) \xrightarrow{\Delta} 6CO_2(g) + 6H_2O(g) + \text{energy}$

8.27 **a.** $CH_3—CH_2—CH_2—CH_2—CH_3$
b. $CH_3—CH_2—CH_2—CH_3$
c. $CH_3—CH_2—CH_2—CH_2—CH_2—CH_3$

8.29 **a.** $2NO(g) + O_2(g) \rightarrow 2NO_2(g)$
b. combination

8.31 **a.** $2NI_3(s) \rightarrow N_2(g) + 3I_2(g)$
b. decomposition

8.33 **a.** $2Cl_2(g) + O_2(g) \rightarrow 2OCl_2(g)$
b. combination

8.35 **a.** 1, 1, 2; combination
b. 2, 2, 1; decomposition

8.37 **a.** carboxylic acid
b. alkene
c. ester
d. amine
e. aldehyde

8.39 **a.** An alcohol contains a hydroxyl group bonded to a carbon.
b. An alkene contains one or more carbon–carbon double bonds.
c. An aldehyde is an organic compound in which the carbon of a carbonyl group is bonded to a hydrogen.
d. Alkanes are hydrocarbons that contain only carbon–carbon single bonds.
e. A carboxylic acid is an organic compound in which the carbon of a carbonyl group is bonded to a hydroxyl group.

8.41 **a.** aromatic, aldehyde
b. aromatic, alkene, aldehyde
c. ketone

8.43 amine, carboxylic acid, amide, aromatic, ester

8.45 **a.** combination
b. combustion
c. double replacement
d. decomposition
e. single replacement

8.47 **a.** $NH_3(g) + HCl(g) \rightarrow NH_4Cl(s)$ combination
b. $Fe_3O_4(s) + 4H_2(g) \rightarrow 3Fe(s) + 4H_2O(g)$ single replacement
c. $2Sb(s) + 3Cl_2(g) \rightarrow 2SbCl_3(s)$ combination
d. $2NI_3(s) \rightarrow N_2(g) + 3I_2(g)$ decomposition
e. $2KBr(aq) + Cl_2(aq) \rightarrow 2KCl(aq) + Br_2(l)$ single replacement
f. $2Fe(s) + 3H_2SO_4(aq) \rightarrow Fe_2(SO_4)_3(aq) + 3H_2(g)$ single replacement
g. $Al_2(SO_4)_3(aq) + 6NaOH(aq) \rightarrow 3Na_2SO_4(aq) + 2Al(OH)_3(s)$ double replacement

8.49 **a.** $Zn(s) + 2HCl(aq) \rightarrow ZnCl_2(aq) + H_2(g)$
b. $BaCO_3(s) \xrightarrow{\Delta} BaO(s) + CO_2(g)$
c. $NaOH(aq) + HCl(aq) \rightarrow NaCl(aq) + H_2O(l)$
d. $2Al(s) + 3F_2(g) \rightarrow 2AlF_3(s)$

8.51 **a.** $4Na(s) + O_2(g) \rightarrow 2Na_2O(s)$ combination
b. $NaCl(aq) + AgNO_3(aq) \rightarrow AgCl(s) + NaNO_3(aq)$ double replacement

8.53 **a.** $3Pb(NO_3)_2(aq) + 2Na_3PO_4(aq) \rightarrow Pb_3(PO_4)_2(s) + 6NaNO_3(aq)$ double replacement
b. $4Ga(s) + 3O_2(g) \xrightarrow{\Delta} 2Ga_2O_3(s)$ combination
c. $2NaNO_3(s) \xrightarrow{\Delta} 2NaNO_2(s) + O_2(g)$ decomposition

8.55 **a.** Reactants: X and Y_2; Products: XY_3
b. $2X + 3Y_2 \rightarrow 2XY_3$
c. combination

8.57 **a.** Moles of C $= 0.40 \cancel{\text{mol } CO_2} \times \dfrac{1 \text{ mol C}}{1 \cancel{\text{mol } CO_2}} = 0.40$ mol of C

Moles of H $= 0.60 \cancel{\text{mol } H_2O} \times \dfrac{2 \text{ mol H}}{1 \cancel{\text{mol } H_2O}} = 1.2$ mol of H

b. $\dfrac{0.40 \text{ mol C}}{0.40} = 1.0$ mol of C $\quad \dfrac{1.2 \text{ mol H}}{0.40} = 3.0$ mol of H (2 SFs)

Empirical formula $= C_{1.0}H_{3.0} = CH_3$

c. Since the empirical formula is CH_3 and one molecule of the compound contains 6 H atoms, the molecular formula must be $2(CH_3) = C_2H_6$.

d. $2C_2H_6(g) + 7O_2(g) \xrightarrow{\Delta} 4CO_2(g) + 6H_2O(g) +$ energy

8.59 aromatic, ketone, alkene, amine, carboxylic acid

8.61 **a.** alcohol
b. amine
c. ketone
d. alkene
e. carboxylic acid

9

Chemical Quantities in Reactions

Study Goals

- Given a quantity in moles of reactant or product, use a mole–mole factor from the balanced equation to calculate the number of moles of another substance in the reaction.
- Given the mass in grams of a substance in a reaction, calculate the mass in grams of another substance in the reaction.
- Identify a limiting reactant when given the quantities of two reactants; calculate the amount of product formed from the limiting reactant.
- Given the actual quantity of product, determine the percent yield for a reaction.
- Given the heat of reaction (enthalpy change), calculate the loss or gain of heat for an exothermic or endothermic reaction.
- Describe the role of ATP in providing energy for the cells of the body.

Chapter Outline

Answers and Solutions to Text Problems

9.1 **a.** (1) Two molecules of sulfur dioxide gas react with one molecule of oxygen gas to produce two molecules of sulfur trioxide gas.
(2) Two mol of sulfur dioxide gas react with 1 mol of oxygen gas to produce 2 mol of of sulfur trioxide gas.

b. (1) Four atoms of solid phosphorus react with five molecules of oxygen gas to produce two molecules of solid diphosphorus pentoxide.
(2) Four mol of solid phosphorus react with 5 mol of oxygen gas to produce 2 mol of solid diphosphorus pentoxide.

9.3 **a.** Reactants: 2 mol of SO_2 and 1 mol of O_2 $= 2\ \cancel{mol}\ (64.07\ g/\cancel{mol}) + 1\ \cancel{mol}\ (32.00\ g/\cancel{mol})$
$= 128.14\ g + 32.00\ g = 160.14$ g of reactants
Products: 2 mol of SO_3 $= 2\ \cancel{mol}\ (80.07\ g/\cancel{mol}) = 160.14$ g of products

b. Reactants: 4 mol of P and 5 mol of O_2 $= 4\ \cancel{mol}\ (30.97\ g/\cancel{mol}) + 5\ \cancel{mol}\ (32.00\ g/\cancel{mol})$
$= 123.88\ g + 160.00\ g = 283.88$ g of reactants
Products: 2 mol of P_2O_5 $= 2\ \cancel{mol}\ (141.94\ g/\cancel{mol}) = 283.88$ g of products

9.5 **a.** $\frac{2\ mol\ SO_2}{1\ mol\ O_2}$ and $\frac{1\ mol\ O_2}{2\ mol\ SO_2}$; $\frac{2\ mol\ SO_2}{2\ mol\ SO_3}$ and $\frac{2\ mol\ SO_3}{2\ mol\ SO_2}$; $\frac{1\ mol\ O_2}{2\ mol\ SO_3}$ and $\frac{2\ mol\ SO_3}{1\ mol\ O_2}$

b. $\frac{4\ mol\ P}{5\ mol\ O_2}$ and $\frac{5\ mol\ O_2}{4\ mol\ P}$; $\frac{4\ mol\ P}{2\ mol\ P_2O_5}$ and $\frac{2\ mol\ P_2O_5}{4\ mol\ P}$; $\frac{5\ mol\ O_2}{2\ mol\ P_2O_5}$ and $\frac{2\ mol\ P_2O_5}{5\ mol\ O_2}$

9.7 **a.** $2.0\ \cancel{mol\ H_2} \times \frac{1\ mol\ O_2}{2\ \cancel{mol\ H_2}} = 1.0$ mol of O_2 (2 SFs)

b. $5.0\ \cancel{mol\ O_2} \times \frac{2\ mol\ H_2}{1\ \cancel{mol\ O_2}} = 10.$ mol of H_2 (2 SFs)

c. $2.5\ \cancel{mol\ O_2} \times \frac{2\ mol\ H_2O}{1\ \cancel{mol\ O_2}} = 5.0$ mol of H_2O (2 SFs)

9.9 **a.** $0.500\ \cancel{mol\ SO_2} \times \frac{5\ mol\ C}{2\ \cancel{mol\ SO_2}} = 1.25$ mol of C (3 SFs)

b. $1.2\ \cancel{mol\ C} \times \frac{4\ mol\ CO}{5\ \cancel{mol\ C}} = 0.96$ mol of CO (2 SFs)

c. $0.50\ \cancel{mol\ CS_2} \times \frac{2\ mol\ SO_2}{1\ \cancel{mol\ CS_2}} = 1.0$ mol of SO_2 (2 SFs)

d. $2.5\ \cancel{mol\ C} \times \frac{1\ mol\ CS_2}{5\ \cancel{mol\ C}} = 0.50$ mol of CS_2 (2 SFs)

9.11 **a.** $2.50\ \cancel{mol\ Na} \times \frac{2\ \cancel{mol\ Na_2O}}{4\ \cancel{mol\ Na}} \times \frac{61.98\ g\ Na_2O}{1\ \cancel{mol\ Na_2O}} = 77.5$ g of Na_2O (3 SFs)

b. $18.0\ \cancel{g\ Na} \times \frac{1\ \cancel{mol\ Na}}{22.99\ \cancel{g\ Na}} \times \frac{1\ \cancel{mol\ O_2}}{4\ \cancel{mol\ Na}} \times \frac{32.00\ g\ O_2}{1\ \cancel{mol\ O_2}} = 6.26$ g of O_2 (3 SFs)

c. $75.0\ \cancel{g\ Na_2O} \times \frac{1\ \cancel{mol\ Na_2O}}{61.98\ \cancel{g\ Na_2O}} \times \frac{1\ \cancel{mol\ O_2}}{2\ \cancel{mol\ Na_2O}} \times \frac{32.00\ g\ O_2}{1\ \cancel{mol\ O_2}} = 19.4$ g of O_2 (3 SFs)

9.13 **a.** $8.00\ \cancel{mol\ NH_3} \times \frac{3\ \cancel{mol\ O_2}}{4\ \cancel{mol\ NH_3}} \times \frac{32.00\ g\ O_2}{1\ \cancel{mol\ O_2}} = 192$ g of O_2 (3 SFs)

b. $6.50\ \cancel{g\ O_2} \times \frac{1\ \cancel{mol\ O_2}}{32.00\ \cancel{g\ O_2}} \times \frac{2\ \cancel{mol\ N_2}}{3\ \cancel{mol\ O_2}} \times \frac{28.02\ g\ N_2}{1\ \cancel{mol\ N_2}} = 3.79$ g of N_2 (3 SFs)

c. $34.0\ \cancel{g\ NH_3} \times \frac{1\ \cancel{mol\ NH_3}}{17.03\ \cancel{g\ NH_3}} \times \frac{6\ \cancel{mol\ H_2O}}{4\ \cancel{mol\ NH_3}} \times \frac{18.02\ g\ H_2O}{1\ \cancel{mol\ H_2O}} = 54.0$ g of H_2O (3 SFs)

9.15 **a.** $28.0\ \cancel{g\ NO_2} \times \frac{1\ \cancel{mol\ NO_2}}{46.01\ \cancel{g\ NO_2}} \times \frac{1\ \cancel{mol\ H_2O}}{3\ \cancel{mol\ NO_2}} \times \frac{18.02\ g\ H_2O}{1\ \cancel{mol\ H_2O}} = 3.66$ g of H_2O (3 SFs)

b. $15.8\ \cancel{g\ NO_2} \times \frac{1\ \cancel{mol\ NO_2}}{46.01\ \cancel{g\ NO_2}} \times \frac{1\ \cancel{mol\ NO}}{3\ \cancel{mol\ NO_2}} \times \frac{30.01\ g\ NO}{1\ \cancel{mol\ NO}} = 3.44$ g of NO (3 SFs)

c. $8.25\ \cancel{g\ NO_2} \times \frac{1\ \cancel{mol\ NO_2}}{46.01\ \cancel{g\ NO_2}} \times \frac{2\ \cancel{mol\ HNO_3}}{3\ \cancel{mol\ NO_2}} \times \frac{63.02\ g\ HNO_3}{1\ \cancel{mol\ HNO_3}} = 7.53$ g of HNO_3 (3 SFs)

9.17 **a.** $2PbS(s) + 3O_2(g) \rightarrow 2PbO(s) + 2SO_2(g)$

b. $0.125\ \cancel{mol\ PbS} \times \frac{3\ \cancel{mol\ O_2}}{2\ \cancel{mol\ PbS}} \times \frac{32.00\ g\ O_2}{1\ \cancel{mol\ O_2}} = 6.00\ g\ of\ O_2$ (3 SFs)

c. $65.0\ \cancel{g\ PbS} \times \frac{1\ \cancel{mol\ PbS}}{239.3\ \cancel{g\ PbS}} \times \frac{2\ \cancel{mol\ SO_2}}{2\ \cancel{mol\ PbS}} \times \frac{64.07\ g\ SO_2}{1\ \cancel{mol\ SO_2}} = 17.4\ g\ of\ SO_2$ (3 SFs)

d. $128\ \cancel{g\ PbO} \times \frac{1\ \cancel{mol\ PbO}}{223.2\ \cancel{g\ PbO}} \times \frac{2\ \cancel{mol\ PbS}}{2\ \cancel{mol\ PbO}} \times \frac{239.3\ g\ PbS}{1\ \cancel{mol\ PbS}} = 137\ g\ of\ PbS$ (3 SFs)

9.19 **a.** The limiting factor is the number of drivers: with only 8 drivers available, only 8 taxis can be used to pick up passengers.

b. The limiting factor is the number of taxis: only seven taxis are in working condition to be driven.

9.21 **a.** $3.0\ \cancel{mol\ N_2} \times \frac{2\ mol\ NH_3}{1\ \cancel{mol\ N_2}} = 6.0\ mol\ of\ NH_3$

$5.0\ \cancel{mol\ H_2} \times \frac{2\ mol\ NH_3}{3\ \cancel{mol\ H_2}} = 3.3\ mol\ of\ NH_3$ (smaller number of moles)

The limiting reactant is 5.0 mol of H_2. (2 SFs)

b. $8.0\ \cancel{mol\ N_2} \times \frac{2\ mol\ NH_3}{1\ \cancel{mol\ N_2}} = 16\ mol\ of\ NH_3$

$4.0\ \cancel{mol\ H_2} \times \frac{2\ mol\ NH_3}{3\ \cancel{mol\ H_2}} = 2.7\ mol\ of\ NH_3$ (smaller number of moles)

The limiting reactant is 4.0 mol of H_2. (2 SFs)

c. $3.0\ \cancel{mol\ N_2} \times \frac{2\ mol\ NH_3}{1\ \cancel{mol\ N_2}} = 6.0\ mol\ of\ NH_3$ (smaller number of moles)

$12.0\ \cancel{mol\ H_2} \times \frac{2\ mol\ NH_3}{3\ \cancel{mol\ N_2}} = 8.0\ mol\ of\ NH_3$

The limiting reactant is 3.0 mol of N_2. (2 SFs)

9.23 **a.** $2.00\ \cancel{mol\ SO_2} \times \frac{2\ mol\ SO_3}{2\ \cancel{mol\ SO_2}} = 2.00\ mol\ of\ SO_3$ (smaller number of moles)

$2.00\ \cancel{mol\ O_2} \times \frac{2\ mol\ SO_3}{1\ \cancel{mol\ O_2}} = 4.00\ mol\ of\ SO_3$

$\therefore$ 2.00 mol of SO_3 produced. (3 SFs)

b. $2.00\ \cancel{mol\ Fe} \times \frac{1\ mol\ Fe_3O_4}{3\ \cancel{mol\ Fe}} = 0.667\ mol\ of\ Fe_3O_4$

$2.00\ \cancel{mol\ H_2O} \times \frac{1\ mol\ Fe_3O_4}{4\ \cancel{mol\ H_2O}} = 0.500\ mol\ of\ Fe_3O_4$ (smaller number of moles)

$\therefore$ 0.500 mol of Fe_3O_4 produced. (3 SFs)

c. $2.00\ \cancel{mol\ C_7H_{16}} \times \frac{7\ mol\ CO_2}{1\ \cancel{mol\ C_7H_{16}}} = 14.0\ mol\ of\ CO_2$

$2.00\ \cancel{mol\ O_2} \times \frac{7\ mol\ CO_2}{11\ \cancel{mol\ O_2}} = 1.27\ mol\ of\ CO_2$ (smaller number of moles)

$\therefore$ 1.27 mol of CO_2 produced. (3 SFs)

9.25 **a.** $20.0\ \cancel{g\ Al} \times \frac{1\ \cancel{mol\ Al}}{26.98\ \cancel{g\ Al}} \times \frac{2\ mol\ AlCl_3}{2\ \cancel{mol\ Al}} = 0.741$ mol of $AlCl_3$

$20.0\ \cancel{g\ Cl_2} \times \frac{1\ \cancel{mol\ Cl_2}}{70.90\ \cancel{g\ Cl_2}} \times \frac{2\ mol\ AlCl_3}{3\ \cancel{mol\ Cl_2}} = 0.188$ mol of $AlCl_3$ (smaller number of moles)

$0.188\ \cancel{mol\ AlCl_3} \times \frac{133.33\ g\ AlCl_3}{1\ \cancel{mol\ AlCl_3}} = 25.1$ g of $AlCl_3$ (3 SFs)

b. $20.0\ \cancel{g\ NH_3} \times \frac{1\ \cancel{mol\ NH_3}}{17.03\ \cancel{g\ NH_3}} \times \frac{6\ mol\ H_2O}{4\ \cancel{mol\ NH_3}} = 1.76$ mol of H_2O

$20.0\ \cancel{g\ O_2} \times \frac{1\ \cancel{mol\ O_2}}{32.00\ \cancel{g\ O_2}} \times \frac{6\ mol\ H_2O}{5\ \cancel{mol\ O_2}} = 0.750$ mol of H_2O (smaller number of moles)

$0.750\ \cancel{mol\ H_2O} \times \frac{18.02\ g\ H_2O}{1\ \cancel{mol\ H_2O}} = 13.5$ g of H_2O (3 SFs)

c. $20.0\ \cancel{g\ CS_2} \times \frac{1\ \cancel{mol\ CS_2}}{76.15\ \cancel{g\ CS_2}} \times \frac{2\ mol\ SO_2}{1\ \cancel{mol\ CS_2}} = 0.525$ mol of SO_2

$20.0\ \cancel{g\ O_2} \times \frac{1\ \cancel{mol\ O_2}}{32.00\ \cancel{g\ O_2}} \times \frac{2\ mol\ SO_2}{3\ \cancel{mol\ O_2}} = 0.417$ mol of SO_2 (smaller number of moles)

$0.417\ \cancel{mol\ SO_2} \times \frac{64.07\ g\ SO_2}{1\ \cancel{mol\ SO_2}} = 26.7$ g of SO_2 (3 SFs)

9.27 **a.** Theoretical yield of CS_2:

$40.0\ \cancel{g\ C} \times \frac{1\ \cancel{mol\ C}}{12.01\ \cancel{g\ C}} \times \frac{1\ \cancel{mol\ CS_2}}{5\ \cancel{mol\ C}} \times \frac{76.15\ g\ CS_2}{1\ \cancel{mol\ CS_2}} = 50.7$ g of CS_2

Percent yield: $\frac{36.0\ \cancel{g\ CS_2}\ \text{(actual)}}{50.7\ \cancel{g\ CS_2}\ \text{(theoretical)}} \times 100\% = 71.0\%$ (3 SFs)

b. Theoretical yield of CS_2:

$32.0\ \cancel{g\ SO_2} \times \frac{1\ \cancel{mol\ SO_2}}{64.07\ \cancel{g\ SO_2}} \times \frac{1\ \cancel{mol\ CS_2}}{2\ \cancel{mol\ SO_2}} \times \frac{76.15\ g\ CS_2}{1\ \cancel{mol\ CS_2}} = 19.0$ g of CS_2

Percent yield: $\frac{12.0\ \cancel{g\ CS_2}\ \text{(actual)}}{19.0\ \cancel{g\ CS_2}\ \text{(theoretical)}} \times 100\% = 63.2\%$ (3 SFs)

9.29 Theoretical yield of Al_2O_3:

$50.0\ \cancel{g\ Al} \times \frac{1\ \cancel{mol\ Al}}{26.98\ \cancel{g\ Al}} \times \frac{2\ \cancel{mol\ Al_2O_3}}{4\ \cancel{mol\ Al}} \times \frac{101.96\ g\ Al_2O_3}{1\ \cancel{mol\ Al_2O_3}} = 94.5$ g of Al_2O_3

Use the percent yield to convert theoretical to actual:

$94.5\ \cancel{g\ Al_2O_3} \times \frac{75.0\ g\ Al_2O_3}{100\ \cancel{g\ Al_2O_3}} = 70.9$ g of Al_2O_3 (actual) (3 SFs)

9.31 Theoretical yield of CO_2:

$30.0\ \cancel{g\ C} \times \frac{1\ \cancel{mol\ C}}{12.01\ \cancel{g\ C}} \times \frac{2\ \cancel{mol\ CO}}{3\ \cancel{mol\ C}} \times \frac{28.01\ g\ CO}{1\ \cancel{mol\ CO}} = 46.6$ g of CO

Percent yield: $\frac{28.2\ \cancel{g\ CO}\ \text{(actual)}}{46.6\ \cancel{g\ CO}\ \text{(theoretical)}} \times 100\% = 60.5\%$ (3 SFs)

9.33 In exothermic reactions, the energy of the products is less than that of the reactants.

9.35 **a.** An exothermic reaction releases energy.
b. An endothermic reaction has a higher energy level for the products than the reactants.
c. Metabolism is an exothermic reaction, providing energy for the body.

9.37 **a.** Heat is released, which makes the reaction exothermic with $\Delta H = -890$ kJ.
b. Heat is absorbed, which makes the reaction endothermic with $\Delta H = +65.3$ kJ.
c. Heat is released, which makes the reaction exothermic with $\Delta H = -850$ kJ.

9.39 $125\ \cancel{\text{g Cl}_2} \times \dfrac{1\ \cancel{\text{mol Cl}_2}}{70.90\ \cancel{\text{g Cl}_2}} \times \dfrac{657\ \text{kJ}}{2\ \cancel{\text{mol Cl}_2}} = 579\ \text{kJ released (3 SFs)}$

9.41 ATP is the abbreviation for adenosine triphosphate.

9.43 Glucose ($C_6H_{12}O_6$) is the primary fuel used by the body to provide energy.

9.45 The ΔH for the breakdown of ATP is -7.3 kcal/mol (-31 kJ/mol), which makes the reaction exothermic.

9.47 ATP is an energy-rich compound because it contains high-energy phosphate bonds that can be cleaved to free a phosphate group from ATP; enough energy is released to "drive" energy-requiring processes in the cell.

9.49 Fats can be stored in unlimited amounts in adipose tissue, so they are the major form of stored energy in the body.

9.51 **a.** $2NO(g) + O_2(g) \rightarrow 2NO_2(g)$ **b.** NO is the limiting reactant.

9.53 **a.** $N_2(g) + 3H_2(g) \rightarrow 2NH_3(g)$ **b.** diagram A; N_2 is the excess reactant.

9.55 **a.** $2NI_3(g) \rightarrow N_2(g) + 3I_2(g)$
b. Theoretical yield: from the $6NI_3$ we could obtain $3N_2$ and $9I_2$.
Actual yield: in the actual products, we obtain $2N_2$ and $6I_2$.

Percent yield: $\dfrac{2N_2 \text{ and } 6I_2 \text{ (actual)}}{3N_2 \text{ and } 9I_2 \text{ (theoretical)}} \times 100\%$

$= \dfrac{2\ \cancel{(N_2 \text{ and } 3I_2)} \text{ (actual)}}{3\ \cancel{(N_2 \text{ and } 3I_2)} \text{ (theoretical)}} \times 100\% = 67\%$

9.57 **a.** $124\ \cancel{\text{g C}_2\text{H}_6\text{O}} \times \dfrac{1\ \cancel{\text{mol C}_2\text{H}_6\text{O}}}{46.07\ \cancel{\text{g C}_2\text{H}_6\text{O}}} \times \dfrac{1\ \text{mol C}_6\text{H}_{12}\text{O}_6}{2\ \cancel{\text{mol C}_2\text{H}_6\text{O}}} = 1.35\ \text{mol of glucose (C}_6\text{H}_{12}\text{O}_6\text{) (3 SFs)}$

b. $0.240\ \cancel{\text{kg C}_6\text{H}_{12}\text{O}_6} \times \dfrac{1000\ \cancel{\text{g}}}{1\ \cancel{\text{kg}}} \times \dfrac{1\ \cancel{\text{mol C}_6\text{H}_{12}\text{O}_6}}{180.16\ \cancel{\text{g C}_6\text{H}_{12}\text{O}_6}} \times \dfrac{2\ \cancel{\text{mol C}_2\text{H}_6\text{O}}}{1\ \cancel{\text{mol C}_6\text{H}_{12}\text{O}_6}} \times \dfrac{46.07\ \text{g C}_2\text{H}_6\text{O}}{1\ \cancel{\text{mol C}_2\text{H}_6\text{O}}}$
$= 123\ \text{g of ethanol (C}_2\text{H}_6\text{O) (3 SFs)}$

9.59 **a.** $2NH_3(g) + 5F_2(g) \rightarrow N_2F_4(g) + 6HF(g)$

b. $4.00\ \cancel{\text{mol HF}} \times \dfrac{2\ \text{mol NH}_3}{6\ \cancel{\text{mol HF}}} = 1.33\ \text{mol of NH}_3\ \text{(3 SFs)}$

$4.00\ \cancel{\text{mol HF}} \times \dfrac{5\ \text{mol F}_2}{6\ \cancel{\text{mol HF}}} = 3.33\ \text{mol of F}_2\ \text{(3 SFs)}$

c. $1.50\ \cancel{\text{mol NH}_3} \times \dfrac{5\ \cancel{\text{mol F}_2}}{2\ \cancel{\text{mol NH}_3}} \times \dfrac{38.00\ \text{g F}_2}{1\ \cancel{\text{mol F}_2}} = 143\ \text{g of F}_2\ \text{(3 SFs)}$

d. $3.40\ \cancel{\text{g NH}_3} \times \dfrac{1\ \cancel{\text{mol NH}_3}}{17.03\ \cancel{\text{g NH}_3}} \times \dfrac{1\ \cancel{\text{mol N}_2\text{F}_4}}{2\ \cancel{\text{mol NH}_3}} \times \dfrac{104.02\ \text{g N}_2\text{F}_4}{1\ \cancel{\text{mol N}_2\text{F}_4}} = 10.4\ \text{g of N}_2\text{F}_4\ \text{(3 SFs)}$

9.61 $12.8\ \cancel{g\ Na} \times \frac{1\ \cancel{mol\ Na}}{22.99\ \cancel{g\ Na}} \times \frac{2\ mol\ NaCl}{2\ \cancel{mol\ Na}} = 0.557$ mol of NaCl

$10.2\ \cancel{g\ Cl_2} \times \frac{1\ \cancel{mol\ Cl_2}}{70.90\ \cancel{g\ Cl_2}} \times \frac{2\ mol\ NaCl}{1\ \cancel{mol\ Cl_2}} = 0.288$ mol of NaCl (smaller number of moles)

$0.288\ \cancel{mol\ NaCl} \times \frac{58.44\ g\ NaCl}{1\ \cancel{mol\ NaCl}} = 16.8$ g of NaCl

$\therefore$ 16.8 g of NaCl produced. (3 SFs)

9.63 **a.** $4.0\ \cancel{mol\ H_2O} \times \frac{1\ \cancel{mol\ C_5H_{12}}}{6\ \cancel{mol\ H_2O}} \times \frac{72.15\ g\ C_5H_{12}}{1\ \cancel{mol\ C_5H_{12}}} = 48$ g of C_5H_{12} (2 SFs)

b. $32.0\ \cancel{g\ O_2} \times \frac{1\ \cancel{mol\ O_2}}{32.00\ \cancel{g\ O_2}} \times \frac{5\ \cancel{mol\ CO_2}}{8\ \cancel{mol\ O_2}} \times \frac{44.01\ g\ CO_2}{1\ \cancel{mol\ CO_2}} = 27.5$ g of CO_2 (3 SFs)

c. $44.5\ \cancel{g\ C_5H_{12}} \times \frac{1\ \cancel{mol\ C_5H_{12}}}{72.15\ \cancel{g\ C_5H_{12}}} \times \frac{5\ mol\ CO_2}{1\ \cancel{mol\ C_5H_{12}}} = 3.08$ mol of CO_2

$108\ \cancel{g\ O_2} \times \frac{1\ \cancel{mol\ O_2}}{32.00\ \cancel{g\ O_2}} \times \frac{5\ mol\ CO_2}{8\ \cancel{mol\ O_2}} = 2.11$ mol of CO_2 (smaller number of moles)

$2.11\ \cancel{mol\ CO_2} \times \frac{44.01\ g\ CO_2}{1\ \cancel{mol\ CO_2}} = 92.9$ g of CO_2

$\therefore$ 92.9 g of CO_2 produced. (3 SFs)

9.65 Theoretical yield of C_2H_6:

$28.0\ \cancel{g\ C_2H_2} \times \frac{1\ \cancel{mol\ C_2H_2}}{26.04\ \cancel{g\ C_2H_2}} \times \frac{1\ \cancel{mol\ C_2H_6}}{1\ \cancel{mol\ C_2H_2}} \times \frac{30.07\ g\ C_2H_6}{1\ \cancel{mol\ C_2H_6}} = 32.3$ g of C_2H_6

Percent yield: $\frac{24.5\ \cancel{g\ C_2H_6}\ (actual)}{32.3\ \cancel{g\ C_2H_6}\ (theoretical)} \times 100\% = 75.9\%$ (3 SFs)

9.67 Theoretical yield of NH_3 would be:

$30.0\ \cancel{g\ NH_3}\ (actual) \times \frac{100.\ g\ NH_3\ (theoretical)}{65.0\ \cancel{g\ NH_3}\ (actual)} = 46.2$ g of NH_3 (theoretical)

$46.2\ \cancel{g\ NH_3} \times \frac{1\ \cancel{mol\ NH_3}}{17.03\ \cancel{g\ NH_3}} \times \frac{1\ \cancel{mol\ N_2}}{2\ \cancel{mol\ NH_3}} \times \frac{28.02\ g\ N_2}{1\ \cancel{mol\ N_2}} = 38.0$ g of N_2 (reacted) (3 SFs)

9.69 **a.** $22.0\ \cancel{g\ C_2H_2} \times \frac{1\ \cancel{mol\ C_2H_2}}{26.04\ \cancel{g\ C_2H_2}} \times \frac{5\ \cancel{mol\ O_2}}{2\ \cancel{mol\ C_2H_2}} \times \frac{6.022 \times 10^{23}\ molecules\ O_2}{1\ \cancel{mol\ O_2}}$

$= 1.27 \times 10^{24}$ molecules of O_2

b. Theoretical yield of CO_2:

$22.0\ \cancel{g\ C_2H_2} \times \frac{1\ \cancel{mol\ C_2H_2}}{26.04\ \cancel{g\ C_2H_2}} \times \frac{4\ \cancel{mol\ CO_2}}{2\ \cancel{mol\ C_2H_2}} \times \frac{44.01\ g\ CO_2}{1\ \cancel{mol\ CO_2}} = 74.4$ g of CO_2

c. Percent yield: $\frac{64.0\ \cancel{g\ CO_2}\ (actual)}{74.4\ \cancel{g\ CO_2}\ (theoretical)} \times 100\% = 86.0\%$

9.71 **a.** $3.00\ \cancel{g\ NO} \times \frac{1\ \cancel{mol\ NO}}{30.01\ \cancel{g\ NO}} \times \frac{90.2\ kJ}{2\ \cancel{mol\ NO}} = 4.51$ kJ

b. $2NO(g) \rightarrow N_2(g) + O_2(g) + 90.2\ kJ$

c. $5.00\ \cancel{g\ NO} \times \frac{1\ \cancel{mol\ NO}}{30.01\ \cancel{g\ NO}} \times \frac{90.2\ kJ}{2\ \cancel{mol\ NO}} = 7.51$ kJ

9.73 **a.** Heat is released, which makes the reaction exothermic.

b. Heat is absorbed, which makes the reaction endothermic.

9.75 Reaction (1) is an exothermic reaction that releases a large amount of energy (14.8 kcal), which provides the energy required for reaction (2).

9.77 $$1.0\ \text{mol } C_{16}H_{32}O_2 \times \frac{129\ \text{mol ATP}}{1\ \text{mol } C_{16}H_{32}O_2} \times \frac{7.3\ \text{kcal}}{1\ \text{mol ATP}} = 940\ \text{kcal (2 SFs)}$$

9.79 **a.** $$50.0\ \text{g CO} \times \frac{1\ \text{mol CO}}{28.01\ \text{g CO}} \times \frac{1\ \text{mol } CH_3OH}{1\ \text{mol CO}}$$

$= 1.79$ mol of CH_3OH (smaller number of moles)

$$10.0\ \text{g } H_2 \times \frac{1\ \text{mol } H_2}{2.016\ \text{g } H_2} \times \frac{1\ \text{mol } CH_3OH}{2\ \text{mol } H_2} = 2.48\ \text{mol of } CH_3OH$$

CO is the limiting reactant.

b. H_2 is the excess reactant.

c. $$1.79\ \text{mol } CH_3OH \times \frac{32.04\ \text{g } CH_3OH}{1\ \text{mol } CH_3OH} = 57.4\ \text{g of } CH_3OH$$

57.4 g of methanol (CH_3OH) can be produced. (3 SFs)

d. Amount of H_2 reacted:

$$50.0\ \text{g CO} \times \frac{1\ \text{mol CO}}{28.01\ \text{g CO}} \times \frac{2\ \text{mol } H_2}{1\ \text{mol CO}} \times \frac{2.016\ \text{g } H_2}{1\ \text{mol } H_2} = 7.20\ \text{g of } H_2\ \text{(3 SFs)}$$

10.0 g (initial) − 7.20 g (reacted) = 2.8 g of H_2 left over. (2 SFs)

9.81 **a.** $4Cr(s) + 3O_2(g) \rightarrow 2Cr_2O_3(s)$

b. combination reaction

c. $$4.50\ \text{mol Cr} \times \frac{3\ \text{mol } O_2}{4\ \text{mol Cr}} = 3.38\ \text{mol of } O_2\ \text{(3 SFs)}$$

d. $$24.8\ \text{g Cr} \times \frac{1\ \text{mol Cr}}{52.00\ \text{g Cr}} \times \frac{2\ \text{mol } Cr_2O_3}{4\ \text{mol Cr}} \times \frac{152.0\ \text{g } Cr_2O_3}{1\ \text{mol } Cr_2O_3} = 36.2\ \text{g of } Cr_2O_3\ \text{(3 SFs)}$$

e. $$0.500\ \text{mol Cr} \times \frac{2\ \text{mol } Cr_2O_3}{4\ \text{mol Cr}} = 0.250\ \text{mol of } Cr_2O_3$$

$$8.00\ \text{g } O_2 \times \frac{1\ \text{mol } O_2}{32.00\ \text{g } O_2} \times \frac{2\ \text{mol } Cr_2O_3}{3\ \text{mol } O_2} = 0.167\ \text{mol of } Cr_2O_3\ \text{(smaller number of moles)}$$

$$0.167\ \text{mol } Cr_2O_3 \times \frac{152.0\ \text{g } Cr_2O_3}{1\ \text{mol } Cr_2O_3} = 25.4\ \text{g of } Cr_2O_3\ \text{(3 SFs)}$$

f. $$74.0\ \text{g Cr} \times \frac{1\ \text{mol Cr}}{52.00\ \text{g Cr}} \times \frac{2\ \text{mol } Cr_2O_3}{4\ \text{mol Cr}} = 0.712\ \text{mol of } Cr_2O_3\ \text{(smaller number of moles)}$$

$$62.0\ \text{g } O_2 \times \frac{1\ \text{mol } O_2}{32.00\ \text{g } O_2} \times \frac{2\ \text{mol } Cr_2O_3}{3\ \text{mol } O_2} = 1.29\ \text{mol of } Cr_2O_3$$

$$0.712\ \text{mol } Cr_2O_3 \times \frac{152.0\ \text{g } Cr_2O_3}{1\ \text{mol } Cr_2O_3} = 108\ \text{g of } Cr_2O_3\ \text{(theoretical)}$$

Use the percent yield to convert theoretical to actual:

$$108\ \text{g } Cr_2O_3 \times \frac{70.0\ \text{g } Cr_2O_3}{100.\ \text{g } Cr_2O_3} = 75.6\ \text{g of } Cr_2O_3\ \text{(actual) (3 SFs)}$$

9.83 **a.** The heat of reaction (ΔH) is negative, so the reaction is exothermic.

b. $$1.5\ \text{mol S} \times \frac{790\ \text{kJ}}{2\ \text{mol S}} = 590\ \text{kJ released (2 SFs)}$$

c. $$125\ \text{g } SO_3 \times \frac{1\ \text{mol } SO_3}{80.07\ \text{g } SO_3} \times \frac{790\ \text{kJ}}{2\ \text{mol } SO_3} = 620\ \text{kJ released (2 SFs)}$$

d. $\Delta H = +790$ kJ

e. The heat of reaction (ΔH) is positive, so the reaction is endothermic.

9.85 **a.** $24\,\cancel{h} \times \frac{60\,\cancel{min}}{1\,\cancel{h}} \times \frac{60\,\cancel{s}}{1\,\cancel{min}} \times \frac{2 \times 10^{6}\,\cancel{molecules\,ATP}}{1\,\cancel{s}\,\cancel{cell}} \times 10^{13}\,\cancel{cells}$

$\times \frac{1\,\cancel{mol\,ATP}}{6.022 \times 10^{23}\,\cancel{molecules\,ATP}} \times \frac{7.3\text{ kcal}}{1\,\cancel{mol\,ATP}} = 21\text{ kcal (2 SFs)}$

b. $24\,\cancel{h} \times \frac{60\,\cancel{min}}{1\,\cancel{h}} \times \frac{60\,\cancel{s}}{1\,\cancel{min}} \times \frac{2 \times 10^{6}\,\cancel{molecules\,ATP}}{1\,\cancel{s}\,\cancel{cell}} \times 10^{13}\,\cancel{cells}$

$\times \frac{1\,\cancel{mol\,ATP}}{6.022 \times 10^{23}\,\cancel{molecules\,ATP}} \times \frac{507\text{ g ATP}}{1\,\cancel{mol\,ATP}} = 1500\text{ g of ATP (2 SFs)}$

Structures of Solids and Liquids

Study Goals

- Draw the electron-dot formulas for covalent compounds or polyatomic ions with multiple bonds and show resonance structures.

- Predict the three-dimensional structure of a molecule or polyatomic ion and classify it as polar or nonpolar.

- Use electronegativity to determine the polarity of a bond or a molecule.

- Describe the attractive forces between ions, polar covalent molecules, and nonpolar covalent molecules.

- Describe the changes of state between solids, liquids, and gases; calculate the energy involved.

Chapter Outline

Chapter Opener: Pharmacist

10.1 Electron-Dot Formulas

10.2 Shapes of Molecules and Ions (VSEPR Theory)

10.3 Electronegativity and Polarity

10.4 Attractive Forces in Compounds

Chemistry and Health: Attractive Forces in Biological Compounds

10.5 Changes of State

Chemistry and Health: Steam Burns
Answers

Answers and Solutions to Text Problems

10.1 **a.** $2\,H(1\,e^-) + 1\,S(6\,e^-) = 2 + 6 = 8$ valence electrons
 b. $2\,I(7\,e^-) = 14$ valence electrons
 c. $1\,C(4\,e^-) + 4\,Cl(7\,e^-) = 4 + 28 = 32$ valence electrons
 d. $1\,O(6\,e^-) + 1\,H(1\,e^-) + 1\,e^-\text{(negative charge)} = 6 + 1 + 1 = 8$ valence electrons

10.3 **a.** $1\,H(1\,e^-) + 1\,F(7\,e^-) = 1 + 7 = 8$ valence electrons

$$ H\!:\!\ddot{\underset{\cdot\cdot}{F}}\!: \qquad or \qquad H\!-\!\ddot{\underset{\cdot\cdot}{F}}\!: $$

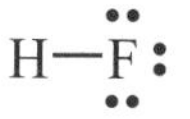

 b. $1\,S(6\,e^-) + 2\,F(7\,e^-) = 6 + 14 = 20$ valence electrons

$$:\!\ddot{\underset{\cdot\cdot}{F}}\!:\!\ddot{\underset{\cdot\cdot}{S}}\!:\!\ddot{\underset{\cdot\cdot}{F}}\!: \qquad or \qquad :\!\ddot{\underset{\cdot\cdot}{F}}\!-\!\ddot{\underset{\cdot\cdot}{S}}\!-\!\ddot{\underset{\cdot\cdot}{F}}\!: $$

 c. $1\,N(5\,e^-) + 3\,Br(7\,e^-) = 5 + 21 = 26$ valence electrons

$$ \begin{array}{c} :\!\ddot{Br}\!: \\[2pt] :\!\ddot{Br}\!:\!N\!:\!\ddot{Br}\!: \end{array} \qquad or \qquad \begin{array}{c} :\!\ddot{Br}\!: \\ | \\ :\!\ddot{Br}\!-\!N\!-\!\ddot{Br}\!: \end{array} $$

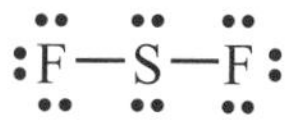

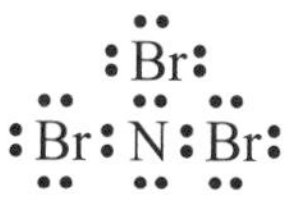

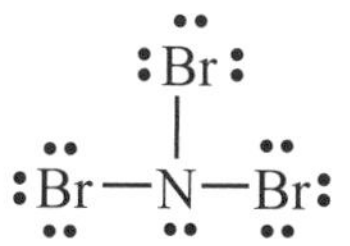

d. 1 B(3 e^-) + 4 H(1 e^-) + 1 e^-(negative charge) = 3 + 4 + 1 = 8 valence electrons

$$\left[\begin{array}{c} H \\ H:\ddot{B}:H \\ H \end{array}\right]^- \quad \text{or} \quad \left[\begin{array}{c} H \\ | \\ H-B-H \\ | \\ H \end{array}\right]^-$$

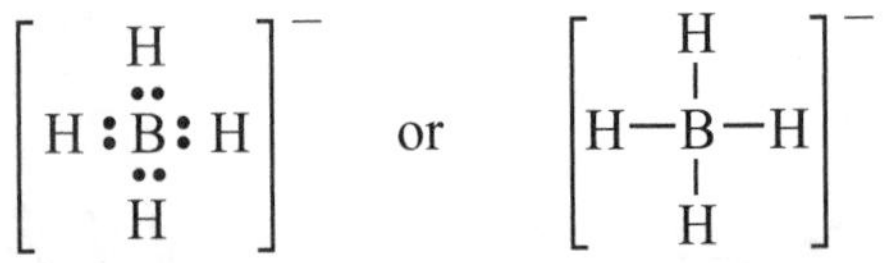

e. 1 C(4 e^-) + 4 H(1 e^-) + 1 O(6 e^-) = 4 + 4 + 6 = 14 valence electrons

$$\begin{array}{c} H \\ H:\ddot{C}:\ddot{O}:H \\ H \end{array} \quad \text{or} \quad \begin{array}{c} H \\ | \\ H-C-\ddot{O}-H \\ | \\ H \end{array}$$

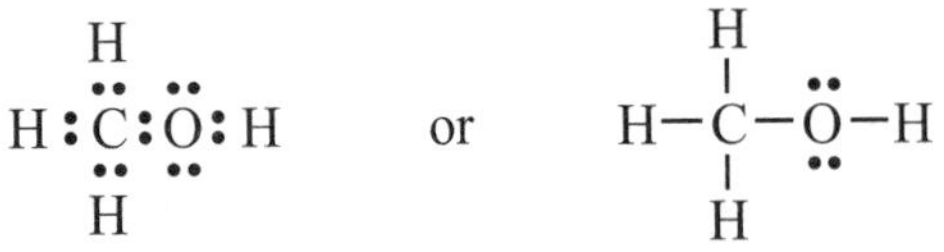

f. 2 N(5 e^-) + 4 H(1 e^-) = 10 + 4 = 14 valence electrons

$$\begin{array}{cc} H & H \\ H:\ddot{N}:\ddot{N}:H \end{array} \quad \text{or} \quad \begin{array}{cc} H & H \\ | & | \\ H-\ddot{N}-\ddot{N}-H \end{array}$$

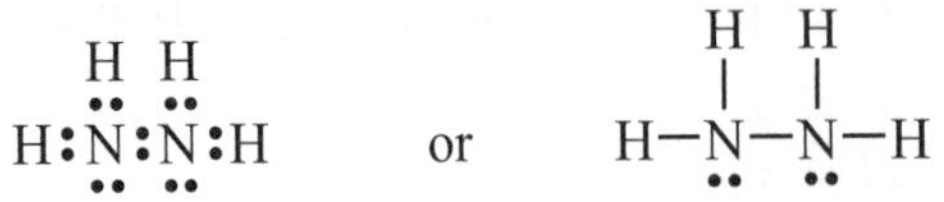

10.5 If complete octets cannot be formed when using only single bonds between atoms, it is necessary to draw multiple bonds.

10.7 Resonance occurs when we can draw two or more electron-dot formulas for the same molecule or ion.

10.9 a. 1 C(4 e^-) + 1 O(6 e^-) = 4 + 6 = 10 valence electrons

$$:C:::O: \quad \text{or} \quad :C\equiv O:$$

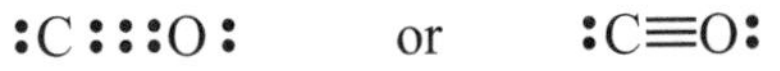

b. 2 C(4 e^-) + 4 H(1 e^-) = 8 + 4 = 12 valence electrons

$$\begin{array}{cc} H & H \\ H:\ddot{C}::\ddot{C}:H \end{array} \quad \text{or} \quad \begin{array}{cc} H & H \\ | & | \\ H-C=C-H \end{array}$$

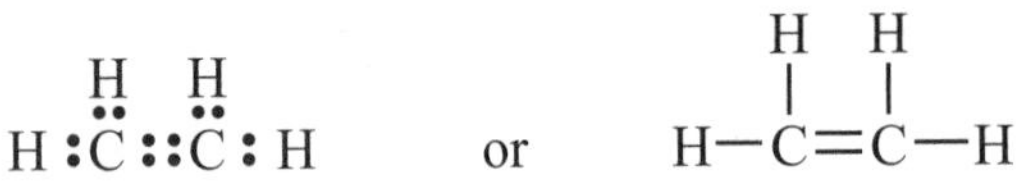

c. 2 H(1 e^-) + 1 C(4 e^-) + 1 O(6 e^-) = 2 + 4 + 6 = 12 valence electrons

$$\begin{array}{c} :\ddot{O}: \\ H:C:H \end{array} \quad \text{or} \quad \begin{array}{c} :O: \\ || \\ H-C-H \end{array}$$

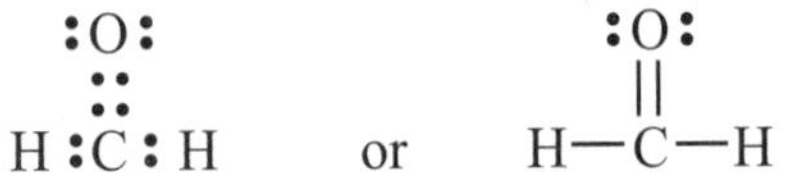

10.11 a. 1 Cl(7 e^-) + 1 N(5 e^-) + 2 O(6 e^-) = 7 + 5 + 12 = 24 valence electrons

$$\begin{array}{c} :O: \\ || \\ :\ddot{C}l-N-\ddot{O}: \end{array} \longleftrightarrow \begin{array}{c} :\ddot{O}: \\ | \\ :\ddot{C}l-N=\ddot{O}: \end{array}$$

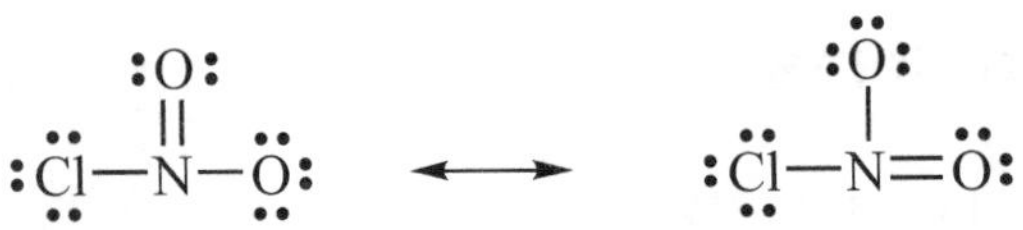

b. 1 O(6 e^-) + 1 C(4 e^-) + 1 N(5 e^-) + 1 e^-(negative charge)
= 6 + 4 + 5 + 1 = 16 valence electrons

$$\left[:O\equiv C-\ddot{N}:\right]^- \longleftrightarrow \left[:\ddot{O}=C=\ddot{N}:\right]^- \longleftrightarrow \left[:\ddot{O}-C\equiv N:\right]^-$$

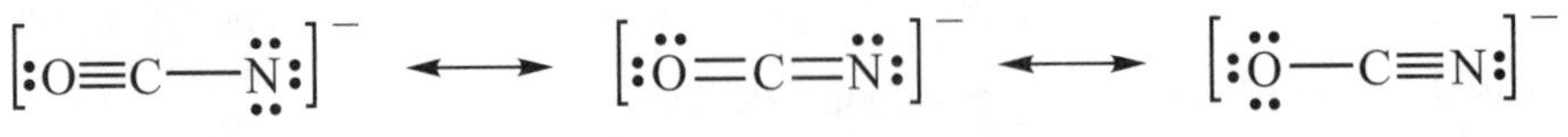

10.13 a. With two bonded atoms and no lone pairs on the central atom, a molecule would have a linear shape.

b. Four electron groups around a central atom have a tetrahedral electron arrangement. With three bonded atoms and one lone pair, the shape is trigonal pyramidal.

10.15 The four electron groups in PCl_3 have a tetrahedral arrangement, but three bonded atoms and one lone pair around a central atom give a trigonal pyramidal shape. The arrangement of electron pairs determines the angles between the pairs, whereas the number of bonded atoms determines the shape of the molecule.

10.17 In BF_3, the central atom B has three bonded atoms and no lone pairs, which gives BF_3 a trigonal planar shape. In NF_3, the central atom N has three bonded atoms and one lone pair, which gives NF_3 a trigonal pyramidal shape.

10.19 **a.** The central atom Ga has three electron groups bonded to three H atoms; GaH_3 has a trigonal planar shape.

b. The central O atom has four electron pairs, but only two are bonded to fluorine atoms. Its shape is bent with bond angles $< 109.5°$.

c. The central atom C has two electron groups bonded to two atoms; HCN is linear.

d. The central atom C has four electron pairs bonded to four chlorine atoms; CCl_4 has a tetrahedral shape.

e. The central atom Se has three electron groups with two bonded atoms and a lone pair, which gives SeO_2 a bent shape with bond angles $< 120°$.

10.21 To find the total valence electrons for an ion, add the total valence electrons for each atom and add the number of electrons indicated by a negative charge.

a. $1\ C(4\ e^-) + 3\ O(6\ e^-) + 2\ e^-$(negative charge) $= 4 + 18 + 2 = 24$ valence electrons

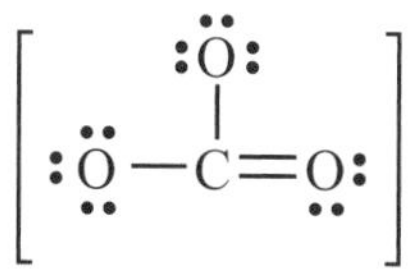

three electron groups around C bonded to three atoms; trigonal planar shape

b. $1\ S(6\ e^-) + 4\ O(6\ e^-) + 2\ e^-$(negative charge) $= 6 + 24 + 2 = 32$ valence electrons

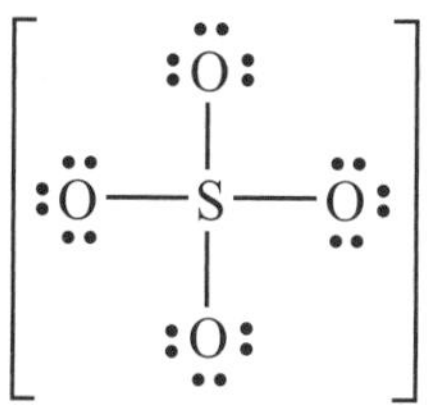

four electron groups around S bonded to four atoms; tetrahedral shape

c. $1\ B(3\ e^-) + 4\ H(1\ e^-) + 1\ e^-$(negative charge) $= 3 + 4 + 1 = 8$ valence electrons

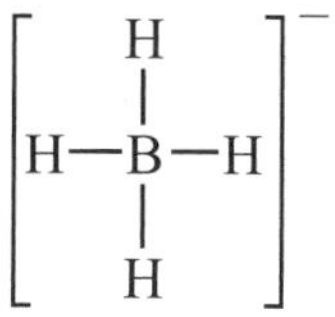

four electron groups around B bonded to four atoms; tetrahedral shape

d. $1\ N(5\ e^-) + 2\ O(6\ e^-) - 1\ e^-$(positive charge) $= 5 + 12 - 1 = 16$ valence electrons

two electron groups around N bonded to two atoms; linear shape

10.23 The electronegativity values increase going from left to right across a period.

10.25 A nonpolar covalent bond would have an electronegativity difference in the range of 0.0 to 0.4.

10.27 **a.** Electronegativity increases going up a group: K, Na, Li

b. Electronegativity increases going left to right across a period: Na, P, Cl

c. Electronegativity increases going across a period and at the top of a group: Ca, Br, O

10.29 A dipole arrow points from the atom with the lower electronegativity value (more positive) to the atom in the bond that has the higher electronegativity value (more negative).

a.

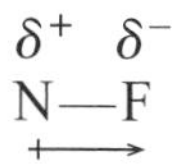

b. $\delta^+ \; \delta^-$
Si—P
$+\!\!\longrightarrow$

c. $\delta^+ \; \delta^-$
C—O
$+\!\!\longrightarrow$

d. $\delta^+ \; \delta^-$
P—Br
$+\!\!\longrightarrow$

e. $\delta^+ \; \delta^-$
B—Cl
$+\!\!\longrightarrow$

10.31 **a.** Si—Br electronegativity difference 2.8 − 1.8 = 1.0, polar covalent
b. Li—F electronegativity difference 4.0 − 1.0 = 3.0, ionic
c. Br—F electronegativity difference 4.0 − 2.8 = 1.2, polar covalent
d. Br—Br electronegativity difference 2.8 − 2.8 = 0.0, nonpolar covalent
e. N—P electronegativity difference 3.0 − 2.1 = 0.9, polar covalent
f. C—P electronegativity difference 2.5 − 2.1 = 0.4, nonpolar covalent

10.33 Electrons are shared equally between two identical F atoms, but unequally between nonidentical atoms H and F.

10.35 **a.** The molecule CS_2 contains two nonpolar covalent C—S bonds and has a linear shape. With only nonpolar bonds, CS_2 is a nonpolar molecule.
b. The molecule NF_3 contains three polar covalent N—F bonds and a lone pair on the central N atom. This asymmetric trigonal pyramidal shape makes NF_3 a polar molecule.
c. The molecule Br_2 contains only a nonpolar covalent Br—Br bond; it is a nonpolar molecule.
d. The molecule SO_3 contains three polar covalent S—O bonds and no lone pair on the central S atom. This symmetrical trigonal planar geometry allows the dipoles to cancel, making SO_3 a nonpolar molecule.

10.37 In the molecule CO_2, the two C—O dipoles cancel, resulting in a nonpolar molecule; in CO, there is only one dipole, making it a polar molecule.

10.39 **a.** BrF is a polar molecule. An attraction between the positive end of one polar molecule and the negative end of another polar molecule is called dipole–dipole attraction.
b. An ionic bond is an attraction between a positive and negative ion, as in KCl.
c. CCl_4 is a nonpolar molecule. The weak attractions that occur between temporary dipoles in nonpolar molecules are called dispersion forces.
d. NF_3 is a polar molecule. An attraction between the positive end of one polar molecule and the negative end of another polar molecule is called dipole–dipole attraction.
e. Cl_2 is a nonpolar molecule. The weak attractions that occur between temporary dipoles in nonpolar molecules are dispersion forces.

10.41 **a.** Hydrogen bonds are strong dipole–dipole attractions that occur between a partially positive hydrogen atom of one molecule and one of the strongly electronegative atoms F, O, or N in another.
b. H_2S is a polar molecule. Dipole–dipole attractions occur between dipoles in polar molecules.
c. CO is a polar molecule. Dipole–dipole attractions occur between dipoles in polar molecules.
d. CF_4 is a nonpolar molecule. Dispersion forces occur between temporary dipoles in nonpolar molecules.
e. $CH_3—CH_2—CH_3$ is a nonpolar molecule. Dispersion forces occur between temporary dipoles in nonpolar molecules.

10.43 **a.** $65.0\ \cancel{\text{g ice}} \times \dfrac{334\ \text{J}}{1\ \cancel{\text{g ice}}} = 21\,700\ \text{J}$ (3 SFs); heat is absorbed

b. $17.0\ \cancel{\text{g ice}} \times \dfrac{334\ \text{J}}{1\ \cancel{\text{g ice}}} = 5680\ \text{J}$ (3 SFs); heat is absorbed

c. $225\ \cancel{\text{g water}} \times \dfrac{334\ \cancel{\text{J}}}{1\ \cancel{\text{g water}}} \times \dfrac{1\ \text{kJ}}{1000\ \cancel{\text{J}}} = 75.2\ \text{kJ}$ (3 SFs); heat is released

d. $50.0\ \cancel{\text{g water}} \times \dfrac{334\ \cancel{\text{J}}}{1\ \cancel{\text{g water}}} \times \dfrac{1\ \text{kJ}}{1000\ \cancel{\text{J}}} = 16.7\ \text{kJ}$ (3 SFs); heat is released

10.45 **a.** $10.0\ \cancel{\text{g water}} \times \dfrac{2260\ \text{J}}{1\ \cancel{\text{g water}}} = 22\,600\ \text{J}$ (3 SFs); heat is absorbed

b. $50.0\ \cancel{\text{g water}} \times \dfrac{2260\ \cancel{\text{J}}}{1\ \cancel{\text{g water}}} \times \dfrac{1\ \text{kJ}}{1000\ \cancel{\text{J}}} = 113\ \text{kJ}$ (3 SFs); heat is absorbed

c. $8.00\ \cancel{\text{kg steam}} \times \dfrac{1000\ \cancel{\text{g}}}{1\ \cancel{\text{kg}}} \times \dfrac{2260\ \text{J}}{1\ \cancel{\text{g steam}}} = 1.81 \times 10^7\ \text{J}$ (3 SFs); heat is released

d. $175\ \cancel{\text{g steam}} \times \dfrac{2260\ \cancel{\text{J}}}{1\ \cancel{\text{g steam}}} \times \dfrac{1\ \text{kJ}}{1000\ \cancel{\text{J}}} = 396\ \text{kJ}$ (3 SFs); heat is released

10.47 **a.** water 15 °C → 72 °C: $\Delta T = 72\ °\text{C} - 15\ °\text{C} = 57\ °\text{C}$;

$20.0\ \cancel{\text{g}} \times 57\ \cancel{°\text{C}} \times \dfrac{4.184\ \text{J}}{\cancel{\text{g}}\ \cancel{°\text{C}}}$

$= 4800\ \text{J}$ (2 SFs)

b. Two calculations are needed:

(1) ice 0 °C → water 0 °C: $50.0\ \cancel{\text{g ice}} \times \dfrac{334\ \text{J}}{1\ \cancel{\text{g ice}}} = 16\,700\ \text{J}$

(2) water 0 °C → 65.0 °C: $\Delta T = 65.0\ °\text{C} - 0\ °\text{C} = 65.0\ °\text{C}$;

$50.0\ \cancel{\text{g}} \times 65.0\ \cancel{°\text{C}} \times \dfrac{4.184\ \text{J}}{\cancel{\text{g}}\ \cancel{°\text{C}}} = 13\,600\ \text{J}$

∴ Total heat needed = 16 700 J + 13 600 J = 30 300 J (3 SFs)

c. Two calculations are needed:

(1) steam 100 °C → water 100 °C: $15.0\ \cancel{\text{g steam}} \times \dfrac{2260\ \cancel{\text{J}}}{1\ \cancel{\text{g steam}}} \times \dfrac{1\ \text{kJ}}{1000\ \cancel{\text{J}}} = 33.9\ \text{kJ}$

(2) water 100 °C → 0 °C: $15.0\ \cancel{\text{g water}} \times 100.\ \cancel{°\text{C}} \times \dfrac{4.184\ \cancel{\text{J}}}{\cancel{\text{g}}\ \cancel{°\text{C}}} \times \dfrac{1\ \text{kJ}}{1000\ \cancel{\text{J}}} = 6.28\ \text{kJ}$

∴ Total heat released = 33.9 kJ + 6.28 kJ = 40.2 kJ (3 SFs)

d. Three calculations are needed:

(1) ice 0 °C → water 0 °C: $24.0\ \cancel{\text{g ice}} \times \dfrac{334\ \cancel{\text{J}}}{1\ \cancel{\text{g ice}}} \times \dfrac{1\ \text{kJ}}{1000\ \cancel{\text{J}}} = 8.02\ \text{kJ}$

(2) water 0 °C → 100 °C: $24.0\ \cancel{\text{g water}} \times 100.\ \cancel{°\text{C}} \times \dfrac{4.184\ \cancel{\text{J}}}{\cancel{\text{g}}\ \cancel{°\text{C}}} \times \dfrac{1\ \text{kJ}}{1000\ \cancel{\text{J}}} = 10.0\ \text{kJ}$

(3) water 100 °C → steam 100 °C: $24.0\ \cancel{\text{g water}} \times \dfrac{2260\ \cancel{\text{J}}}{1\ \cancel{\text{g water}}} \times \dfrac{1\ \text{kJ}}{1000\ \cancel{\text{J}}} = 54.2\ \text{kJ}$

∴ Total heat needed = 8.02 kJ + 10.0 kJ + 54.2 kJ = 72.2 kJ (3 SFs)

10.49 Two calculations are needed:

(1) ice 0 °C → water 0 °C: $275\ \cancel{\text{g ice}} \times \dfrac{334\ \cancel{\text{J}}}{1\ \cancel{\text{g ice}}} \times \dfrac{1\ \text{kJ}}{1000\ \cancel{\text{J}}} = 91.9\ \text{kJ}$

(2) water 0 °C → 24.0 °C: $\Delta T = 24.0\ °\text{C} - 0\ °\text{C} = 24.0\ °\text{C}$;

$275\ \cancel{\text{g water}} \times 24.0\ \cancel{°\text{C}} \times \dfrac{4.184\ \cancel{\text{J}}}{\cancel{\text{g}}\ \cancel{°\text{C}}} \times \dfrac{1\ \text{kJ}}{1000\ \cancel{\text{J}}} = 27.6\ \text{kJ}$

∴ Total heat absorbed = 91.9 kJ + 27.6 kJ = 119.5 kJ

10.51 **a.** SiH_4 is a nonpolar molecule. Dispersion forces occur between temporary dipoles in nonpolar molecules.

b. NO_2 is a polar molecule. Dipole–dipole attractions occur between dipoles in polar molecules.

c. Hydrogen bonds are strong dipole–dipole attractions that occur between a partially positive hydrogen atom of one molecule and one of the strongly electronegative atoms F, O, or N in another.

d. Dispersion forces occur between temporary dipoles in Ar atoms.

10.53 In the molecule BCl_3, the central B atom makes three polar covalent bonds with no lone pair on the B. The trigonal planar geometry makes all B—Cl dipoles cancel, and BCl_3 is a nonpolar molecule. In PCl_3, the central P atom also makes three polar covalent bonds to Cl but there is a lone pair on the P atom. This gives an asymmetric trigonal pyramidal shape, meaning the three dipoles do not cancel, and PCl_3 is a polar molecule.

10.55 **a.** The heat from the skin is used to evaporate the water (perspiration). Therefore, the skin is cooled.

b. On a hot day, there are more liquid water molecules in the damp towels that have sufficient energy to become water vapor. Thus, water evaporates from the towels more readily on a hot day.

c. In a closed plastic bag, some water molecules evaporate, but they cannot escape and will condense back to liquid; the clothes will not dry.

10.57

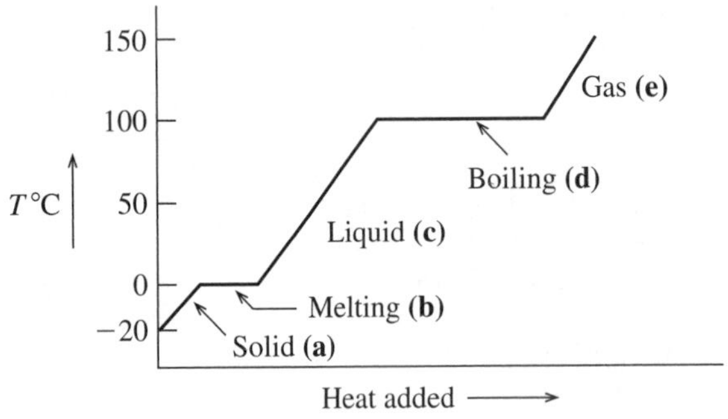

10.59 **a.** The melting point of chloroform is about −60 °C.

b. The boiling point of chloroform is about 60 °C.

c. A represents the solid state. B represents the change from solid to liquid, or melting of the substance. C represents the liquid state as temperature increases. D represents the change from liquid to gas, or boiling of the liquid. E represents the gas state.

d. at −80 °C, solid; at −40 °C, liquid; at 25 °C, liquid; at 80 °C, gas

10.61 To find the total valence electrons for an ion, add the total valence electrons for each atom and add the number of valence electrons indicated by a negative charge. If charge is positive, remove one or more valence electrons.

a. $1\ C(4\ e^-) + 2\ S(6\ e^-) = 4 + 12 = 16$ valence electrons

b. $2\ C(4\ e^-) + 4\ H(1\ e^-) + 1\ O(6\ e^-) = 8 + 4 + 6 = 18$ valence electrons

c. $1\ P(5\ e^-) + 4\ H(1\ e^-) - 1\ e^-$(positive charge) $= 5 + 4 - 1 = 8$ valence electrons

d. $1\ B(3\ e^-) + 3\ Cl(7\ e^-) = 3 + 21 = 24$ valence electrons

e. $1\ S(6\ e^-) + 3\ O(6\ e^-) + 2\ e^-$(negative charge) $= 6 + 18 + 2 = 26$ valence electrons

10.63 **a.** $1\ B(3\ e^-) + 4\ F(7\ e^-) + 1\ e^-$(negative charge) $= 3 + 28 + 1 = 32$ valence electrons

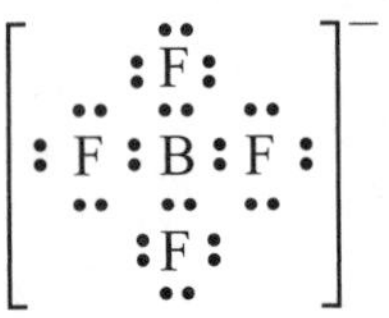

or

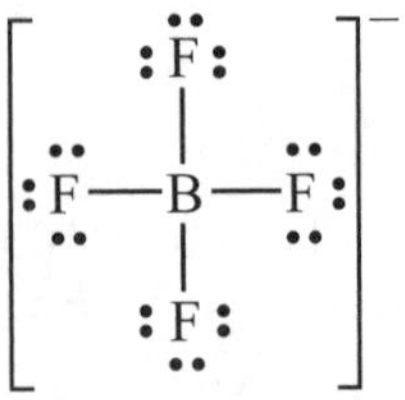

b. 2 Cl(7 e^-) + 1 O(6 e^-) = 14 + 6 = 20 valence electrons

$$:\ddot{C}l:\ddot{O}:\ddot{C}l: \qquad \text{or} \qquad :\ddot{C}l-\ddot{O}-\ddot{C}l:$$

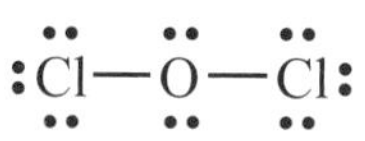

c. 1 N(5 e^-) + 3 H(1 e^-) + 1 O(6 e^-) = 5 + 3 + 6 = 14 valence electrons

$$\begin{array}{c} H \\ H:\!N\!:\!\ddot{O}\!:\!H \end{array} \qquad \text{or} \qquad \begin{array}{c} H \\ | \\ H-N-\ddot{O}-H \end{array}$$

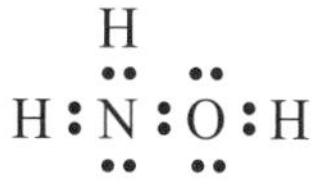

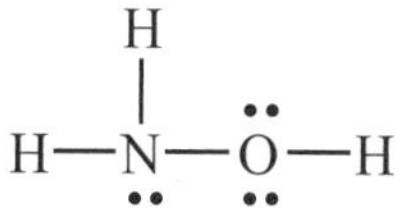

d. 1 N(5 e^-) + 2 O(6 e^-) − 1 e^-(positive charge) = 5 + 12 − 1 = 16 valence electrons

$$\left[:\ddot{O}::N::\ddot{O}:\right]^+ \qquad \text{or} \qquad \left[:\ddot{O}=N=\ddot{O}:\right]^+$$

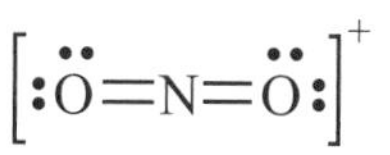

e. 2 C(4 e^-) + 2 H(1 e^-) + 2 Cl(7 e^-) = 8 + 2 + 14 = 24 valence electrons

$$\begin{array}{cc} H & :\ddot{C}l: \\ H:C::C:\ddot{C}l: \end{array} \qquad \text{or} \qquad \begin{array}{cc} H & :\ddot{C}l: \\ | & | \\ H-C=C-\ddot{C}l: \end{array}$$

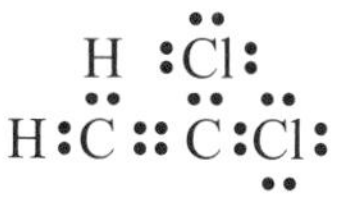

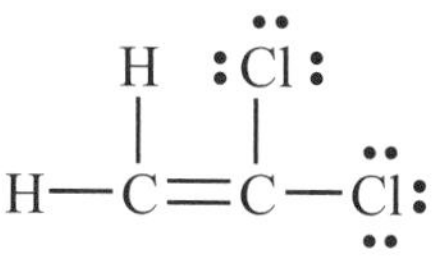

10.65 a. 3 N(5 e^-) + 1 e^-(negative charge) = 15 + 1 = 16 valence electrons

$$\left[:N\equiv N-\ddot{N}:\right]^- \longleftrightarrow \left[:\ddot{N}=N=\ddot{N}:\right]^- \longleftrightarrow \left[:\ddot{N}-N\equiv N:\right]^-$$

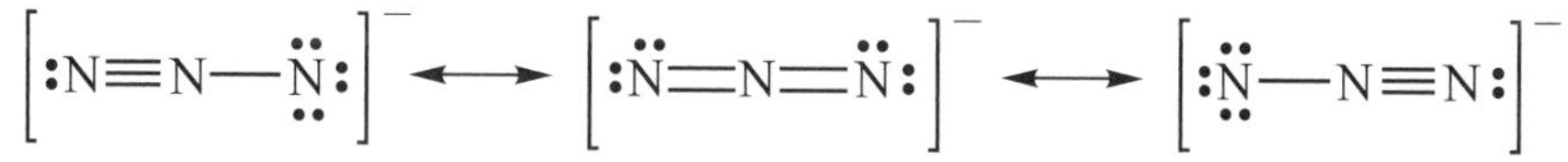

b. 1 N(5 e^-) + 2 O(6 e^-) − 1 e^-(positive charge) = 5 + 12 − 1 = 16 valence electrons

$$\left[:O\equiv N-\ddot{O}:\right]^+ \longleftrightarrow \left[:\ddot{O}=N=\ddot{O}:\right]^+ \longleftrightarrow \left[:\ddot{O}-N\equiv O:\right]^+$$

c. 1 C(4 e^-) + 1 N(5 e^-) + 1 S(6 e^-) + 1 e^-(negative charge)
= 4 + 5 + 6 + 1 = 16 valence electrons

$$\left[:C\equiv N-\ddot{S}:\right]^- \longleftrightarrow \left[:\ddot{C}=N=\ddot{S}:\right]^- \longleftrightarrow \left[:\ddot{C}-N\equiv S:\right]^-$$

10.67 a. Electronegativity is higher at the top of a group: I, Cl, F
b. Electronegativity increases left to right across a period and at the top of a group: K, Li, S, Cl
c. Electronegativity is higher at the top of a group: Ba, Sr, Mg, Be

10.69 Determine the difference in electronegativity values:
a. C—O (3.5 − 2.5 = 1.0) is more polar than C—N (3.5 − 3.0 = 0.5)
b. N—F (4.0 − 3.0 = 1.0) is more polar than N—Br (3.0 − 2.8 = 0.2)
c. S—Cl (3.0 − 2.5 = 0.5) is more polar than Br—Cl (3.0 − 2.8 = 0.2)
d. Br—I (2.8 − 2.5 = 0.3) is more polar than Br—Cl (3.0 − 2.8 = 0.2)
e. N—S (3.0 − 2.5 = 0.5) has the same degree of polarity as N—O (3.5 − 3.0 = 0.5)

10.71 A dipole arrow points from the atom with the lower electronegativity value (more positive) to the atom in the bond that has the higher electronegativity value (more negative).
a. Si has the lower electronegativity value of 1.8, making Si the positive end of the dipole. The Cl atom has a higher electronegativity value of 3.0.

Si—Cl
⟵

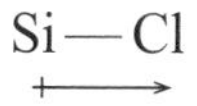

b. C has the lower electronegativity value of 2.5, making C the positive end of the dipole. The N atom has a higher electronegativity value of 3.0.

C—N
⟵

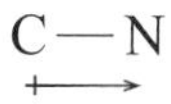

c. Cl has the lower electronegativity value of 3.0, making Cl the positive end of the dipole. The F atom has a higher electronegativity value of 4.0.

 F—Cl
 $\longleftarrow\!\!+$

 d. C has the lower electronegativity value of 2.5, making C the positive end of the dipole. The F atom has a higher electronegativity value of 4.0.

 C—F
 $+\!\!\longrightarrow$

 e. N has the lower electronegativity value of 3.0, making N the positive end of the dipole. The O atom has a higher electronegativity value of 3.5.

 N—O
 $+\!\!\longrightarrow$

10.73 a. Si—Cl electronegativity difference $3.0 - 1.8 = 1.2$, polar covalent
 b. C—C electronegativity difference $2.5 - 2.5 = 0.0$, nonpolar covalent
 c. Na—Cl electronegativity difference $3.0 - 0.9 = 2.1$, ionic
 d. C—H electronegativity difference $2.5 - 2.1 = 0.4$, nonpolar covalent
 e. F—F electronegativity difference $4.0 - 4.0 = 0.0$, nonpolar covalent

10.75 a. 1 N($5\ e^-$) + 3 F($7\ e^-$) = $5 + 21 = 26$ valence electrons

 :F̈—N̈—F̈: four electron groups around N (3 bonded atoms and 1 lone pair);
 | trigonal pyramidal shape
 :F̈:

 b. 1 Si($4\ e^-$) + 4 Br($7\ e^-$) = $4 + 28 = 32$ valence electrons

 :B̈r:
 |
 :B̈r—Si—B̈r: four electron groups around Si bonded to four atoms; tetrahedral shape
 |
 :B̈r:

 c. 1 Be($2\ e^-$) + 2 Cl($7\ e^-$) = $2 + 14 = 16$ valence electrons

 :C̈l—Be—C̈l: two electron groups around Be bonded to two atoms; linear shape

 d. 1 S($6\ e^-$) + 2 O($6\ e^-$) = $6 + 12 = 18$ valence electrons

 :Ö=S̈—Ö: \longleftrightarrow :Ö—S̈=Ö: three electron groups around S

 (2 bonded atoms and 1 lone pair); bent shape ($<120°$)

10.77 a. 1 Br($7\ e^-$) + 2 O($6\ e^-$) + 1 e^-(negative charge) = $7 + 12 + 1 = 20$ valence electrons

 $\left[\ :\ddot{O}—\ddot{Br}—\ddot{O}:\ \right]^-$ four electron groups around Br (2 bonded atoms and 2 lone pairs);
 bent shape ($<109.5°$)

 b. 1 O($6\ e^-$) + 2 H($1\ e^-$) = $6 + 2 = 8$ valence electrons
 ($<109.5°$)
 H
 |
 :Ö—H four electron groups around O (2 bonded atoms and 2 lone pairs); bent shape

 c. 1 C($4\ e^-$) + 3 O($6\ e^-$) + 2 e^-(negative charge) = $4 + 18 + 2 = 24$ valence electrons

 $\left[\ \begin{matrix} :\ddot{O}: \\ | \\ :\ddot{O}=C—\ddot{O}: \end{matrix}\ \right]^{2-}$ three electron groups around C bonded to three atoms; trigonal planar shape

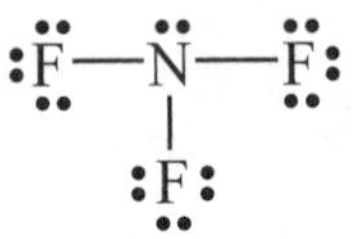

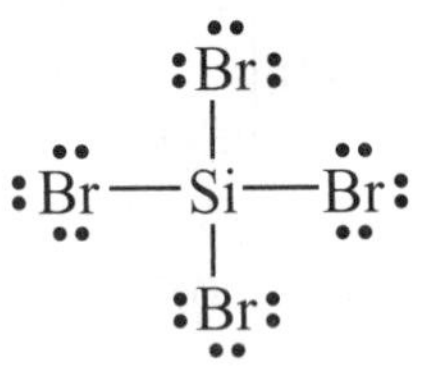

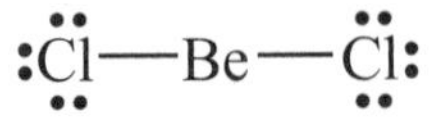

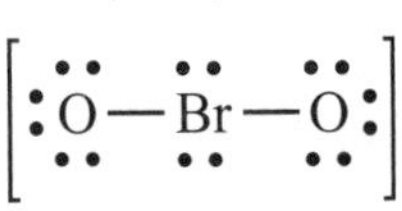

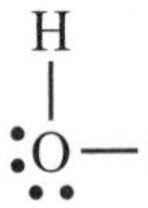

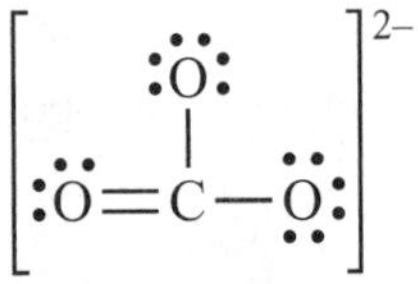

d. 1 C(4 e^-) + 4 F(7 e^-) = 4 + 28 = 32 valence electrons

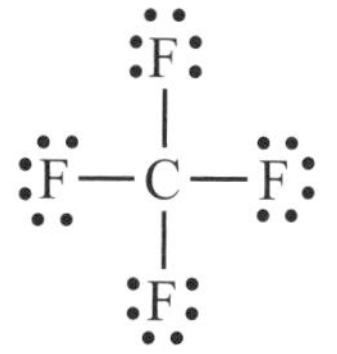
four electron groups around C bonded to four atoms; tetrahedral shape

e. 1 C(4 e^-) + 2 S(6 e^-) = 4 + 12 = 16 valence electrons

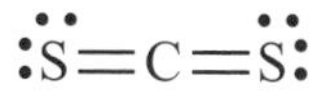
two electron groups around C bonded to two atoms; linear shape

f. 1 P(5 e^-) + 3 O(6 e^-) + 3 e^-(negative charge) = 5 + 18 + 3 = 26 valence electrons

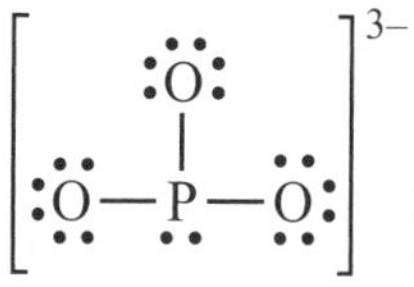
four electron groups around P (3 bonded atoms and one lone pair); trigonal pyramidal shape

10.79 a. The molecule HBr contains only a single polar covalent H—Br bond; it is a polar molecule.

b. The molecule SiO_2 contains two polar covalent Si—O bonds and has a linear shape. The two equal dipoles directed away from each other at 180° will cancel, resulting in a nonpolar molecule.

c. NCl_3 is a nonpolar molecule. There are three nonpolar N—Cl bonds (no dipoles) in a trigonal pyramidal shape.

d. The molecule CH_3Cl has a tetrahedral shape but the atoms bonded to the central atom, C, are not identical. The three C—H bonds are nonpolar, but the C—Cl bond is polar. The nonpolar C—H bonds do not cancel the dipole from the polar C—Cl bond, making it a polar molecule.

e. NI_3 is a polar molecule. The dipoles from the three polar covalent N—I bonds do not cancel because of the asymmetric trigonal pyramidal shape of the molecule.

f. The molecule H_2O contains two polar covalent O—H bonds and two lone pairs on the central O atom, resulting in a bent shape. The dipoles do not cancel, making H_2O a polar molecule.

10.81 a. A molecule that has a central atom with three bonded atoms and no lone pairs will have a trigonal planar shape. Since the bonded atoms are identical, the symmetrical shape allows the dipoles to cancel, making the molecule nonpolar.

b. A molecule that has a central atom with two bonded atoms and one lone pair will have a bent shape. This asymmetric shape means the dipoles do not cancel, and the molecule will be polar.

c. A molecule that has a central atom with two bonded atoms and no lone pairs will have a linear shape. Since the bonded atoms are identical, the symmetrical shape allows the dipoles to cancel, making the molecule nonpolar.

10.83 a. Hydrogen bonding (3) involves strong dipole–dipole attractions that occur between a partially positive hydrogen atom of one molecule and one of the strongly electronegative atoms F, O, or N in another.

b. Dispersion forces (4) occur between temporary dipoles in Kr atoms.

c. CH_4 is a nonpolar molecule. Dispersion forces (4) occur between temporary dipoles in nonpolar molecules.

d. $CHCl_3$ is a polar molecule. Dipole–dipole attractions (2) occur between dipoles in polar molecules.

e. Hydrogen bonding (3) involves strong dipole–dipole attractions that occur between a partially positive hydrogen atom of one molecule and one of the strongly electronegative atoms F, O, or N in another.

f. Ionic bonds (1) are strong attractions between positive and negative ions, as in LiCl.

10.85 When water vapor condenses or liquid water freezes, heat is released, which warms the air.

10.87 Two calculations are required:

(1) water 25 °C → 0 °C: $325\ \cancel{g} \times 25\ \cancel{°C} \times \dfrac{4.184\ \cancel{J}}{\cancel{g}\ \cancel{°C}} \times \dfrac{1\ kJ}{1000\ \cancel{J}} = 34\ kJ$

(2) water 0 °C → ice 0 °C: $325\ \cancel{g} \times \dfrac{334\ \cancel{J}}{1\ \cancel{g}} \times \dfrac{1\ kJ}{1000\ \cancel{J}} = 109\ kJ$

∴ Total heat removed = 34 kJ + 109 kJ = 143 kJ removed

10.89 **a.**

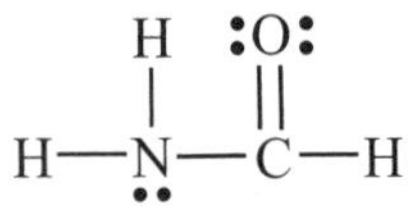

b.

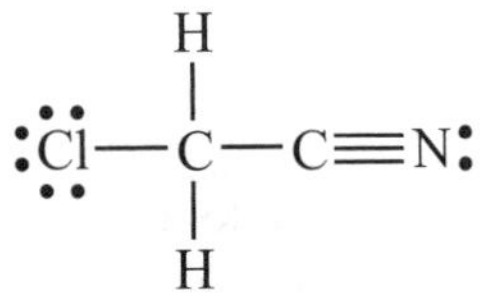

c.

d.

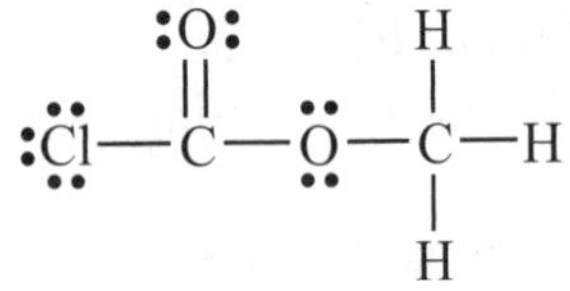

10.91 **a.** The central atom N has four electron groups with three bonded atoms and a lone pair, which gives NH_2Cl a trigonal pyramidal shape.

b. The central atom P has four electron pairs bonded to four H atoms; PH_4^+ has a tetrahedral shape.

c. The central atom C has two electron groups bonded to two atoms and no lone pairs, which gives SCN^- a linear shape.

d. The central atom S has three electron groups bonded to three O atoms; SO_3 has a trigonal planar shape.

10.93 Two calculations are required:

(1) lead 300. °C → 0 °C:

Heat = mass × $\Delta T \times SH$ = $3.0\ \cancel{kg} \times \dfrac{1000\ \cancel{g}}{1\ \cancel{kg}} \times 300.\ \cancel{°C} \times \dfrac{0.13\ J}{\cancel{g}\ \cancel{°C}} = 1.2 \times 10^5\ J$

(2) ice 0 °C → water 0 °C:

Heat = mass × heat of fusion

$\therefore \text{Mass}_{\text{ice}} = \dfrac{\text{heat}}{\text{heat of fusion}} = \dfrac{1.2 \times 10^5\ \cancel{J}}{\left(\dfrac{334\ \cancel{J}}{1\ g}\right)} = 360\ g$ of ice will be melted (2 SFs)

10.95 Two calculations are required:

(1) ice 0 °C → water 0 °C:

Heat = mass × heat of fusion = $45.0\ \cancel{g} \times \dfrac{334\ J}{\cancel{g}} = 15\,000\ J$

(2) water 8.0 °C → 0.0 °C:

Heat = mass × $\Delta T \times SH$

$\therefore \text{Mass}_{\text{water}} = \dfrac{\text{heat}}{\Delta T \times SH} = \dfrac{15\,000\ \cancel{J}}{(8.0\ \cancel{°C})\left(\dfrac{4.184\ \cancel{J}}{1\ g\ \cancel{°C}}\right)} = 450\ g$ of water (2 SFs)

Answers to Combining Ideas from Chapters 8 to 10

CI.13 a. $8.56\ \cancel{g\ Cu} \times \frac{1\ cm^3}{8.94\ \cancel{g\ Cu}} = 0.957\ cm^3$ (3 SFs)

b. $8.56\ \cancel{g\ Cu} \times \frac{1\ \cancel{mol\ Cu}}{63.55\ \cancel{g\ Cu}} \times \frac{6.022 \times 10^{23}\ Cu\ atoms}{1\ \cancel{mol\ Cu}} = 8.11 \times 10^{22}$ atoms of Cu (3 SFs)

c. $2Cu(s) + O_2(g) \rightarrow 2CuO(s)$

d. The reaction of the elements copper and oxygen to form copper(II) oxide is a combination reaction.

e. Use the following as the correct answer for CI.13e text answer.

$8.56\ \cancel{g\ Cu} \times \frac{1\ \cancel{mol\ Cu}}{63.55\ \cancel{g\ Cu}} \times \frac{1\ \cancel{mol\ O_2}}{2\ \cancel{mol\ Cu}} \times \frac{32.00\ g\ O_2}{1\ \cancel{mol\ O_2}} = 2.16$ g of O_2 (3 SFs)

f. From part **e**, we know that 8.56 g of Cu reacts with 2.16 g of O_2
∴ 3.72 g of O_2 is excess O_2; Cu is the limiting reactant.

Molar mass of CuO = 1 mol of Cu + 1 mol of O = 63.55 g + 16.00 g = 79.55 g

$8.56\ \cancel{g\ Cu} \times \frac{1\ \cancel{mol\ Cu}}{63.55\ \cancel{g\ Cu}} \times \frac{2\ \cancel{mol\ CuO}}{2\ \cancel{mol\ Cu}} \times \frac{79.55\ g\ CuO}{1\ \cancel{mol\ CuO}} = 10.7$ g of CuO (3 SFs)

g. Theoretical yield of CuO (from part f) = 10.7 g of CuO
Use the percent yield to convert theoretical to actual:

$10.7\ \cancel{g\ CuO} \times \frac{85.0\ g\ CuO}{100\ \cancel{g\ CuO}} = 9.10$ g of CuO (actual) (3 SFs)

CI.15 a. The formula of sodium hypochlorite is NaClO.

Molar mass of NaClO
= 1 mol of Na + 1 mol of Cl + 1 mol of O = 22.99 g + 35.45 g + 16.00 g = 74.44 g

b. $\left[:\ddot{\underset{\cdot\cdot}{Cl}}-\ddot{\underset{\cdot\cdot}{O}}:\right]^-$

c. Given gallons of bleach **Need** number of ClO^- ions
Plan gal → qt → mL → g of solution → g of NaClO → mol of NaClO → mol of ClO^-
→ number of ClO^- ions

Set Up $1.00\ \cancel{gal} \times \frac{4\ \cancel{qt}}{1\ \cancel{gal}} \times \frac{946.3\ \cancel{mL}}{1\ \cancel{qt}} \times \frac{1.08\ \cancel{g\ solution}}{1\ \cancel{mL\ solution}} \times \frac{5.25\ \cancel{g\ NaClO}}{100\ \cancel{g\ solution}}$
$\times \frac{1\ \cancel{mol\ NaClO}}{74.44\ \cancel{g\ NaClO}} \times \frac{1\ \cancel{mol\ ClO^-}}{1\ \cancel{mol\ NaClO}} \times \frac{6.022 \times 10^{23}\ ClO^-\ ions}{1\ \cancel{mol\ ClO^-}}$
$= 1.74 \times 10^{24}\ ClO^-$ ions (3 SFs)

d. $2NaOH(aq) + Cl_2(g) \rightarrow NaClO(aq) + NaCl(aq) + H_2O(l)$

e. Given gallons of bleach **Need** g of NaOH
Plan gal → qt → mL → g of solution → g of NaClO → mol of NaClO → mol of NaOH
→ g of NaOH

Set Up $1.00\ \cancel{gal} \times \frac{4\ \cancel{qt}}{1\ \cancel{gal}} \times \frac{946.3\ \cancel{mL}}{1\ \cancel{qt}} \times \frac{1.08\ \cancel{g\ solution}}{1\ \cancel{mL\ solution}} \times \frac{5.25\ \cancel{g\ NaClO}}{100\ \cancel{g\ solution}}$
$\times \frac{1\ \cancel{mol\ NaClO}}{74.44\ \cancel{g\ NaClO}} \times \frac{2\ \cancel{mol\ NaOH}}{1\ \cancel{mol\ NaClO}} \times \frac{40.00\ g\ NaOH}{1\ \cancel{mol\ NaOH}}$
= 231 g of NaOH (3 SFs)

f. $165\ \cancel{g\ Cl_2} \times \frac{1\ \cancel{mol\ Cl_2}}{70.90\ \cancel{g\ Cl_2}} \times \frac{1\ mol\ NaClO}{1\ \cancel{mol\ Cl_2}} = 2.33$ mol of NaClO (smaller number of moles)

$275\ \cancel{g\ NaOH} \times \frac{1\ \cancel{mol\ NaOH}}{40.00\ \cancel{g\ NaOH}} \times \frac{1\ mol\ NaClO}{2\ \cancel{mol\ NaOH}} = 3.44$ mol of NaClO

∴ Chlorine is the limiting reactant and

$$2.33 \text{ mol NaClO} \times \frac{74.44 \text{ g NaClO}}{1 \text{ mol NaClO}} = 173 \text{ g of NaClO produced (theoretical)}$$

$$\therefore \text{ Percent yield} = \frac{162 \text{ g NaClO (actual)}}{173 \text{ g NaClO (theoretical)}} \times 100\% = 93.6\% \text{ yield (3 SFs)}$$

CI.17 a. chloral hydrate

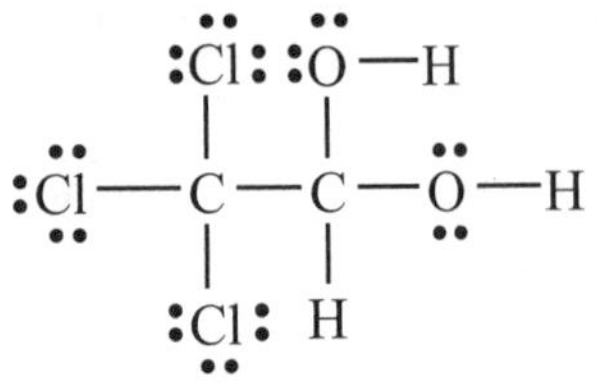

chloral

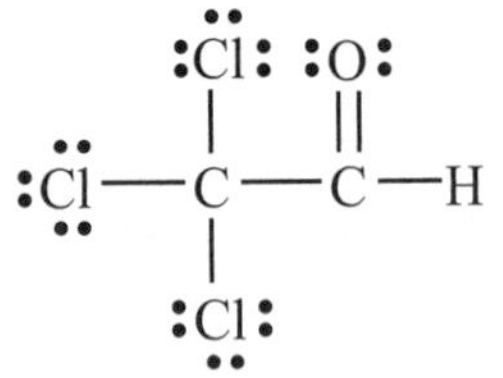

b. Chloral hydrate has two hydroxyl groups, and chloral has an aldehyde (carbonyl) group.

c. Chloral hydrate, $C_2H_3O_2Cl_3 \rightarrow C_{(2\div1)}H_{(3\div1)}O_{(2\div1)}Cl_{(3\div1)} = C_2H_3O_2Cl_3$ (empirical formula)

Chloral, $C_2HOCl_3 \rightarrow C_{(2\div1)}H_{(1\div1)}O_{(1\div1)}Cl_{(3\div1)} = C_2HOCl_3$ (empirical formula)

d.
$$2 \text{ mol C} \times \frac{12.01 \text{ g C}}{1 \text{ mol C}} = 24.02 \text{ g of C}$$

$$3 \text{ mol H} \times \frac{1.008 \text{ g H}}{1 \text{ mol H}} = 3.024 \text{ g of H}$$

$$2 \text{ mol O} \times \frac{16.00 \text{ g O}}{1 \text{ mol O}} = 32.00 \text{ g of O}$$

$$3 \text{ mol Cl} \times \frac{35.45 \text{ g Cl}}{1 \text{ mol Cl}} = 106.4 \text{ g of Cl}$$

$$\therefore \text{ Molar mass of } C_2H_3O_2Cl_3 = 165.4 \text{ g}$$

$$\text{Mass \% Cl} = \frac{106.4 \text{ g Cl}}{165.4 \text{ g } C_2H_3O_2Cl_3} \times 100\% = 64.33\% \text{ Cl (4 SFs)}$$

CI.19 a.

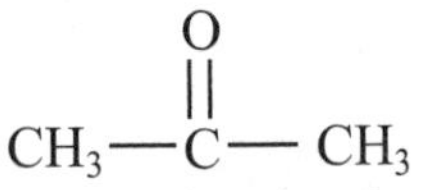

b. The molecular formula of acetone is C_3H_6O.

Molar mass of C_3H_6O

$= 3$ mol of C $+ 6$ mol of H $+ 1$ mol of O $= 3(12.01 \text{ g}) + 6(1.008 \text{ g}) + 16.00 \text{ g} = 58.08 \text{ g}$

c. $C_3H_6O(l) + 4O_2(g) \xrightarrow{\Delta} 3CO_2(g) + 3H_2O(g) + 1790 \text{ kJ}$

d. Combustion is an exothermic reaction.

e.
$$2.58 \text{ g } C_3H_6O \times \frac{1 \text{ mol } C_3H_6O}{58.08 \text{ g } C_3H_6O} \times \frac{1790 \text{ kJ}}{1 \text{ mol } C_3H_6O} = 79.5 \text{ kJ (3 SFs)}$$

f.
$$15.0 \text{ mL } C_3H_6O \times \frac{0.786 \text{ g } C_3H_6O}{1 \text{ mL } C_3H_6O} \times \frac{1 \text{ mol } C_3H_6O}{58.08 \text{ g } C_3H_6O} \times \frac{4 \text{ mol } O_2}{1 \text{ mol } C_3H_6O} \times \frac{32.00 \text{ g } O_2}{1 \text{ mol } O_2}$$

$= 26.0$ g of O_2 (3 SFs)

11
Gases

Study Goals

- Use the kinetic molecular theory of gases to describe the properties of gases.
- Describe the units of measurement used for pressure, and change from one unit to another.
- Use the pressure–volume relationship (Boyle's law) to determine the new pressure or volume of a certain amount of gas at a constant temperature.
- Use the temperature–volume relationship (Charles's law) to determine the new temperature or volume of a certain amount of gas at a constant pressure.
- Use the temperature–pressure relationship (Gay-Lussac's law) to determine the new temperature or pressure of a certain amount of gas at a constant volume.
- Use the combined gas law to find the new pressure, volume, or temperature of a gas when changes in two of these properties are given.
- Use Avogadro's law to describe the relationship between the amount of a gas and its volume, and use this relationship in calculations.
- Use the ideal gas law to solve for P, V, T, or n of a gas when given three of the four values in the ideal gas law.
- Calculate density, molar mass, or volume of a gas in a chemical reaction.
- Determine the mass or volume of a gas that reacts or forms in a chemical reaction.
- Use partial pressures to calculate the total pressure of a mixture of gases.

Chapter Outline

11.10 Partial Pressures (Dalton's Law)
Chemistry and Health: Blood Gases
Chemistry and Health: Hyperbaric Chambers
Answers

Answers and Solutions to Text Problems

11.1 **a.** At a higher temperature, gas particles have greater kinetic energy, which makes them move faster.
b. Because there are great distances between the particles of a gas, they can be pushed closer together and still remain a gas.
c. Gas particles are very far apart, which means that the mass of a gas in a certain volume is very small, resulting in a low density.

11.3 **a.** The temperature of a gas can be expressed in kelvins.
b. The space occupied by a gas is its volume.
c. The amount of a gas can be expressed in grams.
d. Pressure results from the force of gas particles striking the walls of the container.

11.5 Some units used to describe the pressure of a gas are atmospheres (abbreviated atm), mmHg, torr, pounds per square inch (lb/in.^{2} or psi), pascals, kilopascals, and in. Hg.

11.7 **a.** $2.00\ \cancel{\text{atm}} \times \frac{760\ \text{torr}}{1\ \cancel{\text{atm}}} = 1520\ \text{torr (3 SFs)}$

b. $2.00\ \cancel{\text{atm}} \times \frac{14.7\ \text{lb/in.}^2}{1\ \cancel{\text{atm}}} = 29.4\ \text{lb/in.}^2\ \text{(3 SFs)}$

c. $2.00\ \cancel{\text{atm}} \times \frac{760\ \text{mmHg}}{1\ \cancel{\text{atm}}} = 1520\ \text{mmHg (3 SFs)}$

d. $2.00\ \cancel{\text{atm}} \times \frac{101.325\ \text{kPa}}{1\ \cancel{\text{atm}}} = 203\ \text{kPa (3 SFs)}$

11.9 As a diver ascends to the surface, external pressure decreases. If the air in the lungs were not exhaled, its volume would expand and severely damage the lungs. The pressure in the lungs must adjust to changes in the external pressure.

11.11 **a.** Inspiration begins when the diaphragm flattens, causing the lungs to expand. The increased volume of the thoracic cavity reduces the pressure in the lungs such that air flows into the lungs.
b. Expiration occurs as the diaphragm relaxes causing a decrease in the volume of the lungs. The pressure of the air in the lungs increases and air flows out of the lungs.
c. Inspiration occurs when the pressure in the lungs is less than the pressure of the air in the atmosphere.

11.13 **a.** The pressure is greater in cylinder A. According to Boyle's law, a decrease in volume pushes the gas particles closer together, which will cause an increase in the pressure.

b.

Property	Conditions 1	Conditions 2	Know	Predict
Pressure (P)	$P_1 = 650$ mmHg	$P_2 = 1.2$ atm	P increases	
Volume (V)	$V_1 = 220$ mL	$V_2 = ?$		V decreases

According to Boyle's law, $P_1V_1 = P_2V_2$, then

$$V_2 = V_1 \times \frac{P_1}{P_2} = 220\ \text{mL} \times \frac{650\ \cancel{\text{mmHg}}}{1.2\ \cancel{\text{atm}}} \times \frac{1\ \cancel{\text{atm}}}{760\ \cancel{\text{mmHg}}} = 160\ \text{mL (2 SFs)}$$

11.15 a. The pressure of the gas doubles when the volume is halved.

b. The pressure falls to one-third the initial pressure when the volume expands to three times its initial value.

c. The pressure increases to ten times the original pressure when the volume decreases to one-tenth of the initial volume.

11.17 From Boyle's law, we know that pressure is inversely related to volume (e.g. pressure increases when volume decreases).

a. Volume increases; pressure must decrease.

$$P_2 = P_1 \times \frac{V_1}{V_2} = 655 \text{ mmHg} \times \frac{10.0 \text{ L}}{20.0 \text{ L}} = 328 \text{ mmHg (3 SFs)}$$

b. Volume decreases; pressure must increase.

$$P_2 = P_1 \times \frac{V_1}{V_2} = 655 \text{ mmHg} \times \frac{10.0 \text{ L}}{2.50 \text{ L}} = 2620 \text{ mmHg (3 SFs)}$$

c. The mL units must be converted to L for unit cancellation in the calculation, and because the volume increases, pressure must decrease.

$$P_2 = P_1 \times \frac{V_1}{V_2} = 655 \text{ mmHg} \times \frac{10.0 \text{ L}}{13\,800 \text{ mL}} \times \frac{1000 \text{ mL}}{1 \text{ L}} = 475 \text{ mmHg (3 SFs)}$$

d. The mL units must be converted to L for unit cancellation in the calculation, and because the volume decreases, pressure must increase.

$$P_2 = P_1 \times \frac{V_1}{V_2} = 655 \text{ mmHg} \times \frac{10.0 \text{ L}}{1250 \text{ mL}} \times \frac{1000 \text{ mL}}{1 \text{ L}} = 5240 \text{ mmHg (3 SFs)}$$

11.19 Pressure decreases; volume must increase.

$$V_2 = V_1 \times \frac{P_1}{P_2} = 5.0 \text{ L} \times \frac{5.0 \text{ atm}}{1.0 \text{ atm}} = 25 \text{ L of cyclopropane (2 SFs)}$$

11.21 From Boyle's law, we know that pressure is inversely related to volume.

a. Pressure increases; volume must decrease.

$$V_2 = V_1 \times \frac{P_1}{P_2} = 50.0 \text{ L} \times \frac{760. \text{ mmHg}}{1500 \text{ mmHg}} = 25 \text{ L (2 SFs)}$$

b. The mmHg units must be converted to atm for unit cancellation in the calculation, and because the pressure increases, volume must decrease.

$$P_1 = 760. \text{ mmHg} \times \frac{1 \text{ atm}}{760 \text{ mmHg}} = 1.00 \text{ atm}$$

$$V_2 = V_1 \times \frac{P_1}{P_2} = 50.0 \text{ L} \times \frac{1.00 \text{ atm}}{2.0 \text{ atm}} = 25 \text{ L (2 SFs)}$$

c. The mmHg units must be converted to atm for unit cancellation in the calculation, and because the pressure decreases, volume must increase.

$$P_1 = 760. \text{ mmHg} \times \frac{1 \text{ atm}}{760 \text{ mmHg}} = 1.00 \text{ atm}$$

$$V_2 = V_1 \times \frac{P_1}{P_2} = 50.0 \text{ L} \times \frac{1.00 \text{ atm}}{0.500 \text{ atm}} = 100. \text{ L (3 SFs)}$$

d. The mmHg units must be converted to torr for unit cancellation in the calculation, and because the pressure increases, volume must decrease.

$$P_1 = 760. \text{ mmHg} \times \frac{760 \text{ torr}}{760 \text{ mmHg}} = 760. \text{ torr}$$

$$V_2 = V_1 \times \frac{P_1}{P_2} = 50.0 \text{ L} \times \frac{760. \text{ torr}}{850 \text{ torr}} = 45 \text{ L (2 SFs)}$$

11.23 According to Charles's law, there is a direct relationship between Kelvin temperature and volume (e.g. volume increases when temperature increases, if the pressure and amount of gas remain constant).

a. Diagram C shows an increased volume corresponding to an increase in temperature.

b. Diagram A shows a decreased volume corresponding to a decrease in temperature.

c. Diagram B shows no change in volume, which corresponds to no net change in temperature.

11.25 According to Charles's law, a change in the volume of a gas is directly proportional to the change in its Kelvin temperature. In all gas law computations, temperatures must be in kelvins. (Temperatures in °C are converted to K by the addition of 273.) The initial temperature for all cases here is $T_1 = 15\ °\text{C} + 273 = 288\ \text{K}$.

a. Volume increases; temperature must have increased.

$$T_2 = T_1 \times \frac{V_2}{V_1} = 288\ \text{K} \times \frac{5.00\ \cancel{\text{L}}}{2.50\ \cancel{\text{L}}} = 576\ \text{K}$$

$576\ \text{K} - 273 = 303\ °\text{C}$ (3 SFs)

b. Volume decreases; temperature must have decreased.

$$T_2 = T_1 \times \frac{V_2}{V_1} = 288\ \text{K} \times \frac{1250\ \cancel{\text{mL}}}{2.50\ \cancel{\text{L}}} \times \frac{1\ \cancel{\text{L}}}{1000\ \cancel{\text{mL}}} = 144\ \text{K}$$

$144\ \text{K} - 273 = -129\ °\text{C}$ (3 SFs)

c. Volume increases; temperature must have increased.

$$T_2 = T_1 \times \frac{V_2}{V_1} = 288\ \text{K} \times \frac{7.50\ \cancel{\text{L}}}{2.50\ \cancel{\text{L}}} = 864\ \text{K}$$

$864\ \text{K} - 273 = 591\ °\text{C}$ (3 SFs)

d. Volume increases; temperature must have increased.

$$T_2 = T_1 \times \frac{V_2}{V_1} = 288\ \text{K} \times \frac{3550\ \cancel{\text{mL}}}{2.50\ \cancel{\text{L}}} \times \frac{1\ \cancel{\text{L}}}{1000\ \cancel{\text{mL}}} = 409\ \text{K}$$

$409\ \text{K} - 273 = 136\ °\text{C}$ (3 SFs)

11.27 According to Charles's law, a change in the volume of a gas is directly proportional to the change in its Kelvin temperature. In all gas law computations, temperatures must be in kelvins. (Temperatures in °C are converted to K by the addition of 273.) The initial temperature for all cases here is $T_1 = 75\ °\text{C} + 273 = 348\ \text{K}$.

a. When temperature decreases, volume must also decrease.

$T_2 = 55\ °\text{C} + 273 = 328\ \text{K}$

$$V_2 = V_1 \times \frac{T_2}{T_1} = 2500\ \text{mL} \times \frac{328\ \cancel{\text{K}}}{348\ \cancel{\text{K}}} = 2400\ \text{mL (2 SFs)}$$

b. When temperature increases, volume must also increase.

$$V_2 = V_1 \times \frac{T_2}{T_1} = 2500\ \text{mL} \times \frac{680.\ \cancel{\text{K}}}{348\ \cancel{\text{K}}} = 4900\ \text{mL (2 SFs)}$$

c. When temperature decreases, volume must also decrease.

$T_2 = -25\ °\text{C} + 273 = 248\ \text{K}$

$$V_2 = V_1 \times \frac{T_2}{T_1} = 2500\ \text{mL} \times \frac{248\ \cancel{\text{K}}}{348\ \cancel{\text{K}}} = 1800\ \text{mL (2 SFs)}$$

d. When temperature decreases, volume must also decrease.

$$V_2 = V_1 \times \frac{T_2}{T_1} = 2500\ \text{mL} \times \frac{240.\ \cancel{\text{K}}}{348\ \cancel{\text{K}}} = 1700\ \text{mL (2 SFs)}$$

11.29 According to Gay-Lussac's law, temperature is directly related to pressure. For example, temperature increases when the pressure increases. In all gas law computations, temperatures must be in kelvins. (Temperatures in °C are converted to K by the addition of 273.)

a. $T_1 = 155\ °C + 273 = 428\ K \qquad T_2 = 0\ °C + 273 = 273\ K$

$$P_2 = P_1 \times \frac{T_2}{T_1} = 1200\ \cancel{torr} \times \frac{273\ \cancel{K}}{428\ \cancel{K}} \times \frac{760\ mmHg}{760\ \cancel{torr}} = 770\ mmHg\ (2\ SFs)$$

b. $T_1 = 12\ °C + 273 = 285\ K \qquad T_2 = 35\ °C + 273 = 308\ K$

$$P_2 = P_1 \times \frac{T_2}{T_1} = 1.40\ \cancel{atm} \times \frac{308\ \cancel{K}}{285\ \cancel{K}} \times \frac{760\ mmHg}{1\ \cancel{atm}} = 1150\ mmHg\ (3\ SFs)$$

11.31 According to Gay-Lussac's law, temperature is directly related to pressure. For example, temperature increases when the pressure increases. In all gas law computations, temperatures must be in kelvins. (Temperatures in °C are converted to K by the addition of 273.)

a. $T_1 = 25\ °C + 273 = 298\ K$

$$T_2 = T_1 \times \frac{P_2}{P_1} = 298\ K \times \frac{620\ \cancel{mmHg}}{740\ \cancel{mmHg}} = 250.\ K \qquad 250.\ K - 273 = -23\ °C\ (2\ SFs)$$

b. $T_1 = -18\ °C + 273 = 255\ K$

$$T_2 = T_1 \times \frac{P_2}{P_1} = 255\ K \times \frac{1250\ \cancel{torr}}{0.950\ \cancel{atm}} \times \frac{1\ \cancel{atm}}{760\ \cancel{torr}} = 441\ K \qquad 441\ K - 273 = 168\ °C\ (3\ SFs)$$

11.33 **a.** The boiling point is the temperature at which bubbles of vapor appear within the liquid.

b. Vapor pressure is the pressure exerted by a gas above the surface of its liquid.

c. Atmospheric pressure is the pressure exerted on Earth by the particles in the air.

d. The boiling point is the temperature at which the vapor pressure of a liquid becomes equal to the external pressure.

11.35 **a.** On the top of a mountain, water boils below 100 °C because the atmospheric (external) pressure is less than 1 atm. The boiling point is the temperature at which the vapor pressure of a liquid becomes equal to the external (in this case, atmospheric) pressure.

b. Because the pressure inside a pressure cooker is greater than 1 atm, water boils above 100 °C. The higher temperature of the boiling water allows food to cook more quickly.

11.37 $\frac{P_1V_1}{T_1} = \frac{P_2V_2}{T_2}$ Boyle's, Charles's, and Gay-Lussac's laws are combined to make this law.

11.39 $T_1 = 25\ °C + 273 = 298\ K;\quad V_1 = 6.50\ L;\quad P_1 = 845\ mmHg$

a. $T_2 = 325\ K;\quad V_2 = 1850\ mL = 1.85\ L;\quad P_2 = ?$

$$P_2 = P_1 \times \frac{V_1}{V_2} \times \frac{T_2}{T_1} = 845\ \cancel{mmHg} \times \frac{6.50\ \cancel{L}}{1.85\ \cancel{L}} \times \frac{325\ \cancel{K}}{298\ \cancel{K}} \times \frac{1\ atm}{760\ \cancel{mmHg}} = 4.26\ atm\ (3\ SFs)$$

b. $T_2 = 12\ °C + 273 = 285\ K;\quad V_2 = 2.25\ L;\quad P_2 = ?$

$$P_2 = P_1 \times \frac{V_1}{V_2} \times \frac{T_2}{T_1} = 845\ \cancel{mmHg} \times \frac{6.50\ \cancel{L}}{2.25\ \cancel{L}} \times \frac{285\ \cancel{K}}{298\ \cancel{K}} \times \frac{1\ atm}{760\ \cancel{mmHg}} = 3.07\ atm\ (3\ SFs)$$

c. $T_2 = 47\ °C + 273 = 320\ K;\quad V_2 = 12.8\ L;\quad P_2 = ?$

$$P_2 = P_1 \times \frac{V_1}{V_2} \times \frac{T_2}{T_1} = 845\ \cancel{mmHg} \times \frac{6.50\ \cancel{L}}{12.8\ \cancel{L}} \times \frac{320\ \cancel{K}}{298\ \cancel{K}} \times \frac{1\ atm}{760\ \cancel{mmHg}} = 0.606\ atm\ (3\ SFs)$$

11.41 $T_1 = 212\ °C + 273 = 485\ K;\quad V_1 = 124\ mL;\quad P_1 = 1.80\ atm$

$T_2 = ?;\quad V_2 = 138\ mL;\quad P_2 = 0.800\ atm$

$$T_2 = T_1 \times \frac{V_2}{V_1} \times \frac{P_2}{P_1} = 485\ K \times \frac{138\ \cancel{mL}}{124\ \cancel{mL}} \times \frac{0.800\ \cancel{atm}}{1.80\ \cancel{atm}} = 240.\ K$$

$240.\ K - 273 = -33\ °C\ (2\ SFs)$

11.43 The volume increases because the number of gas particles in the tire or basketball is increased.

11.45 According to Avogadro's law, a change in the volume of a gas is directly proportional to the change in the number of moles of gas. $n_1 = 1.50$ mol of Ne; $V_1 = 8.00$ L

a. $V_2 = V_1 \times \frac{n_2}{n_1} = 8.00 \text{ L} \times \frac{\frac{1}{2}(1.50)\ \cancel{\text{mol Ne}}}{1.50\ \cancel{\text{mol Ne}}} = 4.00 \text{ L (3 SFs)}$

b. $n_2 = 1.50 \text{ mol Ne} + 3.50 \text{ mol Ne} = 5.00 \text{ mol of Ne}$

$V_2 = V_1 \times \frac{n_2}{n_1} = 8.00 \text{ L} \times \frac{5.00\ \cancel{\text{mol Ne}}}{1.50\ \cancel{\text{mol Ne}}} = 26.7 \text{ L (3 SFs)}$

c. $25.0\ \cancel{\text{g Ne}} \times \frac{1 \text{ mol Ne}}{20.18\ \cancel{\text{g Ne}}} = 1.24 \text{ mol of Ne added}$

$n_2 = 1.50 \text{ mol Ne} + 1.24 \text{ mol Ne} = 2.74 \text{ mol of Ne}$

$V_2 = V_1 \times \frac{n_2}{n_1} = 8.00 \text{ L} \times \frac{2.74\ \cancel{\text{mol Ne}}}{1.50\ \cancel{\text{mol Ne}}} = 14.6 \text{ L (3 SFs)}$

11.47 At STP, 1 mol of any gas occupies a volume of 22.4 L.

a. $44.8\ \cancel{\text{L O}_2} \times \frac{1 \text{ mol O}_2}{22.4\ \cancel{\text{L O}_2 \text{ (STP)}}} = 2.00 \text{ mol of O}_2 \text{ (3 SFs)}$

b. $4.00\ \cancel{\text{L CO}_2} \times \frac{1 \text{ mol CO}_2}{22.4\ \cancel{\text{L CO}_2 \text{ (STP)}}} = 0.179 \text{ mol of CO}_2 \text{ (3 SFs)}$

c. $6.40\ \cancel{\text{g O}_2} \times \frac{1\ \cancel{\text{mol O}_2}}{32.00\ \cancel{\text{g O}_2}} \times \frac{22.4 \text{ L (STP)}}{1\ \cancel{\text{mol O}_2}} = 4.48 \text{ L (3 SFs)}$

d. $50.0\ \cancel{\text{g Ne}} \times \frac{1\ \cancel{\text{mol Ne}}}{20.18\ \cancel{\text{g Ne}}} \times \frac{22.4\ \cancel{\text{L (STP)}}}{1\ \cancel{\text{mol Ne}}} \times \frac{1000 \text{ mL}}{1\ \cancel{\text{L}}} = 55\,500 \text{ mL (3 SFs)}$

11.49 At STP, 1 mol of any gas occupies a volume of 22.4 L.

a. For F_2, molar mass $= 2(19.00 \text{ g}) = 38.00$ g

$$\text{Density of F}_2 = \frac{\text{molar mass}}{\text{molar volume}} = \frac{\frac{38.00 \text{ g F}_2}{1\ \cancel{\text{mol F}_2}}}{\frac{22.4 \text{ L F}_2}{1\ \cancel{\text{mol F}_2}}} = 1.70 \text{ g/L (3 SFs)}$$

b. For CH_4, molar mass $= 12.01 \text{ g} + 4(1.008 \text{ g}) = 16.04$ g

$$\text{Density of CH}_4 = \frac{\text{molar mass}}{\text{molar volume}} = \frac{\frac{16.04 \text{ g CH}_4}{1\ \cancel{\text{mol CH}_4}}}{\frac{22.4 \text{ L CH}_4}{1\ \cancel{\text{mol CH}_4}}} = 0.716 \text{ g/L (3 SFs)}$$

c. For Ne, molar mass $= 20.18$ g

$$\text{Density of Ne} = \frac{\text{molar mass}}{\text{molar volume}} = \frac{\frac{20.18 \text{ g Ne}}{1\ \cancel{\text{mol Ne}}}}{\frac{22.4 \text{ L Ne}}{1\ \cancel{\text{mol Ne}}}} = 0.901 \text{ g/L (3 SFs)}$$

d. For SO_2, molar mass $= 32.07 \text{ g} + 2(16.00 \text{ g}) = 64.07$ g

$$\text{Density of SO}_2 = \frac{\text{molar mass}}{\text{molar volume}} = \frac{\frac{64.07 \text{ g SO}_2}{1\ \cancel{\text{mol SO}_2}}}{\frac{22.4 \text{ L SO}_2}{1\ \cancel{\text{mol SO}_2}}} = 2.86 \text{ g/L (3 SFs)}$$

11.51 $T = 27\text{ °C} + 273 = 300.\text{ K}$ $\qquad PV = nRT$

$$P = \frac{nRT}{V} = \frac{(2.00\ \cancel{\text{mol}})\left(\dfrac{0.0821\ \cancel{\text{L}}\cdot\text{atm}}{\cancel{\text{mol}}\cdot\cancel{\text{K}}}\right)(300.\ \cancel{\text{K}})}{(10.0\ \cancel{\text{L}})} = 4.93\text{ atm (3 SFs)}$$

11.53 $T = 22\text{ °C} + 273 = 295\text{ K}$ $\qquad PV = nRT$

$$n = \frac{PV}{RT} = \frac{(845\ \cancel{\text{mmHg}})(20.0\ \cancel{\text{L}})}{\left(\dfrac{62.4\ \cancel{\text{L}}\cdot\cancel{\text{mmHg}}}{\cancel{\text{mol}}\cdot\cancel{\text{K}}}\right)(295\ \cancel{\text{K}})} \times \frac{32.00\text{ g }O_2}{1\ \cancel{\text{mol }O_2}} = 29.4\text{ g of }O_2\text{ (3 SFs)}$$

11.55 $n = 25.0\ \cancel{\text{g }N_2} \times \dfrac{1\text{ mol }N_2}{28.02\ \cancel{\text{g }N_2}} = 0.892\text{ mol of }N_2$ $\qquad PV = nRT$

$$T = \frac{PV}{nR} = \frac{(630.\ \cancel{\text{mmHg}})(50.0\ \cancel{\text{L}})}{(0.892\ \cancel{\text{mol}})\left(\dfrac{62.4\ \cancel{\text{L}}\cdot\cancel{\text{mmHg}}}{\cancel{\text{mol}}\cdot\text{K}}\right)} = 566\text{ K} - 273 = 293\text{ °C (3 SFs)}$$

11.57 **a.** $n = 450\ \cancel{\text{mL}} \times \dfrac{1\ \cancel{\text{L}}}{1000\ \cancel{\text{mL}}} \times \dfrac{1\text{ mol}}{22.4\ \cancel{\text{L (STP)}}} = 0.020\text{ mol}$

$$\text{Molar mass} = \frac{\text{mass}}{\text{moles}} = \frac{0.84\text{ g}}{0.020\text{ mol}} = 42\text{ g/mol (2 SFs)}$$

b. $n = 1.00\ \cancel{\text{L}} \times \dfrac{1\text{ mol}}{22.4\ \cancel{\text{L (STP)}}} = 0.0446\text{ mol}$

$$\text{Molar mass} = \frac{\text{mass}}{\text{moles}} = \frac{1.28\text{ g}}{0.0446\text{ mol}} = 28.7\text{ g/mol (3 SFs)}$$

c. $T = 22\text{ °C} + 273 = 295\text{ K}$ $\qquad PV = nRT$

$$n = \frac{PV}{RT} = \frac{(685\ \cancel{\text{mmHg}})(1.00\ \cancel{\text{L}})}{\left(\dfrac{62.4\ \cancel{\text{L}}\cdot\cancel{\text{mmHg}}}{\text{mol}\cdot\cancel{\text{K}}}\right)(295\ \cancel{\text{K}})} = 0.0372\text{ mol}$$

$$\text{Molar mass} = \frac{\text{mass}}{\text{moles}} = \frac{1.48\text{ g}}{0.0372\text{ mol}} = 39.8\text{ g/mol (3 SFs)}$$

d. $T = 24\text{ °C} + 273 = 297\text{ K}$ $\qquad PV = nRT$

$$n = \frac{PV}{RT} = \frac{(0.95\ \cancel{\text{atm}})(2.30\ \cancel{\text{L}})}{\left(\dfrac{0.0821\ \cancel{\text{L}}\cdot\cancel{\text{atm}}}{\text{mol}\cdot\cancel{\text{K}}}\right)(297\ \cancel{\text{K}})} = 0.090\text{ mol}$$

$$\text{Molar mass} = \frac{\text{mass}}{\text{moles}} = \frac{2.96\text{ g}}{0.090\text{ mol}} = 33\text{ g/mol (2 SFs)}$$

11.59 **a.** $8.25\ \cancel{\text{g Mg}} \times \dfrac{1\ \cancel{\text{mol Mg}}}{24.31\ \cancel{\text{g Mg}}} \times \dfrac{1\ \cancel{\text{mol }H_2}}{1\ \cancel{\text{mol Mg}}} \times \dfrac{22.4\text{ L (STP)}}{1\ \cancel{\text{mol }H_2}} = 7.60\text{ L of }H_2\text{ released at STP}$ (3 SFs)

b. $T = 18\text{ °C} + 273 = 291\text{ K}$

$$n = \frac{PV}{RT} = \frac{(735\ \cancel{\text{mmHg}})(5.00\ \cancel{\text{L}})}{\left(\dfrac{62.4\ \cancel{\text{L}}\cdot\cancel{\text{mmHg}}}{\text{mol}\cdot\cancel{\text{K}}}\right)(291\ \cancel{\text{K}})} = 0.202\text{ mol of }H_2$$

$$0.202\ \cancel{\text{mol }H_2} \times \frac{1\ \cancel{\text{mol Mg}}}{1\ \cancel{\text{mol }H_2}} \times \frac{24.31\text{ g Mg}}{1\ \cancel{\text{mol Mg}}} = 4.92\text{ g of Mg (3 SFs)}$$

11.61 $55.2\ \cancel{g\ C_4H_{10}} \times \frac{1\ \cancel{mol\ C_4H_{10}}}{58.12\ \cancel{g\ C_4H_{10}}} \times \frac{13\ mol\ O_2}{2\ \cancel{mol\ C_4H_{10}}} = 6.17\ mol\ of\ O_2$

$T = 25\ °C + 273 = 298\ K \qquad PV = nRT$

$$V = \frac{nRT}{P} = \frac{(6.17\ \cancel{mol})\left(\frac{0.0821\ L \cdot \cancel{atm}}{\cancel{mol} \cdot \cancel{K}}\right)(298\ \cancel{K})}{(0.850\ \cancel{atm})} = 178\ L\ of\ O_2\ (3\ SFs)$$

11.63 $5.4\ \cancel{g\ Al} \times \frac{1\ \cancel{mol\ Al}}{26.98\ \cancel{g\ Al}} \times \frac{3\ mol\ O_2}{4\ \cancel{mol\ Al}} = 0.15\ mol\ of\ O_2$

$0.15\ \cancel{mol\ O_2} \times \frac{22.4\ L\ (STP)}{1\ \cancel{mol\ O_2}} = 3.4\ L\ of\ O_2\ (2\ SFs)$

11.65 **a.** $P_{O_2} = P_{total} - P_{H_2O} = 765\ mmHg - 22\ mmHg = 743\ mmHg\ (3\ SFs)$

b. $V = 256\ \cancel{mL} \times \frac{1\ L}{1000\ \cancel{mL}} = 0.256\ L;\ T = 24\ °C + 273 = 297\ K$

$$n = \frac{PV}{RT} = \frac{(743\ \cancel{mmHg})(0.256\ \cancel{L})}{\left(\frac{62.4\ \cancel{L} \cdot \cancel{mmHg}}{mol \cdot \cancel{K}}\right)(297\ \cancel{K})} = 0.0103\ mol\ of\ O_2\ (3\ SFs)$$

11.67 In a gas mixture, the pressure that each gas exerts as part of the total pressure is called the partial pressure of that gas. Because the air sample is a mixture of gases, the total pressure is the sum of the partial pressures of each gas in the sample.

11.69 To obtain the total pressure in a gas mixture, add up all of the partial pressures using the same pressure unit.

$P_{total} = P_{Nitrogen} + P_{Oxygen} + P_{Helium} = 425\ torr + 115\ torr + 225\ torr = 765\ torr\ (3\ SFs)$

11.71 Because the total pressure in a gas mixture is the sum of the partial pressures using the same pressure unit, addition and subtraction are used to obtain the "missing" partial pressure.

$P_{Nitrogen} = P_{total} - (P_{Oxygen} + P_{Helium})$

$= 925\ torr - (425\ torr + 75\ torr) = 425\ torr\ (3\ SFs)$

11.73 **a.** 2 The fewest number of gas particles will exert the lowest pressure.
b. 1 The greatest number of gas particles will exert the highest pressure.

11.75 **a.** A Volume decreases when temperature decreases.
b. C Volume increases when pressure decreases.
c. A Volume decreases when the number of moles of gas decreases.
d. B Doubling the Kelvin temperature would double the volume, but when half of the gas escapes, the volume would decrease by half. These two opposing effects cancel each other, and there is no overall change in the volume.
e. C Increasing the moles of gas causes an increase in the volume to keep T and P constant.

11.77 $T_1 = 24\ °C + 273 = 297\ K \qquad T_2 = -95\ °C + 273 = 178\ K$

$V_2 = V_1 \times \frac{P_1}{P_2} \times \frac{T_2}{T_1} = 425\ mL \times \frac{745\ \cancel{mmHg}}{0.115\ \cancel{atm}} \times \frac{1\ \cancel{atm}}{760\ \cancel{mmHg}} \times \frac{178\ \cancel{K}}{297\ \cancel{K}} = 2170\ mL\ (3\ SFs)$

11.79 $T_1 = -8\ °C + 273 = 265\ K;\ P_1 = 658\ mmHg;\ V_1 = 31\,000\ L$

$T_2 = 0\ °C + 273 = 273\ K;\ P_2 = 760\ mmHg;\ V_2 = ?$

$V_2 = V_1 \times \frac{P_1}{P_2} \times \frac{T_2}{T_1} = 31\,000\ L \times \frac{658\ \cancel{mmHg}}{760\ \cancel{mmHg}} \times \frac{273\ \cancel{K}}{265\ \cancel{K}} = 28\,000\ L\ (STP)$

$28\,000\ \cancel{L} \times \frac{1\ \cancel{mol\ H_2}}{22.4\ \cancel{L}} \times \frac{2.016\ \cancel{g\ H_2}}{1\ \cancel{mol\ H_2}} \times \frac{1\ kg\ H_2}{1000\ \cancel{g\ H_2}} = 2.5\ kg\ of\ H_2\ (2\ SFs)$

11.81 $T_1 = 15\ °C + 273 = 288\ K;\ P_1 = 745\ mmHg;\ V_1 = 4250\ mL$

$$T_2 = ?;\ P_2 = 1.20\ \cancel{atm} \times \frac{760\ mmHg}{1\ \cancel{atm}} = 912\ mmHg;\ V_2 = 2.50\ \cancel{L} \times \frac{1000\ mL}{1\ \cancel{L}} = 2.50 \times 10^3\ mL$$

$$T_2 = T_1 \times \frac{V_2}{V_1} \times \frac{P_2}{P_1} = 288\ K \times \frac{2500\ \cancel{mL}}{4250\ \cancel{mL}} \times \frac{912\ \cancel{mmHg}}{745\ \cancel{mmHg}} = 207\ K - 273 = -66\ °C\ (2\ SFs)$$

11.83 $T = 18\ °C + 273 = 291\ K \qquad PV = nRT$

$$n = \frac{PV}{RT} = \frac{(2500\ \cancel{mmHg})(2.00\ \cancel{L})}{\left(\frac{62.4\ \cancel{L} \cdot \cancel{mmHg}}{mol \cdot \cancel{K}}\right)(291\ \cancel{K})} = 0.28\ mol\ of\ CH_4$$

$$0.28\ \cancel{mol\ CH_4} \times \frac{16.04\ g\ CH_4}{1\ \cancel{mol\ CH_4}} = 4.5\ g\ of\ CH_4\ (2\ SFs)$$

11.85 $T = 20.0\ °C + 273 = 293\ K \qquad PV = nRT$

$$V = 941\ \cancel{mL} \times \frac{1\ L}{1000\ \cancel{mL}} = 0.941\ L$$

$$n = \frac{PV}{RT} = \frac{(748\ \cancel{torr})(0.941\ \cancel{L})}{\left(\frac{62.4\ \cancel{L} \cdot \cancel{torr}}{mol \cdot \cancel{K}}\right)(293\ \cancel{K})} = 0.0385\ mol$$

$$\text{Molar mass} = \frac{mass}{moles} = \frac{1.62\ g}{0.0385\ mol} = 42.1\ g/mol\ (3\ SFs)$$

11.87 $T = 23\ °C + 273 = 296\ K;\ V = 782\ \cancel{mL} \times \frac{1\ L}{1000\ \cancel{mL}} = 0.782\ L$

$$n = \frac{PV}{RT} = \frac{(752\ \cancel{mmHg})(0.782\ \cancel{L})}{\left(\frac{62.4\ \cancel{L} \cdot \cancel{mmHg}}{mol \cdot \cancel{K}}\right)(296\ \cancel{K})} = 0.0318\ mol$$

$$\text{Molar mass} = \frac{mass}{moles} = \frac{2.23\ g}{0.0318\ mol} = 70.1\ g/mol$$

Empirical formula mass of $CH_2 = 12.01\ g + 2(1.008\ g) = 14.03\ g$

$$\text{Small integer} = \frac{\text{molar mass}}{\text{empirical formula mass of } CH_2} = \frac{70.1\ \cancel{g}}{14.03\ \cancel{g}} = 5$$

$\therefore$ Molecular formula $= C_{(1\times5)}H_{(2\times5)} = C_5H_{10}$

11.89 $T = 5\ °C + 273 = 278\ K \qquad PV = nRT$

$$n = \frac{PV}{RT} = \frac{(1.2\ \cancel{atm})(35.0\ \cancel{L})}{\left(\frac{0.0821\ \cancel{L} \cdot \cancel{atm}}{mol \cdot \cancel{K}}\right)(278\ \cancel{K})} = 1.8\ mol\ of\ CO_2$$

$$1.8\ \cancel{mol\ CO_2} \times \frac{6.022 \times 10^{23}\ molecules\ CO_2}{1\ \cancel{mol\ CO_2}} = 1.1 \times 10^{24}\ molecules\ of\ CO_2\ (2\ SFs)$$

11.91 $$25.0\ \cancel{g\ Zn} \times \frac{1\ \cancel{mol\ Zn}}{65.41\ \cancel{g\ Zn}} \times \frac{1\ \cancel{mol\ H_2}}{1\ \cancel{mol\ Zn}} \times \frac{22.4\ L\ (STP)}{1\ \cancel{mol\ H_2}} = 8.56\ L\ of\ H_2\ produced\ (3\ SFs)$$

11.93 a. $2.5 \times 10^{23}\ \cancel{\text{molecules NO}_2} \times \dfrac{1\ \cancel{\text{mol NO}_2}}{6.022 \times 10^{23}\ \cancel{\text{molecules NO}_2}} \times \dfrac{7\ \cancel{\text{mol O}_2}}{4\ \cancel{\text{mol NO}_2}}$

$\times \dfrac{22.4\ \text{L O}_2\ \text{(STP)}}{1\ \cancel{\text{mol O}_2}}$

$= 16\ \text{L of O}_2$ (2 SFs)

b. $T = 375\ °\text{C} + 273 = 648\ \text{K}$ $\qquad PV = nRT$

$$n = \frac{PV}{RT} = \frac{(725\ \cancel{\text{mmHg}})(5.00\ \cancel{\text{L}})}{\left(\dfrac{62.4\ \cancel{\text{L}} \cdot \cancel{\text{mmHg}}}{\text{mol} \cdot \cancel{\text{K}}}\right)(648\ \cancel{\text{K}})} = 0.0896\ \text{mol of H}_2\text{O}$$

$$0.0896\ \cancel{\text{mol H}_2\text{O}} \times \frac{4\ \cancel{\text{mol NH}_3}}{6\ \cancel{\text{mol H}_2\text{O}}} \times \frac{17.03\ \text{g NH}_3}{1\ \cancel{\text{mol NH}_3}} = 1.02\ \text{g of NH}_3\ \text{(3 SFs)}$$

11.95 The partial pressure of each gas is proportional to the number of particles of each type of gas that is present. Thus, a ratio of partial pressure to total pressure is equal to the ratio of moles of that gas to the total number of moles of gases that are present:

$$\frac{P_{\text{Helium}}}{P_{\text{total}}} = \frac{n_{\text{Helium}}}{n_{\text{total}}}$$

Solving the equation for the partial pressure of helium yields:

$$P_{\text{Helium}} = P_{\text{total}} \times \frac{n_{\text{Helium}}}{n_{\text{total}}} = 2400\ \text{torr} \times \frac{2.0\ \cancel{\text{mol}}}{8.0\ \cancel{\text{mol}}} = 600\ \text{torr}$$

$$P_{\text{Oxygen}} = P_{\text{total}} \times \frac{n_{\text{Oxygen}}}{n_{\text{total}}} = 2400\ \text{torr} \times \frac{6.0\ \cancel{\text{mol}}}{8.0\ \cancel{\text{mol}}} = 1800\ \text{torr}$$

11.97 Because the partial pressure of nitrogen is to be reported in torr, the atm and mmHg units (for oxygen and argon, respectively) must be converted to torr, as follows:

$$P_{\text{Oxygen}} = 0.60\ \cancel{\text{atm}} \times \frac{760\ \text{torr}}{1\ \cancel{\text{atm}}} = 460\ \text{torr} \quad \text{and} \quad P_{\text{Argon}} = 425\ \cancel{\text{mmHg}} \times \frac{1\ \text{torr}}{1\ \cancel{\text{mmHg}}} = 425\ \text{torr}$$

$\therefore P_{\text{Nitrogen}} = P_{\text{total}} - (P_{\text{Oxygen}} + P_{\text{Argon}})$

$= 1250\ \text{torr} - (460\ \text{torr} + 425\ \text{torr}) = 370\ \text{torr}$ (2 SFs)

11.99 a. $P_{\text{H}_2} = P_{\text{total}} - P_{\text{H}_2\text{O}} = 755\ \text{mmHg} - 21\ \text{mmHg} = 734\ \text{mmHg}$

b. $T = 23\ °\text{C} + 273 = 296\ \text{K};\ V = 415\ \cancel{\text{mL}} \times \dfrac{1\ \text{L}}{1000\ \cancel{\text{mL}}} = 0.415\ \text{L}$ $\qquad PV = nRT$

$$n = \frac{PV}{RT} = \frac{(734\ \cancel{\text{mmHg}})(0.415\ \cancel{\text{L}})}{\left(\dfrac{62.4\ \cancel{\text{L}} \cdot \cancel{\text{mmHg}}}{\text{mol} \cdot \cancel{\text{K}}}\right)(296\ \cancel{\text{K}})} = 0.0165\ \text{mol of H}_2\ \text{(3 SFs)}$$

c. $0.0165\ \cancel{\text{mol H}_2} \times \dfrac{2\ \cancel{\text{mol Al}}}{3\ \cancel{\text{mol H}_2}} \times \dfrac{26.98\ \text{g Al}}{1\ \cancel{\text{mol Al}}} = 0.297\ \text{g of Al}$ (3 SFs)

11.101 a. False. The flask containing helium has more moles of helium and thus more helium atoms.

b. False. There are different numbers of moles in the flasks, which means the pressures are different.

c. True. There are more moles of helium, which makes the pressure of helium greater than that of neon.

d. True. The mass and volume of each are the same, which means the mass/volume ratio or density is the same in both flasks.

11.103 $n = 762 \cancel{\text{mL}} \times \dfrac{1 \cancel{\text{L}}}{1000 \cancel{\text{mL}}} \times \dfrac{1 \text{ mol}}{22.4 \cancel{\text{L}} \text{ (STP)}} = 0.0340 \text{ mol}$

$\text{Molar mass} = \dfrac{\text{mass}}{\text{moles}} = \dfrac{1.02 \text{ g}}{0.0340 \text{ mol}} = 30.0 \text{ g/mol}$

$9.60 \cancel{\text{g C}} \times \dfrac{1 \text{ mol C}}{12.01 \cancel{\text{g C}}} = 0.799 \text{ mol C (smaller number of moles)}$

$2.42 \cancel{\text{g H}} \times \dfrac{1 \text{ mol H}}{1.008 \cancel{\text{g H}}} = 2.40 \text{ mol H}$

$\dfrac{0.799 \text{ mol C}}{0.799} = 1.00 \text{ mol of C} \qquad \dfrac{2.40 \text{ mol H}}{0.799} = 3.00 \text{ mol of H}$

$\therefore$ Empirical formula $= C_{1.00}H_{3.00} \rightarrow CH_3$

Empirical formula mass of $CH_3 = 12.01 \text{ g} + 3(1.008 \text{ g}) = 15.03 \text{ g}$

$\text{Small integer} = \dfrac{\text{molar mass}}{\text{empirical formula mass of } CH_2} = \dfrac{30.0 \cancel{\text{g}}}{15.03 \cancel{\text{g}}} = 2$

$\therefore$ Molecular formula $= C_{(1\times2)}H_{(3\times2)} = C_2H_6$

11.105 $T = 37\ °C + 273 = 310 \text{ K}$

$n = \dfrac{PV}{RT} = \dfrac{(1.00 \cancel{\text{atm}})(7.50 \cancel{\text{L}})}{\left(\dfrac{0.0821 \cancel{\text{L}} \cdot \cancel{\text{atm}}}{\text{mol} \cdot \cancel{\text{K}}}\right)(310 \cancel{\text{K}})}$

$= 0.295 \cancel{\text{mol } O_2} \times \dfrac{6 \text{ mol } H_2O}{6 \cancel{\text{mol } O_2}} = 0.295 \text{ mol of } H_2O \text{ (smaller number of moles)}$

$18.0 \cancel{\text{g } C_6H_{12}O_6} \times \dfrac{1 \text{ mol } C_6H_{12}O_6}{180.16 \cancel{\text{g } C_6H_{12}O_6}}$

$= 0.100 \cancel{\text{mol } C_6H_{12}O_6} \times \dfrac{6 \text{ mol } H_2O}{1 \cancel{\text{mol } C_6H_{12}O_6}} = 0.599 \text{ mol of } H_2O$

$\therefore$ O_2 is the limiting reactant.

$0.295 \cancel{\text{mol } H_2O} \times \dfrac{18.02 \text{ g } H_2O}{1 \cancel{\text{mol } H_2O}} = 5.32 \text{ g of } H_2O \text{ can be produced (3 SFs)}$

12
Solutions

Study Goals

- Identify the solute and solvent in a solution. Describe hydrogen bonding in water. Describe the formation of a solution.
- Identify solutes as electrolytes or nonelectrolytes.
- Define solubility; distinguish between an unsaturated and a saturated solution. Identify an insoluble salt.
- Calculate the percent concentration of a solute in a solution; use percent concentration to calculate the amount of solute or solution.
- Calculate the molarity of a solution; use molarity as a conversion factor to calculate the moles of solute or the volume needed to prepare a solution. Describe the dilution of a solution.
- Given the volume and molarity of a solution, calculate the amount of another reactant or product in the reaction.
- Identify a mixture as a solution, a colloid, or a suspension. Describe how particles of a solution affect the freezing point, boiling point, and osmotic pressure of a solution. Describe osmosis and dialysis.

Chapter Outline

Answers and Solutions to Text Problems

12.1 The component present in the smaller amount is the solute; the larger amount is the solvent.
a. NaCl, solute; water, solvent
b. water, solute; ethanol, solvent
c. oxygen, solute; nitrogen, solvent

12.3 **a.** Sodium nitrate, $NaNO_3$, (an ionic solute) would be soluble in water (a polar solvent).
b. Iodine, I_2, (a nonpolar solute) would be soluble in CCl_4 (a nonpolar solvent).
c. Sucrose (a polar solute) would be soluble in water (a polar solvent).
d. Octane (a nonpolar solute) would be soluble in CCl_4 (a nonpolar solvent).

12.5 The K^+ and I^- ions at the surface of the solid are pulled into solution by the polar water molecules, where the hydration process surrounds separate ions with water molecules.

12.7 The strong electrolyte KF completely dissociates into K^+ and F^- ions when it dissolves in water. When it dissolves in water, the weak electrolyte HF exists as mostly HF molecules along with a few H^+ ions and a few F^- ions.

12.9 Strong electrolytes dissociate completely into ions.

a. $KCl(s) \xrightarrow{H_2O} K^+(aq) + Cl^-(aq)$

b. $CaCl_2(s) \xrightarrow{H_2O} Ca^{2+}(aq) + 2Cl^-(aq)$

c. $K_3PO_4(s) \xrightarrow{H_2O} 3K^+(aq) + PO_4^{3-}(aq)$

d. $Fe(NO_3)_3(s) \xrightarrow{H_2O} Fe^{3+}(aq) + 3NO_3^-(aq)$

12.11 **a.** $CH_3{-}COOH(l) \overset{H_2O}{\rightleftharpoons} H^+(aq) + CH_3{-}COO^-(aq)$ mostly molecules, a few ions
An aqueous solution of a weak electrolyte like acetic acid will contain mostly $CH_3{-}COOH$ molecules with a few H^+ ions and a few $CH_3{-}COO^-$ ions.

b. $NaBr(s) \xrightarrow{H_2O} Na^+(aq) + Br^-(aq)$ ions only
An aqueous solution of a strong electrolyte like NaBr will contain only the ions Na^+ and Br^-.

c. $C_6H_{12}O_6(s) \xrightarrow{H_2O} C_6H_{12}O_6(aq)$ molecules only
An aqueous solution of a nonelectrolyte like fructose will contain only $C_6H_{12}O_6$ molecules.

12.13 **a.** Strong electrolyte because only ions are present in the K_2SO_4 solution.
b. Weak electrolyte because mostly NH_3 molecules and a few ions are present in the solution.
c. Nonelectrolyte because only $C_6H_{12}O_6$ molecules are present in the solution.

12.15 **a.** The solution must be saturated because no additional solute dissolves.
b. The solution was unsaturated because the sugar cube dissolves completely.

12.17 **a.** At 20 °C, KCl has a solubility of 34 g of KCl in 100 g of H_2O. Because 25 g of KCl is less than the maximum amount that can dissolve in 100 g of H_2O at 20 °C, the KCl solution is unsaturated.
b. At 20 °C, $NaNO_3$ has a solubility of 88 g of $NaNO_3$ in 100 g of H_2O. Using the solubility as a conversion factor, we can calculate the maximum amount of $NaNO_3$ that can dissolve in 25 g of H_2O:

$$25\ \cancel{g\ H_2O} \times \frac{88\ g\ NaNO_3}{100\ \cancel{g\ H_2O}} = 22\ g\ of\ NaNO_3\ (2\ SFs)$$

Because 11 g of $NaNO_3$ is less than the maximum amount that can dissolve in 25 g of H_2O at 20 °C, the $NaNO_3$ solution is unsaturated.
c. At 20 °C, sucrose has a solubility of 204 g of sucrose in 100 g of H_2O. Using the solubility as a conversion factor, we can calculate the maximum amount of sucrose that can dissolve in 125 g of H_2O:

$$125\ \cancel{g\ H_2O} \times \frac{204\ g\ sugar}{100\ \cancel{g\ H_2O}} = 255\ g\ of\ sugar\ (3\ SFs)$$

Because 400. g of sucrose exceeds the maximum amount that can dissolve in 125 g of H_2O at 20 °C, the sucrose solution is saturated, and excess undissolved sucrose will be present on the bottom of the container.

12.19 a. At 20 °C, KCl has a solubility of 34 g of KCl in 100 g of H_2O.
∴ 200. g of H_2O will dissolve:

$$200.\ \cancel{\text{g H}_2\text{O}} \times \frac{34\text{ g KCl}}{100\ \cancel{\text{g H}_2\text{O}}} = 68\text{ g of KCl (2 SFs)}$$

At 20 °C, 68 g of KCl will remain in solution.

b. Since 80. g of KCl dissolves at 50 °C and 68 g remains in solution at 20 °C, the mass of solid KCl that crystallizes after cooling is (80. g KCl − 68 g KCl) = 12 g of KCl. (2 SFs)

12.21 a. In general, the solubility of solid solutes (like sucrose) increases as temperature is increased.

b. The solubility of a gaseous solute (CO_2) is less at a higher temperature.

c. The solubility of a gaseous solute is less at a higher temperature, and the CO_2 pressure in the can is increased. When the can of warm soda is opened, more CO_2 is released, producing more spray.

12.23 a. Salts containing Li^+ are soluble.

b. The Cl^- salt containing Ag^+ is insoluble.

c. Salts containing CO_3^{2-} ions are usually insoluble.

d. Salts containing K^+ ions are soluble.

e. Salts containing NO_3^- ions are soluble.

12.25 a. No solid forms; salts containing K^+ and Na^+ are soluble.

b. Solid silver sulfide (Ag_2S) forms:

$$2AgNO_3(aq) + K_2S(aq) \rightarrow Ag_2S(s) + 2KNO_3(aq)$$

$$2Ag^+(aq) + \cancel{2NO_3^-(aq)} + \cancel{2K^+(aq)} + S^{2-}(aq) \rightarrow Ag_2S(s) + \cancel{2K^+(aq)} + \cancel{2NO_3^-(aq)}$$

$$2Ag^+(aq) + S^{2-}(aq) \rightarrow Ag_2S(s) \quad \text{Net ionic equation}$$

c. Solid calcium sulfate ($CaSO_4$) forms:

$$CaCl_2(aq) + Na_2SO_4(aq) \rightarrow CaSO_4(s) + 2NaCl(aq)$$

$$Ca^{2+}(aq) + \cancel{2Cl^-(aq)} + \cancel{2Na^+(aq)} + SO_4^{2-}(aq) \rightarrow CaSO_4(s) + \cancel{2Na^+(aq)} + \cancel{2Cl^-(aq)}$$

$$Ca^{2+}(aq) + SO_4^{2-}(aq) \rightarrow CaSO_4(s) \quad \text{Net ionic equation}$$

d. Solid copper phosphate ($Cu_3(PO_4)_2$) forms:

$$3CuCl_2(aq) + 2Li_3PO_4(aq) \rightarrow Cu_3(PO_4)_2(s) + 6LiCl(aq)$$

$$3Cu^{2+}(aq) + \cancel{6Cl^-(aq)} + \cancel{6Li^+(aq)} + 2PO_4^{2-}(aq) \rightarrow Cu_3(PO_4)_2(s) + \cancel{6Li^+(aq)} + \cancel{6Cl^-(aq)}$$

$$3Cu^{2+}(aq) + 2PO_4^{2-}(aq) \rightarrow Cu_3(PO_4)_2(s) \quad \text{Net ionic equation}$$

12.27 The 5.00% (m/m) concentration indicates that there are 5.00 g of glucose in every 100 g of solution. We can use the mass percent as a conversion factor to determine the total mass of glucose needed to prepare 250. g of solution:

$$250.\ \cancel{\text{g solution}} \times \frac{5.00\text{ g glucose}}{100\ \cancel{\text{g solution}}} = 12.5\text{ g of glucose (3 SFs)}$$

So the procedure will be:
Weigh out 12.5 g of glucose and add 237.5 g of water to make a total of 250. g of solution. This will provide a 5.00% (m/m) glucose solution.

12.29 a. mass of solution = 25 g of KCl + 125 g of H_2O = 150. g of solution

$$\frac{25\text{ g KCl}}{150.\text{ g solution}} \times 100\% = 17\%\text{ (m/m) KCl solution (2 SFs)}$$

b.
$$\frac{8.0\text{ g CaCl}_2}{80.0\text{ g solution}} \times 100\% = 10.\%\text{ (m/m) CaCl}_2\text{ solution (2 SFs)}$$

c.
$$\frac{12\text{ g sucrose}}{225\text{ g solution}} \times 100\% = 5.3\%\text{ (m/m) sucrose solution (2 SFs)}$$

12.31 a. $$50.0\ \cancel{\text{g solution}} \times \frac{5.0\ \text{g KCl}}{100\ \cancel{\text{g solution}}} = 2.5\ \text{g of KCl (2 SFs)}$$

b. $$1250\ \cancel{\text{g solution}} \times \frac{4.0\ \text{g NH}_4\text{Cl}}{100\ \cancel{\text{g solution}}} = 50.\ \text{g of NH}_4\text{Cl (2 SFs)}$$

c. $$250\ \cancel{\text{mL solution}} \times \frac{10.0\ \text{mL acetic acid}}{100\ \cancel{\text{mL solution}}} = 25\ \text{mL of acetic acid (2 SFs)}$$

12.33 $$355\ \cancel{\text{mL solution}} \times \frac{22.5\ \text{mL alcohol}}{100\ \cancel{\text{mL solution}}} = 79.9\ \text{mL of alcohol (3 SFs)}$$

12.35 a. $$5.0\ \cancel{\text{g LiNO}_3} \times \frac{100\ \text{g solution}}{25\ \cancel{\text{g LiNO}_3}} = 20.\ \text{g of solution (2 SFs)}$$

b. $$40.0\ \cancel{\text{g KOH}} \times \frac{100\ \text{g solution}}{10.0\ \cancel{\text{g KOH}}} = 400.\ \text{g of solution (3 SFs)}$$

c. $$2.0\ \cancel{\text{mL formic acid}} \times \frac{100\ \text{mL solution}}{10.0\ \cancel{\text{mL formic acid}}} = 20.\ \text{mL of solution (2 SFs)}$$

12.37 $$\text{molarity (M)} = \frac{\text{moles of solute}}{\text{liters of solution}}$$

a. $$\frac{2.00\ \text{mol glucose}}{4.00\ \text{L solution}} = 0.500\ \text{M glucose solution (3 SFs)}$$

b. $$\frac{5.85\ \cancel{\text{g NaCl}}}{40.0\ \cancel{\text{mL solution}}} \times \frac{1\ \text{mol NaCl}}{58.44\ \cancel{\text{g NaCl}}} \times \frac{1000\ \cancel{\text{mL solution}}}{1\ \text{L solution}} = 2.50\ \text{M NaCl solution (3 SFs)}$$

c. $$\frac{4.00\ \cancel{\text{g KOH}}}{2.00\ \text{L solution}} \times \frac{1\ \text{mol KOH}}{56.11\ \cancel{\text{g KOH}}} = 0.0356\ \text{M KOH solution (3 SFs)}$$

12.39 a. $$2.00\ \cancel{\text{L solution}} \times \frac{1.50\ \cancel{\text{mol NaOH}}}{1\ \cancel{\text{L solution}}} \times \frac{40.00\ \text{g NaOH}}{1\ \cancel{\text{mol NaOH}}} = 120.\ \text{g of NaOH (3 SFs)}$$

b. $$125\ \cancel{\text{mL solution}} \times \frac{1\ \cancel{\text{L solution}}}{1000\ \cancel{\text{mL solution}}} \times \frac{0.200\ \cancel{\text{mol KCl}}}{1\ \cancel{\text{L solution}}} \times \frac{74.55\ \text{g KCl}}{1\ \cancel{\text{mol KCl}}} = 1.86\ \text{g of KCl (3 SFs)}$$

c. $$25.0\ \cancel{\text{mL solution}} \times \frac{1\ \cancel{\text{L solution}}}{1000\ \cancel{\text{mL solution}}} \times \frac{3.50\ \cancel{\text{mol HCl}}}{1\ \cancel{\text{L solution}}} \times \frac{36.46\ \text{g HCl}}{1\ \cancel{\text{mol HCl}}} = 3.19\ \text{g of HCl (3 SFs)}$$

12.41 a. $$12.5\ \cancel{\text{g Na}_2\text{CO}_3} \times \frac{1\ \text{mol Na}_2\text{CO}_3}{105.99\ \cancel{\text{g Na}_2\text{CO}_3}} = 0.118\ \text{mol of Na}_2\text{CO}_3$$

$$0.118\ \cancel{\text{mol Na}_2\text{CO}_3} \times \frac{1\ \cancel{\text{L solution}}}{0.120\ \cancel{\text{mol Na}_2\text{CO}_3}} \times \frac{1000\ \text{mL solution}}{1\ \cancel{\text{L solution}}}$$
$$= 983\ \text{mL of Na}_2\text{CO}_3\ \text{solution (3 SFs)}$$

b. $$0.850\ \cancel{\text{mol NaNO}_3} \times \frac{1\ \cancel{\text{L solution}}}{0.500\ \cancel{\text{mol NaNO}_3}} \times \frac{1000\ \text{mL solution}}{1\ \cancel{\text{L solution}}}$$
$$= 1700\ \text{mL}\ (1.70 \times 10^3\ \text{mL})\ \text{of NaNO}_3\ \text{solution (3 SFs)}$$

c. $$30.0\ \cancel{\text{g LiOH}} \times \frac{1\ \text{mol LiOH}}{23.95\ \cancel{\text{g LiOH}}} = 1.25\ \text{mol of LiOH}$$

$$1.25\ \cancel{\text{mol LiOH}} \times \frac{1\ \cancel{\text{L solution}}}{2.70\ \cancel{\text{mol LiOH}}} \times \frac{1000\ \text{mL solution}}{1\ \cancel{\text{L solution}}} = 464\ \text{mL of LiOH solution (3 SFs)}$$

12.43 $M_1V_1 = M_2V_2$

a. $M_2 = M_1 \times \frac{V_1}{V_2} = 6.00\text{ M} \times \frac{0.150\text{ L}}{0.500\text{ L}} = 1.80\text{ M HCl solution (3 SFs)}$

b. $V_2 = 0.250\text{ L} \times \frac{1000\text{ mL}}{1\text{ L}} = 250.\text{ mL}$

$M_2 = M_1 \times \frac{V_1}{V_2} = 2.50\text{ M} \times \frac{10.0\text{ mL}}{250.\text{ mL}} = 0.100\text{ M KCl solution (3 SFs)}$

c. $M_2 = M_1 \times \frac{V_1}{V_2} = 12.0\text{ M} \times \frac{0.250\text{ L}}{1.00\text{ L}} = 3.00\text{ M KBr solution (3 SFs)}$

12.45 $M_1V_1 = M_2V_2$

a. $V_2 = V_1 \times \frac{M_1}{M_2} = 50.0\text{ mL} \times \frac{12.0\text{ M}}{2.00\text{ M}} = 300.\text{ mL (3 SFs)}$

b. $V_2 = V_1 \times \frac{M_1}{M_2} = 18.0\text{ mL} \times \frac{15.0\text{ M}}{1.50\text{ M}} = 180.\text{ mL (3 SFs)}$

c. $V_2 = V_1 \times \frac{M_1}{M_2} = 4.50\text{ mL} \times \frac{18.0\text{ M}}{2.50\text{ M}} = 32.4\text{ mL (3 SFs)}$

12.47 $M_1V_1 = M_2V_2$

a. $V_1 = V_2 \times \frac{M_2}{M_1} = 255\text{ mL} \times \frac{0.200\text{ M}}{4.00\text{ M}} = 12.8\text{ mL of the HNO}_3\text{ solution (3 SFs)}$

b. $V_1 = V_2 \times \frac{M_2}{M_1} = 715\text{ mL} \times \frac{0.100\text{ M}}{6.00\text{ M}} = 11.9\text{ mL of the MgCl}_2\text{ solution (3 SFs)}$

c. $V_2 = 0.100\text{ L} \times \frac{1000\text{ mL}}{1\text{ L}} = 100.\text{ mL}$

$V_1 = V_2 \times \frac{M_2}{M_1} = 100.\text{ mL} \times \frac{0.150\text{ M}}{8.00\text{ M}} = 1.88\text{ mL of the KCl solution (3 SFs)}$

12.49 $M_1V_1 = M_2V_2 \quad V_2 = V_1 \times \frac{M_1}{M_2} = 25.0\text{ mL} \times \frac{3.00\text{ M}}{0.150\text{ M}} = 500.\text{ mL (3 SFs)}$

12.51 a. $50.0\text{ mL solution} \times \frac{1\text{ L solution}}{1000\text{ mL solution}} \times \frac{1.50\text{ mol KCl}}{1\text{ L solution}} = 0.0750\text{ mol of KCl}$

$0.0750\text{ mol KCl} \times \frac{1\text{ mol PbCl}_2}{2\text{ mol KCl}} \times \frac{278.1\text{ g PbCl}_2}{1\text{ mol PbCl}_2} = 10.4\text{ g of PbCl}_2\text{ (3 SFs)}$

b. $50.0\text{ mL solution} \times \frac{1\text{ L solution}}{1000\text{ mL solution}} \times \frac{1.50\text{ mol KCl}}{1\text{ L solution}} = 0.0750\text{ mol of KCl}$

$0.0750\text{ mol of KCl} \times \frac{1\text{ mol Pb(NO}_3)_2}{2\text{ mol KCl}} \times \frac{1\text{ L solution}}{2.00\text{ mol Pb(NO}_3)_2} \times \frac{1000\text{ mL solution}}{1\text{ L solution}}$

$= 18.8\text{ mL of Pb(NO}_3)_2\text{ solution (3 SFs)}$

c. $30.0\text{ mL solution} \times \frac{1\text{ L solution}}{1000\text{ mL solution}} \times \frac{0.400\text{ mol Pb(NO}_3)_2}{1\text{ L solution}} \times \frac{2\text{ mol KCl}}{1\text{ mol Pb(NO}_3)_2}$

$= 0.0240\text{ mol of KCl}$

$20.0\text{ mL solution} \times \frac{1\text{ L solution}}{1000\text{ mL solution}} = 0.0200\text{ L of solution}$

$\text{molarity (M)} = \frac{\text{moles of solute}}{\text{liters of solution}} = \frac{0.0240\text{ mol KCl}}{0.0200\text{ L solution}} = 1.20\text{ M KCl solution (3 SFs)}$

12.53 a. $15.0\ \cancel{\text{g Mg}} \times \frac{1\ \cancel{\text{mol Mg}}}{24.31\ \cancel{\text{g Mg}}} \times \frac{2\ \cancel{\text{mol HCl}}}{1\ \cancel{\text{mol Mg}}} \times \frac{1\ \cancel{\text{L solution}}}{6.00\ \cancel{\text{mol HCl}}} \times \frac{1000\ \text{mL solution}}{1\ \cancel{\text{L solution}}}$

$= 206$ mL of HCl solution (3 SFs)

b. $0.500\ \cancel{\text{L solution}} \times \frac{2.00\ \cancel{\text{mol HCl}}}{1\ \cancel{\text{L solution}}} \times \frac{1\ \cancel{\text{mol H}_2}}{2\ \cancel{\text{mol HCl}}} \times \frac{22.4\ \text{L H}_2}{1\ \cancel{\text{mol H}_2}} = 11.2$ L of H_2 gas at STP (3 SFs)

c. $n = \frac{PV}{RT} = \frac{(735\ \cancel{\text{mmHg}})(5.20\ \cancel{\text{L}})}{\left(\frac{62.4\ \cancel{\text{L}} \cdot \cancel{\text{mmHg}}}{\text{mol} \cdot \cancel{\text{K}}}\right)(298\ \cancel{\text{K}})} = 0.206$ mol of H_2 gas

$0.206\ \cancel{\text{mol H}_2} \times \frac{2\ \text{mol HCl}}{1\ \cancel{\text{mol H}_2}} = 0.412$ mol of HCl

$45.2\ \cancel{\text{mL solution}} \times \frac{1\ \text{L solution}}{1000\ \cancel{\text{mL solution}}} = 0.0452$ L of solution

$\text{molarity (M)} = \frac{\text{moles of solute}}{\text{liters of solution}} = \frac{0.412\ \text{mol HCl}}{0.0452\ \text{L solution}} = 9.12$ M HCl solution (3 SFs)

12.55 a. A solution cannot be separated by a semipermeable membrane.

b. A suspension settles out upon standing.

12.57 a. The strong electrolyte NaCl dissolves in water producing 2 mol of Na^+ and Cl^- ions per mole of NaCl. Since freezing-point lowering depends on the number of particles in solution, only 0.60 mol of NaCl would be needed to have the same effect in 1 kg of water as 1.2 mol of the nonelectrolyte ethylene glycol.

b. The strong electrolyte K_3PO_4 dissolves in water producing 4 mol of K^+ and PO_4^{3-} ions per mole of K_3PO_4. Since freezing-point lowering depends on the number of particles in solution, only 0.30 mol of K_3PO_4 would be needed to have the same effect in 1 kg of water as 1.2 mol of the nonelectrolyte ethylene glycol.

12.59 a. $325\ \cancel{\text{g CH}_3\text{—OH}} \times \frac{1\ \text{mol CH}_3\text{—OH}}{32.04\ \cancel{\text{g CH}_3\text{—OH}}} = 10.1$ mol of CH_3—OH

$455\ \cancel{\text{g H}_2\text{O}} \times \frac{1\ \text{kg H}_2\text{O}}{1000\ \cancel{\text{g H}_2\text{O}}} = 0.455$ kg of H_2O

$\text{molality } (m) = \frac{\text{moles of solute}}{\text{kilograms of water}} = \frac{10.1\ \text{mol CH}_3\text{—OH}}{0.455\ \text{kg H}_2\text{O}}$

$= 22.2\ m$ CH_3—OH solution (3 SFs)

b. $640.\ \cancel{\text{g C}_3\text{H}_8\text{O}_2} \times \frac{1\ \text{mol C}_3\text{H}_8\text{O}_2}{76.09\ \cancel{\text{g C}_3\text{H}_8\text{O}_2}} = 8.41$ mol of $C_3H_8O_2$

$\text{molality } (m) = \frac{\text{moles of solute}}{\text{kilograms of water}} = \frac{8.41\ \text{mol C}_3\text{H}_8\text{O}_2}{1.22\ \text{kg H}_2\text{O}} = 6.89\ m$ $C_3H_8O_2$ solution (3 SFs)

12.61 a. $22.2\ m$ CH_3—OH solution; freezing-point lowering: $\Delta T_f = m\ K_f$;
boiling-point elevation: $\Delta T_b = m\ K_b$

$\Delta T_f = m\ K_f = 22.2\ \cancel{m} \times \frac{1.86\ °\text{C}}{1\ \cancel{m}} = 41.3\ °\text{C}$ (3 SFs)

freezing point $= 0.0\ °\text{C} - 41.3\ °\text{C} = -41.3\ °\text{C}$

$\Delta T_b = m\ K_b = 22.2\ \cancel{m} \times \frac{0.52\ °\text{C}}{1\ \cancel{m}} = 11.5\ °\text{C}$ (3 SFs)

boiling point $= 100.0\ °\text{C} + 11.5\ °\text{C} = 111.5\ °\text{C}$

b. 6.89 m $C_3H_8O_2$ solution; freezing-point lowering: $\Delta T_f = m\,K_f$;
boiling-point elevation: $\Delta T_b = m\,K_b$

$$\Delta T_f = m\,K_f = 6.89\,\cancel{m} \times \frac{1.86\ ^\circ\text{C}}{1\,\cancel{m}} = 12.8\ ^\circ\text{C (3 SFs)}$$

freezing point = 0.0 °C − 12.8 °C = −12.8 °C

$$\Delta T_b = m\,K_b = 6.89\,\cancel{m} \times \frac{0.52\ ^\circ\text{C}}{1\,\cancel{m}} = 3.58\ ^\circ\text{C (3 SFs)}$$

boiling point = 100.0 °C + 3.58 °C = 103.6 °C

12.63 Water will flow from a region of higher solvent concentration (which corresponds to a lower solute concentration) to a region of lower solvent concentration (which corresponds to a higher solute concentration).

a. The volume of compartment B will rise as water flows into compartment B, which contains the 10% (m/v) starch solution.

b. The volume of compartment B will rise as water flows into compartment B, which contains the 8% (m/v) albumin solution.

c. The volume of compartment A will rise as water flows into compartment A, which contains the 10% (m/v) sucrose solution.

12.65 **a.** 3 (no dissociation)

b. 1 (some dissociation, a few ions)

c. 2 (all ionized)

12.67 **a.** 2 To halve the mass percent (m/m), the volume would double.

b. 3 To go to one-fourth the mass percent (m/m), the volume would be four times the initial volume.

12.69 **a.** Beaker 3; Solid silver chloride (AgCl) will precipitate when the two solutions are mixed.

b. $NaCl(aq) + AgNO_3(aq) \rightarrow AgCl(s) + NaNO_3(aq)$

$\cancel{Na^+(aq)} + Cl^-(aq) + Ag^+(aq) + \cancel{NO_3^-(aq)} \rightarrow AgCl(s) + \cancel{Na^+(aq)} + \cancel{NO_3^-(aq)}$

c. $Ag^+(aq) + Cl^-(aq) \rightarrow AgCl(s)$ Net ionic equation

12.71 A "brine" saltwater solution has a high concentration of Na^+Cl^-, which is hypertonic to the cucumber. The skin of the cucumber acts like a semipermeable membrane; therefore, water flows from the more dilute solution inside the cucumber into the more concentrated brine solution that surrounds it. The loss of water causes the cucumber to become a wrinkled pickle.

12.73 Because iodine is a nonpolar molecule, it will dissolve in hexane, a nonpolar solvent. Iodine does not dissolve in water because water is a polar solvent.

12.75 $$80.0\ \cancel{\text{g NaCl}} \times \frac{100\ \text{g water}}{36.0\ \cancel{\text{g NaCl}}} = 222\ \text{g of water needed (3 SFs)}$$

12.77 At 20 °C, KNO_3 has a solubility of 32 g of KNO_3 in 100 g of H_2O.

a. 200. g of H_2O will dissolve:

$$200.\ \cancel{\text{g H}_2\text{O}} \times \frac{32\ \text{g KNO}_3}{100\ \cancel{\text{g H}_2\text{O}}} = 64\ \text{g of KNO}_3\ \text{(2 SFs)}$$

Because 32 g of KNO_3 is less than the maximum amount that can dissolve in 200. g of H_2O at 20 °C, the KNO_3 solution is unsaturated.

b. 50. g of H_2O will dissolve:

$$50.\ \cancel{\text{g H}_2\text{O}} \times \frac{32\ \text{g KNO}_3}{100\ \cancel{\text{g H}_2\text{O}}} = 16\ \text{g of KNO}_3\ \text{(2 SFs)}$$

Because 19 g of KNO_3 exceeds the maximum amount that can dissolve in 50. g of H_2O at 20 °C, the KNO_3 solution is saturated, and excess undissolved KNO_3 will be present on the bottom of the container.

c. 150. g of H_2O will dissolve:

$$150.\ \cancel{g\ H_2O} \times \frac{32\ g\ KNO_3}{100\ \cancel{g\ H_2O}} = 48\ g\ of\ KNO_3\ (2\ SFs)$$

Because 68 g of KNO_3 exceeds the maximum amount that can dissolve in 150. g of H_2O at 20 °C, the KNO_3 solution is saturated, and excess undissolved KNO_3 will be present on the bottom of the container.

12.79 When solutions of $NaNO_3$ and KCl are mixed, no insoluble products are formed; all of the possible combinations of salts are soluble. When KCl and $Pb(NO_3)_2$ solutions are mixed, the insoluble salt $PbCl_2$ forms.

12.81 **a.** Solid silver chloride (AgCl) forms:

$AgNO_3(aq) + LiCl(aq) \rightarrow AgCl(s) + LiNO_3(aq)$

$Ag^+(aq) + \cancel{NO_3^-(aq)} + \cancel{Li^+(aq)} + Cl^-(aq) \rightarrow AgCl(s) + \cancel{Li^+(aq)} + \cancel{NO_3^-(aq)}$

$Ag^+(aq) + Cl^-(aq) \rightarrow AgCl(s)$ Net ionic equation

b. none (no solid forms; salts containing K^+ and Na^+ are soluble)

c. Solid barium sulfate ($BaSO_4$) forms:

$Na_2SO_4(aq) + BaCl_2(aq) \rightarrow BaSO_4(s) + 2NaCl(aq)$

$\cancel{2Na^+(aq)} + SO_4^{2-}(aq) + Ba^{2+}(aq) + \cancel{2Cl^-(aq)} \rightarrow BaSO_4(s) + \cancel{2Na^+(aq)} + \cancel{2Cl^-(aq)}$

$Ba^{2+}(aq) + SO_4^{2-}(aq) \rightarrow BaSO_4(s)$ Net ionic equation

12.83 $$4.5\ \cancel{mL\ propyl\ alcohol} \times \frac{100\ mL\ solution}{12\ \cancel{mL\ propyl\ alcohol}} = 38\ mL\ of\ propyl\ alcohol\ solution\ needed\ (2\ SFs)$$

12.85 mass of solution: 70.0 g of HNO_3 + 130.0 g of H_2O = 200.0 g of solution

a. $$\frac{70.0\ g\ HNO_3}{200.0\ g\ solution} \times 100\% = 35.0\%\ (m/m)\ HNO_3\ solution\ (3\ SFs)$$

b. $$200.0\ \cancel{g\ solution} \times \frac{1\ mL\ solution}{1.21\ \cancel{g\ solution}} = 165\ mL\ of\ solution\ (3\ SFs)$$

c. $$\frac{70.0\ \cancel{g\ HNO_3}}{165\ \cancel{mL\ solution}} \times \frac{1\ mol\ HNO_3}{63.02\ \cancel{g\ HNO_3}} \times \frac{1000\ \cancel{mL\ solution}}{1\ L\ solution} = 6.73\ M\ HNO_3\ solution\ (3\ SFs)$$

12.87 $$60.0\ \cancel{g\ KNO_3} \times \frac{1\ \cancel{mol\ KNO_3}}{101.11\ \cancel{g\ KNO_3}} \times \frac{1\ L\ solution}{2.50\ \cancel{mol\ KNO_3}} = 0.237\ L\ of\ KNO_3\ solution\ (3\ SFs)$$

12.89 $$250.\ \cancel{mL\ solution} \times \frac{1\ \cancel{L\ solution}}{1000\ \cancel{mL\ solution}} \times \frac{2.00\ \cancel{mol\ KCl}}{1\ \cancel{L\ solution}} \times \frac{74.55\ g\ KCl}{1\ \cancel{mol\ KCl}} = 37.3\ g\ of\ KCl\ (3\ SFs)$$

To make a 2.00 M KCl solution, weigh out 37.3 g of KCl (0.500 mol) and place in a volumetric flask. Add enough water to dissolve the KCl and give a final volume of 250. mL.

12.91 **a.** $$2.52\ \cancel{L\ solution} \times \frac{3.00\ \cancel{mol\ KNO_3}}{1\ \cancel{L\ solution}} \times \frac{101.11\ g\ KNO_3}{1\ \cancel{mol\ KNO_3}} = 764\ g\ of\ KNO_3\ (3\ SFs)$$

b. $$75.0\ \cancel{mL\ solution} \times \frac{1\ \cancel{L\ solution}}{1000\ \cancel{mL\ solution}} \times \frac{0.506\ \cancel{mol\ Na_2SO_4}}{1\ \cancel{L\ solution}} \times \frac{142.05\ g\ Na_2SO_4}{1\ \cancel{mol\ Na_2SO_4}}$$

$= 5.39$ g of Na_2SO_4 (3 SFs)

c. $$45.2\ \cancel{mL\ solution} \times \frac{1\ \cancel{L\ solution}}{1000\ \cancel{mL\ solution}} \times \frac{1.80\ \cancel{mol\ HCl}}{1\ \cancel{L\ solution}} \times \frac{36.46\ g\ HCl}{1\ \cancel{mol\ HCl}}$$

$= 2.97$ g of HCl (3 SFs)

12.93 $$60.0\ \cancel{mL\ solution} \times \frac{1\ \cancel{L\ solution}}{1000\ \cancel{mL\ solution}} \times \frac{2.00\ \cancel{mol\ Al(OH)_3}}{1\ \cancel{L\ solution}} \times \frac{3\ \cancel{mol\ HCl}}{1\ \cancel{mol\ Al(OH)_3}} \times \frac{1\ \cancel{L\ solution}}{6.00\ \cancel{mol\ HCl}}$$

$$\times \frac{1000\ mL\ HCl\ solution}{1\ \cancel{L\ solution}} = 60.0\ mL\ of\ HCl\ solution\ (3\ SFs)$$

12.95 $n = \dfrac{PV}{RT} = \dfrac{(745\ \cancel{\text{mmHg}})(4.20\ \cancel{\text{L}})}{\left(\dfrac{62.4\ \cancel{\text{L}}\cdot\cancel{\text{mmHg}}}{\text{mol}\cdot\cancel{\text{K}}}\right)(308\ \cancel{\text{K}})} = 0.163 \text{ mol of } H_2 \text{ gas}$

$0.163\ \cancel{\text{mol } H_2} \times \dfrac{2 \text{ mol HCl}}{1\ \cancel{\text{mol } H_2}} = 0.326 \text{ mol of HCl}$

$355\ \cancel{\text{mL solution}} \times \dfrac{1 \text{ L solution}}{1000\ \cancel{\text{mL solution}}} = 0.355 \text{ L of solution}$

$\text{molarity (M)} = \dfrac{\text{moles of solute}}{\text{liters of solution}} = \dfrac{0.326 \text{ mol HCl}}{0.355 \text{ L solution}} = 0.918 \text{ M HCl solution (3 SFs)}$

12.97 $M_1V_1 = M_2V_2$

a. $M_2 = M_1 \times \dfrac{V_1}{V_2} = 0.200 \text{ M} \times \dfrac{25.0\ \cancel{\text{mL}}}{50.0\ \cancel{\text{mL}}} = 0.100 \text{ M NaBr solution (3 SFs)}$

b. $M_2 = M_1 \times \dfrac{V_1}{V_2} = 1.20 \text{ M} \times \dfrac{15.0\ \cancel{\text{mL}}}{40.0\ \cancel{\text{mL}}} = 0.450 \text{ M } K_2SO_4 \text{ solution (3 SFs)}$

c. $M_2 = M_1 \times \dfrac{V_1}{V_2} = 6.00 \text{ M} \times \dfrac{75.0\ \cancel{\text{mL}}}{255\ \cancel{\text{mL}}} = 1.76 \text{ M NaOH solution (3 SFs)}$

12.99 $M_1V_1 = M_2V_2$

a. $V_2 = V_1 \times \dfrac{M_1}{M_2} = 25.0 \text{ mL} \times \dfrac{5.00\ \cancel{\text{M}}}{2.50\ \cancel{\text{M}}} = 50.0 \text{ mL (3 SFs)}$

b. $V_2 = V_1 \times \dfrac{M_1}{M_2} = 25.0 \text{ mL} \times \dfrac{5.00\ \cancel{\text{M}}}{1.00\ \cancel{\text{M}}} = 125 \text{ mL (3 SFs)}$

c. $V_2 = V_1 \times \dfrac{M_1}{M_2} = 25.0 \text{ mL} \times \dfrac{5.00\ \cancel{\text{M}}}{0.500\ \cancel{\text{M}}} = 250. \text{ mL (3 SFs)}$

12.101 A solution with a high salt (solute) concentration is hypertonic to the cells of the flowers. Water (solvent) will flow out of the cells of the flowers into the hypertonic salt solution that surrounds them, resulting in "dried" flowers.

12.103 **a.** Na^+ salts are soluble.
b. The halide salts containing Pb^{2+} are insoluble.
c. Most halide salts are soluble.
d. A salt containing the NH_4^+ ion is soluble.
e. The salt containing Mg^{2+} and CO_3^{2-} is insoluble.
f. The salt containing Fe^{3+} and PO_4^{3-} is insoluble.

12.105 **a.** mass of NaCl: 25.50 g − 24.10 g = 1.40 g of NaCl
mass of solution: 36.15 g − 24.10 g = 12.05 g of solution

mass percent (m/m): $\dfrac{1.40 \text{ g NaCl}}{12.05 \text{ g solution}} \times 100\% = 11.6\% \text{ (m/m) NaCl solution (3 SFs)}$

b. molarity (M): $\dfrac{1.40\ \cancel{\text{g NaCl}}}{10.0\ \cancel{\text{mL solution}}} \times \dfrac{1 \text{ mol NaCl}}{58.44\ \cancel{\text{g NaCl}}} \times \dfrac{1000\ \cancel{\text{mL solution}}}{1 \text{ L solution}}$
$= 2.40 \text{ M NaCl solution (3 SFs)}$

c. $M_1V_1 = M_2V_2 \quad M_2 = M_1 \times \dfrac{V_1}{V_2} = 2.40 \text{ M} \times \dfrac{10.0\ \cancel{\text{mL}}}{60.0\ \cancel{\text{mL}}} = 0.400 \text{ M NaCl solution (3 SFs)}$

12.107 $15.2\ \cancel{\text{g LiCl}} \times \dfrac{1\ \cancel{\text{mol LiCl}}}{42.39\ \cancel{\text{g LiCl}}} \times \dfrac{1\ \cancel{\text{L solution}}}{1.75\ \cancel{\text{mol LiCl}}} \times \dfrac{1000 \text{ mL solution}}{1\ \cancel{\text{L solution}}}$
$= 205 \text{ mL of LiCl solution (3 SFs)}$

12.109 $4.20\ \cancel{\text{L H}_2} \times \dfrac{1\ \cancel{\text{mol H}_2}}{22.4\ \cancel{\text{L H}_2}} \times \dfrac{2\ \text{mol HCl}}{1\ \cancel{\text{mol H}_2}} = 0.375\ \text{mol of HCl}$

$250.\ \cancel{\text{mL solution}} \times \dfrac{1\ \text{L solution}}{1000\ \cancel{\text{mL solution}}} = 0.250\ \text{L of solution}$

$\text{molarity (M)} = \dfrac{\text{moles of solute}}{\text{liters of solution}} = \dfrac{0.375\ \text{mol HCl}}{0.250\ \text{L solution}} = 1.50\ \text{M HCl solution (3 SFs)}$

12.111 **a.** The boiling point of the NaCl solution is given as 101.04 °C. Since the boiling point for pure water is 100.0 °C, the ΔT_b is 101.04 °C − 100.0 °C = 1.04 °C.

Rearranging $\Delta T_b = m\,K_b$ to solve for m:

$m = \dfrac{\Delta T_b}{K_b} = \dfrac{1.04\ \cancel{^\circ\text{C}}}{0.52\ \cancel{^\circ\text{C}}/m} = 2.0\ m\ \text{NaCl solution (2 SFs)}$

b. $\Delta T_f = m\,K_f = 2.0\ \cancel{m} \times \dfrac{1.86\ ^\circ\text{C}}{1\ \cancel{m}} = 3.7\ ^\circ\text{C (2 SFs)}$

freezing point = 0.0 °C − 3.7 °C = −3.7 °C

13
Chemical Equilibrium

Study Goals

- Describe how temperature, concentration, and catalysts affect the rate of a reaction.
- Use the concept of reversible reactions to explain chemical equilibrium.
- Write the equilibrium constant expression for a reaction.
- Calculate the equilibrium constant for a reversible reaction given the concentrations of reactants and products at equilibrium.
- Use an equilibrium constant to predict the extent of reaction and to calculate equilibrium concentrations.
- Use Le Châtelier's principle to describe the changes made in equilibrium concentrations when reaction conditions change.
- Write the solubility product constant expression for a slightly soluble salt and calculate the K_{sp}; use the K_{sp} to determine the solubility of a slightly soluble salt.

Chapter Outline

Answers and Solutions to Text Problems

13.1 **a.** The rate of a reaction indicates how fast the products form or how fast the reactants are used up.
b. At room temperature, more of the reactants will have the energy necessary to proceed to products (the activation energy) than at the lower temperature of the refrigerator, so the rate of formation of bread mold will be faster.

13.3 Adding $Br_2(g)$ molecules increases the concentration of reactants, which increases the number of collisions that take place between the reactants.

13.5 **a.** Adding more reactant increases the number of collisions that take place between the reactants, which increases the reaction rate.

b. Raising the temperature increases the kinetic energy of the reactant molecules, which increases the number of collisions and makes more collisions effective. The rate of reaction will be increased.

c. Adding a catalyst lowers the energy of activation, which increases the reaction rate.

d. Removing a reactant decreases the number of collisions that take place between the reactants, which decreases the reaction rate.

13.7 A reversible reaction is one in which a forward reaction converts reactants to products, while a reverse reaction converts products to reactants.

13.9 **a.** Broken glass cannot be put back together; this process is not reversible.

b. In this physical process, heat melts the solid form of water, while removing heat can change liquid water back to solid. This process is reversible.

c. A pan is warmed when heated and cooled when heat is removed; this process is reversible.

13.11 In the expression for K_c, the products are divided by the reactants, with each concentration raised to a power equal to its coefficient in the balanced chemical equation:

a. $K_c = \dfrac{[CS_2][H_2]^4}{[CH_4][H_2S]^2}$ **b.** $K_c = \dfrac{[N_2][O_2]}{[NO]^2}$ **c.** $K_c = \dfrac{[CS_2][O_2]^4}{[SO_3]^2[CO_2]}$

13.13 **a.** only one state (gas) is present; homogeneous equilibrium

b. solid and gaseous states present; heterogeneous equilibrium

c. only one state (gas) is present; homogeneous equilibrium

d. gaseous and liquid states present; heterogeneous equilibrium

13.15 **a.** $K_c = \dfrac{[O_2]^3}{[O_3]^2}$ **b.** $K_c = [CO_2][H_2O]$

c. $K_c = \dfrac{[H_2]^3[CO]}{[CH_4][H_2O]}$ **d.** $K_c = \dfrac{[Cl_2]^2}{[HCl]^4[O_2]}$

13.17 $K_c = \dfrac{[NO_2]^2}{[N_2O_4]} = \dfrac{[0.21]^2}{[0.030]} = 1.5 \text{ (2 SFs)}$

13.19 $K_c = \dfrac{[CH_4][H_2O]}{[CO][H_2]^3} = \dfrac{[1.8][2.0]}{[0.51][0.30]^3} = 260 \text{ (2 SFs)}$

13.21 **a.** A large K_c value indicates that the equilibrium mixture contains mostly products.

b. A large K_c value indicates that the equilibrium mixture contains mostly products.

c. A small K_c value indicates that the equilibrium mixture contains mostly reactants.

13.23 $K_c = \dfrac{[HI]^2}{[H_2][I_2]} = 54$

Rearrange the K_c expression to solve for $[H_2]$ and substitute in known values.

$$[H_2] = \frac{[HI]^2}{K_c[I_2]} = \frac{[0.030]^2}{54[0.015]} = 1.1 \times 10^{-3}\text{ M (2 SFs)}$$

13.25 $K_c = \dfrac{[NO]^2[Br_2]}{[NOBr]^2} = 2.0$

Rearrange the K_c expression to solve for [NOBr] and substitute in known values.

$$[NOBr]^2 = \frac{[NO]^2[Br_2]}{K_c} = \frac{[2.0]^2[1.0]}{2.0} = 2.0$$

Take the square root of both sides of the equation.

$[NOBr] = \sqrt{2.0} = 1.4$ M (2 SFs)

13.27 **a.** When more reactant is added to an equilibrium mixture, the product/reactant ratio is initially less than K_c.

b. According to Le Châtelier's principle, equilibrium is reestablished when the forward reaction forms more products to make the product/reactant ratio equal the K_c again.

13.29 **a.** Adding more reactant shifts equilibrium toward product.

b. Adding more product shifts equilibrium toward reactant.

c. Raising the temperature of an endothermic reaction shifts equilibrium toward product.

d. Decreasing volume favors the side of the reaction with fewer moles of gas, so there is a shift toward product.

e. No shift in equilibrium occurs when a catalyst is added.

13.31 **a.** Adding more reactant shifts equilibrium toward product.

b. Raising the temperature of an endothermic reaction shifts equilibrium toward product.

c. Removing product shifts equilibrium toward product.

d. No shift in equilibrium occurs when a catalyst is added.

e. Removing reactant shifts equilibrium toward reactants.

13.33 **a.** $MgCO_3(s) \rightleftarrows Mg^{2+}(aq) + CO_3^{2-}(aq)$; $K_{sp} = [Mg^{2+}][CO_3^{2-}]$

b. $CaF_2(s) \rightleftarrows Ca^{2+}(aq) + 2F^-(aq)$; $K_{sp} = [Ca^{2+}][F^-]^2$

c. $Ag_3PO_4(s) \rightleftarrows 3Ag^+(aq) + PO_4^{3-}(aq)$; $K_{sp} = [Ag^+]^3[PO_4^{3-}]$

13.35 $BaSO_4(s) \rightleftarrows Ba^{2+}(aq) + SO_4^{2-}(aq)$;

$K_{sp} = [Ba^{2+}][SO_4^{2-}] = [1 \times 10^{-5}][1 \times 10^{-5}] = 1 \times 10^{-10}$ (1 SF)

13.37 $Ag_2CO_3(s) \rightleftarrows 2Ag^+(aq) + CO_3^{2-}(aq)$;

$K_{sp} = [Ag^+]^2[CO_3^{2-}] = [2.6 \times 10^{-4}]^2[1.3 \times 10^{-4}] = 8.8 \times 10^{-12}$ (2 SFs)

13.39 $CuI(s) \rightleftarrows Cu^+(aq) + I^-(aq)$; $K_{sp} = [Cu^+][I^-]$

Substitute S for the molarity of each ion into the K_{sp} expression.

$K_{sp} = S \times S = S^2 = 1 \times 10^{-12}$

Calculate the molar solubility (S) by taking the square root of both sides of the equation.

$S = \sqrt{1 \times 10^{-12}} = 1 \times 10^{-6}$ M (1 SF)

$\therefore [Cu^+] = [I^-] = 1 \times 10^{-6}$ M

13.41 $AgCl(s) \rightleftarrows Ag^+(aq) + Cl^-(aq)$; $K_{sp} = [Ag^+][Cl^-]$

Rearrange the K_{sp} expression to solve for $[Cl^-]$ and substitute in known values.

$$[Cl^-] = \frac{K_{sp}}{[Ag^+]} = \frac{1.8 \times 10^{-10}}{[2.0 \times 10^{-3}]} = 9.0 \times 10^{-8} \text{ M (2 SFs)}$$

13.43 **a.** $K_c = \dfrac{[CO_2][H_2O]^2}{[CH_4][O_2]^2}$

b. $K_c = \dfrac{[N_2]^2[H_2O]^6}{[NH_3]^4[O_2]^3}$

c. $K_c = \dfrac{[CH_4]}{[H_2]^2}$

13.45 There are mostly products and a few reactants, so the reaction would have a large value of the equilibrium constant.

13.47 **a.** T_2 is lower than T_1. This would cause the exothermic reaction shown to shift toward products.
b. Because the equilibrium mixture at T_2 has more products, the K_c value at T_2 is larger than the K_c value for the equilibrium mixture at T_1.

13.49 **a.** Raising the temperature of an exothermic reaction shifts equilibrium toward reactants.
b. Decreasing volume favors the side of the reaction with fewer moles of gas, so there is a shift toward product.
c. Adding a catalyst does not shift equilibrium.
d. Adding more reactant shifts equilibrium toward product.

13.51 **a.** A large K_c value indicates that the equilibrium mixture contains mostly products.
b. A K_c value close to 1 indicates that the equilibrium mixture contains both products and reactants.
c. A small K_c value indicates that the equilibrium mixture contains mostly reactants.

13.53 The numerator in the K_c expression gives the products in the equation, and the denominator gives the reactants.
a. $SO_2Cl_2(g) \rightleftarrows SO_2(g) + Cl_2(g)$
b. $Br_2(g) + Cl_2(g) \rightleftarrows 2BrCl(g)$
c. $CO(g) + 3H_2(g) \rightleftarrows CH_4(g) + H_2O(g)$
d. $2O_2(g) + 2NH_3(g) \rightleftarrows N_2O(g) + 3H_2O(g)$

13.55 **a.** $K_c = \dfrac{[N_2][H_2]^3}{[NH_3]^2}$

b. $K_c = \dfrac{[3.0][0.50]^3}{[0.20]^2} = 9.4 \text{ (2 SFs)}$

13.57 $K_c = \dfrac{[N_2O_4]}{[NO_2]^2} = 5.0$

Rearrange the K_c expression to solve for $[N_2O_4]$ and substitute in known values.

$[N_2O_4] = K_c[NO_2]^2 = 5.0[0.50]^2 = 1.3 \text{ M (2 SFs)}$

13.59 **a.** When the reactant $[O_2]$ increases, the rate of the forward reaction increases to shift the equilibrium toward the products.
b. When the product $[O_2]$ increases, the rate of the reverse reaction increases to shift the equilibrium toward the reactants.
c. When the reactant $[O_2]$ increases, the rate of the forward reaction increases to shift the equilibrium toward the products.
d. When the product $[O_2]$ increases, the rate of the reverse reaction increases to shift the equilibrium toward the reactants.

13.61 Decreasing the volume of an equilibrium mixture shifts the equilibrium toward the side of the reaction that has the fewer number of moles of gas. No shift occurs when there are an equal number of moles of gas on both sides of the equation.

a. With 3 mol of gas on the reactant side and 2 mol of gas on the product side, decreasing the volume will shift equilibrium toward products.

b. With 2 mol of gas on the reactant side and 3 mol of gas on the product side, decreasing the volume will shift equilibrium toward reactants.

c. With 6 mol of gas on reactant side and 0 mol of gas on the product side, decreasing the volume will shift equilibrium toward products.

d. With 4 mol of gas on the reactant side and 5 mol of gas on the product side, decreasing the volume will shift equilibrium toward reactants.

13.63 a. $CuCO_3(s) \rightleftarrows Cu^{2+}(aq) + CO_3^{2-}(aq)$; $K_{sp} = [Cu^{2+}][CO_3^{2-}]$

b. $PbF_2(s) \rightleftarrows Pb^{2+}(aq) + 2F^-(aq)$; $K_{sp} = [Pb^{2+}][F^-]^2$

c. $Fe(OH)_3(s) \rightleftarrows Fe^{3+}(aq) + 3OH^-(aq)$; $K_{sp} = [Fe^{3+}][OH^-]^3$

13.65 $FeS(s) \rightleftarrows Fe^{2+}(aq) + S^{2-}(aq)$; $K_{sp} = [Fe^{2+}][S^{2-}]$

$K_{sp} = [7.7 \times 10^{-10}][7.7 \times 10^{-10}] = 5.9 \times 10^{-19}$ (2 SFs)

13.67 $Mn(OH)_2(s) \rightleftarrows Mn^{2+}(aq) + 2OH^-(aq)$; $K_{sp} = [Mn^{2+}][OH^-]^2$

$K_{sp} = [3.7 \times 10^{-5}][7.4 \times 10^{-5}]^2 = 2.0 \times 10^{-13}$ (2 SFs)

13.69 $CdS(s) \rightleftarrows Cd^{2+}(aq) + S^{2-}(aq)$; $K_{sp} = [Cd^{2+}][S^{2-}]$

Substitute S for the molarity of each ion into the K_{sp} expression.

$K_{sp} = S \times S = 1.0 \times 10^{-24}$

Calculate the molar solubility (S) by taking the square root of both sides of the equation.

$S = \sqrt{1.0 \times 10^{-24}} = 1.0 \times 10^{-12}$ M (2 SFs)

$\therefore [Cd^{2+}] = [S^{2-}] = 1 \times 10^{-12}$ M

13.71 $BaSO_4(s) \rightleftarrows Ba^{2+}(aq) + SO_4^{2-}(aq)$; $K_{sp} = [Ba^{2+}][SO_4^{2-}]$

Rearrange the K_{sp} expression to solve for $[SO_4^{2-}]$ and substitute in known values.

$$[SO_4^{2-}] = \frac{K_{sp}}{[Ba^{2+}]} = \frac{1.1 \times 10^{-10}}{[1.0 \times 10^{-3}]} = 1.1 \times 10^{-7} \text{ M (2 SFs)}$$

13.73 a. A small K_c value indicates that the equilibrium mixture contains mostly reactants.

b. A large K_c value indicates that the equilibrium mixture contains mostly products.

13.75 a. $K_c = \dfrac{[NO]^2[Br_2]}{[NOBr]^2}$

b. When the concentrations are substituted into the expression, the result is 1.0, which is not equal to K_c (2.0). Therefore, the system is not at equilibrium.

$$K_c = \frac{[NO]^2[Br_2]}{[NOBr]^2} = \frac{[1.0]^2[1.0]}{[1.0]^2} = 1.0 \text{ (2 SFs)}$$

c. Since the calculated value in part **b** is less than K_c, the rate of the forward reaction will initially increase.

d. When the system has reestablished equilibrium, the $[Br_2]$ and [NO] will have increased, and the [NOBr] will have decreased.

13.77 $Mg(OH)_2(s) \rightleftarrows Mg^{2+}(aq) + 2OH^-(aq)$; $K_{sp} = [Mg^{2+}][OH^-]^2$

$$\frac{9.7 \times 10^{-3}\ \cancel{\text{g Mg(OH)}_2}}{1\ \text{L solution}} \times \frac{1\ \text{mol Mg(OH)}_2}{58.33\ \cancel{\text{g Mg(OH)}_2}} = 1.7 \times 10^{-4}\ \text{M Mg(OH)}_2\ \text{solution}$$

$$\frac{1.7 \times 10^{-4}\ \cancel{\text{mol Mg(OH)}_2}}{1\ \text{L solution}} \times \frac{1\ \text{mol Mg}^{2+}}{1\ \cancel{\text{mol Mg(OH)}_2}} = 1.7 \times 10^{-4}\ \text{M Mg}^{2+} = [\text{Mg}^{2+}]$$

$$\frac{1.7 \times 10^{-4}\ \cancel{\text{mol Mg(OH)}_2}}{1\ \text{L solution}} \times \frac{2\ \text{mol OH}^-}{1\ \cancel{\text{mol Mg(OH)}_2}} = 3.4 \times 10^{-4}\ \text{M OH}^- = [\text{OH}^-]$$

$K_{sp} = [Mg^{2+}][OH^-]^2 = [1.7 \times 10^{-4}][3.4 \times 10^{-4}]^2 = 2.0 \times 10^{-11}$ (2 SFs)

14
Acids and Bases

Study Goals

- Describe and name Arrhenius, Brønsted–Lowry, and organic acids and bases.
- Identify conjugate acid–base pairs for Brønsted–Lowry acids and bases.
- Write equations for the dissociation of strong and weak acids; identify the direction of reaction.
- Write the expression for the dissociation constant of a weak acid or weak base.
- Use the ion product of water to calculate the $[H_3O^+]$ and $[OH^-]$ in an aqueous solution.
- Calculate pH from $[H_3O^+]$; given the pH, calculate $[H_3O^+]$ and $[OH^-]$ of a solution.
- Write balanced equations for reactions of acids with metals, carbonates, and bases.
- Calculate the molarity or volume of an acid or base from titration information.
- Predict whether a salt will form an acidic, basic, or neutral solution.
- Describe the role of buffers in maintaining the pH of a solution.

Chapter Outline

Answers and Solutions to Text Problems

14.1 **a.** Acids taste sour. **b.** Acids neutralize bases.
c. Acids produce H^+ ions in water. **d.** Barium hydroxide is the name of a base.

14.3 Acids containing a simple nonmetal anion use the prefix *hydro*, followed by the name of the anion with its *ide* ending changed to *ic acid*. When the anion is an oxygen-containing polyatomic ion, the *ate* ending of the polyatomic anion is replaced with *ic acid*. Acids with one oxygen less than the common *ic acid* name are named as *ous acids*. Carboxylic acids are named using IUPAC rules where the alkane name of the longest carbon chain containing the carboxyl functional group has the *e* ending replaced with *oic acid*; many carboxylic acids also have common names derived from their natural sources. Bases are named as ionic compounds containing hydroxide anions.
a. hydrochloric acid **b.** calcium hydroxide **c.** carbonic acid
d. nitric acid **e.** sulfurous acid **f.** bromous acid
g. ethanoic acid (IUPAC); acetic acid (common)

14.5 **a.** $Mg(OH)_2$ **b.** HF **c.** HCOOH
d. LiOH **e.** NH_4OH **f.** HIO_4

14.7 A Brønsted–Lowry acid donates a proton (H^+), whereas a Brønsted–Lowry base accepts a proton.
a. HI is the acid (proton donor); H_2O is the base (proton acceptor).
b. H_2O is the acid (proton donor); F^- is the base (proton acceptor).
c. H_2S is the acid (proton donor); $CH_3—CH_2—NH_2$ is the base (proton acceptor).

14.9 To form the conjugate base, remove a proton (H^+) from the acid.
a. F^- **b.** OH^- **c.** HPO_3^{2-}
d. SO_4^{2-} **e.** ClO_2^-

14.11 To form the conjugate acid, add a proton (H^+) to the base.
a. HCO_3^- **b.** H_3O^+ **c.** H_3PO_4
d. HBr **e.** $HClO_4$

14.13 The conjugate acid is a proton donor, and the conjugate base is a proton acceptor.
a. acid H_2CO_3, conjugate base HCO_3^-; base H_2O, conjugate acid H_3O^+
b. acid NH_4^+, conjugate base NH_3; base H_2O, conjugate acid H_3O^+
c. acid HCN, conjugate base CN^-; base NO_2^-, conjugate acid HNO_2
d. acid HF, conjugate base F^-; base $CH_3—COO^-$, conjugate acid $CH_3—COOH$

14.15 $NH_4^+(aq) + H_2O(l) \rightleftharpoons NH_3(aq) + H_3O^+(aq)$

14.17 A strong acid is a good proton donor, whereas its conjugate base is a poor proton acceptor.

14.19 Use Table 14.4 to answer (the stronger acid will be closer to the top of the table).
a. HBr is the stronger acid. **b.** HSO_4^- is the stronger acid. **c.** H_2CO_3 is the stronger acid.

14.21 Use Table 14.4 to answer (the weaker acid will be closer to the bottom of the table).
a. HSO_4^- is the weaker acid. **b.** HNO_2 is the weaker acid. **c.** HCO_3^- is the weaker acid.

14.23 **a.** From Table 14.4, we see that H_2CO_3 is a weaker acid than H_3O^+ and that H_2O is a weaker base than HCO_3^-. Thus, the reactants are favored.
b. From Table 14.4, we see that NH_4^+ is a weaker acid than H_3O^+ and that H_2O is a weaker base than NH_3. Thus, the reactants are favored.
c. From Table 14.4, we see that NH_4^+ is a weaker acid than HCl and that Cl^- is a weaker base than NH_3. Thus, the products are favored.
d. From Table 14.4, we see that $CH_3—NH_3^+$ is a weaker acid than $CH_3—COOH$ and that $CH_3—COO^-$ is a weaker base than $CH_3—NH_2$. Thus, the reactants are favored.

14.25 This equilibrium favors the reactants because NH_4^+ is a weaker acid than HSO_4^- and SO_4^{2-} is a weaker base than NH_3.

$$NH_4^+(aq) + SO_4^{2-}(aq) \rightleftarrows NH_3(aq) + HSO_4^-(aq)$$

14.27 The smaller the K_a value, the weaker the acid. The weaker acid has the stronger conjugate base.

a. H_2SO_3, which has a larger K_a value than HS^-, is the stronger acid.
b. The conjugate base HSO_3^- is formed by removing a proton from the acid H_2SO_3.
c. The stronger acid, H_2SO_3, has the weaker conjugate base, HSO_3^-.
d. The weaker acid, HS^-, has the stronger conjugate base, S^{2-}.
e. The stronger acid, H_2SO_3, dissociates more and produces more ions.

14.29 $H_3PO_4(aq) + H_2O(l) \rightleftarrows H_3O^+(aq) + H_2PO_4^-(aq)$
The K_a expression is the ratio of the [products] divided by the [reactants] with $[H_2O]$ considered constant and part of the K_a:

$$K_a = \frac{[H_3O^+][H_2PO_4^-]}{[H_3PO_4]} = 7.5 \times 10^{-3}$$

14.31 In pure water, a small fraction of the water molecules break apart to form H^+ and OH^-. The H^+ combines with H_2O to form H_3O^+. Every time a H^+ is formed, a OH^- is also formed. Therefore, the concentration of the two must be equal in pure water.

14.33 In an acidic solution, the $[H_3O^+]$ is greater than the $[OH^-]$, which means that the $[H_3O^+]$ is greater than 1.0×10^{-7} M and the $[OH^-]$ is less than 1.0×10^{-7} M.

14.35 The value of $K_w = [H_3O^+][OH^-] = 1.0 \times 10^{-14}$ at 25 °C.

If $[H_3O^+]$ needs to be calculated from $[OH^-]$, then rearranging the K_w for $[H_3O^+]$ gives

$$[H_3O^+] = \frac{1.0 \times 10^{-14}}{[OH^-]}.$$

If $[OH^-]$ needs to be calculated from $[H_3O^+]$, then rearranging the K_w for $[OH^-]$ gives

$$[OH^-] = \frac{1.0 \times 10^{-14}}{[H_3O^+]}.$$

A neutral solution has $[OH^-] = [H_3O^+] = 1.0 \times 10^{-7}$ M. If the $[OH^-] > [H_3O^+]$, the solution is basic; if the $[H_3O^+] > [OH^-]$, the solution is acidic.

a. $[OH^-] = \dfrac{1.0 \times 10^{-14}}{[H_3O^+]} = \dfrac{1.0 \times 10^{-14}}{[2.0 \times 10^{-5}]} = 5.0 \times 10^{-10}$ M;
since $[H_3O^+] > [OH^-]$, the solution is acidic.

b. $[OH^-] = \dfrac{1.0 \times 10^{-14}}{[H_3O^+]} = \dfrac{1.0 \times 10^{-14}}{[1.4 \times 10^{-9}]} = 7.1 \times 10^{-6}$ M;
since $[OH^-] > [H_3O^+]$, the solution is basic.

c. $[H_3O^+] = \dfrac{1.0 \times 10^{-14}}{[OH^-]} = \dfrac{1.0 \times 10^{-14}}{[8.0 \times 10^{-3}]} = 1.3 \times 10^{-12}$ M;
since $[OH^-] > [H_3O^+]$, the solution is basic.

d. $[H_3O^+] = \dfrac{1.0 \times 10^{-14}}{[OH^-]} = \dfrac{1.0 \times 10^{-14}}{[3.5 \times 10^{-10}]} = 2.9 \times 10^{-5}$ M;

since $[H_3O^+] > [OH^-]$, the solution is acidic.

14.37 The value of $K_w = [H_3O^+][OH^-] = 1.0 \times 10^{-14}$ at 25 °C.

When $[OH^-]$ is known, the $[H_3O^+]$ can be calculated by rearranging the K_w for $[H_3O^+]$:

$$[H_3O^+] = \frac{1.0 \times 10^{-14}}{[OH^-]}$$

a. $[H_3O^+] = \dfrac{1.0 \times 10^{-14}}{[OH^-]} = \dfrac{1.0 \times 10^{-14}}{[1.0 \times 10^{-9}]} = 1.0 \times 10^{-5}$ M (2 SFs)

b. $[H_3O^+] = \dfrac{1.0 \times 10^{-14}}{[OH^-]} = \dfrac{1.0 \times 10^{-14}}{[1.0 \times 10^{-6}]} = 1.0 \times 10^{-8}$ M (2 SFs)

c. $[H_3O^+] = \dfrac{1.0 \times 10^{-14}}{[OH^-]} = \dfrac{1.0 \times 10^{-14}}{[2.0 \times 10^{-5}]} = 5.0 \times 10^{-10}$ M (2 SFs)

d. $[H_3O^+] = \dfrac{1.0 \times 10^{-14}}{[OH^-]} = \dfrac{1.0 \times 10^{-14}}{[4.0 \times 10^{-13}]} = 2.5 \times 10^{-2}$ M (2 SFs)

14.39 The value of $K_w = [H_3O^+][OH^-] = 1.0 \times 10^{-14}$ at 25 °C.
When $[H_3O^+]$ is known, the $[OH^-]$ can be calculated by rearranging the K_w for $[OH^-]$:

$$[OH^-] = \frac{1.0 \times 10^{-14}}{[H_3O^+]}$$

a. $[OH^-] = \dfrac{1.0 \times 10^{-14}}{[H_3O^+]} = \dfrac{1.0 \times 10^{-14}}{[1.0 \times 10^{-3}]} = 1.0 \times 10^{-11}$ M (2 SFs)

b. $[OH^-] = \dfrac{1.0 \times 10^{-14}}{[H_3O^+]} = \dfrac{1.0 \times 10^{-14}}{[5.0 \times 10^{-6}]} = 2.0 \times 10^{-9}$ M (2 SFs)

c. $[OH^-] = \dfrac{1.0 \times 10^{-14}}{[H_3O^+]} = \dfrac{1.0 \times 10^{-14}}{[1.8 \times 10^{-12}]} = 5.6 \times 10^{-3}$ M (2 SFs)

d. $[OH^-] = \dfrac{1.0 \times 10^{-14}}{[H_3O^+]} = \dfrac{1.0 \times 10^{-14}}{[4.0 \times 10^{-13}]} = 2.5 \times 10^{-2}$ M (2 SFs)

14.41 In a neutral solution, the $[H_3O^+] = 1.0 \times 10^{-7}$ M.

$pH = -\log[H_3O^+] = -\log[1.0 \times 10^{-7}] = 7.00$. The pH value contains two *decimal places*, which represent the two significant figures in the coefficient 1.0.

14.43 An acidic solution has a pH less than 7.0 (or a pOH greater than 7.0). A basic solution has a pH greater than 7.0 (or a pOH less than 7.0). A neutral solution has a pH (or pOH) equal to 7.0.

a. basic (pH $7.38 > 7.0$) **b.** acidic (pH $2.8 < 7.0$) **c.** basic (pOH $2.8 < 7.0$)
d. acidic (pH $5.52 < 7.0$) **e.** acidic (pH $4.2 < 7.0$) **f.** basic (pH $7.6 > 7.0$)

14.45 Since pH is a logarithmic scale, an increase or decrease of 1 pH unit changes the $[H_3O^+]$ by a factor of 10. Thus, a pH of 3 ($[H_3O^+] = 10^{-3}$ M, or 0.001 M) is ten times more acidic than a pH of 4 ($[H_3O^+] = 10^{-4}$ M, or 0.0001 M).

14.47 $pH = -\log[H_3O^+]$

Since the value of $K_w = [H_3O^+][OH^-] = 1.0 \times 10^{-14}$ at 25 °C, if $[H_3O^+]$ needs to be calculated from $[OH^-]$, rearranging the K_w for $[H_3O^+]$ gives $[H_3O^+] = \dfrac{1.0 \times 10^{-14}}{[OH^-]}$.

a. $pH = -\log[H_3O^+] = -\log[1.0 \times 10^{-4}] = 4.00$ (2 SFs on the right of the decimal point)

b. $pH = -\log[H_3O^+] = -\log[3.0 \times 10^{-9}] = 8.52$ (2 SFs on the right of the decimal point)

c. $[H_3O^+] = \dfrac{1.0 \times 10^{-14}}{[1.0 \times 10^{-5}]} = 1.0 \times 10^{-9}\text{ M}$

$\text{pH} = -\log[1.0 \times 10^{-9}] = 9.00$ (2 SFs on the right of the decimal point)

d. $[H_3O^+] = \dfrac{1.0 \times 10^{-14}}{[2.5 \times 10^{-11}]} = 4.0 \times 10^{-4}\text{ M}$

$\text{pH} = -\log[4.0 \times 10^{-4}] = 3.40$ (2 SFs on the right of the decimal point)

e. $\text{pH} = -\log[H_3O^+] = -\log[6.7 \times 10^{-8}] = 7.17$ (2 SFs on the right of the decimal point)

f. $[H_3O^+] = \dfrac{1.0 \times 10^{-14}}{[8.2 \times 10^{-4}]} = 1.2 \times 10^{-11}\text{ M}$

$\text{pH} = -\log[1.2 \times 10^{-11}] = 10.92$ (2 SFs on the right of the decimal point)

14.49 On a calculator, pH is calculated by entering $-$*log*, followed by the coefficient *EE (EXP)* key and the power of 10 followed by the change sign (+/−) key. On some calculators, the concentration is entered first (coefficient *EXP*–power) followed by *log* and +/− key.

$[H_3O^+] = \dfrac{1.0 \times 10^{-14}}{[OH^-]}$; $[OH^-] = \dfrac{1.0 \times 10^{-14}}{[H_3O^+]}$; $\text{pH} = -\log[H_3O^+]$; $\text{pOH} = -\log[OH^-]$

$[H_3O^+]$	$[OH^-]$	pH	pOH	Acidic, Basic, or Neutral?
1.0×10^{-8} M	1.0×10^{-6} M	8.00	6.00	Basic
3.2×10^{-4} M	3.1×10^{-11} M	3.49	10.51	Acidic
2.8×10^{-5} M	3.6×10^{-10} M	4.55	9.45	Acidic
1.0×10^{-12} M	1.0×10^{-2} M	12.00	2.00	Basic

$[H_3O^+]$ and $[OH^-]$ all have 2 SFs here, so all pH/pOH values have 2 decimal places on the right of the decimal point.

14.51 Acids react with active metals to form $H_2(g)$ and a salt of the metal. The reaction of acids with carbonates yields CO_2, H_2O, and a salt. In a neutralization reaction, an acid and a base react to form a salt and H_2O.

a. $ZnCO_3(s) + 2HBr(aq) \rightarrow CO_2(g) + H_2O(l) + ZnBr_2(aq)$
b. $Zn(s) + 2HCl(aq) \rightarrow H_2(g) + ZnCl_2(aq)$
c. $HCl(aq) + NaHCO_3(s) \rightarrow CO_2(g) + H_2O(l) + NaCl(aq)$
d. $H_2SO_4(aq) + Mg(OH)_2(s) \rightarrow MgSO_4(aq) + 2H_2O(l)$

14.53 In balancing a neutralization equation, the number of H^+ and OH^- must be equalized by placing coefficients in front of the formulas for the acid and base.

a. $2HCl(aq) + Mg(OH)_2(s) \rightarrow MgCl_2(aq) + 2H_2O(l)$
b. $H_3PO_4(aq) + 3LiOH(aq) \rightarrow Li_3PO_4(aq) + 3H_2O(l)$

14.55 The products of a neutralization are water and a salt. In balancing a neutralization equation, the number of H^+ and OH^- must be equalized by placing coefficients in front of the formulas for the acid and base.

a. $H_2SO_4(aq) + 2NaOH(aq) \rightarrow Na_2SO_4(aq) + 2H_2O(l)$
b. $3HCl(aq) + Fe(OH)_3(s) \rightarrow FeCl_3(aq) + 3H_2O(l)$
c. $H_2CO_3(aq) + Mg(OH)_2(s) \rightarrow MgCO_3(aq) + 2H_2O(l)$

14.57 To a measured volume of the formic acid solution, add a few drops of indicator. Place a solution of NaOH of known molarity in a buret. Add NaOH to the acid solution until one drop changes the color of the solution. Use the volume and molarity of the NaOH solution to calculate the moles of NaOH added to reach the endpoint. This equals the moles of formic acid in the sample. Then calculate the molarity of the formic acid solution from the calculated moles of formic acid and the known volume you started with.

14.59 In the titration equation, one mole of HCl reacts with one mole of NaOH.

$$28.6\ \cancel{\text{mL NaOH solution}} \times \frac{1\ \cancel{\text{L solution}}}{1000\ \cancel{\text{mL solution}}} \times \frac{0.145\ \cancel{\text{mol NaOH}}}{1\ \cancel{\text{L solution}}} \times \frac{1\text{ mol HCl}}{1\ \cancel{\text{mol NaOH}}}$$

$= 0.004\ 15$ mol of HCl

$$5.00\ \cancel{\text{mL HCl solution}} \times \frac{1\text{ L solution}}{1000\ \cancel{\text{mL solution}}} = 0.005\ 00\text{ L of HCl solution}$$

$$\text{molarity (M) of HCl} = \frac{\text{moles of solute}}{\text{liters of solution}} = \frac{0.004\ 15\text{ mol HCl}}{0.005\ 00\text{ L solution}}$$

$= 0.830$ M HCl solution (3 SFs)

14.61 In the titration equation, one mole of H_2SO_4 reacts with two moles of KOH.

$$38.2\ \cancel{\text{mL KOH solution}} \times \frac{1\ \cancel{\text{L solution}}}{1000\ \cancel{\text{mL solution}}} \times \frac{0.163\ \cancel{\text{mol KOH}}}{1\ \cancel{\text{L solution}}} \times \frac{1\text{ mol } H_2SO_4}{2\ \cancel{\text{mol KOH}}}$$

$= 0.003\ 11$ mol of H_2SO_4

$$25.0\ \cancel{\text{mL } H_2SO_4 \text{ solution}} \times \frac{1\text{ L solution}}{1000\ \cancel{\text{mL solution}}} = 0.0250\text{ L of } H_2SO_4 \text{ solution}$$

$$\text{molarity (M) of } H_2SO_4 = \frac{\text{moles of solute}}{\text{liters of solution}} = \frac{0.003\ 11\text{ mol } H_2SO_4}{0.0250\text{ L solution}}$$

$= 0.124$ M H_2SO_4 solution (3 SFs)

14.63 In the titration equation, one mole of H_3PO_4 reacts with three moles of NaOH.

$$50.0\ \cancel{\text{mL } H_3PO_4 \text{ solution}} \times \frac{1\ \cancel{\text{L } H_3PO_4 \text{ solution}}}{1000\ \cancel{\text{mL } H_3PO_4 \text{ solution}}} \times \frac{0.0224\ \cancel{\text{mol } H_3PO_4}}{1\ \cancel{\text{L } H_3PO_4 \text{ solution}}} \times \frac{3\ \cancel{\text{mol NaOH}}}{1\ \cancel{\text{mol } H_3PO_4}}$$

$$\times \frac{1\ \cancel{\text{L NaOH solution}}}{0.204\ \cancel{\text{mol NaOH}}} \times \frac{1000\text{ mL NaOH solution}}{1\ \cancel{\text{L NaOH solution}}} = 16.5\text{ mL of NaOH solution (3 SFs)}$$

14.65 When a salt contains an anion from a weak acid, the anion removes a proton from H_2O. The resulting solution is basic as long as the cation is from a strong base, which would have no effect on pH.

14.67 A solution of a salt with a cation from a weak base and an anion from a strong acid will be acidic. A solution of a salt with a cation from a strong base and an anion from a weak acid will be basic. Solutions of salts with cations from strong bases and anions from strong acids are neutral.

a. Neutral solution: Mg^{2+} is the cation from the strong base $Mg(OH)_2$, and Cl^- is the anion from the strong acid HCl. Neither ion will affect the pH of the solution.

b. Acidic solution: NO_3^- is the anion from the strong acid HNO_3 and will not affect pH; NH_4^+ is the cation from the weak base NH_3 and donates a proton to water to make the solution acidic.

$$NH_4^+(aq) + H_2O(l) \rightleftharpoons NH_3(aq) + H_3O^+(aq)$$

c. Basic solution: Na^+ is the cation from the strong base NaOH and will not affect pH; CO_3^{2-} is the anion from the weak acid H_2CO_3 and removes a proton from water to make the solution basic.

$$CO_3^{2-}(aq) + H_2O(l) \rightleftharpoons HCO_3^-(aq) + OH^-(aq)$$

d. Basic solution: K^+ is the cation from the strong base KOH and will not affect pH; S^{2-} is the anion from the weak acid H_2S and removes a proton from water to make the solution basic.

$$S^{2-}(aq) + H_2O(l) \rightleftharpoons HS^-(aq) + OH^-(aq)$$

14.69 A buffer system contains a weak acid and a salt containing its conjugate base (or a weak base and a salt containing its conjugate acid).

a. This is not a buffer system because it only contains the strong base NaOH and the neutral salt NaCl.

b. This is a buffer system; it contains the weak acid H_2CO_3 and a salt containing its conjugate base HCO_3^-.

c. This is a buffer system; it contains the weak acid HF and a salt containing its conjugate base F^-.

d. This is not a buffer system because it only contains the neutral salts KCl and NaCl.

14.71 a. A buffer system keeps the pH of a solution constant.

b. The same conjugate base is used to produce the same acid when the base neutralizes added H_3O^+.

c. When H_3O^+ is added to the buffer system, the F^- from the salt NaF reacts with the acid to neutralize it.

$F^-(aq) + H_3O^+(aq) \rightarrow HF(aq) + H_2O(l)$

d. When OH^- is added to the buffer system, the weak acid HF reacts with the OH^- to neutralize it.

$HF(aq) + OH^-(aq) \rightarrow F^-(aq) + H_2O(l)$

14.73 $HNO_2(aq) + H_2O(l) \rightleftarrows NO_2^-(aq) + H_3O^+(aq)$

Rearrange the K_a for $[H_3O^+]$ and use it to calculate the pH.

$$[H_3O^+] = K_a \times \frac{[HNO_2]}{[NO_2^-]} = 4.5 \times 10^{-4} \times \frac{[0.10\ \cancel{M}]}{[0.10\ \cancel{M}]} = 4.5 \times 10^{-4}\ M$$

$pH = -\log[H_3O^+] = -\log[4.5 \times 10^{-4}] = 3.35$ (2 SFs on the right of the decimal point)

14.75 $HF(aq) + H_2O(l) \rightleftarrows F^-(aq) + H_3O^+(aq)$

Rearrange the K_a for $[H_3O^+]$ and use it to calculate the pH.

$$[H_3O^+] = K_a \times \frac{[HF]}{[F^-]} = 7.2 \times 10^{-4} \times \frac{[0.10\ \cancel{M}]}{[0.10\ \cancel{M}]} = 7.2 \times 10^{-4}\ M$$

$pH = -\log[7.2 \times 10^{-4}] = 3.14$ (2 SFs on the right of the decimal point)

$$[H_3O^+] = K_a \times \frac{[HF]}{[F^-]} = 7.2 \times 10^{-4} \times \frac{[0.060\ \cancel{M}]}{[0.120\ \cancel{M}]} = 3.6 \times 10^{-4}\ M$$

$pH = -\log[3.6 \times 10^{-4}] = 3.44$ (2 SFs on the right of the decimal point)

$\therefore$ The solution with 0.10 M HF/0.10 M NaF is more acidic.

14.77 a. This diagram represents a weak acid; only a few HX molecules dissociate into H_3O^+ and X^- ions.

b. This diagram represents a strong acid; all of the HX molecules dissociate into H_3O^+ and X^- ions.

c. This diagram represents a weak acid; only a few HX molecules dissociate into H_3O^+ and X^- ions.

14.79 a. Hyperventilation will lower the CO_2 concentration in the blood, which lowers the $[H_2CO_3]$, which decreases the $[H_3O^+]$ and increases the blood pH.

b. Breathing into a paper bag will increase the CO_2 concentration in the blood, increase the $[H_2CO_3]$, increase $[H_3O^+]$, and lower the blood pH back toward the normal range.

14.81 a. acid; bromous acid

b. base; rubidium hydroxide

c. salt; magnesium nitrate

d. acid; butanoic acid (IUPAC), butyric acid (common)

e. acid; perchloric acid

14.83 An acidic solution has a pH less than 7.0. A neutral solution has a pH equal to 7.0. A basic solution has a pH greater than 7.0.

a. acidic (pH $5.2 < 7.0$)

b. basic (pH $7.5 > 7.0$)

c. acidic (pH $3.8 < 7.0$)

d. acidic (pH $2.5 < 7.0$)

e. basic (pH $12.0 > 7.0$)

14.85 a. $Mg(OH)_2$ is considered a strong base because all the $Mg(OH)_2$ that dissolves is completely dissociated in aqueous solution.

b. $Mg(OH)_2(s) + 2HCl(aq) \rightarrow MgCl_2(aq) + 2H_2O(l)$

14.87 Use Table 14.4 to answer (the stronger acid will be closer to the top of the table).
a. HF is the stronger acid. **b.** H_3O^+ is the stronger acid.
c. HNO_2 is the stronger acid. **d.** HCO_3^- is the stronger acid.

14.89 $[H_3O^+] = \dfrac{1.0 \times 10^{-14}}{[OH^-]}$; $[OH^-] = \dfrac{1.0 \times 10^{-14}}{[H_3O^+]}$;

$pH = -\log[H_3O^+]$; $pOH = -\log[OH^-]$; $pH + pOH = 14.00$

a. $pH = -\log[H_3O^+] = -\log[2.0 \times 10^{-8}] = 7.70$ (2 SFs on the right of the decimal point)
since $pH + pOH = 14.00$, $pOH = 14.00 - pH = 14.00 - 7.70 = 6.30$

b. $pH = -\log[5.0 \times 10^{-2}] = 1.30$ (2 SFs on the right of the decimal point)
since $pH + pOH = 14.00$, $pOH = 14.00 - pH = 14.00 - 1.30 = 12.70$

c. $[H_3O^+] = \dfrac{1.0 \times 10^{-14}}{[3.5 \times 10^{-4}]} = 2.9 \times 10^{-11}$ M

$pH = -\log[2.9 \times 10^{-11}] = 10.54$ (2 SFs on the right of the decimal point)
$pOH = -\log[OH^-] = -\log[3.5 \times 10^{-4}] = 3.46$

d. $[H_3O^+] = \dfrac{1.0 \times 10^{-14}}{[0.0054]} = 1.9 \times 10^{-12}$ M

$pH = -\log[1.9 \times 10^{-12}] = 11.73$ (2 SFs on the right of the decimal point)
$pOH = -\log[OH^-] = -\log[0.0054] = 2.27$

14.91 **a.** basic ($pH > 7.0$)
b. acidic ($pH < 7.0$)
c. basic ($pH > 7.0$)
d. basic ($pH > 7.0$)

14.93 If the pH is given, the $[H_3O^+]$ can be found by using the relationship $[H_3O^+] = 10^{-pH}$.
The $[OH^-]$ can be found by rearranging $K_w = [H_3O^+][OH^-] = 1 \times 10^{-14}$.

a. $pH = 3.00$; $[H_3O^+] = 10^{-pH} = 10^{-3.00} = 1.0 \times 10^{-3}$ M (2 SFs)

$[OH^-] = \dfrac{1.0 \times 10^{-14}}{[H_3O^+]} = \dfrac{1.0 \times 10^{-14}}{[1.0 \times 10^{-3}]} = 1.0 \times 10^{-11}$ M (2 SFs)

b. $pH = 6.48$; $[H_3O^+] = 10^{-pH} = 10^{-6.48} = 3.3 \times 10^{-7}$ M (2 SFs)

$[OH^-] = \dfrac{1.0 \times 10^{-14}}{[H_3O^+]} = \dfrac{1.0 \times 10^{-14}}{[3.3 \times 10^{-7}]} = 3.0 \times 10^{-8}$ M (2 SFs)

c. $pH = 8.85$; $[H_3O^+] = 10^{-pH} = 10^{-8.85} = 1.4 \times 10^{-9}$ M (2 SFs)

$[OH^-] = \dfrac{1.0 \times 10^{-14}}{[H_3O^+]} = \dfrac{1.0 \times 10^{-14}}{[1.4 \times 10^{-9}]} = 7.1 \times 10^{-6}$ M (2 SFs)

d. $pH = 11.00$; $[H_3O^+] = 10^{-pH} = 10^{-11.00} = 1.0 \times 10^{-11}$ M (2 SFs)

$[OH^-] = \dfrac{1.0 \times 10^{-14}}{[H_3O^+]} = \dfrac{1.0 \times 10^{-14}}{[1.0 \times 10^{-11}]} = 1.0 \times 10^{-3}$ M (2 SFs)

14.95 **a.** Solution A, with a pH of 4.5, is more acidic.
b. In solution A, the $[H_3O^+] = 10^{-pH} = 10^{-4.5} = 3 \times 10^{-5}$ M (1 SF)
In solution B, the $[H_3O^+] = 10^{-pH} = 10^{-6.7} = 2 \times 10^{-7}$ M (1 SF)

c. In solution A, the $[OH^-] = \dfrac{1.0 \times 10^{-14}}{[H_3O^+]} = \dfrac{1.0 \times 10^{-14}}{[3 \times 10^{-5}]} = 3 \times 10^{-10}\text{ M (1 SF)}$

In solution B, the $[OH^-] = \dfrac{1.0 \times 10^{-14}}{[H_3O^+]} = \dfrac{1.0 \times 10^{-14}}{[2 \times 10^{-7}]} = 5 \times 10^{-8}\text{ M (1 SF)}$

14.97 The $[OH^-]$ can be calculated from the moles of NaOH (each NaOH produces 1 OH^-) and the volume of the solution (in L).

$$0.225\ \cancel{\text{g NaOH}} \times \frac{1\ \cancel{\text{mol NaOH}}}{40.00\ \cancel{\text{g NaOH}}} \times \frac{1\text{ mol OH}^-}{1\ \cancel{\text{mol NaOH}}} = 0.005\ 63\text{ mol of OH}^-$$

$$[OH^-] = \frac{0.005\ 63\text{ mol OH}^-}{0.250\text{ L solution}} = 0.0225\text{ M (3 SFs)}$$

14.99 $2.5\ \cancel{\text{g HCl}} \times \dfrac{1\ \cancel{\text{mol HCl}}}{36.46\ \cancel{\text{g HCl}}} \times \dfrac{1\text{ mol H}_3\text{O}^+}{1\ \cancel{\text{mol HCl}}} = 0.069\text{ mol of H}_3\text{O}^+\text{ (2 SFs)}$

$$[H_3O^+] = \frac{0.069\text{ mol H}_3\text{O}^+}{0.425\text{ L solution}} = 0.16\text{ M (2 SFs)}$$

$\text{pH} = -\log[H_3O^+] = -\log[0.16] = 0.80$ (2 SFs on the right of the decimal point)

$\text{pOH} = 14.00 - \text{pH} = 14.00 - 0.80 = 13.20$ (2 SFs on the right of the decimal point)

14.101 A solution of a salt with a cation from a weak base and an anion from a strong acid will be acidic. A solution of a salt with a cation from a strong base and an anion from a weak acid will be basic. Solutions of salts with cations from strong bases and anions from strong acids are neutral.

a. Basic solution: K^+ is the cation from the strong base KOH and will not affect pH; F^- is the anion from the weak acid HF and removes a proton from water to make the solution basic.

$$F^-(aq) + H_2O(l) \rightleftarrows HF(aq) + OH^-(aq)$$

b. Basic solution: Na^+ is the cation from the strong base NaOH and will not affect pH; CN^- is the anion from the weak acid HCN and removes a proton from water to make the solution basic.

$$CN^-(aq) + H_2O(l) \rightleftarrows HCN(aq) + OH^-(aq)$$

c. Acidic solution: Cl^- is the anion from the strong acid HCl and will not affect pH; $CH_3NH_3^+$ is the cation from the weak base CH_3NH_2 and donates a proton to water to make the solution acidic.

$$CH_3NH_3^+(aq) + H_2O(l) \rightleftarrows CH_3NH_2(aq) + H_3O^+(aq)$$

d. Neutral solution: Na^+ is the cation from the strong base NaOH, and Br^- is the anion from the strong acid HBr. Neither ion will affect the pH of the solution.

14.103 This buffer solution is made from the weak acid H_3PO_4 and a salt containing its conjugate base $H_2PO_4^-$.

a. Add acid: $H_2PO_4^-(aq) + H_3O^+(aq) \rightarrow H_3PO_4(aq) + H_2O(l)$

b. Add base: $H_3PO_4(aq) + OH^-(aq) \rightarrow H_2PO_4^-(aq) + H_2O(l)$

c. $[H_3O^+] = K_a \times \dfrac{[H_3PO_4]}{[H_2PO_4^-]} = 7.5 \times 10^{-3} \times \dfrac{[0.50\ \cancel{M}]}{[0.20\ \cancel{M}]} = 1.9 \times 10^{-2}\text{ M}$

$\text{pH} = -\log[1.9 \times 10^{-2}] = 1.72$ (2 SFs on the right of the decimal point)

14.105 **a.** In the titration equation, one mole of HCl reacts with one mole of NaOH.

$HCl(aq) + NaOH(aq) \rightarrow NaCl(aq) + H_2O(l)$

$$25.0\ \cancel{\text{mL HCl solution}} \times \frac{1\ \cancel{\text{L HCl solution}}}{1000\ \cancel{\text{mL HCl solution}}} \times \frac{0.288\ \cancel{\text{mol HCl}}}{1\ \cancel{\text{L HCl solution}}} \times \frac{1\ \cancel{\text{mol NaOH}}}{1\ \cancel{\text{mol HCl}}}$$

$$\times \frac{1\ \cancel{\text{L NaOH solution}}}{0.150\ \cancel{\text{mol NaOH}}} \times \frac{1000\text{ mL NaOH solution}}{1\ \cancel{\text{L NaOH solution}}} = 48.0\text{ mL of NaOH solution (3 SFs)}$$

b. In the titration equation, one mole of H_2SO_4 reacts with two moles of NaOH.

$H_2SO_4(aq) + 2NaOH(aq) \rightarrow Na_2SO_4(aq) + 2H_2O(l)$

$$10.0\ \cancel{\text{mL } H_2SO_4 \text{ solution}} \times \frac{1\ \cancel{\text{L } H_2SO_4 \text{ solution}}}{1000\ \cancel{\text{mL } H_2SO_4 \text{ solution}}} \times \frac{0.560\ \cancel{\text{mol } H_2SO_4}}{1\ \cancel{\text{L } H_2SO_4 \text{ solution}}} \times \frac{2\ \cancel{\text{mol NaOH}}}{1\ \cancel{\text{mol } H_2SO_4}}$$

$$\times \frac{1\ \cancel{\text{L NaOH solution}}}{0.150\ \cancel{\text{mol NaOH}}} \times \frac{1000\ \text{mL NaOH solution}}{1\ \cancel{\text{L NaOH solution}}} = 74.7\ \text{mL of NaOH solution (3 SFs)}$$

14.107 In the titration equation, one mole of H_2SO_4 reacts with two moles of NaOH.

$$45.6\ \cancel{\text{mL NaOH solution}} \times \frac{1\ \cancel{\text{L solution}}}{1000\ \cancel{\text{mL solution}}} \times \frac{0.205\ \cancel{\text{mol NaOH}}}{1\ \cancel{\text{L solution}}} \times \frac{1\ \text{mol } H_2SO_4}{2\ \cancel{\text{mol NaOH}}}$$

$= 0.004\ 67$ mol of H_2SO_4

$$20.0\ \cancel{\text{mL } H_2SO_4 \text{ solution}} \times \frac{1\ \text{L solution}}{1000\ \cancel{\text{mL solution}}} = 0.0200\ \text{L of } H_2SO_4 \text{ solution}$$

$$\text{molarity (M) of } H_2SO_4 = \frac{\text{moles of solute}}{\text{liters of solution}} = \frac{0.004\ 67\ \text{mol } H_2SO_4}{0.0200\ \text{L solution}}$$

$= 0.234$ M H_2SO_4 solution (3 SFs)

14.109 **a.** To form the conjugate base, remove a proton (H^+) from the acid.

1. HS^- 2. $H_2PO_4^-$ 3. CO_3^{2-}

b. 1. $\dfrac{[H_3O^+][HS^-]}{[H_2S]}$ 2. $\dfrac{[H_3O^+][H_2PO_4^-]}{[H_3PO_4]}$ 3. $\dfrac{[H_3O^+][CO_3^{2-}]}{[HCO_3^-]}$

c. HCO_3^- (see Table 14.4, the weakest acid will be closest to the bottom of the table)

d. H_3PO_4 (see Table 14.4, the strongest acid will be closest to the top of the table)

14.111 **a.** $ZnCO_3(s) + H_2SO_4(aq) \rightarrow CO_2(g) + H_2O(l) + ZnSO_4(aq)$

b. $2Al(s) + 6HNO_3(aq) \rightarrow 3H_2(g) + 2Al(NO_3)_3(aq)$

c. $2H_3PO_4(aq) + 3Ca(OH)_2(s) \rightarrow Ca_3(PO_4)_2(s) + 6H_2O(l)$

d. $KHCO_3(s) + HNO_3(aq) \rightarrow CO_2(g) + H_2O(l) + KNO_3(aq)$

14.113 KOH (strong base) $\rightarrow K^+(aq) + OH^-(aq)$ (100% dissociation)

$[OH^-] = 0.050\ \text{M} = 5.0 \times 10^{-2}$ M

a. $[H_3O^+] = \dfrac{1.0 \times 10^{-14}}{[OH^-]} = \dfrac{1.0 \times 10^{-14}}{[5.0 \times 10^{-2}]} = 2.0 \times 10^{-13}$ M (2 SFs)

b. $\text{pH} = -\log[H_3O^+] = -\log[2.0 \times 10^{-13}] = 12.70$ (2 SFs on the right of the decimal point)

c. $\text{pOH} = -\log[OH^-] = -\log[5.0 \times 10^{-2}] = 1.30$,

or $\text{pOH} = 14.00 - 12.70 = 1.30$ (2 SFs on the right of the decimal point)

d. $H_3PO_4(aq) + 3KOH(aq) \rightarrow K_3PO_4(aq) + 3H_2O(l)$

e. In the titration equation, one mole of H_2SO_4 reacts with two moles of KOH.

$H_2SO_4(aq) + 2KOH(aq) \rightarrow K_2SO_4(aq) + 2H_2O(l)$

$$40.0\ \cancel{\text{mL } H_2SO_4 \text{ solution}} \times \frac{1\ \cancel{\text{L } H_2SO_4 \text{ solution}}}{1000\ \cancel{\text{mL } H_2SO_4 \text{ solution}}} \times \frac{0.035\ \cancel{\text{mol } H_2SO_4}}{1\ \cancel{\text{L } H_2SO_4 \text{ solution}}} \times \frac{2\ \cancel{\text{mol KOH}}}{1\ \cancel{\text{mol } H_2SO_4}}$$

$$\times \frac{1\ \cancel{\text{L KOH solution}}}{0.050\ \cancel{\text{mol KOH}}} \times \frac{1000\ \text{mL KOH solution}}{1\ \cancel{\text{L KOH solution}}} = 56\ \text{mL of KOH solution (2 SFs)}$$

14.115 **a.** The $[H_3O^+]$ can be found by using the relationship $[H_3O^+] = 10^{-pH}$.

$$[H_3O^+] = 10^{-pH} = 10^{-4.2} = 6 \times 10^{-5}\ \text{M (1 SF)}$$

$$[OH^-] = \frac{1.0 \times 10^{-14}}{[H_3O^+]} = \frac{1.0 \times 10^{-14}}{[6 \times 10^{-5}]} = 2 \times 10^{-10}\ \text{M (1 SF)}$$

b. $[H_3O^+] = 10^{-pH} = 10^{-6.5} = 3 \times 10^{-7}\ \text{M (1 SF)}$

$$[OH^-] = \frac{1.0 \times 10^{-14}}{[H_3O^+]} = \frac{1.0 \times 10^{-14}}{[3 \times 10^{-7}]} = 3 \times 10^{-8}\ \text{M (1 SF)}$$

c. In the titration equation, one mole of $CaCO_3$ reacts with two moles of acid HA.

$$1.0\ \cancel{\text{kL solution}} \times \frac{1000\ \cancel{\text{L solution}}}{1\ \cancel{\text{kL solution}}} \times \frac{6 \times 10^{-5}\ \cancel{\text{mol HA}}}{1\ \cancel{\text{L solution}}} \times \frac{1\ \cancel{\text{mol CaCO}_3}}{2\ \cancel{\text{mol HA}}}$$

$$\times \frac{100.09\ \text{g CaCO}_3}{1\ \cancel{\text{mol CaCO}_3}} = 3\ \text{g of CaCO}_3\ \text{(1 SF)}$$

Answers to Combining Ideas from Chapters 11 to 14

CI.21 **a.** CH_4

b. $7.0 \times 10^6 \cancel{\text{gal}} \times \frac{4 \cancel{\text{qt}}}{1 \cancel{\text{gal}}} \times \frac{946.3 \cancel{\text{mL}}}{1 \cancel{\text{qt}}} \times \frac{0.45 \cancel{\text{g}}}{1 \cancel{\text{mL}}} \times \frac{1 \text{ kg}}{1000 \cancel{\text{g}}} = 1.2 \times 10^7$ kg of LNG (2 SFs)

c. $7.0 \times 10^6 \cancel{\text{gal}} \times \frac{4 \cancel{\text{qt}}}{1 \cancel{\text{gal}}} \times \frac{946.3 \cancel{\text{mL}}}{1 \cancel{\text{qt}}} \times \frac{0.45 \cancel{\text{g}}}{1 \cancel{\text{mL}}} \times \frac{1 \cancel{\text{mol CH}_4}}{16.04 \cancel{\text{g}}} \times \frac{22.4 \text{ L CH}_4}{1 \cancel{\text{mol CH}_4 \text{ (STP)}}}$

$= 1.7 \times 10^{10}$ L of LNG (STP) (2 SFs)

d. $CH_4(g) + 2O_2(g) \xrightarrow{\Delta} CO_2(g) + 2H_2O(g)$

e. $7.0 \times 10^6 \cancel{\text{gal}} \times \frac{4 \cancel{\text{qt}}}{1 \cancel{\text{gal}}} \times \frac{946.3 \cancel{\text{mL}}}{1 \cancel{\text{qt}}} \times \frac{0.45 \cancel{\text{g}}}{1 \cancel{\text{mL}}} \times \frac{1 \cancel{\text{mol CH}_4}}{16.04 \cancel{\text{g}}}$

$\times \frac{2 \cancel{\text{mol O}_2}}{1 \cancel{\text{mol CH}_4}} \times \frac{32.00 \cancel{\text{g O}_2}}{1 \cancel{\text{mol O}_2}} \times \frac{1 \text{ kg O}_2}{1000 \cancel{\text{g O}_2}}$

$= 4.8 \times 10^7$ kg of O_2 (2 SFs)

f. $7.0 \times 10^6 \cancel{\text{gal}} \times \frac{4 \cancel{\text{qt}}}{1 \cancel{\text{gal}}} \times \frac{946.3 \cancel{\text{mL}}}{1 \cancel{\text{qt}}} \times \frac{0.45 \cancel{\text{g}}}{1 \cancel{\text{mL}}} \times \frac{1 \cancel{\text{mol CH}_4}}{16.04 \cancel{\text{g}}} \times \frac{883 \text{ kJ}}{1 \cancel{\text{mol CH}_4}}$

$= 6.6 \times 10^{11}$ kJ (2 SFs)

CI.23 **a.** $[H_2] = \frac{2.02 \cancel{\text{g H}_2}}{10.0 \text{ L}} \times \frac{1 \text{ mol H}_2}{2.016 \cancel{\text{g H}_2}} = 0.100 \text{ M}$

$[S_2] = \frac{10.3 \cancel{\text{g S}_2}}{10.0 \text{ L}} \times \frac{1 \text{ mol S}_2}{64.14 \cancel{\text{g S}_2}} = 0.0161 \text{ M}$

$[H_2S] = \frac{68.2 \cancel{\text{g H}_2\text{S}}}{10.0 \text{ L}} \times \frac{1 \text{ mol H}_2\text{S}}{34.09 \cancel{\text{g H}_2\text{S}}} = 0.200 \text{ M}$

$K_c = \frac{[H_2S]^2}{[H_2]^2[S_2]} = \frac{[0.200]^2}{[0.100]^2[0.0161]} = 248$ (3 SFs)

b. If more H_2 (a reactant) is added, the equilibrium will shift toward the products.

c. If the volume decreases from 10.0 L to 5.00 L (at constant temperature), the equilibrium will shift toward the products (fewer moles of gas).

d. $[H_2] = \frac{0.300 \text{ mol H}_2}{5.00 \text{ L}} = 0.0600 \text{ M}$

$[H_2S] = \frac{2.50 \text{ mol H}_2\text{S}}{5.00 \text{ L}} = 0.500 \text{ M}$

$K_c = \frac{[H_2S]^2}{[H_2]^2[S_2]} = 248$

Rearrange the expression to solve for $[S_2]$ and substitute in known values.

$[S_2] = \frac{[H_2S]^2}{[H_2]^2 K_c} = \frac{[0.500]^2}{[0.0600]^2(248)} = 0.280$ M (3 SFs)

e. Because the reaction is exothermic (heat is a product), an increase in temperature will shift the equilibrium toward the reactants and decrease the value of K_c.

CI.25 a. $2M(s) + 6HCl(aq) \rightarrow 2MCl_3(aq) + 3H_2(g)$

b. $34.8\ \cancel{\text{mL solution}} \times \dfrac{1\ \cancel{\text{L solution}}}{1000\ \cancel{\text{mL solution}}} \times \dfrac{0.520\ \cancel{\text{mol HCl}}}{1\ \cancel{\text{L solution}}} \times \dfrac{3\ \text{mol H}_2}{6\ \cancel{\text{mol HCl}}}$

$= 0.00905$ mol of H_2 (3 SFs)

$T = 24\ °C + 273 = 297\ K$

Rearrange the ideal gas law $PV = nRT$ to solve for V,

$$V = \frac{nRT}{P} = \frac{(0.00905\ \cancel{\text{mol}})\left(\dfrac{62.4\ \cancel{\text{L}} \cdot \cancel{\text{mmHg}}}{\cancel{\text{mol}} \cdot \cancel{\text{K}}}\right)(297\ \cancel{\text{K}})}{(720.\ \cancel{\text{mmHg}})} \times \frac{1000\ \text{mL}}{1\ \cancel{\text{L}}} = 233\ \text{mL of H}_2\ (3\ \text{SFs})$$

c. $34.8\ \cancel{\text{mL solution}} \times \dfrac{1\ \cancel{\text{L solution}}}{1000\ \cancel{\text{mL solution}}} \times \dfrac{0.520\ \cancel{\text{mol HCl}}}{1\ \cancel{\text{L solution}}} \times \dfrac{2\ \text{mol M}}{6\ \cancel{\text{mol HCl}}}$

$= 6.03 \times 10^{-3}$ mol of M (3 SFs)

d. $\dfrac{0.420\ \text{g M}}{6.03 \times 10^{-3}\ \text{mol M}} = 69.7$ g/mol of M (3 SFs); $\therefore$ metal M is gallium

e. $2Ga(s) + 6HCl(aq) \rightarrow 2GaCl_3(aq) + 3H_2(g)$

f. Ga $1s^22s^22p^63s^23p^64s^23d^{10}4p^1$;

Ga^{3+} $1s^22s^22p^63s^23p^63d^{10}$

15

Oxidation and Reduction

Study Goals

- Identify what is oxidized and reduced in an oxidation–reduction reaction.
- Assign and use oxidation numbers to identify elements that are oxidized or reduced, and to balance an oxidation–reduction equation.
- Balance oxidation–reduction equations using the half-reaction method.
- Classify alcohols as primary, secondary, or tertiary; write equations for the oxidation of alcohols.
- Write the half-reactions that occur at the anode and cathode of a voltaic cell; write the shorthand cell notation.
- Describe the half-cell reactions and the overall reactions that occur in electrolytic cells.

Chapter Outline

Chapter Opener: Conservator of Photographs

15.1 Oxidation–Reduction Reactions

15.2 Oxidation Numbers

15.3 Balancing Oxidation–Reduction Equations Using Half-Reactions

15.4 Oxidation of Alcohols
Chemistry and Health: Methanol Poisoning
Chemistry and Health: Oxidation of Alcohol in the Body

15.5 Electrical Energy from Oxidation–Reduction Reactions
Chemistry and the Environment: Corrosion: Oxidation of Metals
Chemistry and the Environment: Fuel Cells: Clean Energy for the Future

15.6 Oxidation–Reduction Reactions That Require Electrical Energy
Answers

Answers and Solutions to Text Problems

15.1 Oxidation is the loss of electrons; reduction is the gain of electrons.
a. Na^+ gains an electron to form Na; this is a reduction.
b. Ni loses electrons to form Ni^{2+}; this is an oxidation.
c. Cr^{3+} gains electrons to form Cr; this is a reduction.
d. $2H^+$ gain electrons to form H_2; this is a reduction.

15.3 An oxidized substance has lost electrons; a reduced substance has gained electrons.
a. Zn loses electrons and is oxidized. Cl_2 gains electrons and is reduced.
b. Br^- (in NaBr) loses electrons and is oxidized. Cl_2 gains electrons and is reduced.
c. Pb loses electrons and is oxidized. O_2 gains electrons and is reduced.
d. Sn^{2+} loses electrons and is oxidized. Fe^{3+} gains electrons and is reduced.

15.5 **a.** An element has an oxidation number of zero (Rule 2); Cu = 0.
b. An element has an oxidation number of zero (Rule 2); F in F_2 = 0.
c. The oxidation number of a monatomic ion is equal to its charge (Rule 3); Fe^{2+} = +2.
d. The oxidation number of a monatomic ion is equal to its charge (Rule 3); Cl^- = −1.

15.7 **a.** Cl has an oxidation number of −1 (Rule 5). Because KCl is neutral, the oxidation number of K is calculated as +1. Oxidation numbers: K = +1, Cl = −1
$K + Cl = 0$
$K + (-1) = 0$
$\therefore K = +1$
b. In MnO_2, the oxidation number of O is −2 (Rule 6). Because MnO_2 is neutral, the oxidation number of Mn is calculated as +4. Oxidation numbers: Mn = +4, O = −2
$1Mn + 2O = 0$
$Mn + 2(-2) = 0$
$\therefore Mn = +4$
c. In CO, the oxidation number of O is −2 (Rule 6). Because CO is neutral, the oxidation number of C is calculated as +2. Oxidation numbers: C = +2, O = −2
$C + O = 0$
$C + (-2) = 0$
$\therefore C = +2$
d. In Mn_2O_3, the oxidation number of O is −2 (Rule 6). Because Mn_2O_3 is neutral, the oxidation number of Mn is calculated as +3. Oxidation numbers: Mn = +3, O = −2
$2Mn + 3O = 0$
$2Mn + 3(-2) = 0$
$2Mn = +6$
$\therefore Mn = +3$

15.9 **a.** $AlPO_4$ is an ionic compound composed of the Al^{3+} and PO_4^{3-} ions. The oxidation number of the monatomic ion Al^{3+} is +3 (Rule 3). In the polyatomic ion PO_4^{3-}, the oxidation number of O is −2 (Rule 6). Because the sum of the oxidation numbers in PO_4^{3-} must equal −3, the oxidation number of P is calculated as +5. Oxidation numbers:
Al = +3, P = +5, O = −2
$P + 4O = -3$
$P + 4(-2) = -3$
$\therefore P = +5$
b. In SO_3^{2-}, the oxidation number of O is −2 (Rule 6). Because the sum of the oxidation numbers in the polyatomic ion SO_3^{2-} must equal −2 (Rule 1), the oxidation number of S is calculated as +4. Oxidation numbers: S = +4, O = −2
$S + 3O = -2$
$S + 3(-2) = -2$
$\therefore S = +4$
c. In Cr_2O_3, the oxidation number of O is −2 (Rule 6). Because Cr_2O_3 is neutral, the oxidation number of Cr is calculated as +3. Oxidation numbers: Cr = +3, O = −2
$2Cr + 3O = 0$
$2Cr + 3(-2) = 0$
$2Cr = +6$
$\therefore Cr = +3$
d. In NO_3^-, the oxidation number of O is −2 (Rule 6). Because the sum of the oxidation numbers in the polyatomic ion NO_3^- must equal −1 (Rule 1), the oxidation number of N is calculated as +5. Oxidation numbers: N = +5, O = −2
$N + 3O = -1$
$N + 3(-2) = -1$
$\therefore N = +5$

15.11 a. In HSO_4^-, the oxidation number of H is +1 (Rule 7) and the oxidation number of O is −2 (Rule 6). Because the sum of the oxidation numbers in the polyatomic ion HSO_4^- must equal −1 (Rule 1), the oxidation number of S is calculated as +6.
Oxidation numbers: H = +1, S = +6, O = −2
$H + S + 4O = -1$
$+1 + S + 4(-2) = -1$
$\therefore S = +6$

b. In H_3PO_3, the oxidation number of H is +1 (Rule 7) and the oxidation number of O is −2 (Rule 6). Because H_3PO_3 is neutral, the oxidation number of P is calculated as +3.
Oxidation numbers: H = +1, P = +3, O = −2
$3H + P + 3O = 0$
$3(+1) + P + 3(-2) = 0$
$\therefore P = +3$

c. In $Cr_2O_7^{2-}$, the oxidation number of O is −2 (Rule 6). Because the sum of the oxidation numbers in the polyatomic ion $Cr_2O_7^{2-}$ must equal −2 (Rule 1), the oxidation number of Cr is calculated as +6. Oxidation numbers: Cr = +6, O = −2
$2Cr + 7O = -2$
$2Cr + 7(-2) = -2$
$2Cr = +12$
$\therefore Cr = +6$

d. Na_2CO_3 is an ionic compound composed of the Na^+ and CO_3^{2-} ions. The oxidation number of the monatomic ion Na^+ is +1 (Rule 3). In the polyatomic ion CO_3^{2-}, the oxidation number of O is −2 (Rule 6). Because the sum of the oxidation numbers in CO_3^{2-} must equal −2, the oxidation number of C is calculated as +4. Oxidation numbers: Na = +1, C = +4, O = −2
$2Na + C + 3O = 0$
$2(+1) + C + 3(-2) = 0$
$\therefore C = +4$

15.13 a. HNO_3 H = +1, O = −2
$HNO_3 \rightarrow (+1) + N + 3(-2) = 0 \quad \therefore N = +5$

b. C_3H_6 H = +1
$C_3H_6 \rightarrow 3C + 6(+1) = 0 \quad \therefore 3C = -6$ and $C = -2$

c. K_3PO_4 K = +1, O = −2
$K_3PO_4 \rightarrow 3(+1) + P + 4(-2) = 0 \quad \therefore P = +5$

d. CrO_4^{2-} O = −2
$CrO_4^{2-} \rightarrow Cr + 4(-2) = -2 \quad \therefore Cr = +6$

15.15 a. A reducing agent is the substance that is oxidized; it provides electrons for reduction.

b. An oxidizing agent is the substance that gains electrons; it accepts the electrons lost in an oxidation, and so is reduced.

15.17 a. Zn is the reducing agent. Cl_2 is the oxidizing agent.

b. Br^- (in NaBr) is the reducing agent. Cl_2 is the oxidizing agent.

c. Pb is the reducing agent. O_2 is the oxidizing agent.

d. Sn^{2+} is the reducing agent. Fe^{3+} is the oxidizing agent.

15.19 Assign oxidation numbers and determine which one increases and which one decreases. The substance with an increase in oxidation number is oxidized and is also the reducing agent. The substance with a decrease in oxidation number is reduced and is also the oxidizing agent.

a. $\underset{+2-2}{2NiS} + \underset{0}{3O_2} \rightarrow \underset{+2-2}{2NiO} + \underset{+4-2}{2SO_2}$

Ni: +2 → +2	No change
S: −2 → +4	Oxidation number increases (oxidation)
O: 0 → −2	Oxidation number decreases (reduction)

$\therefore S^{2-}$ (in NiS) is oxidized; O_2 is reduced. NiS is the reducing agent, and O_2 is the oxidizing agent.

b. $Sn^{2+} + 2Fe^{3+} \rightarrow Sn^{4+} + 2Fe^{2+}$

+2 +3 +4 +2

Sn: $+2 \rightarrow +4$ Oxidation number increases (oxidation)

Fe: $+3 \rightarrow +2$ Oxidation number decreases (reduction)

∴ Sn^{2+} is oxidized; Fe^{3+} is reduced. Sn^{2+} is the reducing agent, and Fe^{3+} is the oxidizing agent.

c. $CH_4 + 2O_2 \rightarrow CO_2 + 2H_2O$

−4+1 0 +4−2 +1−2

C: $-4 \rightarrow +4$ Oxidation number increases (oxidation)

H: $+1 \rightarrow +1$ No change

O: $0 \rightarrow -2$ Oxidation number decreases (reduction)

∴ C (in CH_4) is oxidized; O_2 is reduced. CH_4 is the reducing agent, and O_2 is the oxidizing agent.

d. $2Cr_2O_3 + 3Si \rightarrow 4Cr + 3SiO_2$

+3−2 0 0 +4−2

Cr: $+3 \rightarrow 0$ Oxidation number decreases (reduction)

O: $-2 \rightarrow -2$ No change

Si: $0 \rightarrow +4$ Oxidation number increases (oxidation)

∴ Si is oxidized; Cr^{3+} (in Cr_2O_3) is reduced. Si is the reducing agent, and Cr_2O_3 is the oxidizing agent.

15.21 a. $Cu_2S(s) + H_2(g) \rightarrow Cu(s) + H_2S(g)$

+1−2 0 0 +1−2

2Cu: $+1 \rightarrow 0$ decreases by 1 (reduction) ∴ 2 e^- gained

2H: $0 \rightarrow +1$ increases by 1 (oxidation) ∴ 2 e^- lost

∴ balanced equation is $Cu_2S(s) + H_2(g) \rightarrow 2Cu(s) + H_2S(g)$

b. $Fe(s) + Cl_2(g) \rightarrow FeCl_3(s)$

0 0 +3−1

Fe: $0 \rightarrow +3$ increases by 3 (oxidation) ∴ (3 e^- lost) × 2 = 6 e^- lost

2Cl: $0 \rightarrow -1$ decreases by 1 (reduction) ∴ (2 e^- gained) × 3 = 6 e^- gained

∴ balanced equation is $2Fe(s) + 3Cl_2(g) \rightarrow 2FeCl_3(s)$

c. $Al(s) + H_2SO_4(aq) \rightarrow Al_2(SO_4)_3(aq) + H_2(g)$

0 +1+6−2 +3+6−2 0

Al: $0 \rightarrow +3$ increases by 3 (oxidation) ∴ (3 e^- lost) × 2 = 6 e^- lost

2H: $+1 \rightarrow 0$ decreases by 1 (reduction) ∴ (2 e^- gained) × 3 = 6 e^- gained

∴ balanced equation is $2Al(s) + 3H_2SO_4(aq) \rightarrow Al_2(SO_4)_3(aq) + 3H_2(g)$

15.23 a. $Sn^{2+}(aq) \rightarrow Sn^{4+}(aq)$

Balance charge with e^-: $Sn^{2+}(aq) \rightarrow Sn^{4+}(aq) + 2e^-$

b. $Mn^{2+}(aq) \rightarrow MnO_4^-(aq)$

Balance O with H_2O: $Mn^{2+}(aq) + 4H_2O(l) \rightarrow MnO_4^-(aq)$

Balance H with H^+: $Mn^{2+}(aq) + 4H_2O(l) \rightarrow MnO_4^-(aq) + 8H^+(aq)$

Balance charge with e^-: $Mn^{2+}(aq) + 4H_2O(l) \rightarrow MnO_4^-(aq) + 8H^+(aq) + 5e^-$

c. $NO_2^-(aq) \rightarrow NO_3^-(aq)$

Balance O with H_2O: $NO_2^-(aq) + H_2O(l) \rightarrow NO_3^-(aq)$

Balance H with H^+: $NO_2^-(aq) + H_2O(l) \rightarrow NO_3^-(aq) + 2H^+(aq)$

Balance charge with e^-: $NO_2^-(aq) + H_2O(l) \rightarrow NO_3^-(aq) + 2H^+(aq) + 2e^-$

d. $ClO_3^-(aq) \rightarrow ClO_2(aq)$

Balance O with H_2O: $ClO_3^-(aq) \rightarrow ClO_2(aq) + H_2O(l)$

Balance H with H^+: $ClO_3^-(aq) + 2H^+(aq) \rightarrow ClO_2(aq) + H_2O(l)$

Balance charge with e^-: $ClO_3^-(aq) + 2H^+(aq) + e^- \rightarrow ClO_2(aq) + H_2O(l)$

15.25 Write half-reactions, multiply by small numbers to equalize electrons lost and gained, add together, and combine common species.

a. $Ag(s) \rightarrow Ag^+(aq) + e^-$

$2H^+(aq) + NO_3^-(aq) + e^- \rightarrow NO_2(g) + H_2O(l)$

Overall: $2H^+(aq) + Ag(s) + NO_3^-(aq) \rightarrow Ag^+(aq) + NO_2(g) + H_2O(l)$

b. $2Fe(s) + 3H_2O(l) \rightarrow Fe_2O_3(s) + 6H^+(aq) + 6\,e^-$

$10H^+(aq) + 2CrO_4^{2-}(aq) + 6\,e^- \rightarrow Cr_2O_3(s) + 5H_2O(l)$

Overall (in acid): $4H^+(aq) + 2Fe(s) + 2CrO_4^{2-}(aq) \rightarrow Fe_2O_3(s) + Cr_2O_3(s) + 2H_2O(l)$

$\therefore$ in base: $\underbrace{4H^+(aq) + 4OH^-(aq)}_{4H_2O(l)} + 2Fe(s) + 2CrO_4^{2-}(aq) \rightarrow$

$Fe_2O_3(s) + Cr_2O_3(s) + 2H_2O(l) + 4OH^-(aq)$

Overall (in base): $2H_2O(l) + 2Fe(s) + 2CrO_4^{2-}(aq) \rightarrow Fe_2O_3(s) + Cr_2O_3(s) + 4OH^-(aq)$

c. $[4H^+(aq) + NO_3^-(aq) + 3\,e^- \rightarrow NO(g) + 2H_2O(l)] \times 4$

$[2H_2O(l) + S(s) \rightarrow SO_2(g) + 4H^+(aq) + 4\,e^-] \times 3$

Overall: $4H^+(aq) + 4NO_3^-(aq) + 3S(s) \rightarrow 4NO(g) + 3SO_2(g) + 2H_2O(l)$

d. $2S_2O_3^{2-}(aq) \rightarrow S_4O_6^{2-}(aq) + 2\,e^-$

$Cu^{2+}(aq) + 2\,e^- \rightarrow Cu(s)$

Overall: $2S_2O_3^{2-}(aq) + Cu^{2+}(aq) \rightarrow S_4O_6^{2-}(aq) + Cu(s)$

e. $[PbO_2(s) + 4H^+(aq) + 2\,e^- \rightarrow Pb^{2+}(aq) + 2H_2O(l)] \times 5$

$[Mn^{2+}(aq) + 4H_2O(l) \rightarrow MnO_4^-(aq) + 8H^+(aq) + 5\,e^-] \times 2$

Overall: $4H^+(aq) + 5PbO_2(s) + 2Mn^{2+}(aq) \rightarrow 5Pb^{2+}(aq) + 2MnO_4^-(aq) + 2H_2O(l)$

15.27 **a.** One carbon group attached to the carbon atom bonded to the —OH makes this a primary alcohol.

b. One carbon group attached to the carbon atom bonded to the —OH makes this a primary alcohol.

c. Three carbon groups attached to the carbon atom bonded to the —OH makes this a tertiary alcohol.

15.29 **a.** $CH_3-CH_2-CH_2-CH_2-\overset{\overset{\displaystyle O}{\|}}{C}-H$

b. none

c. $CH_3-\overset{\overset{\displaystyle O}{\|}}{C}-CH_2-\overset{\overset{\displaystyle CH_3}{|}}{CH}-CH_3$

15.31 Oxidation occurs at the anode; reduction occurs at the cathode. In the shorthand cell notation, the oxidation half-cell is written on the left side of the double vertical line, and the reduction half-cell is written on the right side.

a. Anode reaction: $Pb(s)\,|\,Pb^{2+}(aq) = Pb(s) \rightarrow Pb^{2+}(aq) + 2\,e^-$

Cathode reaction: $Cu^{2+}(aq)\,|\,Cu(s) = Cu^{2+}(aq) + 2\,e^- \rightarrow Cu(s)$

Overall cell reaction: $Cu^{2+}(aq) + Pb(s) \rightarrow Cu(s) + Pb^{2+}(aq)$

b. Anode reaction: $Cr(s)\,|\,Cr^{2+}(aq) = [Cr(s) \rightarrow Cr^{2+}(aq) + 2\,e^-] \times 1$

Cathode reaction: $Ag^+(aq)\,|\,Ag(s) = [Ag^+(aq) + e^- \rightarrow Ag(s)] \times 2$

Overall cell reaction: $2Ag^+(aq) + Cr(s) \rightarrow 2Ag(s) + Cr^{2+}(aq)$

15.33 **a.** The anode is a Cd metal electrode in a Cd^{2+} solution. The anode reaction is

$Cd(s) \rightarrow Cd^{2+}(aq) + 2\,e^-$

The cathode is a Sn metal electrode in a Sn^{2+} solution. The cathode reaction is

$Sn^{2+}(aq) + 2\,e^- \rightarrow Sn(s)$

The shorthand notation for this cell is

$Cd(s)\,|\,Cd^{2+}(aq)\,||\,Sn^{2+}(aq)\,|\,Sn(s)$

b. The anode is a Zn metal electrode in a Zn^{2+} solution. The anode reaction is
$Zn(s) \rightarrow Zn^{2+}(aq) + 2e^-$
The cathode is a C(graphite) electrode, where Cl_2 gas is reduced to Cl^-.
The cathode reaction is
$Cl_2(g) + 2e^- \rightarrow 2Cl^-(aq)$
The shorthand notation for this cell is
$Zn(s) \mid Zn^{2+}(aq) \parallel Cl^-(aq), Cl_2(g) \mid C(\text{graphite})$

15.35 a. The Cd(*s*) has lost electrons, which makes the half-reaction an oxidation.
b. Cd metal is oxidized.
c. Because this is oxidation, it takes place at the anode.

15.37 a. The Zn(*s*) has lost electrons, which makes the half-reaction an oxidation.
b. Zn metal is oxidized.
c. Because this is oxidation, it takes place at the anode.

15.39 a. The half-reaction to plate tin is $Sn^{2+}(aq) + 2e^- \rightarrow Sn(s)$.
b. The reduction of Sn^{2+} to Sn occurs at the cathode, which is the iron can.
c. The oxidation of Sn to Sn^{2+} occurs at the anode, which is the tin bar.

15.41 Since Fe is above Sn in the activity series, if the Fe is exposed to air and water, Fe will be oxidized and rust will form. To protect iron, Sn would have to be *more* active than Fe and it is not.

15.43 a. Electrons are lost in an oxidation.
b. An oxidizing agent undergoes reduction.
c. O_2 gains electrons to form OH^-; this is a reduction.
d. Br_2 gains electrons to form $2Br^-$; this is a reduction.
e. Sn^{2+} loses electrons to form Sn^{4+}; this is an oxidation.

15.45 a.	VO_2	$O = -2$	Calculate: $V + 2(-2) = 0$	$\therefore V = +4$
b.	Ag_2CrO_4	$Ag = +1, O = -2$	Calculate: $2(+1) + Cr + 4(-2) = 0$	$\therefore Cr = +6$
c.	$S_2O_8^{2-}$	$O = -2$	Calculate: $2S + 8(-2) = -2$	$\therefore 2S = +14$ and $S = +7$
d.	$FeSO_4$	$Fe = +2, O = -2$	Calculate: $(+2) + S + 4(-2) = 0$	$\therefore S = +6$

15.47 $\underset{+3-2}{Cr_2O_3(s)} + \underset{0}{Si(s)} \rightarrow \underset{0}{Cr(s)} + \underset{+4-2}{SiO_2(s)}$

Cr: $+3 \rightarrow 0$ — Oxidation number decreases (reduction)
Si: $0 \rightarrow +4$ — Oxidation number increases (oxidation)

a. Cr^{3+} (in Cr_2O_3) is reduced.
b. Si is oxidized.
c. Cr_2O_3 is the oxidizing agent.
d. Si is the reducing agent.
e. $[Cr_2O_3(s) + 6H^+(aq) + 6e^- \rightarrow 2Cr(s) + 3H_2O(l)] \times 2$
$[Si(s) + 2H_2O(l) \rightarrow SiO_2(s) + 4H^+(aq) + 4e^-] \times 3$

Overall: $2Cr_2O_3(s) + 3Si(s) \rightarrow 4Cr(s) + 3SiO_2(s)$

15.49 a. One carbon group attached to the carbon atom bonded to the —OH makes this a primary alcohol.
b. Two carbon groups attached to the carbon atom bonded to the —OH makes this a secondary alcohol.
c. One carbon group attached to the carbon atom bonded to the —OH makes this a primary alcohol.

15.51 a.

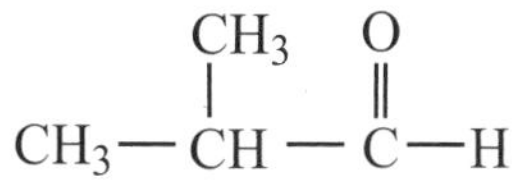

$$CH_3-\overset{\overset{\displaystyle CH_3}{|}}{CH}-\overset{\overset{\displaystyle O}{\|}}{C}-H$$

b.

$$CH_3-CH_2-\overset{\overset{\displaystyle O}{\|}}{C}-CH_3$$

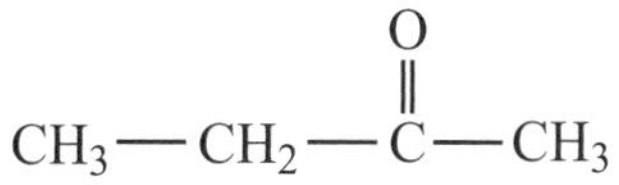

15.53 a. $Fe(s) \rightarrow Fe^{2+}(aq) + 2\,e^-$
b. $Ni^{2+}(aq) + 2\,e^- \rightarrow Ni(s)$
c. Fe is the anode.
d. Ni is the cathode.
e. The electrons flow from Fe to Ni.
f. $Fe(s) + Ni^{2+}(aq) \rightarrow Fe^{2+}(aq) + Ni(s)$
g. $Fe(s) \,|\, Fe^{2+}(aq) \,\|\, Ni^{2+}(aq) \,|\, Ni(s)$

15.55 Reactions **b**, **c**, and **d** all involve loss and gain of electrons; **b**, **c**, and **d** are oxidation–reduction reactions.
a. No. No change in oxidation numbers: $Ag = +1$, $N = +5$, $Na = +1$, $Cl = -1$
b. Yes. Li: $0 \rightarrow +1$ \therefore oxidation \qquad N: $0 \rightarrow -3$ \therefore reduction
c. Yes. Ni: $0 \rightarrow +2$ \therefore oxidation \qquad Pb: $+2 \rightarrow 0$ \therefore reduction
d. Yes. K: $0 \rightarrow +1$ \therefore oxidation \qquad H: $+1 \rightarrow 0$ \therefore reduction

15.57 a. Fe^{3+} gains an electron to form Fe^{2+}; this is a reduction.
b. Fe^{2+} is loses an electron to form Fe^{3+}; this is an oxidation.

15.59 a. Co_2O_3 $\quad O = -2$ \qquad Calculate: $2Co + 3(-2) = 0$ $\therefore 2Co = +6$ \quad and $Co = +3$
b. $KMnO_4$ $\quad K = +1, O = -2$ \quad Calculate: $+1 + Mn + 4(-2) = 0$ $\therefore Mn = +7$
c. $SbCl_5$ $\quad Cl = -1$ \qquad Calculate: $Sb + 5(-1) = 0$ $\therefore Sb = +5$
d. ClO_3^- $\quad O = -2$ \qquad Calculate: $Cl + 3(-2) = -1$ $\therefore Cl = +5$
e. PO_4^{3-} $\quad O = -2$ \qquad Calculate: $P + 4(-2) = -3$ $\therefore P = +5$

15.61 a. $\underset{+2\,-1}{FeCl_2(aq)} + \underset{0}{Cl_2(g)} \rightarrow \underset{+3\,-1}{FeCl_3(aq)}$

Fe: $+2 \rightarrow +3$ \qquad Oxidation number increases (oxidation)
Cl: $0 \rightarrow -1$ \qquad Oxidation number decreases (reduction)
Fe (in $FeCl_2$) is oxidized, and Cl (in Cl_2) is reduced.
\therefore balanced equation is $2FeCl_2(aq) + Cl_2(g) \rightarrow 2FeCl_3(aq)$

b. $\underset{+1\,-2}{H_2S(g)} + \underset{0}{O_2(g)} \rightarrow \underset{+1\,-2}{H_2O(l)} + \underset{+4\,-2}{SO_2(g)}$

H: $+1 \rightarrow +1$ \qquad No change
S: $-2 \rightarrow +4$ \qquad Oxidation number increases (oxidation)
O: $0 \rightarrow -2$ \qquad Oxidation number decreases (reduction)
S (in H_2S) is oxidized, and O (in O_2) is reduced.
\therefore balanced equation is $2H_2S(g) + 3O_2(g) \rightarrow 2H_2O(l) + 2SO_2(g)$

c. $\underset{+5\,-2}{P_2O_5(s)} + \underset{0}{C(s)} \rightarrow \underset{0}{P(s)} + \underset{+2\,-2}{CO(g)}$

P: $+5 \rightarrow 0$ \qquad Oxidation number decreases (reduction)
O: $-2 \rightarrow -2$ \qquad No change
C: $0 \rightarrow +2$ \qquad Oxidation number increases (oxidation)
C is oxidized, and P (in P_2O_5) is reduced.
\therefore balanced equation is $P_2O_5(s) + 5C(s) \rightarrow 2P(s) + 5CO(g)$

15.63 a. Balance charge with e^-: $Zn(s)$ $\rightarrow Zn^{2+}(aq) + 2\,e^-$

 b. Balance O with H_2O: $SnO_2^{2-}(aq) + H_2O(l)$ $\rightarrow SnO_3^{2-}(aq)$

 Balance H with H^+: $SnO_2^{2-}(aq) + H_2O(l)$ $\rightarrow SnO_3^{2-}(aq) + 2H^+(aq)$

 Balance charge with e^-: $SnO_2^{2-}(aq) + H_2O(l)$ $\rightarrow SnO_3^{2-}(aq) + 2H^+(aq) + 2\,e^-$

 c. Balance O with H_2O: $SO_3^{2-}(aq) + H_2O(l)$ $\rightarrow SO_4^{2-}(aq)$

 Balance H with H^+: $SO_3^{2-}(aq) + H_2O(l)$ $\rightarrow SO_4^{2-}(aq) + 2H^+(aq)$

 Balance charge with e^-: $SO_3^{2-}(aq) + H_2O(l)$ $\rightarrow SO_4^{2-}(aq) + 2H^+(aq) + 2\,e^-$

 d. Balance O with H_2O: $NO_3^-(aq)$ $\rightarrow NO(g) + 2H_2O(l)$

 Balance H with H^+: $NO_3^-(aq) + 4H^+(aq)$ $\rightarrow NO(g) + 2H_2O(l)$

 Balance charge with e^-: $NO_3^-(aq) + 4H^+(aq) + 3\,e^- \rightarrow NO(g) + 2H_2O(l)$

15.65 Write half-reactions, multiply by small numbers to equalize electrons lost and gained, add together, and combine common species.

 a. $Zn(s)$ $\rightarrow Zn^{2+}(aq) + 2\,e^-$

 $[NO_3^-(aq) + 2H^+(aq) + e^-$ $\rightarrow NO_2(g) + H_2O(l)] \times 2$

 Overall: $Zn(s) + 2NO_3^-(aq) + 4H^+(aq) \rightarrow Zn^{2+}(aq) + 2NO_2(g) + 2H_2O(l)$

 b. $[MnO_4^-(aq) + 8H^+(aq) + 5\,e^-$ $\rightarrow Mn^{2+}(aq) + 4H_2O(l)] \times 2$

 $[SO_3^{2-}(aq) + H_2O(l)$ $\rightarrow SO_4^{2-}(aq) + 2H^+(aq) + 2\,e^-] \times 5$

 Overall: $2MnO_4^-(aq) + 5SO_3^{2-}(aq) + 6H^+(aq) \rightarrow 2Mn^{2+}(aq) + 5SO_4^{2-}(aq) + 3H_2O(l)$

 c. $[2I^-(aq)$ $\rightarrow I_2(s) + 2\,e^-] \times 3$

 $ClO_3^-(aq) + 6H^+(aq) + 6\,e^-$ $\rightarrow Cl^-(aq) + 3H_2O(l)$

 Overall: $ClO_3^-(aq) + 6I^-(aq) + 6H^+(aq) \rightarrow Cl^-(aq) + 3I_2(s) + 3H_2O(l)$

 d. $[C_2O_4^{2-}(aq)$ $\rightarrow 2CO_2(g) + 2\,e^-] \times 3$

 $Cr_2O_7^{2-}(aq) + 14H^+(aq) + 6\,e^-$ $\rightarrow 2Cr^{3+}(aq) + 7H_2O(l)$

 Overall: $Cr_2O_7^{2-}(aq) + 3C_2O_4^{2-}(aq) + 14H^+(aq) \rightarrow 2Cr^{3+}(aq) + 6CO_2(g) + 7H_2O(l)$

15.67 Primary alcohols oxidize to aldehydes. Secondary alcohols oxidize to ketones.

 a.
$$CH_3-CH_2-\overset{\displaystyle O}{\overset{\displaystyle \|}{C}}-H$$

 b.
$$CH_3-\overset{\displaystyle O}{\overset{\displaystyle \|}{C}}-CH_2-CH_2-CH_3$$

 c.
$$CH_3-CH_2-CH_2-\overset{\displaystyle O}{\overset{\displaystyle \|}{C}}-H$$

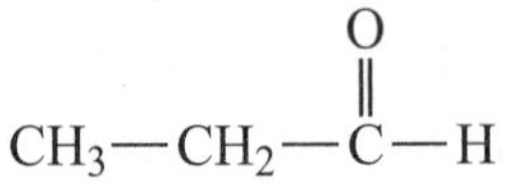

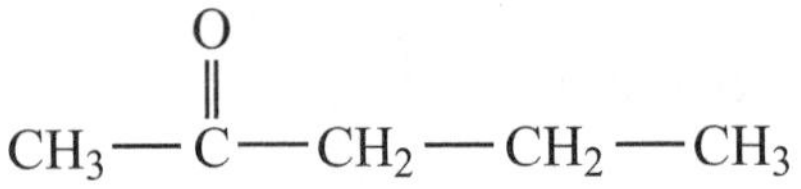

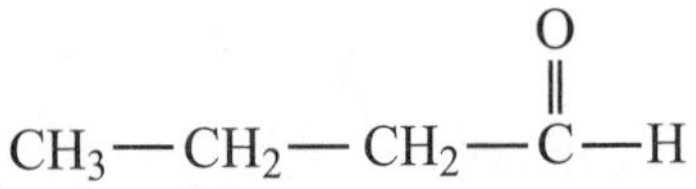

15.69 In the activity series, a metal oxidizes spontaneously when combined with the ions in solutions of any metal below it.

 a. Because Cu is below H_2 in the activity series, the reaction will not be spontaneous.

 b. Because Fe is above Ni in the activity series, the reaction will be spontaneous.

 c. Because Ag is below Cu in the activity series, the reaction will not be spontaneous.

 d. Because Cr is above Ni in the activity series, the reaction will be spontaneous.

 e. Because Zn is above Cu in the activity series, the reaction will be spontaneous.

 f. Because Zn is above Pb in the activity series, the reaction will be spontaneous.

15.71 **a.** The anode is Mg.
b. The cathode is Ni.
c. The half-reaction at the anode is $Mg(s) \rightarrow Mg^{2+}(aq) + 2\,e^-$.
d. The half-reaction at the cathode is $Ni^{2+}(aq) + 2\,e^- \rightarrow Ni(s)$.
e. The overall reaction is $Mg(s) + Ni^{2+}(aq) \rightarrow Mg^{2+}(aq) + Ni(s)$.
f. The shorthand cell notation is $Mg(s)\,|\,Mg^{2+}(aq)\,||\,Ni^{2+}(aq)\,|\,Ni(s)$.

15.73 In the activity series, Fe is above Ni, Pb, and Ag, which means it is oxidized. However, it is below Ca and Al, which means it cannot be oxidized and therefore Fe will not reduce Ca^{2+} or Al^{3+}.
a. $Ca^{2+}(aq)$ will not be reduced by an iron strip.
b. $Ag^{+}(aq)$ will be reduced by an iron strip.
c. $Ni^{2+}(aq)$ will be reduced by an iron strip.
d. $Al^{3+}(aq)$ will not be reduced by an iron strip.
e. $Pb^{2+}(aq)$ will be reduced by an iron strip.

15.75

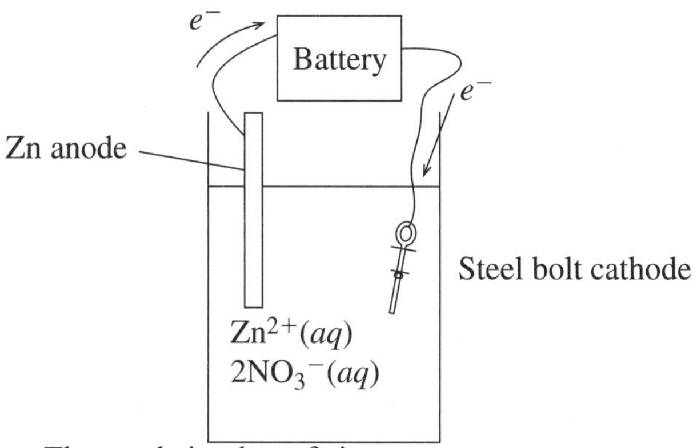

a. The anode is a bar of zinc.
b. The cathode is the steel bolt.
c. $Zn(s) \rightarrow Zn^{2+}(aq) + 2\,e^-$
d. $Zn^{2+}(aq) + 2\,e^- \rightarrow Zn(s)$
e. The purpose of the zinc coating is to prevent rusting of the bolt by H_2O and O_2.

15.77 **a.** $Pb(s) + SO_4^{2-}(aq) \rightarrow PbSO_4(s) + 2\,e^-$
b. $Pb(s)$ loses electrons; it is oxidized.
c. The oxidation half-reaction takes place at the anode.

15.79 **a.**
$$[Ag(s) \rightarrow Ag^{+}(aq) + e^-] \times 3$$
$$NO_3^-(aq) + 4H^{+}(aq) + 3\,e^- \rightarrow NO(g) + 2H_2O(l)$$

Overall: $4H^{+}(aq) + 3Ag(s) + NO_3^-(aq) \rightarrow 3Ag^{+}(aq) + NO(g) + 2H_2O(l)$

b. $$15.0\,\cancel{g\,Ag} \times \frac{1\,\cancel{mol\,Ag}}{107.9\,\cancel{g\,Ag}} \times \frac{1\,\cancel{mol\,NO}}{3\,\cancel{mol\,Ag}} \times \frac{22.4\text{ L (STP)}}{1\,\cancel{mol\,NO}} = 1.04\text{ L of NO}(g)\text{ at STP (3 SFs)}$$

15.81 **a.**
$$[CuS(s) + 4H_2O(l) \rightarrow CuSO_4(aq) + 8H^{+}(aq) + 8\,e^-] \times 3$$
$$[HNO_3(aq) + 3H^{+}(aq) + 3\,e^- \rightarrow NO(g) + 2H_2O(l)] \times 8$$

Overall: $3CuS(s) + 8HNO_3(aq) \rightarrow 3CuSO_4(aq) + 8NO(g) + 4H_2O(l)$

b. $$24.8\,\cancel{g\,CuS} \times \frac{1\,\cancel{mol\,CuS}}{95.62\,\cancel{g\,CuS}} \times \frac{8\,\cancel{mol\,HNO_3}}{3\,\cancel{mol\,CuS}} \times \frac{1000\text{ mL solution}}{16.0\,\cancel{mol\,HNO_3}}$$
$= 43.2$ mL of HNO_3 solution (3 SFs)

15.83 Reactions **a** and **c** both involve loss and gain of electrons; **a** and **c** are oxidation–reduction reactions.

a. Yes. Ca: $0 \rightarrow +2$ ∴ oxidation H: $+1 \rightarrow 0$ ∴ reduction

b. No. No change in oxidation numbers: Ca = +2, C = +4, O = −2

c. Yes. Cl: $0 \rightarrow -1$ ∴ reduction Br: $-1 \rightarrow 0$ ∴ oxidation

d. No. No change in oxidation numbers: Ba = +2, Cl = −1, Na = +1, or S = +6

15.85 a. Br_2 ∴ Br = 0

b. $HBrO_2$ H = +1, O = −2 Calculate: $+1 + Br + 2(-2) = 0$ ∴ Br = +3

c. BrO_3^- O = −2 Calculate: $Br + 3(-2) = -1$ ∴ Br = +5

d. $NaBrO_4$ Na = +1, O = −2 Calculate: $+1 + Br + 4(-2) = 0$ ∴ Br = +7

15.87

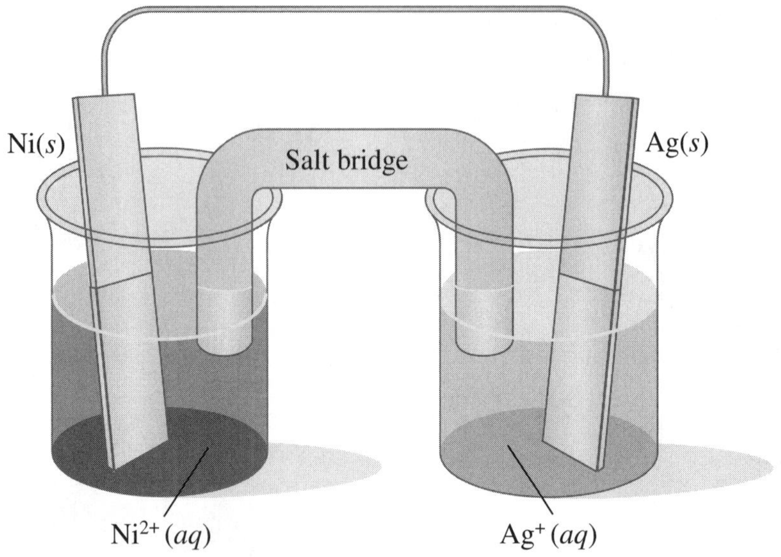

a. Ni(*s*) is the anode.

b. Ag(*s*) is the cathode.

c. $Ni(s) \rightarrow Ni^{2+}(aq) + 2\,e^-$

d. $Ag^+(aq) + e^- \rightarrow Ag(s)$

e. $Ni(s) + 2Ag^+(aq) \rightarrow Ni^{2+}(aq) + 2Ag(s)$

15.89 a. $2Al(s) + 3H_2O(l) \rightarrow Al_2O_3(s) + 6H^+(aq) + 6\,e^-$

b. $2H^+(aq) + 2\,e^- \rightarrow H_2(g)$

c. $2Al(s) + 3H_2O(l) \rightarrow Al_2O_3(s) + 6H^+(aq) + 6\,e^-$

$[2H^+(aq) + 2\,e^- \rightarrow H_2(g)] \times 3$

Overall: $2Al(s) + 3H_2O(l) \rightarrow Al_2O_3(s) + 3H_2(g)$

16
Nuclear Radiation

Study Goals

- Describe alpha, beta, positron, and gamma radiation.
- Write an equation showing mass numbers and atomic numbers for radioactive decay.
- Describe the detection and measurement of radiation.
- Given the half-life of a radioisotope, calculate the amount of radioisotope remaining after one or more half-lives.
- Describe the use of radioisotopes in medicine.
- Describe the processes of nuclear fission and fusion.

Chapter Outline

Answers and Solutions to Text Problems

16.1 **a.** An alpha particle and a helium nucleus both contain two protons and two neutrons. Alpha particles are emitted from unstable nuclei during radioactive decay.
b. α, ${}^{4}_{2}\mathrm{He}$
c. Alpha particles may be emitted from unstable nuclei to form more stable, lower energy nuclei.

16.3 **a.** ${}^{39}_{19}\mathrm{K}$, ${}^{40}_{19}\mathrm{K}$, ${}^{41}_{19}\mathrm{K}$
b. Each isotope has 19 protons and 19 electrons, but they differ in the number of neutrons present. Potassium-39 has $39 - 19 = 20$ neutrons, potassium-40 has $40 - 19 = 21$ neutrons, and potassium-41 has $41 - 19 = 22$ neutrons.

16.5

Medical Use	Atomic Symbol	Mass Number	Number of Protons	Number of Neutrons
Heart imaging	${}^{201}_{81}Tl$	201	81	120
Radiation therapy	${}^{60}_{27}Co$	60	27	33
Abdominal scan	${}^{67}_{31}Ga$	67	31	36
Hyperthyroidism	${}^{131}_{53}I$	131	53	78
Leukemia treatment	${}^{32}_{15}P$	32	15	17

16.7 **a.** α, ${}^{4}_{2}He$ **b.** n, ${}^{1}_{0}n$ **c.** β, ${}^{0}_{-1}e$ **d.** ${}^{15}_{7}N$ **e.** ${}^{125}_{53}I$

16.9 **a.** beta particle (β, ${}^{0}_{-1}e$) **b.** alpha particle (α, ${}^{4}_{2}He$) **c.** neutron (n, ${}^{1}_{0}n$)
d. sodium-24 (${}^{24}_{11}Na$) **e.** carbon-14 (${}^{14}_{6}C$)

16.11 **a.** Because beta particles are so much less massive and move faster than alpha particles, beta radiation can penetrate farther into body tissue.
b. Radiation breaks bonds and forms reactive species that cause undesirable reactions in the cells of the body.
c. Radiation technicians leave the room to increase the distance between them and the radiation. A thick wall of concrete or lead acts to shield them further.
d. Wearing gloves when handling radioisotopes shields the skin from α and β radiation.

16.13 The mass number of the radioactive atom is reduced by 4 when an alpha particle (${}^{4}_{2}He$) is emitted. The unknown product will have an atomic number that is 2 less than the atomic number of the radioactive atom.
a. ${}^{208}_{84}Po \rightarrow {}^{204}_{82}Pb + {}^{4}_{2}He$ **b.** ${}^{232}_{90}Th \rightarrow {}^{228}_{88}Ra + {}^{4}_{2}He$
c. ${}^{251}_{102}No \rightarrow {}^{247}_{100}Fm + {}^{4}_{2}He$ **d.** ${}^{220}_{86}Rn \rightarrow {}^{216}_{84}Po + {}^{4}_{2}He$

16.15 The mass number of the radioactive atom is not changed when a beta particle (${}^{0}_{-1}e$) is emitted. The unknown product will have an atomic number that is 1 higher than the atomic number of the radioactive atom.
a. ${}^{25}_{11}Na \rightarrow {}^{25}_{12}Mg + {}^{0}_{-1}e$ **b.** ${}^{20}_{8}O \rightarrow {}^{20}_{9}F + {}^{0}_{-1}e$
c. ${}^{92}_{38}Sr \rightarrow {}^{92}_{39}Y + {}^{0}_{-1}e$ **d.** ${}^{42}_{19}K \rightarrow {}^{42}_{20}Ca + {}^{0}_{-1}e$

16.17 The mass number of the radioactive atom is not changed when a positron (${}^{0}_{+1}e$) is emitted. The unknown product will have an atomic number that is 1 less than the atomic number of the radioactive atom.
a. ${}^{26}_{14}Si \rightarrow {}^{26}_{13}Al + {}^{0}_{+1}e$ **b.** ${}^{54}_{27}Co \rightarrow {}^{54}_{26}Fe + {}^{0}_{+1}e$
c. ${}^{77}_{37}Rb \rightarrow {}^{77}_{36}Kr + {}^{0}_{+1}e$ **d.** ${}^{93}_{45}Rh \rightarrow {}^{93}_{44}Ru + {}^{0}_{+1}e$

16.19 Balance the mass numbers and the atomic numbers in each nuclear equation.
a. ${}^{28}_{13}Al \rightarrow {}^{28}_{14}Si + {}^{0}_{-1}e$ $? = {}^{28}_{14}Si$ **b.** ${}^{87}_{36}Kr \rightarrow {}^{86}_{36}Kr + {}^{1}_{0}n$ $? = {}^{87}_{36}Kr$
c. ${}^{66}_{29}Cu \rightarrow {}^{66}_{30}Zn + {}^{0}_{-1}e$ $? = {}^{0}_{-1}e$ **d.** ${}^{238}_{92}U \rightarrow {}^{4}_{2}He + {}^{234}_{90}Th$ $? = {}^{238}_{92}U$
e. ${}^{188}_{80}Hg \rightarrow {}^{188}_{79}Au + {}^{0}_{+1}e$ $? = {}^{188}_{79}Au$

16.21 Balance the mass numbers and the atomic numbers in each nuclear equation.

a. ${}^{1}_{0}n + {}^{9}_{4}\text{Be} \rightarrow {}^{10}_{4}\text{Be}$ $\quad ? = {}^{10}_{4}\text{Be}$

b. ${}^{0}_{-1}e + {}^{32}_{16}\text{S} \rightarrow {}^{32}_{15}\text{P}$ $\quad ? = {}^{0}_{-1}e$

c. ${}^{1}_{0}n + {}^{27}_{13}\text{Al} \rightarrow {}^{24}_{11}\text{Na} + {}^{4}_{2}\text{He}$ $\quad ? = {}^{27}_{13}\text{Al}$

d. ${}^{4}_{2}\text{He} + {}^{27}_{13}\text{Al} \rightarrow {}^{30}_{15}\text{P} + {}^{1}_{0}n$ $\quad ? = {}^{30}_{15}\text{P}$

16.23 **a.** When radiation enters the Geiger counter, it ionizes a gas in the detection tube, which produces a burst of electrical current that is detected by the instrument.

b. The becquerel (Bq) is the SI unit for activity. The curie (Ci) is the original unit for activity of radioactive samples.

c. The SI unit for absorbed dose is the gray (Gy). The rad (radiation absorbed dose) is a unit of radiation absorbed per gram of sample. It is the older unit.

d. A kilogray is 1000 Gy, which is equivalent to 100 000 rad.

16.25 $70.0\ \cancel{\text{kg body mass}} \times \dfrac{4.20\ \mu\text{Ci}}{1\ \cancel{\text{kg body mass}}} = 294\ \mu\text{Ci}$ (3 SFs)

16.27 When pilots fly at high altitudes, they are exposed to higher levels of cosmic radiation because there are fewer molecules in the atmosphere to absorb the radiation.

16.29 A half-life is the time required for one-half of a radioactive sample to decay.

16.31 **a.** After one half-life, one-half of the sample would be radioactive: $80.0\text{ mg} \times \frac{1}{2} = 40.0\text{ mg}$ (3 SFs)

b. After two half-lives, one-fourth of the sample would still be radioactive:

$80.0\text{ mg} \times \frac{1}{2} \times \frac{1}{2} = 80.0\text{ mg} \times \frac{1}{4} = 20.0\text{ mg}$ (3 SFs)

c. $18\ \cancel{\text{h}} \times \dfrac{1\text{ half-life}}{6.0\ \cancel{\text{h}}} = 3.0\text{ half-lives} \quad 80.0\text{ mg} \times \frac{1}{2} \times \frac{1}{2} \times \frac{1}{2} = 80.0\text{ mg} \times \frac{1}{8} = 10.0\text{ mg}$ (3 SFs)

d. $24\ \cancel{\text{h}} \times \dfrac{1\text{ half-life}}{6.0\ \cancel{\text{h}}} = 4.0\text{ half-lives} \quad 80.0\text{ mg} \times \frac{1}{2} \times \frac{1}{2} \times \frac{1}{2} \times \frac{1}{2} = 80.0\text{ mg} \times \frac{1}{16} = 5.00\text{ mg}$ (3 SFs)

16.33 The radiation level in a radioactive sample is cut in half with each half-life. We must first determine the number of half-lives: $\frac{1}{4} = \frac{1}{2} \times \frac{1}{2}$ or 2 half-lives

Because each half-life for strontium-85 is 65 days, it will take 130 days for the radiation level of strontium-85 to fall to one-fourth of its original value: $2\ \cancel{\text{half-lives}} \times \dfrac{65\text{ days}}{1\ \cancel{\text{half-life}}} = 130\text{ days}$ (2 SFs)

16.35 **a.** Since the elements calcium and phosphorus are part of the bone, any calcium or phosphorus atom, regardless of isotope, will be carried to and become part of the bony structures of the body. Once there, the radiation emitted by any radioactive isotope can be used for diagnosis or treatment of bone diseases.

b. Strontium (Sr) acts much like calcium (Ca) because both are Group 2A (2) elements. The body will accumulate radioactive strontium in bones in the same way that it incorporates calcium. Radioactive strontium is harmful to children because the radiation it produces causes more damage in cells that are rapidly dividing.

16.37 $4.0\ \cancel{\text{mL solution}} \times \dfrac{45\ \mu\text{Ci}}{1\ \cancel{\text{mL solution}}} = 180\ \mu\text{Ci}$ of selenium-75 (2 SFs)

16.39 Nuclear fission is the splitting of a large atom into smaller fragments with a simultaneous release of large amounts of energy.

16.41 ${}^{1}_{0}n + {}^{235}_{92}\text{U} \rightarrow {}^{131}_{50}\text{Sn} + {}^{103}_{42}\text{Mo} + 2\,{}^{1}_{0}n + \text{energy}$ $\quad ? = {}^{103}_{42}\text{Mo}$

16.43 **a.** Neutrons bombard a nucleus in the fission process.
b. The nuclear process that occurs in the Sun is fusion.
c. Fission is the process where a large nucleus splits into smaller nuclei.
d. Fusion is the process where small nuclei combine to form larger nuclei.

16.45 **a.** ${}^{11}_{6}C$
b.

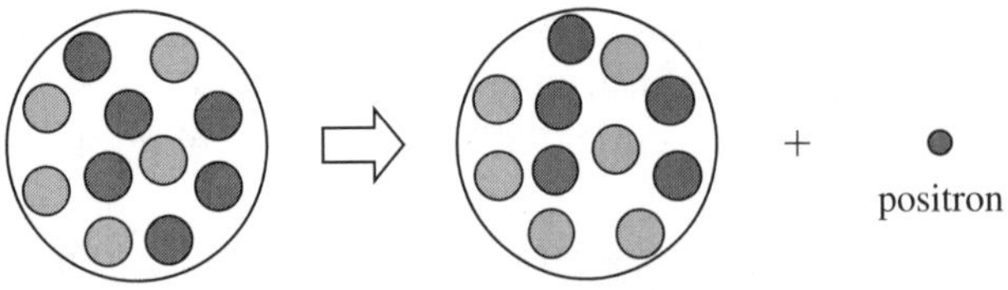

16.47

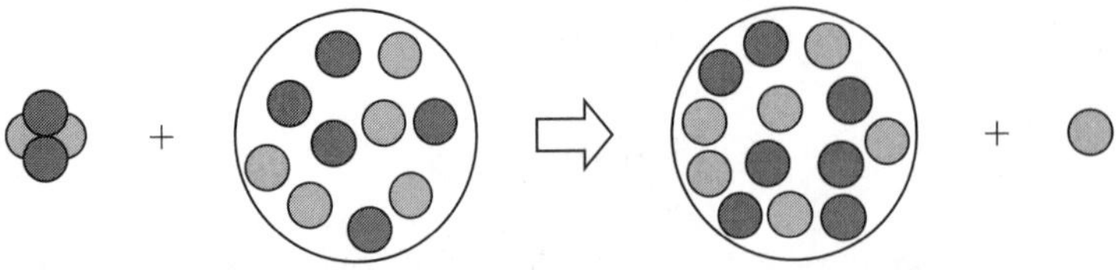

16.49 Half of a radioactive sample decays with each half-life:

$$6.4\ \mu Ci \text{ of } {}^{14}_{6}C \xrightarrow{\text{1 half-life}} 3.2\ \mu Ci \text{ of } {}^{14}_{6}C \xrightarrow{\text{1 half-life}} 1.6\ \mu Ci \text{ of } {}^{14}_{6}C \xrightarrow{\text{1 half-life}} 0.80\ \mu Ci \text{ of } {}^{14}_{6}C$$

$\therefore$ the activity of carbon-14 drops to 0.80 μCi in 3 half-lives or 3×5730 years, which makes the age of the painting 17 200 years.

16.51 **a.** ${}^{25}_{11}Na$ has 11 protons and $25 - 11 = 14$ neutrons.
b. ${}^{61}_{28}Ni$ has 28 protons and $61 - 28 = 33$ neutrons.
c. ${}^{84}_{37}Rb$ has 37 protons and $84 - 37 = 47$ neutrons.
d. ${}^{110}_{47}Ag$ has 47 protons and $110 - 47 = 63$ neutrons.

16.53 **a.** In alpha decay, a helium nucleus is emitted from the nucleus of a radioisotope. In beta decay, a neutron in an unstable nucleus is converted to a proton and an electron, which is emitted as a beta particle. In gamma emission, high-energy radiation is emitted from the nucleus of a radioisotope.
b. Alpha radiation is symbolized by α or ${}^{4}_{2}He$. Beta radiation is symbolized by β or ${}^{0}_{-1}e$. Gamma radiation is symbolized by γ or ${}^{0}_{0}\gamma$.

16.55 **a.** gamma emission
b. positron emission
c. alpha decay

16.57 **a.** ${}^{225}_{90}Th \rightarrow {}^{221}_{88}Ra + {}^{4}_{2}He$
b. ${}^{210}_{83}Bi \rightarrow {}^{206}_{81}Tl + {}^{4}_{2}He$
c. ${}^{137}_{55}Cs \rightarrow {}^{137}_{56}Ba + {}^{0}_{-1}e$
d. ${}^{126}_{50}Sn \rightarrow {}^{126}_{51}Sb + {}^{0}_{-1}e$
e. ${}^{121}_{53}I \rightarrow {}^{121}_{52}Te + {}^{0}_{+1}e$

16.59 **a.** ${}^{4}_{2}He + {}^{14}_{7}N \rightarrow {}^{17}_{8}O + {}^{1}_{1}H$ $\quad ? = {}^{17}_{8}O$
b. ${}^{4}_{2}He + {}^{27}_{13}Al \rightarrow {}^{30}_{14}Si + {}^{1}_{1}H$ $\quad ? = {}^{1}_{1}H$
c. ${}^{1}_{0}n + {}^{235}_{92}U \rightarrow {}^{90}_{38}Sr + 3{}^{1}_{0}n + {}^{143}_{54}Xe$ $\quad ? = {}^{143}_{54}Xe$
d. ${}^{127}_{55}Cs \rightarrow {}^{127}_{54}Xe + {}^{0}_{+1}e$ $\quad ? = {}^{127}_{55}Cs$

16.61 **a.** ${}^{16}_{8}O + {}^{16}_{8}O \rightarrow {}^{28}_{14}Si + {}^{4}_{2}He$
b. ${}^{18}_{8}O + {}^{249}_{98}Cf \rightarrow {}^{263}_{106}Sg + 4{}^{1}_{0}n$
c. ${}^{222}_{86}Rn \rightarrow {}^{218}_{84}Po + {}^{4}_{2}He$
then polonium-218 decays as follows: ${}^{218}_{84}Po \rightarrow {}^{214}_{82}Pb + {}^{4}_{2}He$
d. ${}^{80}_{38}Sr \rightarrow {}^{80}_{37}Rb + {}^{0}_{+1}e$

16.63 Half of a radioactive sample decays with each half-life:

$1.2 \text{ mg of } {}^{32}_{15}\text{P} \xrightarrow{\text{1 half-life}} 0.60 \text{ mg of } {}^{32}_{15}\text{P} \xrightarrow{\text{1 half-life}} 0.30 \text{ mg of } {}^{32}_{15}\text{P}$

∴ 2 half-lives must have elapsed during this time (28 d), yielding the half-life for phosphorus-32:

$$\frac{28 \text{ d}}{2 \text{ half-lives}} = 14 \text{ d/half-life}$$

16.65 **a.** ${}^{47}_{20}\text{Ca} \rightarrow {}^{47}_{21}\text{Sc} + {}^{0}_{-1}e$

b. First, calculate the number of half-lives that have passed:

$$18 \cancel{\text{d}} \times \frac{1 \text{ half-life}}{4.5 \cancel{\text{d}}} = 4.0 \text{ half-lives}$$

Now we can calculate the number of milligrams of calcium-47 that remain:

$16 \text{ mg} \times \frac{1}{2} \times \frac{1}{2} \times \frac{1}{2} \times \frac{1}{2} = 16 \text{ mg} \times \frac{1}{16} = 1.0 \text{ mg of calcium-47 (2 SFs)}$

c. Half of a radioactive sample decays with each half-life:

$4.8 \text{ mg of } {}^{47}_{20}\text{Ca} \xrightarrow{\text{1 half-life}} 2.4 \text{ mg of } {}^{47}_{20}\text{Ca} \xrightarrow{\text{1 half-life}} 1.2 \text{ mg of } {}^{47}_{20}\text{Ca}$

∴ 2 half-lives have passed.

$$2 \cancel{\text{half-lives}} \times \frac{4.5 \text{ d}}{1 \cancel{\text{half-life}}} = 9.0 \text{ d (2 SFs)}$$

16.67 First, calculate the number of half-lives that have passed:

$$24 \cancel{\text{h}} \times \frac{1 \text{ half-life}}{6.0 \cancel{\text{h}}} = 4.0 \text{ half-lives}$$

Now we can calculate the number of milligrams of technicium-99m that remain:

$120 \text{ mg} \times \frac{1}{2} \times \frac{1}{2} \times \frac{1}{2} \times \frac{1}{2} = 120 \text{ mg} \times \frac{1}{16} = 7.5 \text{ mg (2 SFs)}$

16.69 The irradiation of meats, fruits, and vegetables kills bacteria such as *E. coli* that can cause food-borne illnesses. In addition, food spoilage is deterred and shelf life is extended.

16.71 In the fission process, an atom splits into smaller nuclei with a simultaneous release of large amounts of energy. In fusion, two (or more) small nuclei combine (fuse) to form a larger nucleus, with a simultaneous release of large amounts of energy.

16.73 Fusion occurs naturally in the Sun and other stars.

16.75 First, calculate the number of half-lives that have elapsed:

$$27 \cancel{\text{d}} \times \frac{1 \text{ half-life}}{4.5 \cancel{\text{d}}} = 6.0 \text{ half-lives}$$

Because the activity of a radioactive sample is cut in half with each half-life, the activity must have been double its present value before each half-life. For 6.0 half-lives, we need to double the value 6 times:

$1.0\ \mu\text{Ci of } {}^{47}_{20}\text{Ca} \xleftarrow{\text{1 half-life}} 2.0\ \mu\text{Ci of } {}^{47}_{20}\text{Ca} \xleftarrow{\text{1 half-life}} 4.0\ \mu\text{Ci of } {}^{47}_{20}\text{Ca} \xleftarrow{\text{1 half-life}} 8.0\ \mu\text{Ci of } {}^{47}_{20}\text{Ca}$

$\xleftarrow{\text{1 half-life}} 16\ \mu\text{Ci of } {}^{47}_{20}\text{Ca} \xleftarrow{\text{1 half-life}} 32\ \mu\text{Ci of } {}^{47}_{20}\text{Ca} \xleftarrow{\text{1 half-life}} 64\ \mu\text{Ci of } {}^{47}_{20}\text{Ca}$

$1.0\ \mu\text{Ci} \times (2 \times 2 \times 2 \times 2 \times 2 \times 2) = 64\ \mu\text{Ci (2 SFs)}$

16.77 First, calculate the number of half-lives that have passed since the technician was exposed:

$$36 \cancel{\text{h}} \times \frac{1 \text{ half-life}}{12 \cancel{\text{h}}} = 3.0 \text{ half-lives}$$

Because the activity of a radioactive sample is cut in half with each half-life, the activity must have been double its present value before each half-life. For 3.0 half-lives, we need to double the value 3 times:

$2.0\ \mu\text{Ci of } {}^{42}_{19}\text{K} \xleftarrow{\text{1 half-life}} 4.0\ \mu\text{Ci of } {}^{42}_{19}\text{K} \xleftarrow{\text{1 half-life}} 8.0\ \mu\text{Ci of } {}^{42}_{19}\text{K} \xleftarrow{\text{1 half-life}} 16\ \mu\text{Ci of } {}^{42}_{19}\text{K}$

$2.0\ \mu\text{Ci} \times (2 \times 2 \times 2) = 16\ \mu\text{Ci (2 SFs)}$

Answers to Combining Ideas from Chapters 15 and 16

CI.27 a. A charge of −2 is obtained on the product side by adding 2 e^-:

$C_2O_4^{2-}(aq) \rightarrow 2CO_2(g) + 2\,e^-$ balanced oxidation half-reaction

b. $MnO_4^-(aq) \rightarrow Mn^{2+}(aq)$

Add $4H_2O$ to the product side to balance O:

$MnO_4^-(aq) \rightarrow Mn^{2+}(aq) + 4H_2O(l)$

Add $8H^+$ to the reactant side to balance H:

$MnO_4^-(aq) + 8H^+(aq) \rightarrow Mn^{2+}(aq) + 4H_2O(l)$

A charge of +2 is obtained on the reactant side by adding 5 e^-:

$5\,e^- + MnO_4^-(aq) + 8H^+(aq) \rightarrow Mn^{2+}(aq) + 4H_2O(l)$

balanced reduction half-reaction

c.

$$2 \times [5\,e^- + MnO_4^-(aq) + 8H^+(aq) \rightarrow Mn^{2+}(aq) + 4H_2O(l)]$$
$$5 \times [C_2O_4^{2-}(aq) \rightarrow 2CO_2(g) + 2\,e^-]$$
$$2MnO_4^-(aq) + 16H^+(aq) + 5C_2O_4^{2-}(aq) \rightarrow 10CO_2(g) + 2Mn^{2+}(aq) + 8H_2O(l)$$

balanced ionic equation

d.

$$0.758\ \cancel{g\ Na_2C_2O_4} \times \frac{1\ \cancel{mol\ Na_2C_2O_4}}{134.00\ \cancel{g\ Na_2C_2O_4}} \times \frac{1\ \cancel{mol\ C_2O_4^{2-}}}{1\ \cancel{mol\ Na_2C_2O_4}} \times \frac{2\ \cancel{mol\ MnO_4^-}}{5\ \cancel{mol\ C_2O_4^{2-}}}$$
$$\times \frac{1\ mol\ KMnO_4}{1\ \cancel{mol\ MnO_4^-}} \times \frac{1}{24.6\ \cancel{mL\ solution}} \times \frac{1000\ \cancel{mL\ solution}}{1\ L\ solution}$$

= 0.0920 M $KMnO_4$ solution (3 SFs)

CI.29 a.

$$0.121\ \cancel{g\ Mg} \times \frac{1\ \cancel{mol\ Mg}}{24.31\ \cancel{g\ Mg}} \times \frac{1\ mol\ H_2}{1\ \cancel{mol\ Mg}} = 0.004\,98\ mol\ of\ H_2\ \text{(smaller number of moles)}$$

$$50.0\ \cancel{mL\ solution} \times \frac{1\ \cancel{L}}{1000\ \cancel{mL}} \times \frac{1.00\ \cancel{mol\ HCl}}{1\ \cancel{L\ solution}} \times \frac{1\ mol\ H_2}{2\ \cancel{mol\ HCl}} = 0.0250\ mol\ of\ H_2$$

∴ Mg is the limiting reactant.

b. $T = 33.0\ °C + 273 = 306\ K$; $P = 750.\ mmHg$; $n = 0.004\,98$ mol of H_2 (from **a**)

$$V = \frac{nRT}{P} = \frac{(0.004\,98\ \cancel{mol})\left(\dfrac{62.4\ \cancel{L}\cdot\cancel{mmHg}}{\cancel{mol}\cdot\cancel{K}}\right)(306\ \cancel{K})}{750.\ \cancel{mmHg}} \times \frac{1000\ mL}{1\ \cancel{L}} = 127\ mL\ of\ H_2\ \text{(3 SFs)}$$

c.

$$50.0\ \cancel{mL\ solution} \times \frac{1.00\ g}{1\ \cancel{mL}} = 50.0\ g\ of\ solution$$

$$\Delta T = T_{final} - T_{initial} = 33.0\ °C - 22.0\ °C = 11.0\ °C$$

$$50.0\ \cancel{g\ solution} \times \frac{4.184\ J}{\cancel{g}\ \cancel{°C}} \times 11.0\ \cancel{°C} = 2.30 \times 10^3\ J\ \text{(2 SFs)}$$

d.

$$\frac{2.30 \times 10^3\ J}{0.121\ g\ Mg} = 1.90 \times 10^4\ J/g\ of\ Mg\ \text{(3 SFs)}$$

$$\frac{2.30 \times 10^3\ \cancel{J}}{0.121\ \cancel{g\ Mg}} \times \frac{1\ kJ}{1000\ \cancel{J}} \times \frac{24.31\ \cancel{g\ Mg}}{1\ mol\ Mg} = 462\ kJ/mol\ of\ Mg\ \text{(3 SFs)}$$

CI.31 a.

Isotope	Number of Protons	Number of Neutrons	Number of Electrons
$^{27}_{14}\text{Si}$	14	13	14
$^{28}_{14}\text{Si}$	14	14	14
$^{29}_{14}\text{Si}$	14	15	14
$^{30}_{14}\text{Si}$	14	16	14
$^{31}_{14}\text{Si}$	14	17	14

b. Electron configuration of Si: $1s^22s^22p^63s^23p^2$
Abbreviated electron configuration: $[\text{Ne}]3s^23p^2$

c. $^{28}_{14}\text{Si}$ $\quad 27.977 \times \dfrac{92.230}{100} = 25.803$ amu

$^{29}_{14}\text{Si}$ $\quad 28.977 \times \dfrac{4.683}{100} = 1.357$ amu

$^{30}_{14}\text{Si}$ $\quad 29.974 \times \dfrac{3.087}{100} = 0.9253$ amu

Atomic mass of Si = 28.085 amu

d. $^{27}_{14}\text{Si} \rightarrow {}^{27}_{13}\text{Al} + {}^{0}_{+1}e$
$^{31}_{14}\text{Si} \rightarrow {}^{31}_{15}\text{P} + {}^{0}_{-1}e$

e.

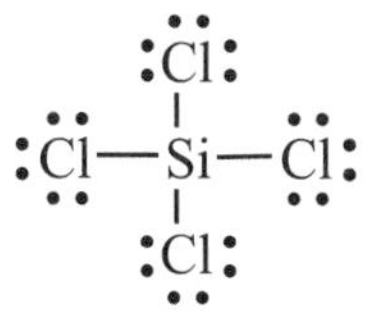

With 4 bonding pairs and no lone pairs, the shape of $SiCl_4$ is tetrahedral.

f. Half of a radioactive sample decays with each half-life:

16 μCi of $^{31}_{14}\text{Si} \xrightarrow{\text{1 half-life}}$ 8.0 μCi of $^{31}_{14}\text{Si} \xrightarrow{\text{1 half-life}}$ 4.0 μCi of $^{31}_{14}\text{Si} \xrightarrow{\text{1 half-life}}$ 2.0 μCi of $^{31}_{14}\text{Si}$

Therefore, 3 half-lives have passed.

$3\ \cancel{\text{half-lives}} \times \dfrac{2.6\ \text{h}}{1\ \cancel{\text{half-life}}} = 7.8\ \text{h (2 SFs)}$

CI.33 a. $^{238}_{92}\text{U} \rightarrow {}^{234}_{90}\text{Th} + {}^{4}_{2}\text{He}$ $\quad ? = {}^{4}_{2}\text{He}$

b. $^{234}_{90}\text{Th} \rightarrow {}^{234}_{91}\text{Pa} + {}^{0}_{-1}e$ $\quad ? = {}^{234}_{91}\text{Pa}$

c. $^{226}_{88}\text{Ra} \rightarrow {}^{222}_{86}\text{Rn} + {}^{4}_{2}\text{He}$ $\quad ? = {}^{226}_{88}\text{Ra}$

17

Organic Chemistry

Study Goals

- Write the IUPAC names and draw the condensed structural formulas for alkanes with substituents.
- Write the IUPAC names and formulas for alkenes and alkynes; draw the condensed structural formulas of the monomers that form a polymer.
- Describe the bonding in benzene; name aromatic compounds, and draw their condensed structural formulas.
- Write the IUPAC and common names of alcohols, phenols, and ethers; draw the condensed structural formulas when given their names.
- Write the IUPAC and common names of aldehydes and ketones; draw the condensed structural formulas when given their names.
- Write the IUPAC and common names of carboxylic acids and esters; draw the condensed structural formulas when given their names.
- Write the IUPAC and common names of amines and amides; draw the condensed structural formulas when given their names.

Chapter Outline

Answers and Solutions to Text Problems

17.1 **a.** Pentane has a carbon chain of five carbon atoms.
b. Hexane has a carbon chain of six carbon atoms.

17.3 **a.** CH_3-CH_3 **b.** $CH_3-CH_2-CH_2-CH_2-CH_3$

17.5 Two structures are isomers if they have the same molecular formula, but different arrangements of atoms.
a. same molecule **b.** isomers of C_6H_{14}

17.7 **a.** 2-fluorobutane **b.** 2,2-dimethylpropane
c. 2-chloro-3-methylpentane

17.9 Draw the main chain with the number of carbon atoms indicated by the alkane name. For example, butane has a main chain of four carbon atoms, and hexane has a main chain of six carbon atoms. Attach substituents on the carbon atoms indicated. For example, in 3-methylpentane, a CH_3- group is bonded to carbon 3 of a five-carbon chain.

$$
\begin{array}{c}
\quad\quad\quad\quad\quad CH_3 \\
\quad\quad\quad\quad\quad | \\
\textbf{a.}\ \text{2-methylbutane}\quad CH_3-CH-CH_2-CH_3
\end{array}
$$

$$
\begin{array}{c}
\quad\quad\quad\quad\quad\quad\quad Cl \\
\quad\quad\quad\quad\quad\quad\quad | \\
CH_3-CH_2-C-CH_2-CH_3 \\
\quad\quad\quad\quad\quad\quad\quad | \\
\textbf{b.}\ \text{3,3-dichloropentane}\quad Cl
\end{array}
$$

$$
\begin{array}{c}
\quad\quad\quad\quad\quad CH_3\ \ CH_3\quad\quad\quad CH_3 \\
\quad\quad\quad\quad\quad |\quad\quad |\quad\quad\quad\quad | \\
\textbf{c.}\ \text{2,3,5-trimethylhexane}\quad CH_3-CH-CH-CH_2-CH-CH_3
\end{array}
$$

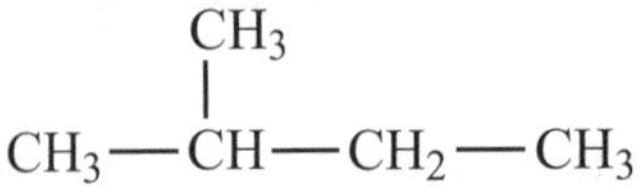

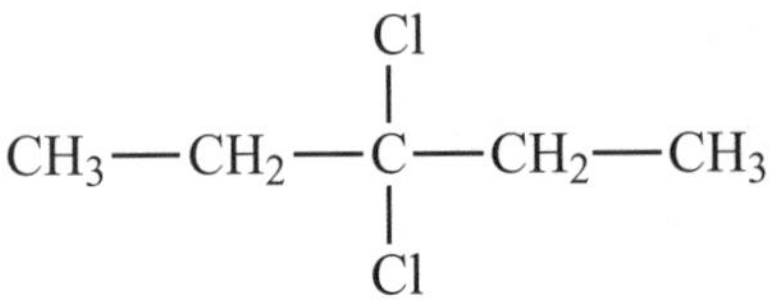

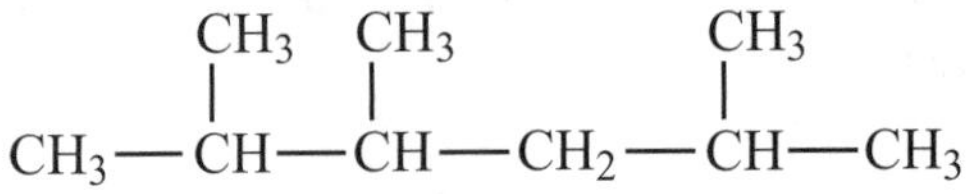

17.11 **a.** An alkene has a carbon–carbon double bond.
b. An alkyne has a carbon–carbon triple bond.

17.13 **a.** The two-carbon compound with a double bond is named ethene.
b. This is a three-carbon alkene with a methyl substituent. The name is 2-methylpropene.
c. The five-carbon compound with a triple bond between carbon 2 and carbon 3 is named 2-pentyne.

17.15 **a.** Propene is the three-carbon alkene. $H_2C=CH-CH_3$
b. 1-Pentene is the five-carbon compound with a double bond between carbon 1 and carbon 2.
$H_2C=CH-CH_2-CH_2-CH_3$
c. 2-Methyl-1-butene has a four-carbon chain with a double bond between carbon 1 and carbon 2

$$
\begin{array}{c}
\quad\quad\quad\quad\quad\quad\quad\quad\quad CH_3 \\
\quad\quad\quad\quad\quad\quad\quad\quad\quad | \\
\text{and a methyl group attached to carbon 2.}\quad H_2C=C-CH_2-CH_3
\end{array}
$$

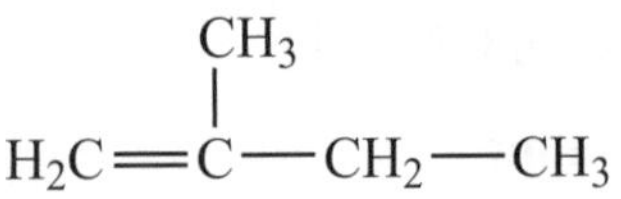

17.17 A polymer is a long-chain molecule consisting of many repeating smaller units. These smaller units are called monomers.

17.19 Polyethylene is a polymer of the monomer ethylene (ethene).

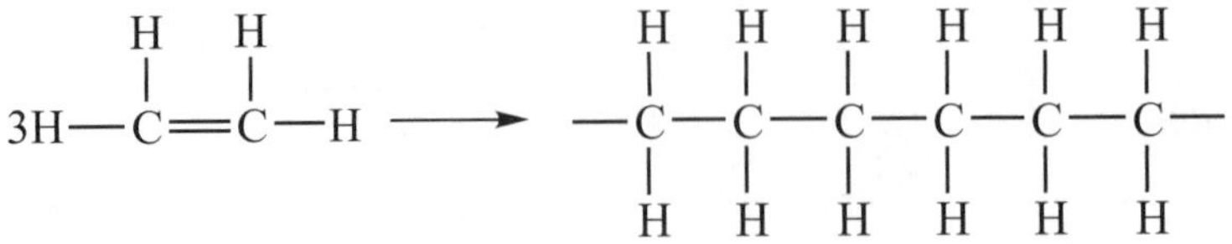

17.21 The six-carbon ring with alternating single and double bonds is benzene.
 a. 2-chlorotoluene
 b. ethylbenzene
 c. 1,3,5-trichlorobenzene

17.23 a.

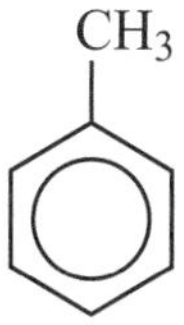

 b.

c.

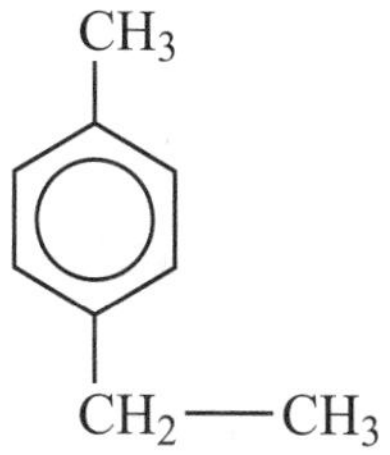

17.25 a. This compound has a two-carbon chain (ethane). The final *e* is dropped, and *ol* added to indicate an alcohol. The IUPAC name is ethanol.
 b. This compound has a four-carbon chain with a hydroxyl group attached to carbon 2. The IUPAC name is 2-butanol.
 c. This compound has a five-carbon chain with a hydroxyl group attached to carbon 2. The IUPAC name is 2-pentanol.
 d. This compound is the six-carbon benzene ring with a hydroxyl group attached. The IUPAC name is phenol.

17.27 a. 1-Propanol has a three-carbon chain with a hydroxyl group attached to carbon 1.
 $CH_3 \!-\! CH_2 \!-\! CH_2 \!-\! OH$
 b. Methyl alcohol has a hydroxyl group attached to a one-carbon alkane. $CH_3 \!-\! OH$
 c. 3-Pentanol has a five-carbon chain with a hydroxyl group attached to carbon 3.

$$CH_3 \!-\! CH_2 \!-\! \overset{\displaystyle \overset{OH}{|}}{CH} \!-\! CH_2 \!-\! CH_3$$

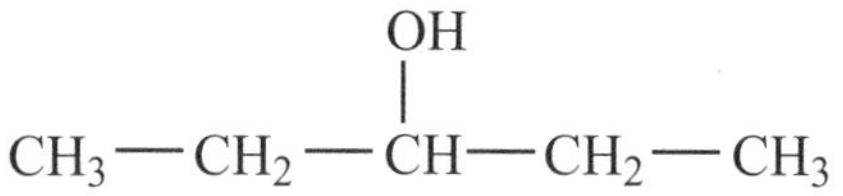

 d. 2-Methyl-2-butanol has a four-carbon chain with a methyl and hydroxyl group attached to carbon 2.

$$CH_3 \!-\! \underset{\displaystyle \underset{CH_3}{|}}{\overset{\displaystyle \overset{OH}{|}}{C}} \!-\! CH_2 \!-\! CH_3$$

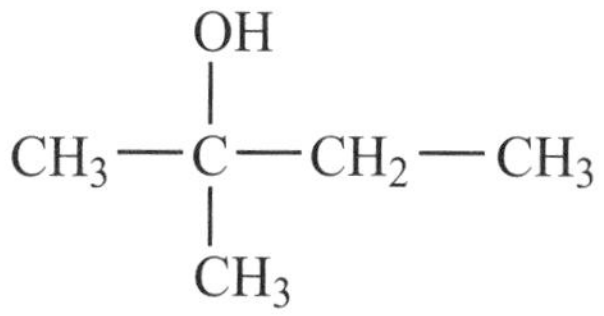

17.29 a. ethyl methyl ether **b.** dipropyl ether **c.** methyl propyl ether

17.31 a. acetaldehyde **b.** methyl propyl ketone **c.** formaldehyde

17.33 a. propanal **b.** 2-methyl-3-pentanone **c.** benzaldehyde

17.35 a. Acetaldehyde is the common name of the aldehyde with two carbons.

$$CH_3-C(=O)-H$$

b. 2-Pentanone has a ketone group on carbon 2 of a five-carbon chain.

$$CH_3-C(=O)-CH_2-CH_2-CH_3$$

c. Butyl methyl ketone has a four-carbon group and a one-carbon group on either side of the carbonyl carbon.

$$CH_3-C(=O)-CH_2-CH_2-CH_2-CH_3$$

17.37 a. Ethanoic acid (acetic acid) is the carboxylic acid with two carbons.

b. Propanoic acid (propionic acid) is the three-carbon carboxylic acid.

c. 4-Hydroxybenzoic acid has a hydroxyl group on carbon 4 of the ring in the aromatic carboxylic acid.

17.39 a. Propionic acid is the common name of the three-carbon carboxylic acid. $CH_3-CH_2-C(=O)-OH$

b. Benzoic acid is a carboxylic acid group attached to a benzene ring.

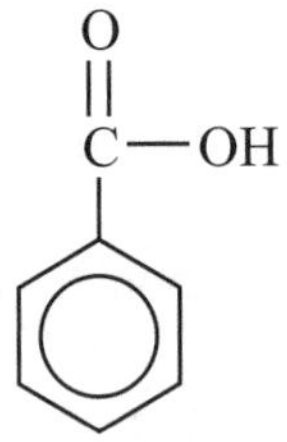

c. 2-Chloroethanoic acid is a carboxylic acid that has a two-carbon chain with a $-Cl$ on carbon 2.

$$Cl-CH_2-C(=O)-OH$$

d. 3-Hydroxypropanoic acid is a carboxylic acid that has a three-carbon chain with a hydroxyl group on carbon 3.

$$HO-CH_2-CH_2-C(=O)-OH$$

17.41 a. The ester has a one-carbon part from the alcohol methanol and a two-carbon part from the carboxylic acid acetic (ethanoic) acid.

$$CH_3-C(=O)-O-CH_3$$

b. The ester has a one-carbon part from the alcohol methanol and a five-carbon part from the carboxylic acid pentanoic acid.

$$CH_3-CH_2-CH_2-CH_2-C(=O)-O-CH_3$$

17.43 A carboxylic acid and an alcohol react to give an ester with the elimination of water.

a. 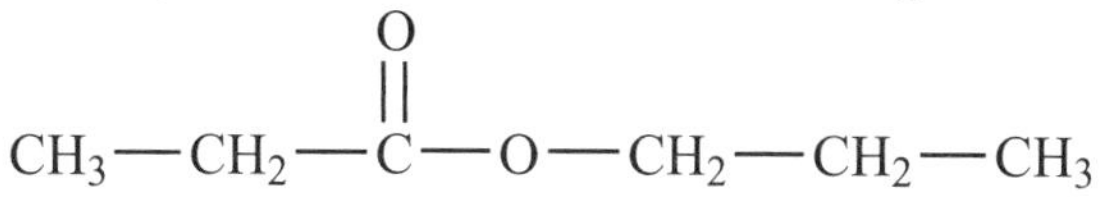

b. 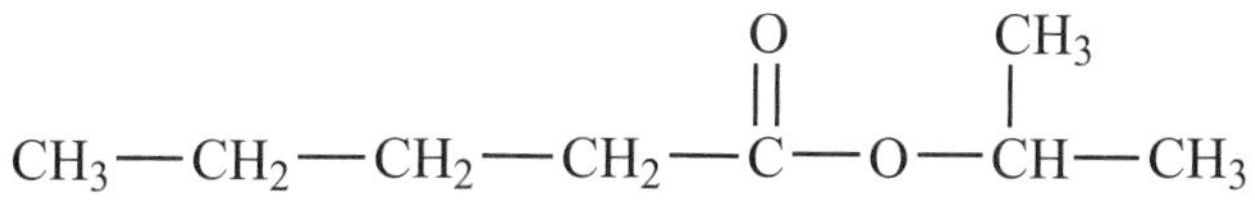

17.45 **a.** The name of this ester is methyl methanoate. The carbonyl portion of the ester contains one carbon; the name is derived from methanoic acid. The alkyl portion has one carbon, which is methyl.

b. The name of this ester is methyl ethanoate. The carbonyl portion of the ester contains two carbons; the name is derived from ethanoic acid. The alkyl portion has one carbon, which is methyl.

c. The name of this ester is methyl butanoate. The carbonyl portion of the ester contains four carbons; the name is derived from butanoic acid. The alkyl portion has one carbon, which is methyl.

17.47 **a.** Acetic acid is the two-carbon carboxylic acid. Methanol gives a one-carbon alkyl group.

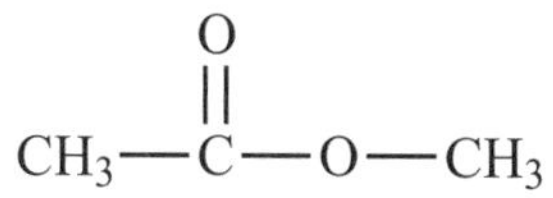

b. Formic acid is the one-carbon carboxylic acid bonded to the four-carbon 1-butanol.

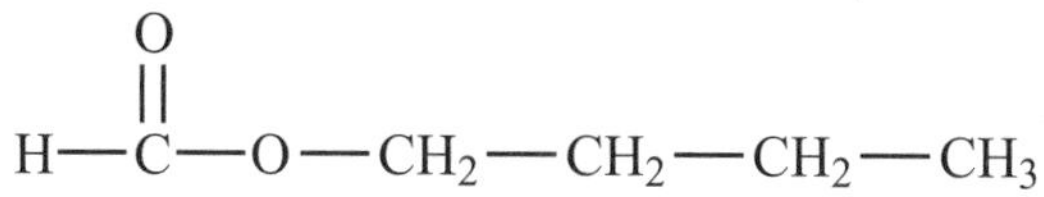

c. Pentanoic acid is the five-carbon carboxylic acid bonded to the two-carbon ethanol.

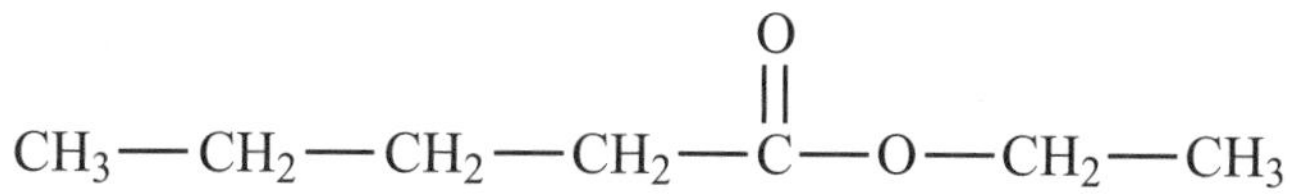

d. Propanoic acid is the three-carbon carboxylic acid bonded to the three-carbon propanol.

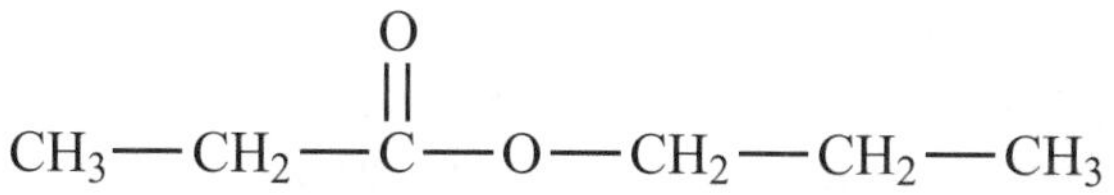

17.49 The common name of an amine consists of naming the alkyl groups bonded to the nitrogen atom in alphabetical order followed by *amine*.

a. A methyl group attached to $—NH_2$ is methylamine.

b. A methyl and a propyl group attached to the nitrogen atom would be named methylpropylamine.

c. One methyl and two ethyl groups attached to the nitrogen atom is named diethylmethylamine.

17.51 **a.** Ethanamide (acetamide); ethanamide tells us that the carbonyl portion has two carbon atoms.

b. 2-Chlorobutanamide (2-chlorobutyramide) is a four-carbon amide with a $-Cl$ on carbon 2.

c. Methanamide (formamide) has only the carbonyl carbon bonded to the amino group.

17.53 **a.** Propionamide is the common name of the amide of propanoic acid, which has three carbon atoms.

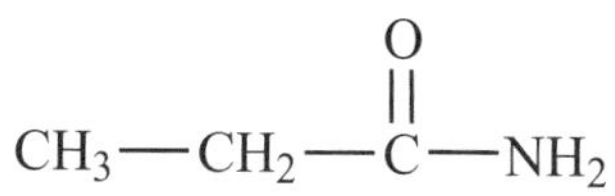

b. 2-Methylpentanamide is an amide of the five-carbon pentanoic acid with a methyl group attached to carbon 2.

$$CH_3-CH_2-CH_2-\overset{\overset{\displaystyle CH_3}{|}}{CH}-\overset{\overset{\displaystyle O}{||}}{C}-NH_2$$

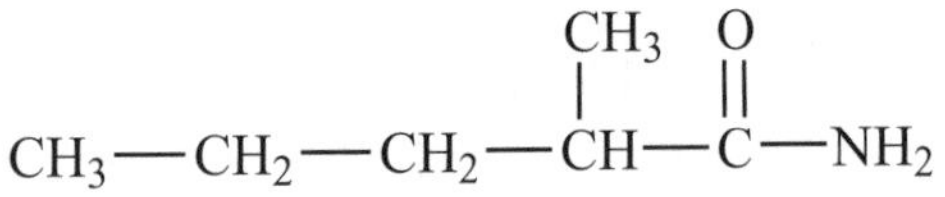

c. Methanamide is the simplest of the amides, with only one carbon atom.

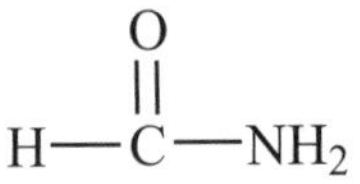

$$H-\overset{\overset{\displaystyle O}{||}}{C}-NH_2$$

17.55

$$-\overset{\overset{\displaystyle F}{|}}{\underset{\underset{\displaystyle F}{|}}{C}}-\overset{\overset{\displaystyle F}{|}}{\underset{\underset{\displaystyle F}{|}}{C}}-\overset{\overset{\displaystyle F}{|}}{\underset{\underset{\displaystyle F}{|}}{C}}-\overset{\overset{\displaystyle F}{|}}{\underset{\underset{\displaystyle F}{|}}{C}}-\overset{\overset{\displaystyle F}{|}}{\underset{\underset{\displaystyle F}{|}}{C}}-\overset{\overset{\displaystyle F}{|}}{\underset{\underset{\displaystyle F}{|}}{C}}-\overset{\overset{\displaystyle F}{|}}{\underset{\underset{\displaystyle F}{|}}{C}}-\overset{\overset{\displaystyle F}{|}}{\underset{\underset{\displaystyle F}{|}}{C}}-$$

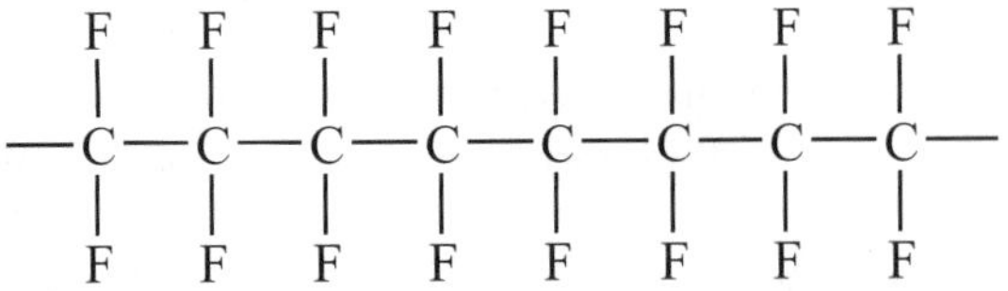

17.57 a. 3,7-dimethyl-6-octenal
 b. The *en* signifies that a double bond is present.
 c. The *al* signifies that an aldehyde is present.

17.59 a.
$$CH_3-CH_2-\overset{\overset{\displaystyle CH_2-CH_3}{|}}{CH}-CH_2-CH_2-CH_3$$

 b.
$$CH_3-\overset{\overset{\displaystyle CH_3}{|}}{CH}-\overset{\overset{\displaystyle CH_3}{|}}{CH}-CH_2-CH_3$$

 c.
$$Cl-CH_2-CH_2-\overset{\overset{\displaystyle CH_3}{|}}{\underset{\underset{\displaystyle Cl}{|}}{C}}-CH_2-CH_2-CH_2-CH_3$$

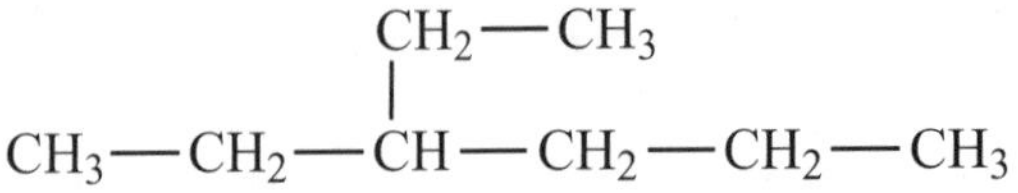

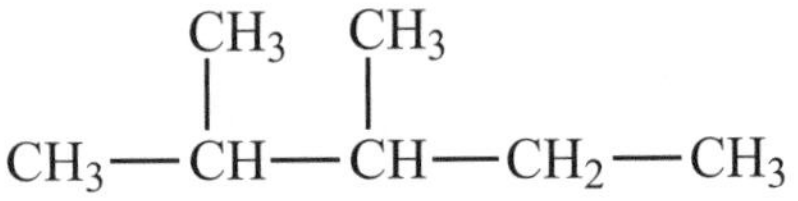

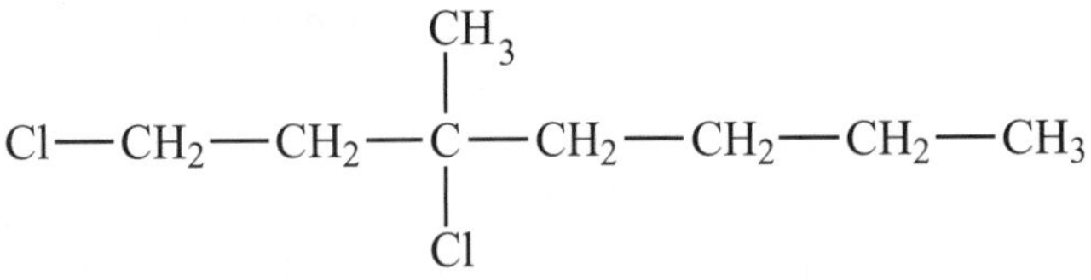

17.61 a. This compound contains a five-carbon chain with a double bond between carbon 1 and carbon 2 and a methyl group attached to carbon 2. The IUPAC name is 2-methyl-1-pentene.
 b. This compound has a four-carbon chain with a triple bond between carbon 1 and carbon 2. The IUPAC name is 1-butyne.
 c. This compound contains a five-carbon chain with a double bond between carbon 2 and carbon 3. The IUPAC name is 2-pentene.

17.63 a. This compound has a methyl group attached to the aromatic ring of benzene and is known as methylbenzene or toluene.
 b. A benzene ring with a $-Cl$ attached to carbon 1 and carbon 2 of the ring is named 1,2-dichlorobenzene.
 c. A benzene ring with an ethyl group on carbon 1 and a methyl group on carbon 4 of the ring is named 1-ethyl-4-methylbenzene; if the name toluene is used as the parent compound, the ethyl group is considered to be on carbon 4 of the ring and the name would be 4-ethyltoluene.

17.65 a.

b.

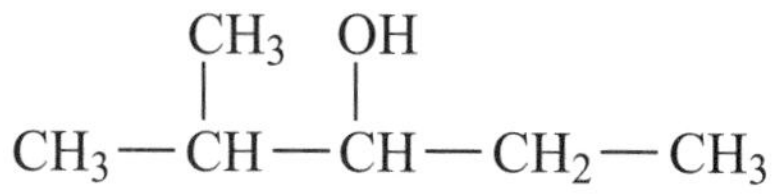

$$CH_3\overset{\overset{\displaystyle CH_3}{|}}{CH}\overset{\overset{\displaystyle OH}{|}}{CH}CH_2-CH_3$$

c.

$$CH_3-CH_2-\overset{\overset{\displaystyle O}{||}}{C}-CH_2-CH_3$$

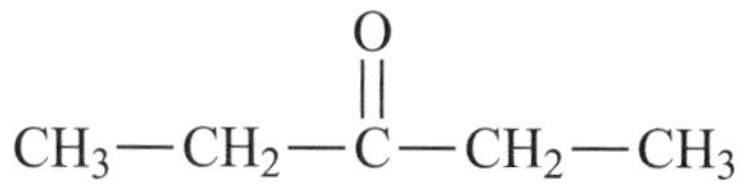

17.67 a. With a hydroxyl group on carbon 3 and a $-Cl$ on carbon 4 of the aromatic aldehyde benzalde-hyde, the compound will be named 4-chloro-3-hydroxybenzaldehyde.

b. With a $-Cl$ attached to carbon 3 of the three-carbon aldehyde, the compound will be named 3-chloropropanal.

c. With a $-Cl$ attached to carbon 2 of the five-carbon ketone, the compound will be named 2-chloro-3-pentanone.

17.69 a. 4-Chlorobenzaldehyde is a benzene with an aldehyde group and a chlorine atom on carbon 4 of the ring.

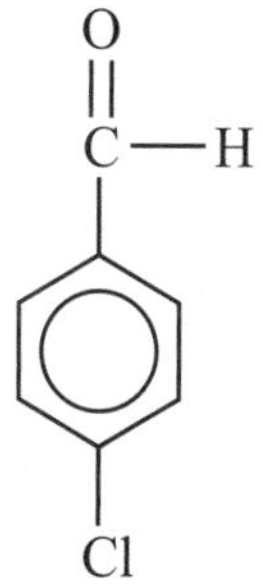

b. 3-Chloropropionaldehyde is a three-carbon aldehyde with a $-Cl$ located two carbons from the carbonyl group.

$$Cl-CH_2-CH_2-\overset{\overset{\displaystyle O}{||}}{C}-H$$

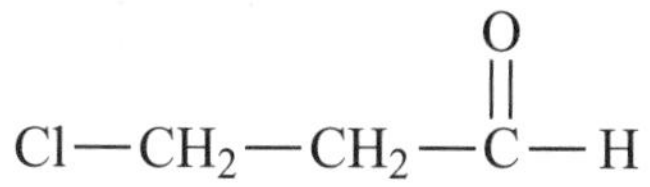

c. Ethyl methyl ketone (butanone) is a four-carbon ketone.

$$CH_3-\overset{\overset{\displaystyle O}{||}}{C}-CH_2-CH_3$$

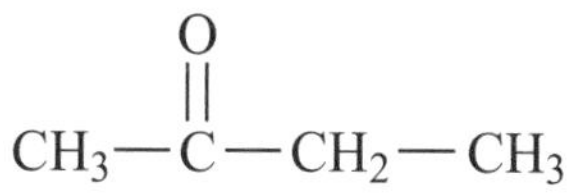

17.71 a. With a methyl group attached to carbon 3 of the four-carbon carboxylic acid, the compound is named 3-methylbutanoic acid.

b. The name of this ester is ethyl benzoate. The carbonyl portion of the ester is derived from the aromatic benzoic acid. The alkyl portion has two carbons, which is ethyl.

c. The name of this ester is ethyl propanoate. The carbonyl portion of the ester contains three carbons; the name is derived from propanoic acid. The alkyl portion has two carbons, which is ethyl.

17.73 a. $CH_3-CH_2-NH_2$

b. $CH_3-NH-CH_3$

c.

$$CH_3-CH_2-\overset{\overset{\displaystyle CH_2-CH_3}{|}}{N}-CH_2-CH_3$$

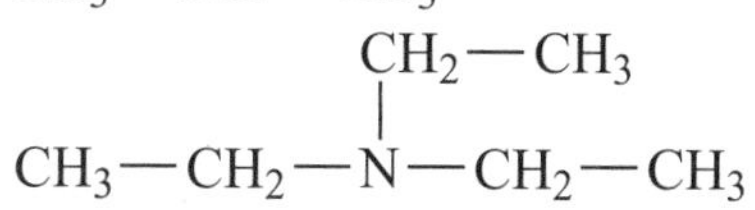

17.75 There are eight alcohols (isomers) that have the molecular formula $C_5H_{12}O$.

$$CH_3-CH_2-CH_2-CH_2-CH_2-OH \qquad \text{1-pentanol}$$

$$\begin{array}{c} OH \\ | \\ CH_3-CH-CH_2-CH_2-CH_3 \end{array} \qquad \text{2-pentanol}$$

$$\begin{array}{c} OH \\ | \\ CH_3-CH_2-CH-CH_2-CH_3 \end{array} \qquad \text{3-pentanol}$$

$$\begin{array}{c} CH_3 \\ | \\ HO-CH_2-CH-CH_2-CH_3 \end{array} \qquad \text{2-methyl-1-butanol}$$

$$\begin{array}{c} CH_3 \\ | \\ HO-CH_2-CH_2-CH-CH_3 \end{array} \qquad \text{3-methyl-1-butanol}$$

$$\begin{array}{c} CH_3 \\ | \\ CH_3-C-CH_2-CH_3 \\ | \\ OH \end{array} \qquad \text{2-methyl-2-butanol}$$

$$\begin{array}{c} OH \quad CH_3 \\ | \qquad | \\ CH_3-CH-CH-CH_3 \end{array} \qquad \text{3-methyl-2-butanol}$$

$$\begin{array}{c} CH_3 \\ | \\ CH_3-C-CH_2-OH \\ | \\ CH_3 \end{array} \qquad \text{2,2-dimethyl-1-propanol}$$

17.77

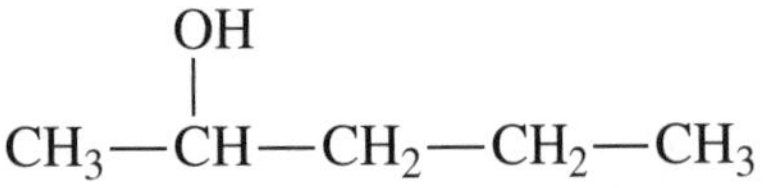

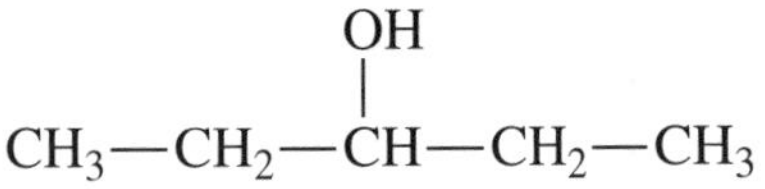

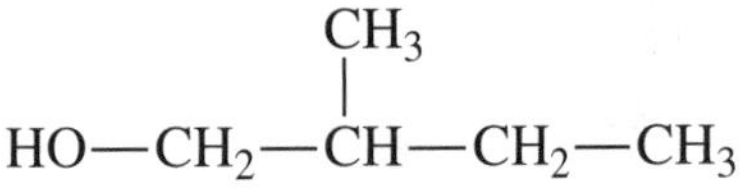

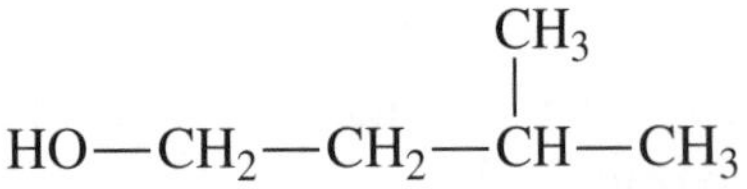

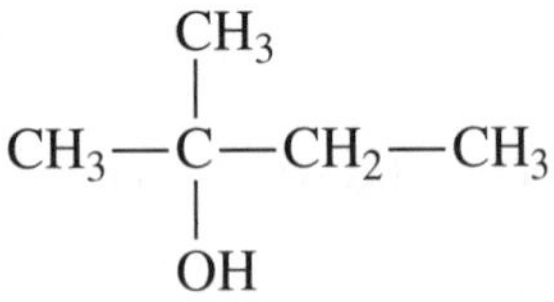

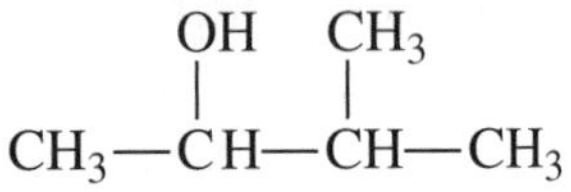

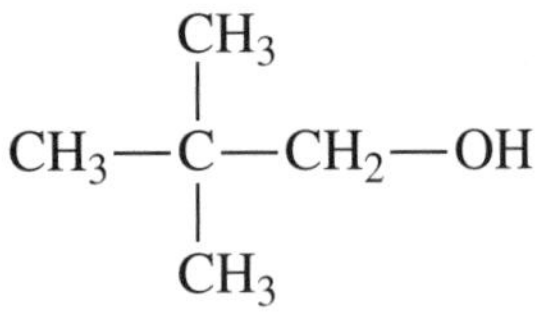

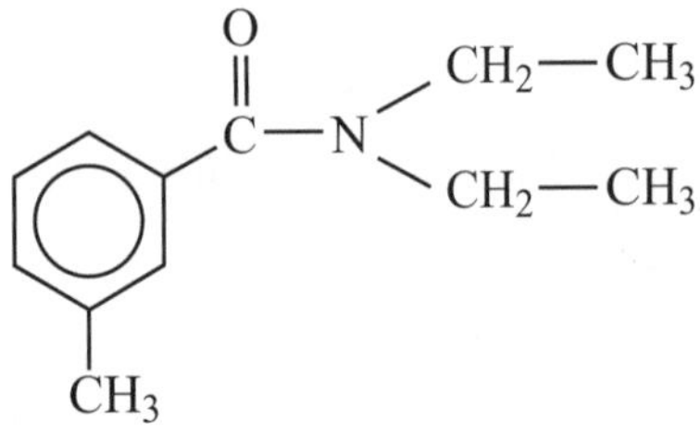

18

Biochemistry

Study Goals

- Classify a carbohydrate as an aldose or ketose; draw the open-chain and cyclic structures for glucose, galactose, and fructose.
- Describe the monosaccharide units and linkages in disaccharides and polysaccharides.
- Describe some properties of lipids.
- Describe protein functions, and draw structures for amino acids and dipeptides.
- Identify the levels of structure of a protein.
- Describe the role of an enzyme in an enzyme-catalyzed reaction.
- Describe the structure of the nucleic acids in DNA and RNA.
- Describe the synthesis of protein from mRNA.

Chapter Outline

Answers and Solutions to Text Problems

18.1 Hydroxyl (—OH) groups are found in all monosaccharides along with a carbonyl (C=O) group on the first or second carbon of the chain.

18.3 A ketopentose contains hydroxyl and ketone functional groups and has five carbon atoms.

18.5 a. Fructose is a ketose; it has a carbonyl group on carbon 2.
 b. Ribose is an aldose; it has a —CHO, an aldehyde group.
 c. Dihydroxyacetone is a ketose; it has a carbonyl group on carbon 2.
 d. Xylose is an aldose; it has a —CHO, an aldehyde group.
 e. Galactose is an aldose; it has a —CHO, an aldehyde group.

18.7 In galactose, the hydroxyl group on carbon 4 extends to the left. In glucose, this hydroxyl goes to the right.

18.9 In the cyclic structure of glucose, there are five carbon atoms and an oxygen atom in the ring portion.

18.11 a. This is the α form because the —OH on carbon 1 is down.
 b. This is the α form because the —OH on carbon 1 is down.

18.13 a. Isomaltose is a disaccharide.
 b. Isomaltose contains two molecules of α-glucose.
 c. The glycosidic link in isomaltose is an α-1,6-glycosidic bond.
 d. The downward position of the —OH on carbon 1 of the second glucose makes it α-isomaltose.

18.15 a. Another name for table sugar is sucrose.
 b. Lactose is the disaccharide found in milk and milk products.
 c. Maltose is also called malt sugar.
 d. Lactose contains the monosaccharides galactose and glucose.

18.17 a. Cellulose is a polysaccharide that is not digestible by humans.
 b. Amylose and amylopectin (in starch) are the storage forms of carbohydrates in plants.
 c. Amylose is the polysaccharide which contains only α-1,4-glycosidic bonds.
 d. Glycogen is the most highly branched polysaccharide.

18.19 a. Lauric acid contains only carbon–carbon single bonds; it is saturated.
 b. Linolenic acid has three carbon–carbon double bonds; it is unsaturated.
 c. Stearic acid contains only carbon–carbon single bonds; it is saturated.

18.21 Glyceryl trimyristate (trimyristin) has three myristic acids (14-carbon saturated fatty acid) forming ester bonds with glycerol.

$$CH_2-O-\overset{\displaystyle O}{\overset{||}{C}}-(CH_2)_{12}-CH_3$$
$$HC-O-\overset{\displaystyle O}{\overset{||}{C}}-(CH_2)_{12}-CH_3$$
$$CH_2-O-\overset{\displaystyle O}{\overset{||}{C}}-(CH_2)_{12}-CH_3$$

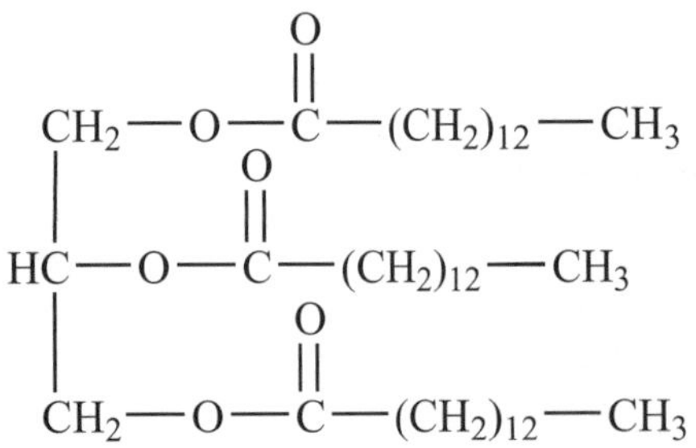

18.23 Safflower oil contains fatty acids with two or three carbon–carbon double bonds; olive oil contains a large amount of oleic acid, which has only one (monounsaturated) carbon–carbon double bond.

18.25 a. Partial hydrogenation means that some of the double bonds in the unsaturated fatty acids have been converted to single bonds by the addition of hydrogen.
 b. Since the margarine has more saturated fatty acids than the original vegetable oil, the fatty acid chains pack together more tightly and remain solid at higher temperatures.

18.27

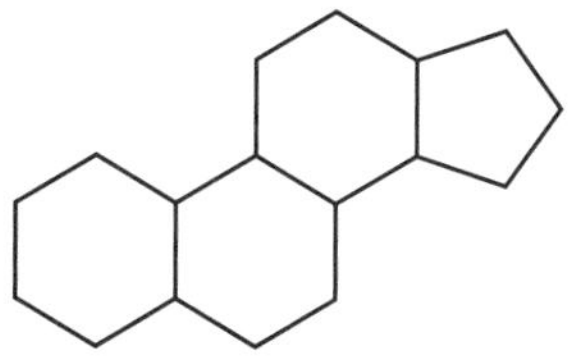

18.29 All amino acids contain a carboxylate group ($-COO^-$) and an ammonium group ($-NH_3^+$) on the alpha carbon.

18.31 a.

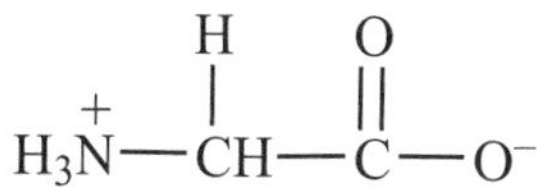

b.

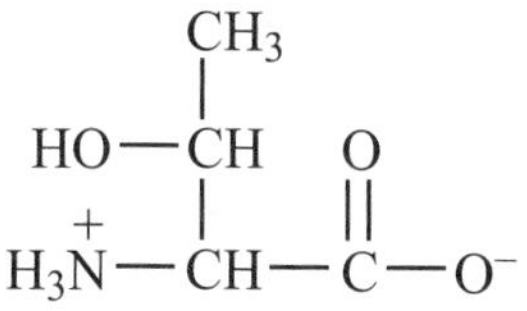

c.

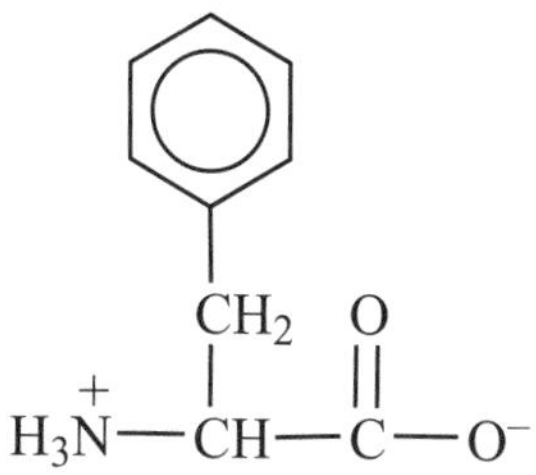

18.33 a. Glycine, which has only a —H as the R group, is hydrophobic (nonpolar).

b. Threonine has an R group that contains the polar —OH group; threonine is hydrophilic (polar, neutral).

c. Phenylalanine has an R group with a nonpolar benzene ring; phenylalanine is hydrophobic (nonpolar).

18.35 The abbreviations of most amino acids are derived from the first three letters in the name.

a. alanine **b.** valine **c.** lysine **d.** cysteine

18.37 In a peptide, the amino acids are joined by peptide bonds (amide bonds). The first amino acid has a free ammonium group and the last one has a free carboxylate group which ionize in biological environments.

a.

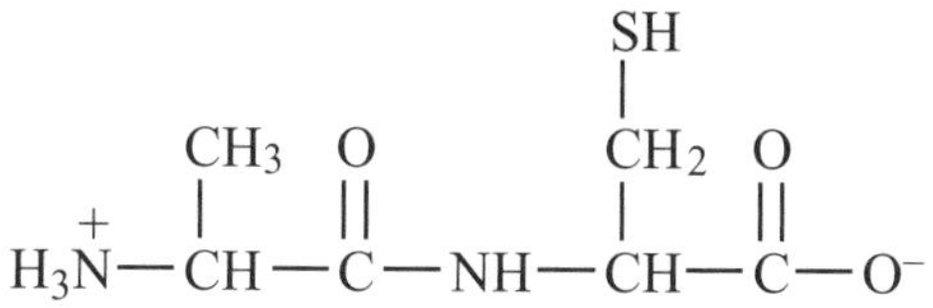

b.

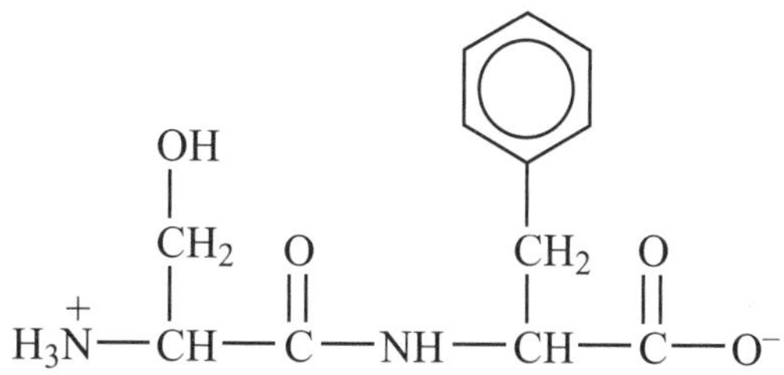

c.

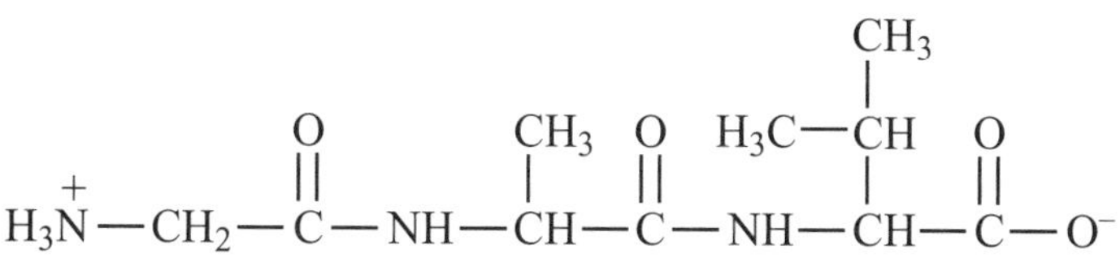

18.39 The possible primary structures of a tripeptide containing one valine and two serines are Val–Ser–Ser, Ser–Val–Ser, and Ser–Ser–Val.

18.41 In the α helix, hydrogen bonds form between the oxygen atom in the carbonyl group and hydrogen in the amide group in the next turn of the helix of the polypeptide chain. In a β-pleated sheet, hydrogen bonds occur between parallel peptides or across sections of a long polypeptide chain.

18.43 **a.** The two cysteine residues have —SH groups, which react to form a disulfide bond.
b. Serine has a polar —OH group that can form a hydrogen bond with the carboxyl group of aspartic acid.
c. Two leucine residues have nonpolar hydrocarbon R groups; they would have a hydrophobic interaction.

18.45 **a.** Cysteine, with its —SH group, can form disulfide cross-links.
b. Leucine and valine are likely to be found on the inside of the protein structure because they have nonpolar side groups and are hydrophobic.
c. The cysteine and aspartic acid would be on the outside of the protein because they have polar R groups.
d. The order of the amino acids (the primary structure) provides the side chains (R groups), whose interactions determine the tertiary structure of the protein.

18.47 **a.** An enzyme (1) has a tertiary structure that recognizes the substrate.
b. The combination of an enzyme and its substrate forms the enzyme–substrate complex (2).
c. The substrate (3) has a structure that fits the active site of the enzyme.

18.49 **a.** The equation for an enzyme–catalyzed reaction is:
$E + S \rightleftarrows ES \rightarrow E + P$
E = enzyme, S = substrate, ES = enzyme–substrate complex, P = products
b. The active site is a region or pocket within the tertiary structure of an enzyme that accepts the substrate, aligns the substrate for reaction, and catalyzes the reaction.

18.51 DNA contains two purines, adenine (A) and guanine (G), and two pyrimidines, cytosine (C) and thymine (T). RNA contains the same bases, except thymine (T) is replaced by the pyrimidine uracil (U).
a. Thymine is present in DNA.
b. Cytosine is present in both DNA and RNA.

18.53 The two DNA strands are held together by hydrogen bonds between the complementary bases in each strand.

18.55 **a.** Since T pairs with A, if one strand of DNA has the sequence —A—A—A—A—A—A—, the second DNA strand would be —T—T—T—T—T—T—.
b. Since C pairs with G, if one strand of DNA has the sequence —G—G—G—G—G—G—, the second DNA strand would be —C—C—C—C—C—C—.
c. Since T pairs with A, and C pairs with G, if one strand of DNA has the sequence —A—G—T—C—C—A—G—G—T—, the second DNA strand would be —T—C—A—G—G—T—C—C—A—.
d. Since T pairs with A and C pairs with G, if one strand of DNA has the sequence —C—T—G—T—A—T—A—C—G—T—T—, the second DNA strand would be —G—A—C—A—T—A—T—G—C—A—A—.

18.57 The two DNA strands separate (in a way that is similar to the unzipping of a zipper) to allow each of the bases to pair with its complementary base (A binds with T, C binds with G), which produces two exact copies of the original DNA.

18.59 The three types of RNA are messenger RNA (mRNA), ribosomal RNA (rRNA), and transfer RNA (tRNA).

18.61 In transcription, the information contained in DNA is transferred to mRNA molecules. Transcription begins when the section of a DNA to be copied unwinds, and each base in the DNA template strand is paired with its complementary base to form the messenger RNA molecule: G with C, C with G, T with A, and A with U (uracil).

18.63 The strand of mRNA would have the following sequence:
—G—G—C—U—U—C—C—A—A—G—U—G—.

18.65 **a.** The codon CUG in mRNA codes for the amino acid leucine (Leu).
b. The codon UCC in mRNA codes for the amino acid serine (Ser).
c. The codon GGU in mRNA codes for the amino acid glycine (Gly).
d. The codon AGG in mRNA codes for the amino acid arginine (Arg).

18.67 **a.** Melezitose is a trisaccharide.
b. Melezitose contains two glucose molecules and a fructose molecule.

18.69

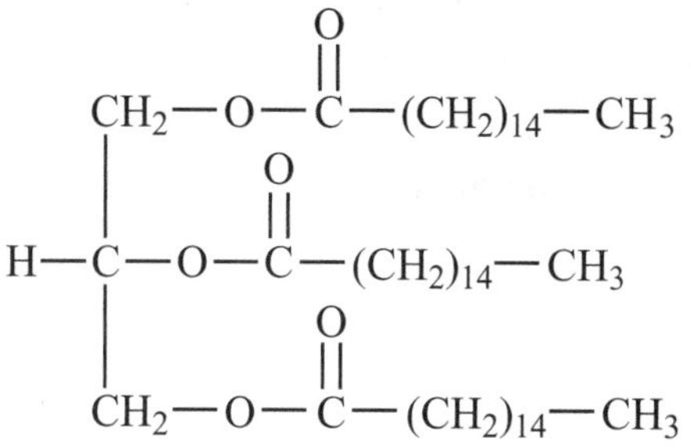

18.71 **a.**

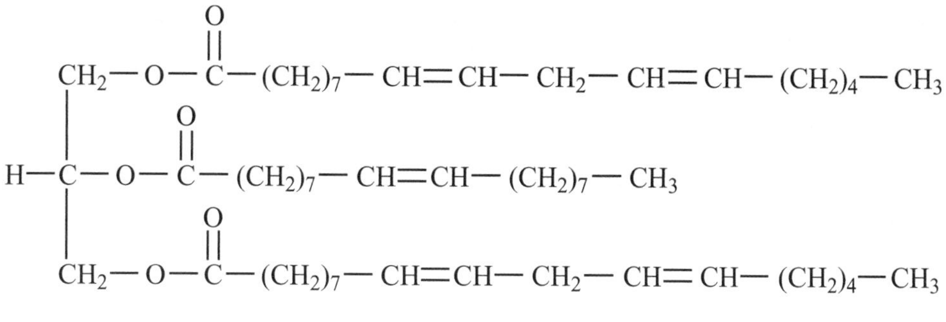

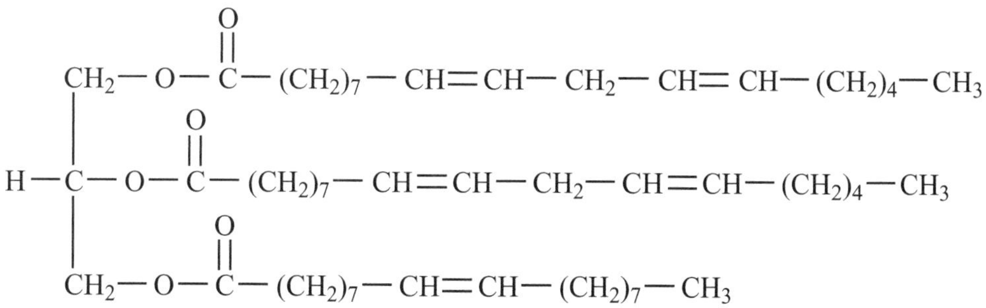

b.

$$CH_2-O-\overset{\displaystyle O}{\overset{\|}{C}}-(CH_2)_7-CH=CH-CH_2-CH=CH-(CH_2)_4-CH_3$$

$$H-\overset{\displaystyle |}{C}-O-\overset{\displaystyle O}{\overset{\|}{C}}-(CH_2)_7-CH=CH-(CH_2)_7-CH_3 \qquad\qquad + 5H_2 \overset{Ni}{\longrightarrow}$$

$$CH_2-O-\overset{\displaystyle O}{\overset{\|}{C}}-(CH_2)_7-CH=CH-CH_2-CH=CH-(CH_2)_4-CH_3$$

$$CH_2-O-\overset{\displaystyle O}{\overset{\|}{C}}-(CH_2)_{16}-CH_3$$

$$CH-O-\overset{\displaystyle O}{\overset{\|}{C}}-(CH_2)_{16}-CH_3$$

$$CH_2-O-\overset{\displaystyle O}{\overset{\|}{C}}-(CH_2)_{16}-CH_3$$

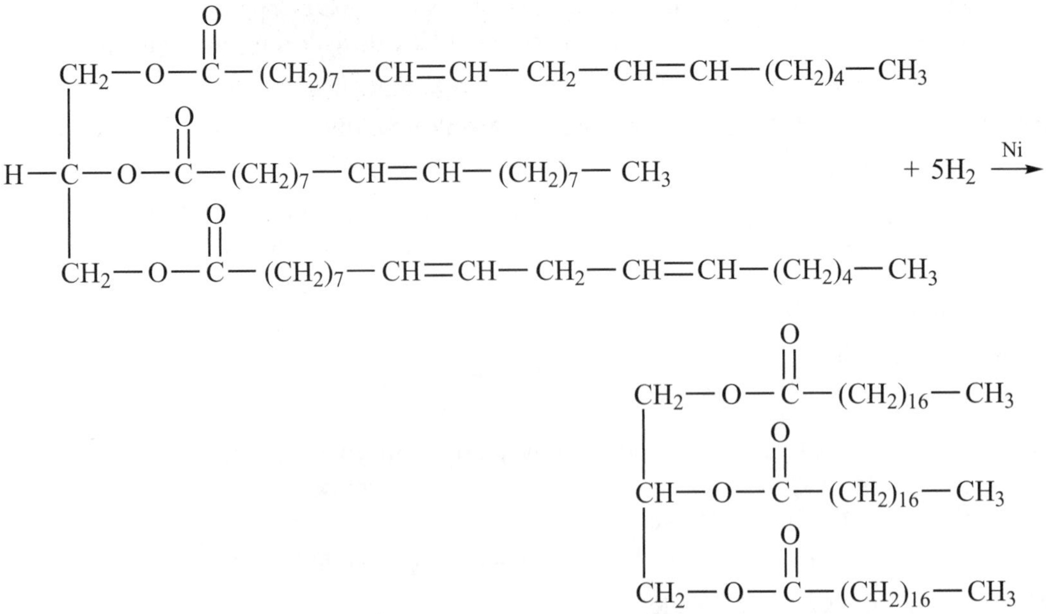

18.73 a. The polar R groups of asparagine and serine would interact by hydrogen bond.

 b. The R groups of aspartic acid and lysine would form a salt bridge (ionic bond).

 c. The R groups of the two cysteines contain —SH groups which can react to form a disulfide bond.

 d. The nonpolar R groups of leucine and alanine would have a hydrophobic interaction.

18.75 Glucose and galactose differ only at carbon 4 where the —OH in glucose is on the right side, and in galactose it is on the left side.

18.77

α-Gulose β-Gulose

18.79

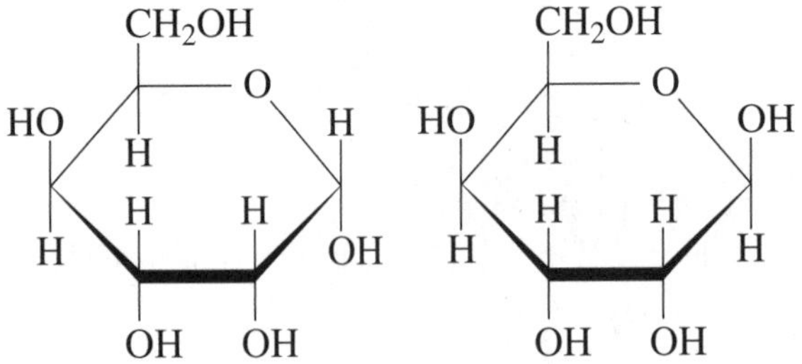

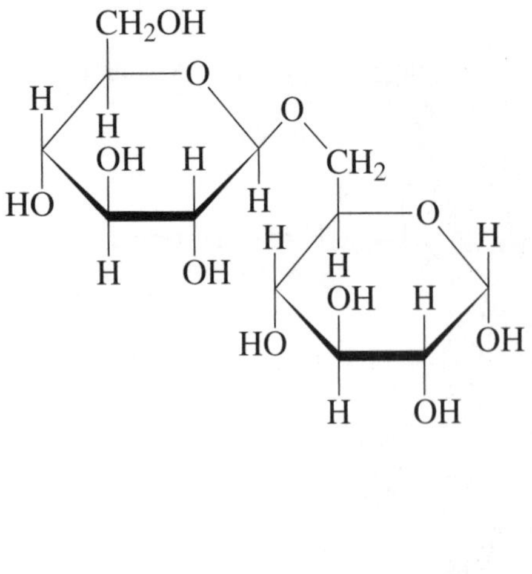

18.81

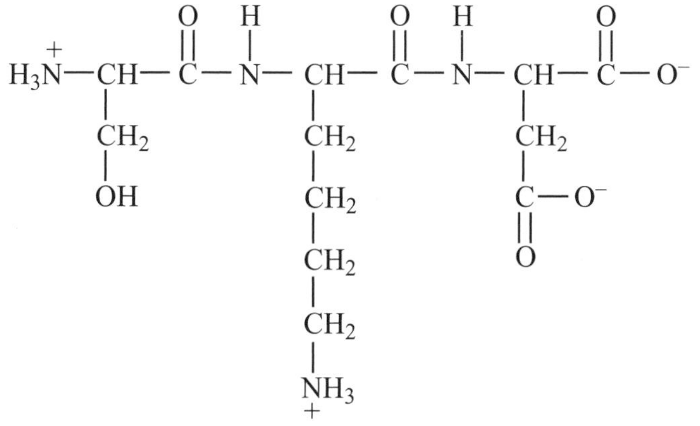

18.83 **a.** thymine and deoxyribose **b.** adenine and ribose
c. cytosine and ribose **d.** guanine and deoxyribose

18.85 **a.** —C—T—G—A—A—T—C—C—G—
b. —A—C—G—T—T—T—G—A—T—C—G—A—
c. —T—A—G—C—T—A—G—C—T—A—G—C—

18.87 **a.** Messenger RNA (mRNA) carries genetic information from the nucleus to the ribosomes.
b. Messenger RNA (mRNA) acts as a template for protein synthesis.

18.89 Raffinose contains the monosaccharides galactose, glucose, and fructose.

18.91 **a.** The secondary structure of a protein depends on hydrogen bonds to form a helix or a pleated sheet. The tertiary structure is determined by the interactions of R groups such as disulfide bonds and salt bridges and determines the three-dimensional structure of the protein.
b. Nonessential amino acids can be synthesized by the body, but essential amino acids must be supplied by the diet.
c. Polar amino acids have hydrophilic R groups, while nonpolar amino acids have hydrophobic R groups.
d. Dipeptides contain two amino acids, whereas tripeptides contain three amino acids.
e. An ionic bond (salt bridge) is an interaction between a basic and an acidic R group; a disulfide bond links the —SH groups of two cysteines.
f. The α helix is the secondary shape like a spiral staircase or corkscrew. The β-pleated sheet is a secondary structure that is formed by many peptide chains side by side.
g. The tertiary structure of a protein is its three-dimensional structure. In the quaternary structure, two or more peptide subunits interact to form a biologically active protein.

18.93 **a.** Because A bonds with T, salmon DNA with 28% A will also have 28% T, which gives a total of 56%. The remaining 44% for the G and C nucleotides is divided as 22% G and 22% C.
b. Because C bonds with G, human DNA with 20% C will also have 20% G, which gives a total of 40%. The remaining 60% for the A and T nucleotides is divided as 30% A and 30% T.

18.95 **a.** The mRNA sequence would be: —CGA—AAA—GUU—UUU—
b. From the table of mRNA codons, the amino acids would be —Arg—Lys—Val—Phe—.

18.97 Using the genetic code, the codons indicate the following amino acid sequence:
START—Tyr—Gly—Gly—Phe—Leu—STOP

Answers to Combining Ideas from Chapters 17 and 18

CI.35 a. The —OH group in BHT is bonded to a carbon atom in an aromatic ring, which means that BHT has a phenol functional group.

b. BHT is referred to as an "antioxidant" because it, rather than the food, reacts with oxygen in the food container, thus preventing or retarding spoilage of food.

c. $15\ \text{oz cereal} \times \dfrac{1\ \text{lb}}{16\ \text{oz}} \times \dfrac{1\ \text{kg}}{2.205\ \text{lb}} \times \dfrac{50.\ \text{mg BHT}}{1\ \text{kg cereal}} = 21\ \text{mg of BHT (2 SFs)}$

CI.37 a. acetylene H—C≡C—H

$2C_2H_2(g) + 5O_2(g) \xrightarrow{\Delta} 4CO_2(g) + 2H_2O(g) + \text{energy}$

b. $8.5\ \text{L}\ C_2H_2 \times \dfrac{1\ \text{mol}\ C_2H_2}{22.4\ \text{L (STP)}} \times \dfrac{5\ \text{mol}\ O_2}{2\ \text{mol}\ C_2H_2} \times \dfrac{32.00\ \text{g}\ O_2}{1\ \text{mol}\ O_2} = 30.\ \text{g of}\ O_2\ \text{(2 SFs)}$

c. $30.0\ \text{g}\ C_2H_2 \times \dfrac{1\ \text{mol}\ C_2H_2}{26.04\ \text{g}\ C_2H_2} \times \dfrac{4\ \text{mol}\ CO_2}{2\ \text{mol}\ C_2H_2} \times \dfrac{22.4\ \text{L (STP)}}{1\ \text{mol}\ CO_2}$

$= 51.6\ \text{L of}\ CO_2\ \text{(STP) (3 SFs)}$

CI.39 a. Molar mass of MTBE ($C_5H_{12}O$)

$= 5\ \text{mol of C} + 12\ \text{mol of H} + 1\ \text{mol of O} = 5(12.01\ \text{g}) + 12(1.008\ \text{g}) + 16.00\ \text{g} = 88.15\ \text{g}$

$100.\ \text{g gasoline} \times \dfrac{2.7\ \text{g O}}{100\ \text{g gasoline}} \times \dfrac{88.15\ \text{g MTBE}}{16.00\ \text{g O}} = 15\ \text{g of MTBE (2 SFs)}$

b. $1.00\ \text{L gasoline} \times \dfrac{1000\ \text{mL}}{1\ \text{L}} \times \dfrac{0.740\ \text{g gasoline}}{1\ \text{mL gasoline}} \times \dfrac{15\ \text{g MTBE}}{100\ \text{g gasoline}}$

$\times \dfrac{1\ \text{mL MTBE}}{0.740\ \text{g MTBE}} \times \dfrac{1\ \text{L}}{1000\ \text{mL}}$

$= 0.15\ \text{L of MTBE (2 SFs)}$

c. $2C_5H_{12}O(l) + 15O_2(g) \xrightarrow{\Delta} 10CO_2(g) + 12H_2O(g) + \text{energy}$

d. $1.00\ \text{L MTBE} \times \dfrac{1000\ \text{mL}}{1\ \text{L}} \times \dfrac{0.740\ \text{g MTBE}}{1\ \text{mL MTBE}} \times \dfrac{1\ \text{mol MTBE}}{88.15\ \text{g MTBE}}$

$\times \dfrac{15\ \text{mol}\ O_2}{2\ \text{mol MTBE}} \times \dfrac{22.4\ \text{L (STP)}}{1\ \text{mol}\ O_2} \times \dfrac{100\ \text{L air}}{21\ \text{L}\ O_2}$

$= 6700\ \text{L}\ (6.7 \times 10^3\ \text{L})\ \text{of air (STP) (2 SFs)}$

CI.41 a. Adding NaOH will saponify the glyceryl tristearate (fat), breaking it up into fatty acid salts and glycerol, which are soluble and will wash down the drain.

b.

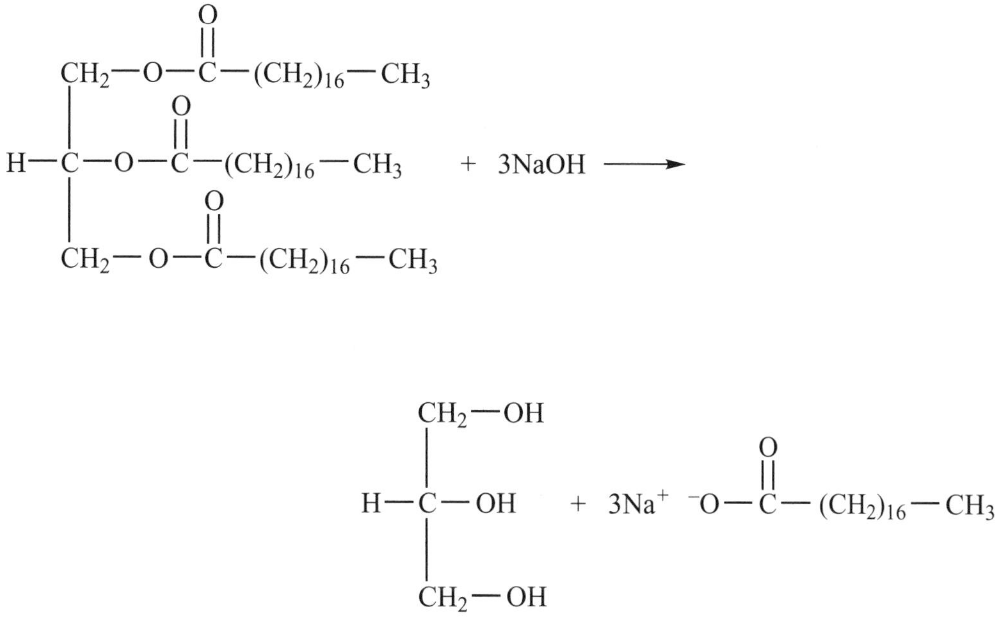

$$CH_2-O-\overset{\overset{\textstyle O}{\|}}{C}-(CH_2)_{16}-CH_3$$

$$H-\overset{\displaystyle |}{C}-O-\overset{\overset{\textstyle O}{\|}}{C}-(CH_2)_{16}-CH_3 \quad + \quad 3NaOH \longrightarrow$$

$$CH_2-O-\overset{\overset{\textstyle O}{\|}}{C}-(CH_2)_{16}-CH_3$$

$$CH_2-OH$$

$$H-\overset{\displaystyle |}{C}-OH \quad + \quad 3Na^+ \; {}^-O-\overset{\overset{\textstyle O}{\|}}{C}-(CH_2)_{16}-CH_3$$

$$CH_2-OH$$

CI.43 a.

$$HO-\overset{\overset{\textstyle O}{\|}}{C}-\underset{}{\bigcirc}-\overset{\overset{\textstyle O}{\|}}{C}-O-CH_2-CH_2-OH$$

b.

$$HO-CH_2-CH_2-O-\overset{\overset{\textstyle O}{\|}}{C}-\underset{}{\bigcirc}-\overset{\overset{\textstyle O}{\|}}{C}-O-CH_2-CH_2-OH$$

c. $1.7 \times 10^9 \; \cancel{\text{lb PETE}} \times \dfrac{1 \text{ kg PETE}}{2.205 \; \cancel{\text{lb PETE}}} = 7.7 \times 10^8$ kg of PETE (2 SFs)

d. $1.7 \times 10^9 \; \cancel{\text{lb PETE}} \times \dfrac{453.6 \; \cancel{\text{g PETE}}}{1 \; \cancel{\text{lb PETE}}} \times \dfrac{1 \; \cancel{\text{mL PETE}}}{1.38 \; \cancel{\text{g PETE}}} \times \dfrac{1 \text{ L PETE}}{1000 \; \cancel{\text{mL PETE}}}$

$= 5.6 \times 10^8$ L of PETE (2 SFs)

e. $5.6 \times 10^8 \; \cancel{\text{L PETE}} \times \dfrac{1 \text{ landfill}}{2.7 \times 10^7 \; \cancel{\text{L PETE}}} = 21$ landfills (2 SFs)

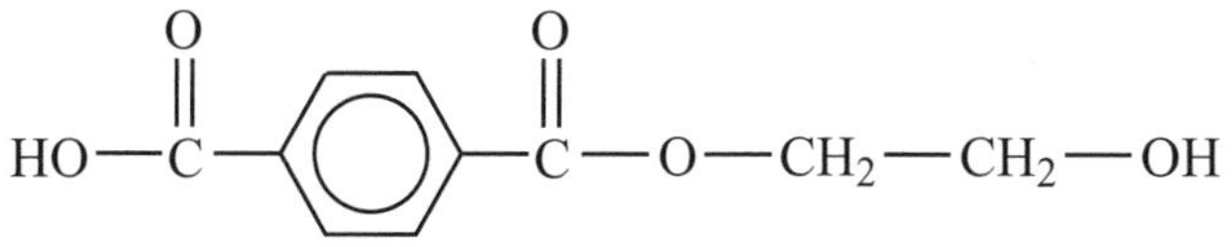

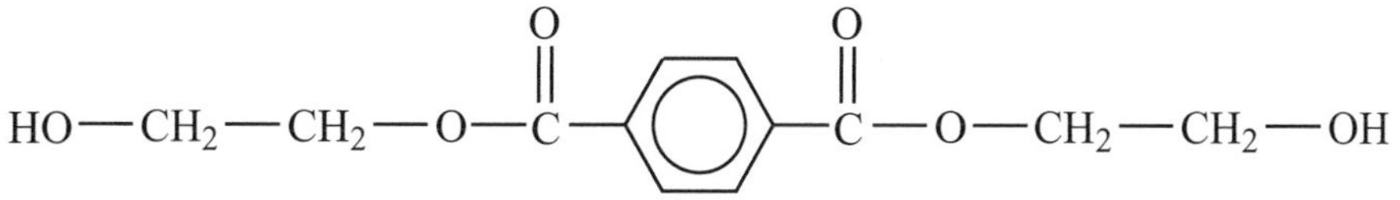